Smart Innovation, Systems and Technologies

Volume 232

The Smart Innovation, Systems and Technologies book series encompasses the topics of knowledge, intelligence, innovation and sustainability. The aim of the series is to make available a platform for the publication of books on all aspects of single and multi-disciplinary research on these themes in order to make the latest results available in a readily-accessible form. Volumes on interdisciplinary research combining two or more of these areas is particularly sought.

The series covers systems and paradigms that employ knowledge and intelligence in a broad sense. Its scope is systems having embedded knowledge and intelligence, which may be applied to the solution of world problems in industry, the environment and the community. It also focusses on the knowledge-transfer methodologies and innovation strategies employed to make this happen effectively. The combination of intelligent systems tools and a broad range of applications introduces a need for a synergy of disciplines from science, technology, business and the humanities. The series will include conference proceedings, edited collections, monographs, handbooks, reference books, and other relevant types of book in areas of science and technology where smart systems and technologies can offer innovative solutions.

High quality content is an essential feature for all book proposals accepted for the series. It is expected that editors of all accepted volumes will ensure that contributions are subjected to an appropriate level of reviewing process and adhere to KES quality principles.

Indexed by SCOPUS, EI Compendex, INSPEC, WTI Frankfurt eG, zbMATH, Japanese Science and Technology Agency (JST), SCImago, DBLP.

All books published in the series are submitted for consideration in Web of Science.

More information about this series at http://www.springer.com/series/8767

Andrey Ronzhin · Vladislav Shishlakov
Editors

Electromechanics and Robotics

Proceedings of 16th International Conference on Electromechanics and Robotics "Zavalishin's Readings" (ER(ZR) 2021), St. Petersburg, Russia, 14–17 April 2021

Editors
Andrey Ronzhin
St. Petersburg Federal Research Center
of the Russian Academy of Sciences
St. Petersburg, Russia

Vladislav Shishlakov
State University of Airspace
Instrumentation
St. Petersburg, Russia

ISSN 2190-3018 ISSN 2190-3026 (electronic)
Smart Innovation, Systems and Technologies
ISBN 978-981-16-2816-0 ISBN 978-981-16-2814-6 (eBook)
https://doi.org/10.1007/978-981-16-2814-6

This Springer imprint is published by the registered company Springer Nature Singapore Pte Ltd.
The registered company address is: 152 Beach Road, #21-01/04 Gateway East, Singapore 189721, Singapore

Organization

General Chair

Yulia Antokhina

Co-chairs

Oleg Baulin
Sergey Emelyanov
Vladislav Shishlakov

Committees

Chair of Program Committee

Andrey Ronzhin

Program Committee

Karsten Berns, Germany
Nikolay Bolotnik, Russia
Yi-Tung Chen, USA
Sergey Chigvincev, Russia
Alexander Danilov, Russia
Vlado Delic, Serbia
Ivan Ermolov, Russia

Naohisa Hashimoto, Japan
Han-Pang Huang, Taiwan
Shu Huang, Taiwan
Viktor Glazunov, Russia
Mehmet Guzey, Turkey
Oliver Jokisch, Germany
Airat Kalimgulov, Russia
Alexey Kashevnik, Russia
Marat Khakimyanov, Russia
Regina Khazieva, Russia
Pavel Khlyupin, Russia
Sergey Konesev, Russia
Eugeni Magid, Russia
Roman Meshcheryakov, Russia
Zuhra Pavlova, Russia
Vladimir Pavlovskiy, Russia
Francesco Pierri, Italy
Yuriy Poduraev, Russia
Mirko Rakovic, Serbia
Raul Rojas, Germany
Jose Rosado, Portugal
Vitali Shabanov, Russia
Hooman Samani, Taiwan
Yulia Sandamirskaya, Switzerland
Jesus Savage, Mexico
Valery Sapelnikov, Russia
Robert Sattarov, Russia
Vladimir Serebrenny, Russia
Michail Sit, Moldova
Lev Stankevich, Russia
Tilo Strutz, Germany
Georgi Vukov, Bulgaria
Sergey Yatsun, Russia
Arkadiy Yuschenko, Russia
Milos Zelezny, Czech Republic
Lyudmila Zinchenko, Russia

Co-chair of Organizing Committee

Pavel Khlyupin, Sergey Solyonyj, Sergey Yatsun, Andrey Ronzhin

Organizing Committee

Marina Astapova, Peter Bezmen, Polina Chernousova, Natalia Dormidontova, Oksana Emelyanova, Natalia Kashina, Regina Khazieva, Marat Khakimyanov, Sergey Konesev, Alena Lopotova, Boris Lushnikov, Anna Motienko, Irina Podnozova, Yevgeny Politov, Alexander Rukavitsyn, Anton Saveliev, Ekaterina Savelyeva, Oksana Solenaya, Anastasia Statkevich, Sergey Timofeev, Irina Vatamaniuk, Elizaveta Usina, Andrey Yatsun

Foreword

Dmitry Aleksandrovich Zavalishin (1900–1968)—a Russian scientist— is a corresponding member of the USSR Academy of Sciences and founder of the school of valve energy converters based on electric machines and valve converters energy. The first conference was organized by the Institute of Innovative Technologies in Electromechanics and Robotics of the St. Petersburg State University of Aerospace Instrumentation in 2006.

The purpose of the conference is the exchange of information and progressive results of scientific research work of scientific and pedagogical workers, young scientists, graduate students, applicants, and students in the field of: automatic control systems, electromechanics, electric power engineering and electrical engineering, mechatronics, robotics, automation, technical physics and management in the electric power industry.

We express our deepest gratitude to all participants for their valuable contribution to the successful organization of ER(ZR)-2021, hope for and look forward to your attention to the next International Conference on Electromechanics and Robotics "Zavalishin's Readings" in 2022. The conference website is located at: http://suai.edu.ru/conference/zav-read/.

May 2021

Prof. Yulia A. Antokhina
General Chair of 16th International Conference on Electromechanics and Robotics "Zavalishin's Readings"—2021
Rector of the St. Petersburg State University of Aerospace Instrumentation
St. Petersburg, Russia

Preface

In this year, the conference, the 16th International Conference on Electromechanics and Robotics "Zavalishin's Readings"—2021, ER(ZR)-2021, was organized with XV International Conference "Vibration-2021. Vibration technologies, mechatronics and controlled machines" and VI International Conference "Electric drive, electrical technology and electrical equipment of enterprises" during April 14–17, 2021, in St. Petersburg, Russia. The conferences were organized by St. Petersburg State University of Aerospace Instrumentation (SUAI, St. Petersburg, Russia), St. Petersburg Federal Research Center of the Russian Academy of Sciences (SPC RAS, St. Petersburg, Russia), Southwest State University (SWSU, Kursk, Russia), and Ufa State Petroleum Technical University (USPTU, Ufa, Russia).

To avoid the COVID-19 pandemic in the world, ER(ZR)-2021 was organized as a virtual conference still. However, the videoconference attracted increased number of participants, decreased costs for travel and accommodation, etc.

During the conference, the invited talks were given by Prof. Abolfazl Vahedi (Iran University of Science and Technology, Tehran, Iran), Assoc. Prof. Pavel Khlyupin (Ufa State Petroleum Technical University, Ufa, Russia), and Prof. Victor Glazunov (Mechanical Engineering Research Institute of the Russian Academy of Sciences). More than 180 papers of authors from Iran, Nigeria, Russia, Turkey, and Uzbekistan were submitted to the conference, and each paper was reviewed by several scientists. Around 25% of the best papers were published in English proceedings by Springer in series Smart Innovation, Systems, and Technologies indexed in SCOPUS, Thomson Reuters (Web of Science), Inspec, etc. Due to great efforts of reviewers, this book was carefully prepared and consists of 43 contributions.

Special thanks are to the members of the local organizing committee for their tireless effort and enthusiasm during the conference organization. Hope for and look forward to your attention to the ER(ZR)-2022. The conference website is located at: http://suai.edu.ru/conference/zav-read/.

May 2020

Prof. Andrey Ronzhin
Chair of Program Committee of 16th International Conference on Electromechanics and Robotics "Zavalishin's Readings"—2021
Director of St. Petersburg Federal Research Center of the Russian Academy of Sciences
St. Petersburg, Russia

Prof. Vladislav Shishlakov
Co-chair of 16th International Conference on Electromechanics and Robotics "Zavalishin's Readings"—2021
Vice-Rector for Educational Technologies and Innovative Activities
St. Petersburg State University of Aerospace Instrumentation
St. Petersburg, Russia

Contents

About the Editors

Prof. Andrey Ronzhin is Director of St. Petersburg Federal Research Center of the Russian Academy of Sciences (SPC RAS) and Head of the Department of Electromechanics and Robotics Systems at St. Petersburg University of Airspace Instrumentation. His research focuses on interaction of autonomous robotic systems and users in a cyber-physical environment. He is a member of Scientific Board of Robotics and Mechatronics of the Russian Academy of Sciences, the Academy of Navigation and Motion Control, Co-chairman of International Conference Interactive Collaborative Robotics (ICR). He is Deputy Editor-in-Chief of Journal "Informatics and Automation".

Prof. Vladislav Shishlakov is Vice-Rector for Educational Technologies and Innovative Activities, St. Petersburg State University of Aerospace Instrumentation (SUAI) and Head of the Department of Management in Technical Systems. He is Honorary Worker at the Ministry of Education and Science of the Russian Federation since 2009. His research interests are related to development of methods of synthesis of nonlinear systems of automatic control systems, which are continuous, and with different types of signal modulation, as well as the development and research of electromechanical and electric power systems and complexes based on the effects of high-temperature superconductivity.

Robotics and Automation

Analysis of Mechanisms with Parallel-Serial Structure 5-DOF and Extended Working Area

Viktor Glazunov, Gleb Filippov, Gagik Rashoyan, Lubov Gavrilina, Konstantin Shalyukhin, and Sergey Skvortsov

Abstract The subject of article is the kinematic and dynamic analysis of the parallel-sequential structure spatial mechanisms. The synthesis of a parallel-sequential structure mechanism for extended objects processing with kinematic chains made according to the l-coordinate scheme was carried out. Its analysis has been carried out. Kinematic and dynamic models of this mechanism have been developed. Mathematical models have been developed to study the kinematics and dynamics of a mechanism with an extended working area. The solution of the inverse problem on positions for this mechanism is given. This solution uses the classic coordinate transformation method. The final system of the constraints equations is obtained. The description of the methodology for solving the problem of velocities, which consists in differentiating the constraints equations (the Angeles-Gosselin method), is given. A system of implicit functions is presented. The solution of the problem of dynamics for a partial planar parallel structure mechanism with three kinematic chains is described. This solution was carried out using a control law that minimizes errors in speed, position, and acceleration. The mutual influence of drive motors on each other, as well as on the law of the executive link motion, is investigated. Full-scale and numerical experiments have been carried out on the developed models, and their results are presented. In a numerical experiment, a situation was simulated in which movement along one coordinate occurs according to a harmonic law, and no movement along the rest.

1 Introduction

The modern development of engineering and technologies imposes increased requirements on process equipment in terms of such indicators as velocity, accuracy, and

Scopus Author ID: 57204705748

V. Glazunov · G. Filippov (✉) · G. Rashoyan · L. Gavrilina · K. Shalyukhin · S. Skvortsov
Mechanical Engineering Research Institute, Russian Academy of Sciences (IMASH RAN), 4, Maly Kharitonyevsky Pereulok, Moscow 101990, Russia

A. Ronzhin and V. Shishlakov (eds.), *Electromechanics and Robotics*, Smart Innovation, Systems and Technologies 232, https://doi.org/10.1007/978-981-16-2814-6_1

load capacity to take external forces [1]. In this case, the equipment for additive technologies, laser systems used in the design and manufacture of parts for new aircraft and rocket engines, is considered.

Extension along one linear coordinate is a distinctive feature of many parts of these objects. For example, it can be an engine turbine blade airfoil, extended body parts. The relatively small size of the working area of the currently existing process equipment is not always able to provide the processing of these parts. Equipment created on the basis of mechanisms of a parallel-serial structure allows the most complete response to the presented requirement.

2 Statement of the Problem

It is known that 5-DOF of the end-effector are sufficient for traditional metalworking methods, as well as additive technologies or laser methods for processing parts, since the laser beam or the extruder of a 3D printer does not require its rotation around its own axis. For a spindle with a fixed blade tool, rotation can always be controlled by a separate motor [2]. Thus, the number of degrees of freedom of the mechanism is reduced by one unit, the control system of the mechanism is simplified, and the number of motors is reduced and, as a consequence, leads to a decrease in the number of kinematic chains in some cases. It should be noted that similar work is being carried out in foreign and domestic scientific centers. An example of such a mechanism is the Neumann's Tricept robot [3] designed for machining complex-shaped parts. According to this principle, for example, a METROM robot-machine of German production was created.

In [4], a robotic manipulator design is proposed, consisting of two serial robots, one of which is transformed into a parallel robot in order to increase the rigidity and area of the working space.

In [5], a 5-axis parallel-serial manipulator is proposed, for which direct and inverse problems of kinematics and dynamics was resolved, the working area was calculated, but its geometric shape was insufficient for extended parts processing.

Also, the domestic design of the manipulator developed by I.A. Nesmiyanov et al. [6, 7] is known. It is designed for processing and transportation of extended parts. The problems of determining the driving forces and moments that implement a given law of motion are solved, and a method for determining the forces in kinematic pairs for the rational design of the manipulator according to the criteria of rigidity and strength is developed. Paper [8] is devoted to the development and application of parallel-serial mechanisms as part of a robotic complex for carrying out process operations in the manufacture of parts of a turbojet engine nozzle.

A distinctive feature of these manipulators is the presence of a "frame structure," which increases the load-bearing capacity and rigidity of the manipulator and provides sufficiently large linear movements for the tool, but limited rotation angles [9].

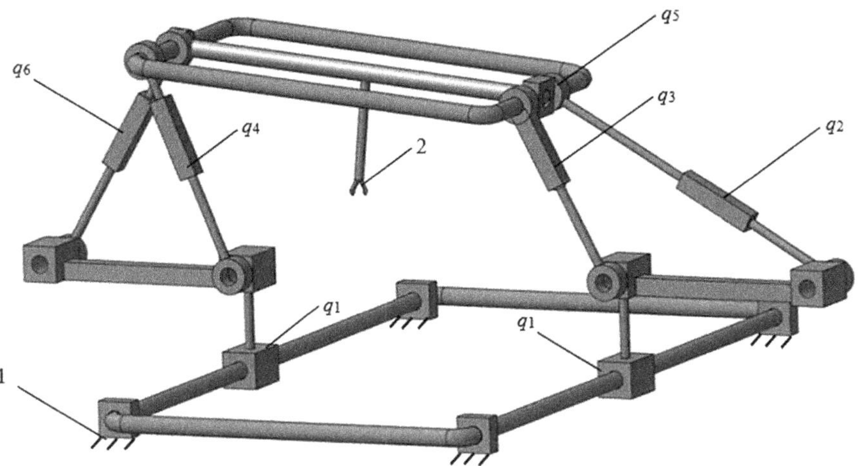

Fig. 1 Parallel-serial mechanism for processing extended objects

The object of the study is the mechanism of the parallel-serial structure shown in Fig. 1. A simplified diagram of the mechanism (Fig. 1) includes a fixed base 1, an output link 2, one common or two synchronous drives q_1, a drive q_5 in series, and drives of a partial mechanism of a parallel structure q_2, q_3, q_4, q_6. The robot under consideration uses linear motors in a partial planar mechanism of a parallel structure. They are made according to the l-coordinate scheme, and engines of this type were widely used by Professor Koliskor. It should be noted that there is an additional kinematic chain with a q6 drive, which is designed to eliminate possible special positions and reduce the load on all motors. An additional kinematic chain with a q6 drive adds additional rigidity to the structure and ensures an even distribution of the load. In this case, the additional kinematic does not affect the total number of degrees of freedom of the output link.

3 Solving the Problem

For the chosen scheme of the mechanism, it is necessary to develop kinematic and dynamic models and carry out computational experiments on the models.

Let us consider the solution of the inverse kinematics for this mechanism.

The inverse position problem is solved on the basis of the equations of relationships between the absolute coordinates of the output link and the generalized coordinates of the mechanism, which follow from the geometric relationships.

3.1 Inverse Kinematics Problem

Let the coordinates of the output link (*x*, *y*, *z*) and the angles of its orientation (α, β) be given in the absolute coordinate system *OXYZ* (Fig. 2). It is required to determine the generalized coordinates ($q_1, q_2, q_3, q_4, q_5, q_6$) (Figs. 1 and 2). First, we determine the coordinates q_1 and q_5. When solving the inverse problem of kinematics, we will use the classical method—the coordinate transformation method. For this, knowing the position and orientation of the tool, it is possible to determine the position of the center point A_1 of the frame and, accordingly, the first and fifth generalized coordinates. To determine the generalized coordinates q_2, q_3, q_4, q_6 which correspond to the partial planar mechanism of the parallel structure, we find the coordinates of points A_2 and A_3. Then the generalized coordinates of the planar mechanism q_2, q_3, q_4, q_6 are determined.

As a result, the following constraint equations were obtained:

$$\begin{cases} q_1 = x + L_0 \cdot \sin\beta; \\ q_2 = \sqrt{\left(B_{2x} - \left(x - L_0 \cdot \sin(\alpha) + \frac{a}{2} \cdot \cos(\alpha)\right)\right)^2 + \left(\left(y + L_0 \cdot \cos(\alpha) + \frac{a}{2} \cdot \sin(\alpha)\right) - B_{2y}\right)^2} - L_2; \\ q_3 = \sqrt{\left(B_{3x} - \left(x - L_0 \cdot \sin(\alpha) + \frac{a}{2} \cdot \cos(\alpha)\right)\right)^2 + \left(\left(y + L_0 \cdot \cos(\alpha) + \frac{a}{2} \cdot \sin(\alpha)\right) - B_{3y}\right)^2} - L_3; \\ q_4 = \sqrt{\left(B_{4x} - \left(x - L_0 \cdot \sin(\alpha) - \frac{a}{2} \cdot \cos(\alpha)\right)\right)^2 + \left(\left(y + L_0 \cdot \cos(\alpha) - \frac{a}{2} \cdot \sin(\alpha)\right) - B_{4y}\right)^2} - L_4; \\ q_5 = \beta; \\ q_6 = \sqrt{\left(B_{6x} - \left(x - L_0 \cdot \sin(\alpha) - \frac{a}{2} \cdot \cos(\alpha)\right)\right)^2 + \left(\left(y + L_0 \cdot \cos(\alpha) - \frac{a}{2} \cdot \sin(\alpha)\right) - B_{6y}\right)^2} - L_6. \end{cases} \tag{1}$$

where $B_2\left(B_{2x}, B_{2y}\right)$, $B_3\left(B_{3x}, B_{3y}\right)$, $B_4\left(B_{4x}, B_{4y}\right)$, and $B_6\left(B_{6x}, B_{6y}\right)$ are the coordinates of the points of intersection of the kinematic chains with the base, L_2, L_3, L_4, L_6 and a are the known distances.

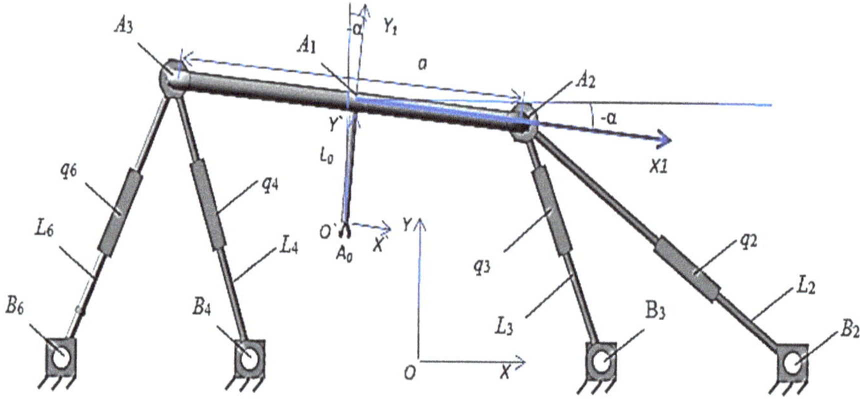

Fig. 2 Partial planar mechanism of parallel structure

To solve the problem of velocities, the method of differentiation of constraint equations is used [10, 11]. The essence of the method is as follows. It is necessary to represent constraint Eq. (1) in the form of implicit functions F_i. In this case, it is necessary to carry out a number of transformations for the convenience of performing the subsequent differentiation operation. They consist, first of all, in reducing the considered expressions to a form in which the degree of the polynomials included in it is equal to one. Unfortunately, this stage of the methodology for analyzing mechanisms of parallel and parallel-sequential structure is rather difficult to algorithmize. The general recommendation comes down to performing classical transformations. If it is impossible to reduce the expressions to a form in which it will subsequently be possible to calculate the partial derivatives of implicit functions in generalized coordinates, the researcher may need to find an alternative solution to the inverse problem of positions, in which the constraint equations will take a form more convenient for further transformations. The constraint Eq. (1) in the form of implicit functions F_i after performing the necessary transformations:

$$\begin{cases} F_1 = x + L_1 \cdot \sin(\beta) - q_1; \\ F_2 = \left(B_{2x} - \left(x - L_0 \cdot \sin(\alpha) + \frac{a}{2} \cdot \cos(\alpha)\right)\right)^2 \\ \quad + \left(\left(y + L_0 \cdot \cos(\alpha) + \frac{a}{2} \cdot \sin(\alpha)\right) - B_{2y}\right)^2 - (L_2 + q_2)^2; \\ F_3 = \left(B_{3x} - \left(x - L_0 \cdot \sin(\alpha) + \frac{a}{2} \cdot \cos(\alpha)\right)\right)^2 \\ \quad + \left(\left(y + L_0 \cdot \cos(\alpha) + \frac{a}{2} \cdot \sin(\alpha)\right) - B_{3y}\right)^2 - (L_3 + q_3)^2; \\ F_4 = \left(B_{4x} - \left(x - L_0 \cdot \sin(\alpha) - \frac{a}{2} \cdot \cos(\alpha)\right)\right)^2 \\ \quad + \left(\left(y + L_0 \cdot \cos(\alpha) - \frac{a}{2} \cdot \sin(\alpha)\right) - B_{4y}\right)^2 - (L_4 + q_4)^2; \\ F_5 = \beta - q_5; \\ F_6 = \left(B_{6x} - \left(x - L_0 \cdot \sin(\alpha) - \frac{a}{2} \cdot \cos(\alpha)\right)\right)^2 \\ \quad + \left(\left(y + L_0 \cdot \cos(\alpha) - \frac{a}{2} \cdot \sin(\alpha)\right) - B_{6y}\right)^2 - (L_6 + q_6)^2. \end{cases}$$

Differentiating the functions F_i by the generalized and absolute coordinates of the output link, it is possible to obtain the relation for the velocities.

$$\begin{pmatrix} \partial F_1/\partial\alpha & \partial F_1/\partial\beta & \partial F_1/\partial x & \partial F_1/\partial y & \partial F_1/\partial z \\ \partial F_2/\partial\alpha & \partial F_2/\partial\beta & \partial F_2/\partial x & \partial F_2/\partial y & \partial F_2/\partial z \\ \partial F_3/\partial\alpha & \partial F_3/\partial\beta & \partial F_3/\partial x & \partial F_3/\partial y & \partial F_3/\partial z \\ \partial F_4/\partial\alpha & \partial F_4/\partial\beta & \partial F_4/\partial x & \partial F_4/\partial y & \partial F_4/\partial z \\ \partial F_5/\partial x & \partial F_5/\partial\beta & \partial F_5/\partial x & \partial F_5/\partial y & \partial F_5/\partial z \end{pmatrix} \begin{pmatrix} \dot{\alpha} \\ \dot{\beta} \\ \dot{x} \\ \dot{y} \\ \dot{z} \end{pmatrix} \tag{2}$$

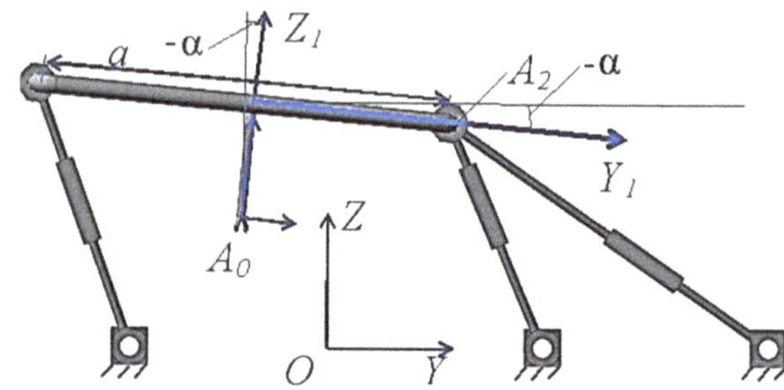

Fig. 3 Partial planar mechanism of a parallel structure with kinematic chains made according to the structure of *l*-coordinates

$$= -\begin{pmatrix} \partial F_1/\partial\alpha & 0 & 0 & 0 & 0 \\ 0 & \partial F_2/\partial\beta & 0 & 0 & 0 \\ 0 & 0 & \partial F_3/\partial x & 0 & 0 \\ 0 & 0 & 0 & \partial F_4/\partial x & 0 \\ 0 & 0 & 0 & 0 & \partial F_5/\partial x \end{pmatrix} \begin{pmatrix} \dot{q}_1 \\ \dot{q}_2 \\ \dot{q}_3 \\ \dot{q}_4 \\ \dot{q}_5 \end{pmatrix}$$

Relation (2) allows us to determine the generalized velocities for the given absolute velocities of the output link of the mechanism.

3.2 Mechanism Dynamics Problem

For the mechanism of a parallel structure with three drives, the inverse problem of dynamics is solved. In this case, the fourth engine performs the function of unloading. The absolute coordinates z, y, α are linear and angular coordinates of the output link, respectively (Fig. 3).

It is necessary to determine the forces in the drives at which the output link would move according to a given law. It was assumed that the entire mass is concentrated on the output link.

To solve the problem, the d'Alembert–Lagrange equations were used:

$$\begin{cases} m\ddot{z} = P_1\partial q_1/\partial z + P_2\partial q_2/\partial z + P_3\partial q_3/\partial z; \\ m\ddot{y} = P_1\partial q_1/\partial y + P_2\partial q_2/\partial y + P_3\partial q_3/\partial y - mg; \\ J\ddot{\alpha} = P_1\partial q_1/\partial\alpha + P_2\partial q_2/\partial\alpha + P_3\partial q_3/\partial\alpha, \end{cases}$$

where m is the mass of the output link; $\ddot{z}$, $\ddot{y}$—accelerations of the center of mass; $\ddot{\alpha}$—angular acceleration; P_i—forces in drives; J—moment of inertia of the output link; $\partial q_1/\partial z,\ldots, \partial q_3/\partial\alpha$—variable coefficients, nonlinearly dependent on the coordinates of the output link; g—acceleration of gravity.

The drives control law is considered, which minimizes errors in velocity, position, and acceleration.

The law of acceleration change has the form:

$$\begin{cases} \ddot{z} = \ddot{z}_T + Hg \cdot (\dot{z}_T - \dot{z}) + Hg_1 \cdot (z_T - z); \\ \ddot{y} = \ddot{y}_T + Hg \cdot (\dot{y}_T - \dot{y}) + Hg_1 \cdot (y_T - y); \\ \ddot{\alpha} = \ddot{\alpha}_T + Hg \cdot (\dot{\alpha}_T - \dot{\alpha}) + Hg_1 \cdot (\alpha_T - \alpha), \end{cases} \tag{3}$$

where Hg, Hg_1 are the coefficients that determine the type and rate of the transient process; $z_T(t)$, $y_T(t)$, $\alpha_T(t)$, $\dot{z}_T(t)$, $\dot{y}_T(t)$, $\dot{\alpha}_T(t)$, $\ddot{z}_T(t)$, $\ddot{y}_T(t)$,$\ddot{\alpha}_T(t)$ are the required coordinates, velocities, and accelerations; $z(t)$, $y(t)$, $\alpha(t)$, $\dot{z}(t)$, $\dot{y}(t)$, $\dot{\alpha}(t)$, $\ddot{z}(t)$, $\ddot{y}(t)$, $\ddot{\alpha}(t)$—actual coordinates, velocities, and accelerations.

In this case, the forces in the drives are determined according to the relations:

$$\begin{cases} P_1 = m \cdot (\ddot{z}_T + Hg \cdot (\dot{z}_T - \dot{z}) + Hg_1 \cdot (z_T - z)) \cdot A_{11} + \\ +m \cdot (\ddot{y}_T + Hg \cdot (\dot{y}_T - \dot{y}) + Hg_1 \cdot (y_T - y)) \cdot A_{22} + \\ +J \cdot (\ddot{\alpha}_T + Hg \cdot (\dot{\alpha}_T - \dot{\alpha}) + Hg_1 \cdot (\alpha_T - \alpha)) \cdot A_{33}; \\ P_2 = m \cdot (\ddot{z}_T + Hg \cdot (\dot{z}_T - \dot{z}) + Hg_1 \cdot (z_T - z)) \cdot B_{11} + \\ +m \cdot (\ddot{y}_T + Hg \cdot (\dot{y}_T - \dot{y}) + Hg_1 \cdot (y_T - y)) \cdot B_{22} + \\ +J \cdot (\ddot{\alpha}_T + Hg \cdot (\dot{\alpha}_T - \dot{\alpha}) + Hg_1 \cdot (\alpha_T - \alpha)) \cdot B_{33}; \\ P_3 = m \cdot (\ddot{z}_T + Hg \cdot (\dot{z}_T - \dot{z}) + Hg_1 \cdot (z_T - z)) \cdot C_{11} + \\ +m \cdot (\ddot{y}_T + Hg \cdot (\dot{y}_T - \dot{y}) + Hg_1 \cdot (y_T - y)) \cdot C_{22} + \\ +J \cdot (\ddot{\alpha}_T + Hg \cdot (\dot{\alpha}_T - \dot{\alpha}) + Hg_1 \cdot (\alpha_T - \alpha)) \cdot C_{33}, \end{cases}$$

where $A_{11}, A_{22}, \ldots, C_{33}$ are variable coefficients that nonlinearly depend on the output link coordinates obtained from the solution of the velocity problem.

4 Computational Experiment

As an example, the case was considered when the required movement is only associated with movement along the Y axis:

$$\begin{cases} z_T(t) = 0; \\ y_T(t) = 0,5 \cdot \sin(10 \cdot t); \\ \alpha_T(t) = 0. \end{cases}$$

Z-axis movement and α rotation must be zero. However, due to the presence of dynamic mutual influence between the drives, motion was obtained in all three coordinates. The previously mentioned control law (3) allows quickly reducing to zero the changes of the coordinates z and α (Fig. 4).

In the considered motion with feedback, we obtain graphs of changes in the velocities of the center of mass: $\dot{z}(t)$, $\dot{y}(t)$, $\dot{\alpha}(t)$ (Fig. 5):

A method of dynamic analysis of parallel structure mechanisms is developed. It allows one to get a solution of the dynamics problem without performing linearization. It is advisable to use this technique to organize the management of parallel structure mechanisms. It turned out that with the corresponding coefficients Hg, Hg_1

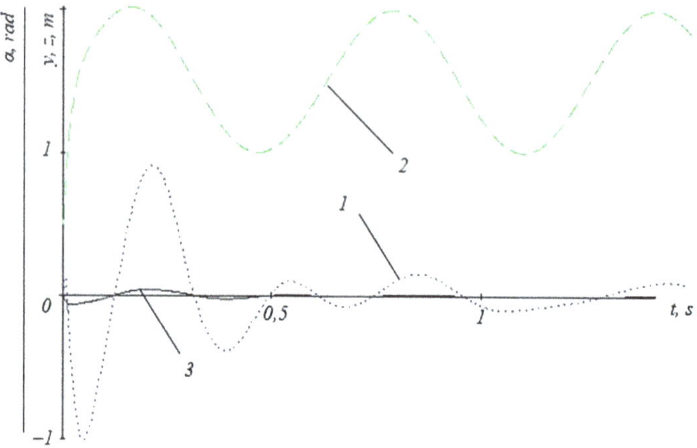

Fig. 4 Stabilization of coordinates z and α of the output link: 1—$z(t)$, 2—$y(t)$, 3—$\alpha(t)$

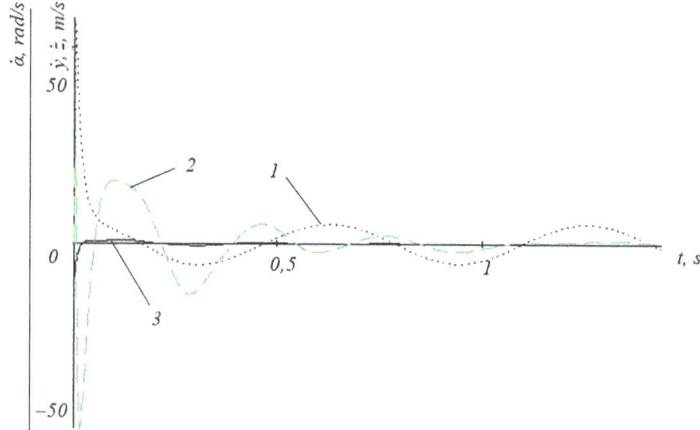

Fig. 5 Changing the velocities of the output link: 1—$\dot{z}(t)$, 2—$\dot{y}(t)$, 3—$\dot{\alpha}(t)$

of the system (3), providing a given transition process, it is possible to obtain an acceptable and stable solution of the problem.

At the same time, one can consider such situations when the model does not correspond to the actual parameters of the mechanism. This algorithm allows you to control the output link with sufficient accuracy and with the preservation of the stability condition of the movement. The developed approach allows one to conduct numerical experiments and select the necessary parameters for the control system. The technique can be modified to apply to other classes of mechanisms.

In order to test the functionality in practice, designs were developed, and the corresponding prototypes were made. Figure 6 shows a spatial 5-DOF mechanism for laser head controlling. The mechanism consists of the base, an output link (laser

Fig. 6 Model of a manipulator with a planar partial mechanism

head), which are connected by four two-wire kinematic circuits. The ability to process elongated parts, in this case, is achieved by moving the mechanism with the laser head along one coordinate by the straight guides. The control of the output link in the plane perpendicular to the axis of the mechanism movement is provided by the coordinated movement of the four linear actuators. The design of this mechanism is protected by a patent (12).

5 Conclusion

In the above study, a current scientific problem of developing a new family of parallel-serial mechanisms with 5-DOF for multi-axis manipulation systems is considered. The methodology of structural synthesis and kinematic analysis of parallel-serial mechanisms has been developed. The considered manipulator has prospects of application in various multi-axis robotic systems.

For the mechanism under consideration, the problems of positions and velocities are solved, and a dynamic analysis is carried out. A design was developed, a prototype was made for the implementation of additive technologies and focused on the manufacture of elongated parts (Fig. 6), the control system of which was based on the method of kinematic and dynamic analysis described in the paper. The maximum deviation from the experimentally reproducible trajectory according to the results of one hundred measurements is no more than 20 microns.

References

1. Glazunov, V.A.: Parallel Structure Mechanisms and Their Application: Robotic, Process, Medical, Educational Systems. Institute for Computer Research, Moscow-Izhevsk (2018)
2. Glazunov, V.A., Filippov, G.S., Rashoyan, G.V., Aleshin, A.K., Shalyukhin, K.A., Skvortsov, S.A., Antonov, A.V., Terekhova, A.N.: Velocity analysis of a spherical parallel robot. J. Phys. Conf. Ser. **1260**(11), 112012 (2019)
3. Neumann, K.E.: Robot. US Patent No. 4,732,525 (1988)
4. Lai, C.Y., Chavez, D.E.V., Ding, S.: Transformable parallel-serial manipulator for robotic machining. Int. J. Adv. Manuf. Technol. **97**(5), 2987–2996 (2018)
5. Zhang, D., Xu, Y., Yao, J., Zhao, Y.: Design of a novel 5-DOF hybrid serial-parallel manipulator and theoretical analysis of its parallel part. Robot. Comput.-Integr. Manuf. **53**, 228–239 (2018)
6. Zhoga, V.V., Dyashkin-Titov, V.V., Nesmiyanov, I.A., Vorobieva, N.S.: The task of positioning a manipulator of a parallel-sequential structure with a controlled gripper. Mechatron. Autom. Control **17**(8), 525–530 (2016)
7. Dyashkin-Titov, V.V., Zhoga, V.V., Nesmiyanov, I.A., Vorob'eva, N.S.: Dynamics of the manipulator parallel-serial structure. In Advances in Mechanical Engineering, pp. 33–43. Springer, Cham (2018)
8. Glazunov, V.A., Filippov, G.S., Lastochkin, A.B.: Development and Application of Robotic Systems Based on Modern Principles for Carrying Out Process Operations in the Manufacture of the Central Body of the Turbojet Engine Nozzle. Perspective methods of processing machine parts. Lenand, Moscow (2019)
9. Glazunov, V.A., Filippov, G.S., Lastochkin, A.B., Ceccarelli, M., Skvortsov, S.A., Rashoyan, G.V., Aleshin, A.K., Shaluhin, K.A.: 5DOF mechanism for vertebral surgery kinematic analysis and velocity calculation. In: IFToMM World Congress on Mechanism and Machine Science, pp. 1741–1749. Springer, Cham (2019)
10. Gosselin, C.M.: Parallel computational algorithms for the kinematics and dynamics. Measur Control **118**(1), 22–28 (1996)
11. Aleshin, A.K., Glazunov, V.A., Shai, O., Rashoyan, G.V., Skvortsov, S.A., Lastochkin, A.B.: Infinitesimal displacement analysis of a parallel manipulator with circular guide via the differentiation of constraint equations. J. Mach. Manuf. Reliab. **45**(5), 398–402 (2016)

Studying of Copying Control System with Nonlinear Measurer

Sergey Jatsun, Andrei Malchikov, Andrey Yatsun, and Ekaterina Saveleva

Abstract The article is devoted to the development and mathematical modeling of the exoskeleton suit femoral link copying control system. The material discusses various approaches to ensuring the master and slave links movement synchronization in a human–machine system and minimizing the efforts of interaction between them. To control the force between the links, an original design measurer with nonlinear mechanical properties is used. On the proposed criterion for assessing the regulation quality basis, an analysis of the applicability of various methods for the copying control task is carried out. To compensate the feedback loop nonlinear effects, it is proposed to use a corrective action block, the parameters of which are determined experimentally. As the numerical simulations results shown, the control system with a correcting device usage makes it possible to obtain the exoskeleton suit drive system movement required parameters executive link. Integral error values less than 0.7% of the amplitude for the system studied parameters.

1 Introduction

The creation of modern robotic devices for medical, industrial, and other applications involves the use of advanced automatic control systems and adaptive, intelligent controllers [1–3]. The construction of complex control systems requires the use of measuring systems with certain properties. The design and adjustment of system properties determine the operational properties of the final device. One of the most interesting tasks in terms of building complex measuring systems is the development of a sensor system in human–machine devices [4–8]. In this case, the measurer is characterized not only by indicators of accuracy, repeatability, range, etc., but also by the influence on the controlled system functioning. The example of human–machine systems are lower extremities active exoskeletons [8–10].

S. Jatsun · A. Malchikov (✉) · A. Yatsun · E. Saveleva
Department of Mechanics, Mechatronics and Robotics, South-West State University, 94, 50 Let Oktyabrya Str, 305040 Kursk, Russia
e-mail: teormeh@inbox.ru

A. Ronzhin and V. Shishlakov (eds.), *Electromechanics and Robotics*, Smart Innovation, Systems and Technologies 232, https://doi.org/10.1007/978-981-16-2814-6_2

Various design schemes of exoskeletons are known. Depending on the field of application and the characteristics of performed tasks, exoskeletons can be passive or active, driven by various types of drives. This paper presents an active exoskeleton, which is an electromechanical unit equipped with a copy control system. This paper considers the active exoskeleton hip joint, which is an electromechanical unit equipped with a copying control system [11]. The source [11] considers the copying system implementation for the exoskeleton executive links movement control using measuring cuffs based on force and displacement measurers. As full-scale experiments shown, such structures, including guides and springs, are difficult for manufacturing and configuration, and they also have significant mass-dimensional characteristics that restrict their application. The paper considers the influence of the measurer nonlinear properties on the copying industrial exoskeleton femoral unit control system functioning. An original method of compensating effect creating, which allows to take into account the measurer nonlinear properties which generates drive system supply voltages, is proposed.

2 The Control Object Description

The electromechanical system considered in the work is presented as a pair of links fixed at one point. The leader link movement is set kinematically, and the follower link movement occurs under the action of the drive force and a visco-elastic element—a measuring cuff which connects the links to each other.

To generate drive control voltages, the information from force sensors is used. The force sensors are located in the cuffs devices for attaching of the exoskeleton links to the operator's thighs.

The design scheme of the investigated man–machine system can be presented in the form shown in Fig. 1.

In this diagram, the exoskeleton link is represented by the segment $O_1 O_{E3}$, fixed in the hinge O_1. The force F_{act} acts from the side of the linear actuator, applied at the point O_{A2} at an angle α_{act}, and the force from the side of the cuff attached to the operator's thigh applied at the point O_{E2} acts on the link. The operator's thigh is represented on the diagram by the segment $O_1 O_{H3}$, fixed at the point O_1. Let us assume that, in addition to gravity, the thigh is acted upon by the force from the exoskeleton applied at the point O_{H2}.

The task of the control system is to provide such a movement of the exoskeleton link, at which the required constant force on the cuff P would be provided (for the copying mode of movement, $P = 0$). An important feature of this work is the use of the original design measuring cuff with nonlinear properties, through which the force interaction between the operator and the exoskeleton is transmitted.

The system motion dynamics can be described by the equations [12]:

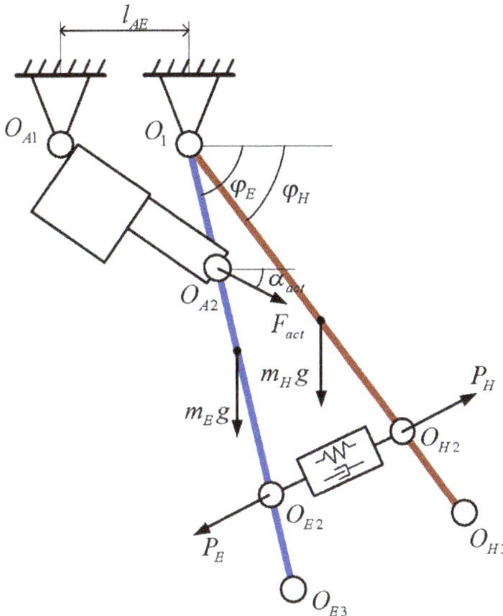

Fig. 1 Device design scheme

$$\phi_E \frac{m_E L_E^2}{3} = L_A F_{\text{act}} \cos\alpha \cdot \sin\phi_E - \frac{L_E}{2} m_E g \cos\phi_E - L_{CE} P_E \cos\left(\frac{\phi_H - \phi_E}{2}\right); \quad (1)$$

$$\phi_H \frac{m_H L_H^2}{3} = L_{CH} P_H \cos\left(\frac{\phi_H - \phi_E}{2}\right) - \frac{L_H}{2} m_H g \cos\phi_H.$$

For these equations, the following designations are accepted: L_E, L_H—length of the exoskeleton's femoral link and the operator's thigh, L_A—the distance from the hinge to the motor mount, L_{CE}, L_{CH}—the distance from the hinge to the cuff attachment for the exoskeleton and the operator, respectively; $m_E m_H$—masses of the exoskeleton link and the operator's hip.

To obtain a mathematical model of a prototype measuring cuff, consider the design diagram shown in Fig. 2.

The device works as follows. The force F_H from the operator 1 through the mount 2 is transmitted to the arm of the cuff, which consists of two plates—deformable 4, which takes up most the load and the force 7 on which the limiter is located 6. The force acting on the measuring plate is recorded by the strain gauge 5 (YZC131), which is interrogated using the HX711 load cell amplifier. Force measurement accuracy is not less than 0.01(H). By adjusting the position of the limiter, you can obtain the required nonlinear elastic characteristic of the sensor.

The operation principle, as well as the bracket numerical parameters, was studied experimentally on a laboratory bench. In the course of the experiments, the characteristics for the cuff without using a mechanical stop were obtained ($s \ll s_L$). The

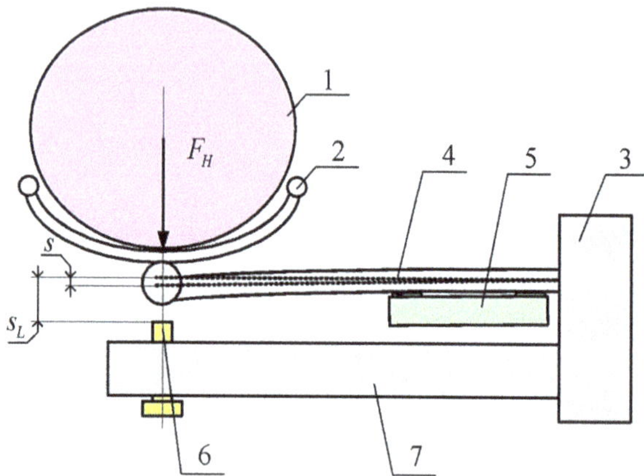

Fig. 2 Measuring cuff scheme. In this diagram, the following designations are adopted: 1—operator's thigh, 2—thigh attachment, 3—the exoskeleton attachment bracket, 4—deformable (measuring) bracket plate, 5—strain gauge sensor, 6—limiter, 7—the bracket force plate

cuff with a mechanical stop was set to a gap of 0.3 mm (s_L= 0.3). The results of both measurements will be shown in the form of graphs (Fig. 3).

As the tests results showed, the suspension final rigidity is approximately 20 (kN/m), upon reaching the limiter it becomes equal to about 300 (kN/m). Thus, the elastic force as a displacement function can be represented as follows:

$$F = c_1 s, if\ |s| < s_L otherwise (c_1 + c_2)s \tag{2}$$

where s—is the amount of deformation (the cuff attachment point displacement), c_1—is the coefficient of the measuring plate elasticity (20 kN/m), and c_2—is the coefficient of the cuff force plate elasticity (300 kN/m). The sensor's sensitive element readings are also of interest, which also has a variable slope of the graph when a load is applied.

$$F_s = cs_1 s if\ |s| < s_L otherwise (cs_1 + cs_2)s \tag{3}$$

cs_1—is the scaling sensor coefficient, and cs_2—is the scaling sensor coefficient when it is fixedly the limiter.

Using the obtained data, you can write an equation for the mathematical model of the measuring cuff as follows:

$$|P_E| = |P_H| = \begin{matrix} c_1 s + \mu\left(\dot{\phi}_H - \dot{\phi}_E\right) \text{ if } |s| < s_L \\ (c_1 + c_2)s + \mu(\dot{\phi}_H - \dot{\phi}_E) \text{ if } |s| \geq s_L \end{matrix}, \tag{4}$$

where μ—viscous coefficient (take equal 20 Ns) and s is the gap size, determined by the angular displacement of the links relative to each other:

$$s = L_H \sqrt{2 - 2\cos\cos\ (\phi_H - \phi_E)}\,.$$

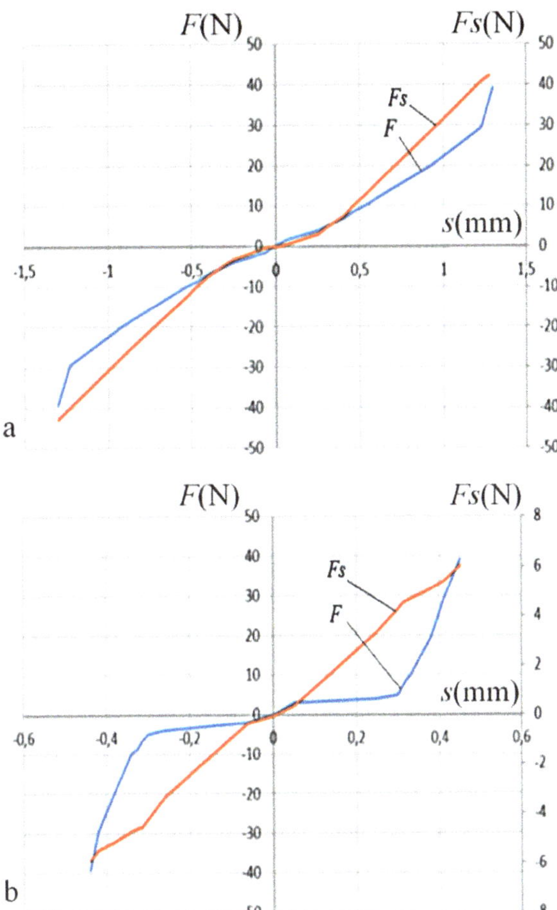

Fig. 3 Dependences of the force value on the displacement of the measuring plate: **a** without a limiter, **b** with a limiter

3 The Control System Methods

Let us consider different ways of determining the exoskeleton drive system supply voltage. Let us set the hip displacement kinematically as follows:

$$\begin{aligned}\phi_H &= A_0 \sin\ (\omega t),\\ \dot{\phi}_H &= A_0 \omega \cos\ (\omega t),\\ \ddot{\phi}_H &= -A_0 \omega^2 \sin\ (\omega t)\end{aligned} \tag{5}$$

From Eq. (1), we can obtain an expression for the interaction force (6):

$$P_H = \left(\frac{L_H}{2} m_H g \sin\phi_H - \ddot{\phi}_H \frac{m_H L_H^2}{3}\right) L_{CH} \cos\left(\frac{\phi_H - \phi_E}{2}\right), \tag{6}$$

Substituting expressions for the defining motion:

$$P_H = (\frac{L_H}{2} m_H g \sin(A_0 \sin(\omega t)) + (A_0 \omega^2 \sin(\omega t))(\frac{m_H L_H^2}{3}) L_C H \cos(\frac{A_0 \sin(\omega t) - \phi_E}{2}) \quad (7)$$

Note that the copying control system task is reduced to minimizing the interaction force, which is determined by the distance between the cuff attachment points $O_{E2}O_{H2}$.

In addition to the P_H force described by Eqs. (4) and (7), the movement of the exoskeleton link is affected by the drive force [13]:

$$F_{\text{act}} = c_M I \cdot k_{pg} k_b \eta \frac{2\pi}{pn},$$

where: c_M—is the electric motor torque coefficient, I—is the motor current, k_{pg}, k_b—is the total gear ratio of the planetary and belt reducers, η—is the reduced total efficiency of the mechanism, p—is the pitch of the ball screw thread, and n—is the number of ball screw starts.

The current strength can be obtained from the Kirchhoff equation for the armature circuit:

$$U = L\dot{I} + RI + k_\omega \dot{\varphi},$$

where L and R are the inductance and active resistance of the armature circuit, k_ω is the motor speed coefficient, and $\dot{\varphi} = \dot{\phi}_E k_{pg} k_b \frac{2\pi}{pn}$—is the shaft rotation speed.

Neglecting the current inductive component, not having a significant effect on the process for a given motion mode, we can obtain the supply voltage value required to create the required force F_{act}:

$$U = \frac{R F_{act}}{c_M k_{pg} k_b \eta \frac{2\pi}{pn}} + k_\omega \dot{\phi}_E k_{pg} k_b \frac{2\pi}{pn}. \quad (8)$$

Further, the drive force value can be obtained from Eq. (8), which ultimately gives an expression for the voltage required to implement the motion mode.

4 Simulation Results

Let us show the results of mathematical modeling of a proportional control system for a device with mechanical stops in the measuring cuff. Movement parameters of the master link according (5) are $A_0 = 0.5$rad, $\omega = 2$ rad/s.

As can be seen from the simulation results, the appearance of nonlinearities, which cannot be taken into account in proportional control, significantly reduce the control quality (Fig. 4).

Since the process is periodic, we take the integral positioning error of the links as a criterion reflecting the regulation quality in the following form:

$$K_I = \frac{1}{T}\int_0^T |\varphi_H - \varphi_E| \mathrm{d}t,$$

where T modeling time.

The integral error for the graph shown is $K_I = 0.0768$.

To ensure the required control quality, it is proposed to use a feedback sensor that measures the force between the operator's leg and the exoskeleton link. Consider

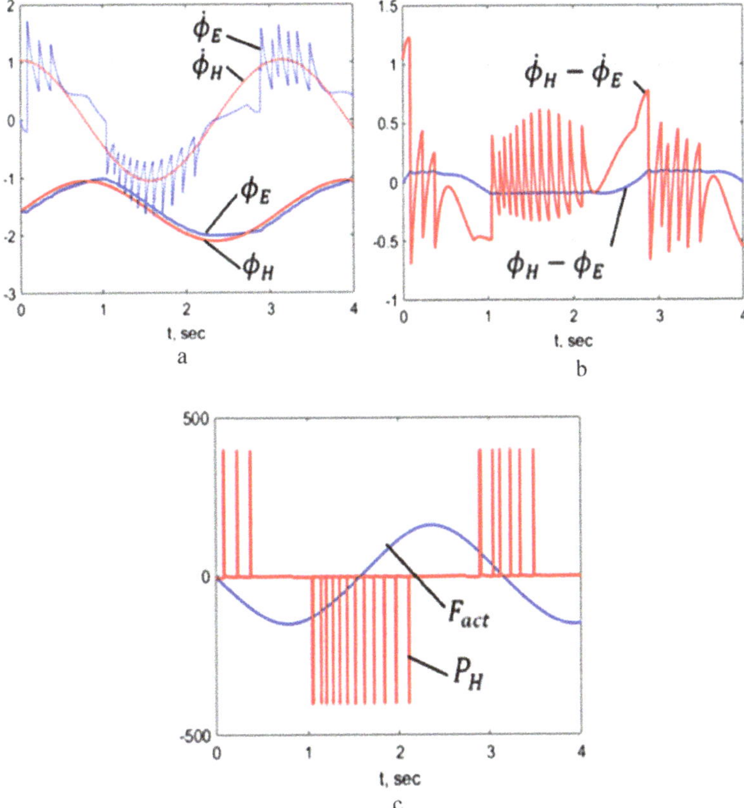

Fig. 4 Simulation results with proportional control and nonlinear cuff: **a** angular displacement (rad) and velocity (rad/s), **b** positioning and velocity error, **c** actuator force (N)

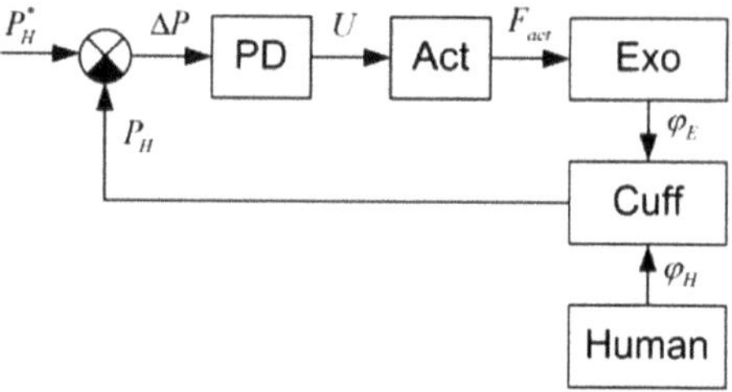

Fig. 5 Structure of the closed control system

the system behavior when controlling the human–machine interaction strength, the structural diagram will be as follows (Fig. 5):

In this case, the drive voltage will be determined by the expression:

$U = k_p\left(P_H^* - P_H\right) - k_d\left(\frac{dP_H^*}{dt} - \frac{dP_H}{dt}\right).$

For the PD controller with parameters $k_p = 0.5, k_d = 0.2$ (parameters were obtained experimentally), the simulation results for nonlinear cuff will look as shown in Fig. 6.

As can be seen from the simulation results, the feedback introduction makes it possible to drive system actuator motion control process more accurately. The integral error is $K_I = 0.0718$. However, as can be seen in Fig. 6, if the cuff movement stop is used, the system demonstrates an oscillatory process with each change of the driving link movement direction.

To take into account the measurer nonlinear properties, a control algorithm is proposed using a corrective action block, the parameters of which are determined experimentally.

The adaptive control system structure is shown in Fig. 7.

In this case, the drive voltage will be determined by the expression:

$U = k_p\left(P_H^* - (P_H + P'_H)\right) - k_d\left(\frac{dP_H^*}{dt} - \frac{d(P_H + P'_H)}{dt}\right),$

where P'_H—is the correcting interaction force value obtained from the block MM equations for which are (2) and (3).

Using the obtained characteristics of the measuring cuff in the corrective block, we simulate the system under the same initial conditions (Fig. 8).

The integral error value for adaptive control is $K_I = 0.0039$. Paying attention to the initial surge on the charts—it is clear that it is duties by the model initialization. At the initial moment of time, all the variables values are equal to zero that means that the model requires several iterations in order to form the corrective action for the PD controller correctly.

5 Conclusions

The paper presents a study of the copying control system. Various approaches to the of the drive system exoskeleton hip joint supply voltage formation are described.

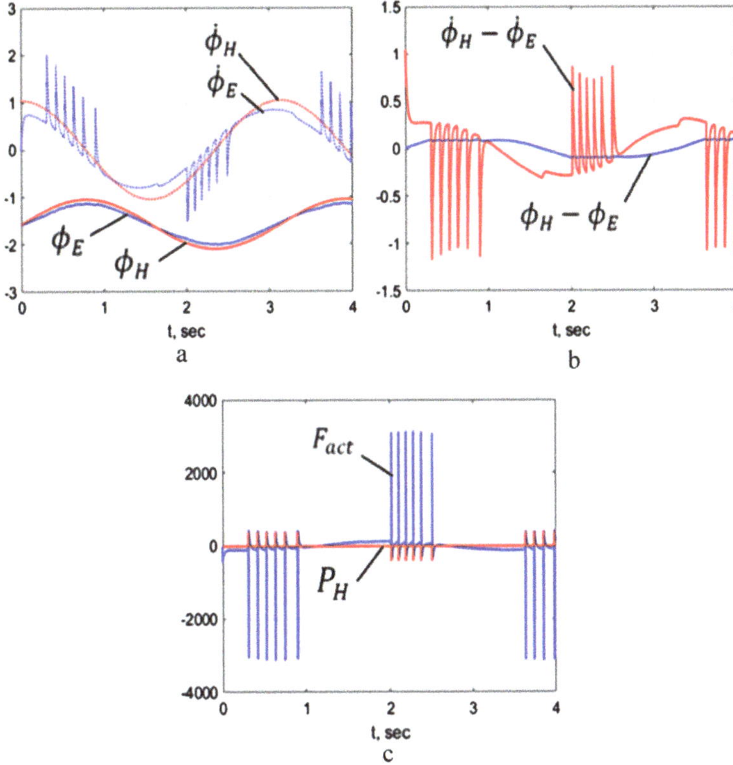

Fig. 6 Modeling of a linear PD controller: **a** angular displacement (rad) and velocity(rad/s), **b** positioning and velocity error, and **c** actuator force (N)

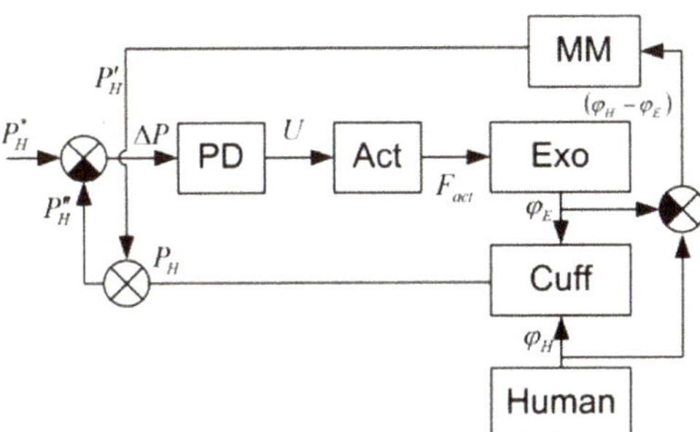

Fig. 7 Control system with a corrective action block MM diagram

The best regulation quality was shown by the copying control system algorithm with the usage of the measuring cuff model. The corrective action block parameters the reflecting its nonlinear properties are obtained experimentally. The proposed copying control system usage made it possible to reduce the integral error values to

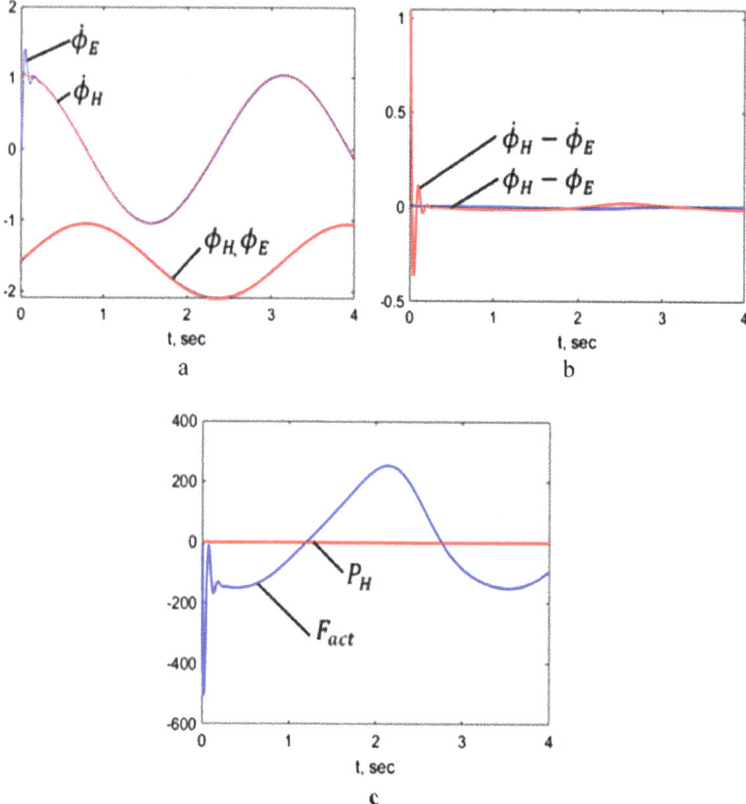

Fig. 8 Results of the control algorithm simulation using the corrective action block

0.0039 rad (0.7% of the amplitude), and to eliminate the effect of hesitation when changing the movement direction.

Acknowledgements The work was supported by RFBR, research project № 19-08-00440, and Andrei Malchikov was supported by the President grant, project MK-780.2020.8

References

1. He, W., Dong, Y.: Adaptive fuzzy neural network control for a constrained robot using impedance learning. IEEE Trans. Neural Netw. Learn. Syst. **29**(4), 1174–1186 (2017)
2. Anam, K., Al-Jumaily, A.A.: Active exoskeleton control systems: state of the art. Procedia Eng. **41**, 988–994 (2012)
3. Kazerooni, H., Racine, J.L., Huang, L., Steger, R.: On the control of the berkeley lower extremity exoskeleton (BLEEX). In: Proceedings of the 2005 IEEE International Conference on Robotics and Automation, pp. 4353–4360 (2005)

4. Yu, Y., Nassar, J., Xu, C., Min, J., Yang, Y., Dai, A., Doshi, R., Huang, A., Song, Y., Gelhar, R., Ames, A.D., Gao, W.: Biofuel-powered soft electronic skin with multiplexed and wireless sensing for human-machine interfaces. Sci. Robot. **5**(41) (2020)
5. Huang, Q., He, S., Wang, Q., Gu, Z., Peng, N., Li, K., Zhang, Y., Shao, M., Li, Y.: An EOG-based human–machine interface for wheelchair control. IEEE Trans. Biomed. Eng. **65**(9), 2023–2032 (2017)
6. Dong, W., Wang, Y., Zhou, Y., Bai, Y., Ju, Z., Guo, J., Gu, G., Bai, K., Ouyang, G., Chen, S., Zhang, Q., Huang, Y.: Soft human–machine interfaces: design, sensing and stimulation. Int. J. Intell. Robot. Appl. **2**(3), 313–338 (2018)
7. De. Santis, A., Siciliano, B., De. Luca, A., Bicchi, A.: An atlas of physical human–robot interaction. Mech. Mach. Theory **43**(3), 253–270 (2008)
8. Jatsun, S., Malchikov, A., Loktionova, O., Yatsun, A.: Modeling of human-machine interaction in an industrial exoskeleton control system. In: International Conference on Interactive Collaborative Robotics, pp. 116–125. Springer, Cham (2020)
9. Li, Z., Huang, B., Ye, Z., Deng, M., Yang, C.: Physical human–robot interaction of a robotic exoskeleton by admittance control. IEEE Trans. Industr. Electron. **65**(12), 9614–9624 (2018)
10. De. Rossi, S.M.M., Vitiello, N., Lenzi, T., Ronsse, R., Koopman, B., Persichetti, A., Carrozza, M.C.: Sensing pressure distribution on a lower-limb exoskeleton physical human-machine interface. Sensors **11**(1), 207–227 (2011)
11. Jatsun, S., Malchikov, A., Yatsun, A.: Comparative analysis of the industrial exoskeleton control systems. In: Proceedings of 14th International Conference on Electromechanics and Robotics "Zavalishin's Readings", pp. 63–74. Springer, Singapore (2020)
12. Lewis, F.L., Dawson, D.M., Abdallah, C.T.: Robot Manipulator Control: Theory and Practice. CRC Press, Boca Raton (2003)
13. Yatsun, A., Karlov, A., Malchikov, A., Jatsun, S.: Investigation of the dynamical characteristics of the lower-limbs exoskeleton actuators. MATEC Web of Conferences, Vol. 161, p. 03008 EDP Sciences (2018)

Determination of Special Positions for Solving the Problem of Joint-Relative Manipulation Mechanisms Kinematic Control

Vasily Pashchenko, Alexey Romanov, Maxim Chaikin, Vladimir Zakharov, Vasily Pashchenko, and Alexey Romanov

Abstract The article deals with topical issues of solving the problem of analyzing the special provisions of the mechanisms of joint-relative manipulation. The analysis of existing methods and approaches for determining the special provisions of the mechanisms is carried out. The approach based on the search for bifurcation conditions is considered in detail. Within the framework of using this approach, special positions of the planar six-branch mechanism are found through the compilation of coupling equations. The working area of the flat six-shaft mechanism is constructed taking into account the permissible and special provisions. The method of finding the working space of the mechanism of joint-relative manipulation using the iterative method is shown. The working area of the mechanism of joint-relative manipulation is constructed, taking into account the permissible and special provisions. The working area of the mechanisms of joint-relative manipulation is investigated, taking into account special positions, and an approach to solving the problem of forming a program trajectory in the space of generalized coordinates is proposed, taking into account the presence of special positions of the mechanism.

1 Introduction

Modern production development requires manufacturers and developers to introduce new robotic mechanisms. In frequency, one of the most promising areas of development of the machine-building industry is the creation of effective technological equipment with high accuracy, productivity, and load capacity [1–4].

V. Pashchenko (✉) · A. Romanov · M. Chaikin · V. Zakharov
Kaluga Branch of the Bauman Moscow State Technical University, Kaluga, Russia
e-mail: pashenkovn@inbox.ru

V. Pashchenko
AO "Kaluga Astral", Kaluga, Russia

A. Romanov
Blagonravov Mechanical Engineering Research Institute of Russian Academy of Sciences, Moscow, Russia

A. Ronzhin and V. Shishlakov (eds.), *Electromechanics and Robotics*, Smart Innovation, Systems and Technologies 232, https://doi.org/10.1007/978-981-16-2814-6_3

For the synthesis of equipment with such requirements, it is possible to use mechanisms of a parallel structure. Such mechanisms have increased performance in terms of carrying capacity, accuracy, and speed of movement of the output link since they perceive the load similarly to spatial trusses [5–7]. However, having a wide range of advantages, they have several disadvantages, including a relatively small working area [8–10] and the presence of special positions (singularities) [11, 12]. Special positions are those positions in which the mobility of the links is lost, which can lead to a loss of the degree of freedom, loss of structural rigidity, uncontrolled movement of both intermediate links, and the output link of the mechanism [13–16].

These drawbacks can be eliminated by organizing joint-relative manipulation. The general structural synthesis of joint-relative manipulation mechanisms (JRMM) is the presence of two output moving links. On the one hand, the organization of such a structure allows both reproducing a given trajectory, orienting the body in a moving coordinate system, and the movement of the coordinate system itself, which makes it possible to eliminate the indicated disadvantages [17]. On the other hand, they complicate the control system. Effective management of the ISM requires knowledge of the structure of the work area (size, availability of special provisions).

The most well-known approaches to determining singular positions are:

1. Method Angeles and Goslin [18] is based on finding the Jacobi matrices, for the recording of which the constraint equations are drawn up, substantiating the relationship between a set of generalized coordinates, as well as the values of coordinates and angles of position of the output link of the mechanism.
2. A method for finding bifurcation conditions for determining singular positions. The method is based on the following statement: the bifurcation conditions characterize the unstable positions of the mechanism in terms of its structural parameters, and the singularities describe the same positions in space [19]. Bifurcation conditions can be found through the compilation of the Jacobi and Hessian matrices [20], through the equilibrium or compatibility equations in closed kinematic structures [21], and also through the analysis of higher-order constraint equations [22].
3. Power screws method [23]. The method is based on the definition of power screws, which are transmitted to the output link by means of kinematic chains. In this case, the condition for degeneracy is the linear dependence of the found power screws.

The aim of the work is to find special positions of the mechanism and to solve the problem of forming the JRMM program trajectory based on the data obtained.

The objectives of this work are to determine singular positions, study the JRMM working area taking into account singular positions, and solve the problem of forming a trajectory in the space of generalized coordinates.

2 Formulation of the Problem

Let us consider the possibilities of using the method for finding the bifurcation conditions to determine the singular positions through the compatibility equations of the mechanism.

When analyzing closed mechanisms with kinematic m pairs, it is always possible to single out n kinematic pairs associated with executive (leading) links and $m - n$ kinematic pairs associated with passive (free) links. The quantities that describe the angle of rotation of the connection associated with the drive kinematic pairs will be designated as control variables $\theta = (\theta_1, \theta_2, \ldots \theta_3)$, and the quantities that describe the position of passive kinematic pairs will be designated as state variables $\xi = (\xi_1, \xi_2, \ldots, \xi_{m-n})$.

In the process of movement of the output link, the mechanism must satisfy the condition of the closed kinematic chain, which is expressed by the compatibility equation, which describes the dependence of geometric parameters on state variables:

$$F(\xi, \theta) = 0.$$

Let us differentiate the compatibility equation with respect to control variables:

$$\sum_{j=1}^{m-n} \frac{\partial F_k}{\partial \xi_j} K_m + \frac{\partial F_k}{\partial \theta_i} = 0, \quad i = 1 \ldots n, \ j = 1 \ldots m - n,$$

where k—the number of compatibility equations, which is equal to the number of state variables,

$$K_m = \frac{\partial \xi_j}{\partial \theta_i}.$$

If the parameter K_{m} is equal to 0 or undefined, then the state variables cease to be dependent on the control variables, which means that bifurcation conditions can be found that lead to the appearance of singular positions.

Let us find the generalized speeds of the mechanism, proceeding from the fact that the state variables in mechanisms with closed kinematic chains are functions of the control variables:

$$\frac{d\xi_j}{\mathrm{d}t} = \frac{\partial \xi_j}{\partial \theta_i} \frac{\mathrm{d}\theta_i}{\mathrm{d}t} = K \frac{\mathrm{d}\theta_i}{\mathrm{d}t}; \tag{1}$$

$$\frac{\mathrm{d}\theta_i}{\mathrm{d}t} = \frac{\partial \theta_i}{\partial \xi_j} \frac{\mathrm{d}\xi_j}{\mathrm{d}t} = \frac{1}{K} \frac{\mathrm{d}\xi_j}{\mathrm{d}t}. \tag{2}$$

Let us present the following classification of condition for the appearance of a singularity:

Fig. 1 Joint-relative manipulation mechanism

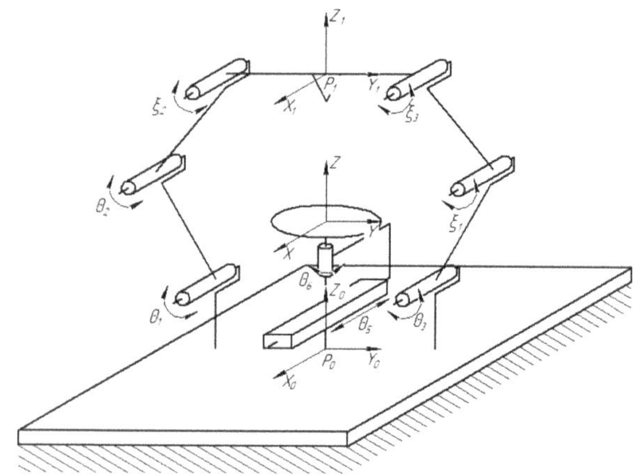

1. $K = 0$—the first type of singularity. Based on Eq. (1), any control action $\frac{d\theta_i}{dt}$ corresponds to a nonzero speed $\frac{d\xi_j}{dt}$. This means that the passive joint loses one or more degrees of mobility and cannot perform the required movement under a given action of the executive joint.
2. $K = \infty$—the second type of singularity. Based on Eq. (2), any output speed $\frac{d\xi_j}{dt}$ corresponds to a control action $\frac{d\theta_i}{dt}$ equal to zero. Consequently, the passive joint receives one or more additional degrees of mobility.
3. $K = 0/0$—the third type of singularity. In this case, the executive joint can move with a passive joint, and the corresponding passive joint can move with a corresponding self-locking executive joint.

Let us solve the set tasks using the example of a JRMM with five degrees of freedom, the structure of which is shown in Fig. 1.

The mechanism consists of a six-link mechanism (drive kinematic pairs $\theta_1, \theta_2, \theta_3$), which has three degrees of freedom and a translational-rotary mechanism, which has two degrees of freedom (drive kinematic pairs θ_4, θ_5). $X_1Y_1Z_1$—coordinate system of the output link of the six-link mechanism. $X_0Y_0Z_0$—coordinate system of the output link of the translational-rotary mechanism, and *XYZ*—the basic coordinate system JRMM.

3 Finding Conditions for the Occurrence of Singular Positions

The translational-rotary mechanism is an open kinematic chain, has two degrees of freedom, and can carry out translational and rotational movements independent of each other, which are characterized by control variables θ_5 and θ_6, respectively.

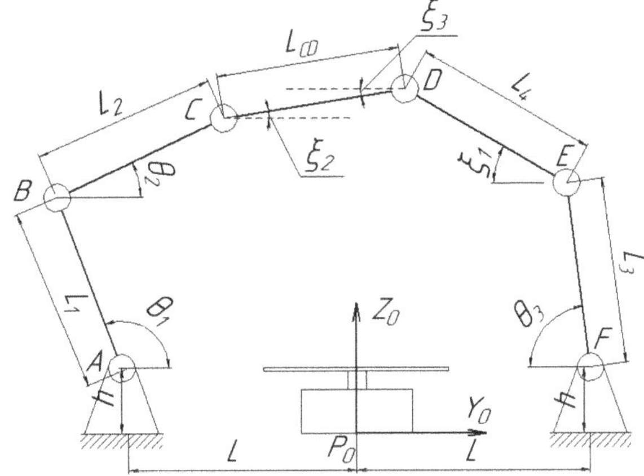

Fig. 2 General configuration of the six-link mechanism JRMM

Therefore, we can conclude that this mechanism does not have singular positions, and its working space is limited only by the size of the forward stroke.

Let us dwell on a detailed examination of the six-link mechanism. We denote the connection points of the intermediate links of the six-link mechanism as *A*, *B*, *C*, *D*, *E*, and *F*, and the links connecting *A* and $B - L_1$,, *C* and $D - L_2$, *E* and $D - L_3$, *F* and $E - L_4$, *C* and $D - L_{CD}$, the distance between the base coordinate system and the center of the base of the six-link mechanism—*L*, and the height between the base of the joint-relative manipulation mechanism to the first rotational kinematic pair of the six-link mechanism *h* (Fig. 2).

Since the state variables of the six-link mechanism ξ_2 and ξ_3 are equal, describe the same position of the output link, and are determined in the same way through the known parameters of the mechanism and control variables [24], we will consider only one state variable of the output link ξ_2. Let us compose compatibility equations for control variables $\theta_1,\ \theta_2,\ \theta_3$, and state variables and ξ_1 и ξ_2:

$$
\begin{aligned}
F_1 &= 2L - L_1 \cos\theta_1 - L_2 \cos\theta_3 - L_3 \cos\theta_3 - L_4 \cos\xi_1 - L_{CD} \cos\xi_2; \\
F_2 &= -L_1 \sin\theta_1 - L_2 \sin\theta_2 + L_3 \sin\theta_3 + L_4 \sin\xi_1 + L_{CD} \sin\xi_2,
\end{aligned}
\tag{3}
$$

where is F_1—the compatibility equation about the *Y*-axis and F_2—compatibility equation about the *Z*-axis.

Let us differentiate the compatibility Eq. (3) with respect to the control variables $\theta_1,\ \theta_2,\ \theta_3$:

$$
\begin{aligned}
\frac{\partial F_1}{\partial \theta_1} &= L_1 \sin\theta_1 + L_4 \sin\xi_1 \frac{\partial \xi_1}{\partial \theta_1} + L_{CD} \sin\xi_2 \frac{\partial \xi_2}{\partial \theta_1} = 0; \\
\frac{\partial F_2}{\partial \theta_1} &= -L_1 \cos\theta_1 + L_4 \cos\xi_1 \frac{\partial \xi_1}{\partial \theta_1} + L_{CD} \cos\xi_2 \frac{\partial \xi_2}{\partial \theta_1} = 0;
\end{aligned}
$$

$$\frac{\partial F_1}{\partial \theta_2} = L_2 \sin\theta_2 + L_4 \sin\xi_1 \frac{\partial \xi_1}{\partial \theta_2} + L_{CD} \sin\xi_2 \frac{\partial \xi_2}{\partial \theta_2} = 0;$$
$$\frac{\partial F_2}{\partial \theta_2} = -L_2 \cos\theta_2 + L_4 \cos\xi_1 \frac{\partial \xi_1}{\partial \theta_2} + L_{CD} \cos\xi_2 \frac{\partial \xi_2}{\partial \theta_2} = 0;$$
$$\frac{\partial F_1}{\partial \theta_3} = L_3 \sin\theta_3 + L_4 \sin\xi_1 \frac{\partial \xi_1}{\partial \theta_3} + L_{CD} \sin\xi_2 \frac{\partial \xi_2}{\partial \theta_3} = 0;$$
$$\frac{\partial F_2}{\partial \theta_3} = L_3 \cos\theta_3 + L_4 \cos\xi_1 \frac{\partial \xi_1}{\partial \theta_3} + L_{CD} \cos\xi_2 \frac{\partial \xi_2}{\partial \theta_3} = 0.$$

Let us find the parameters $K_1 \ldots K_6$ by calculating the dependence of the non-drive pairs on the drive ones:

$$K_1 = \frac{\partial \xi_1}{\partial \theta_1} = -\frac{L_1 \sin(\xi_2 - \theta_1)}{L_4 \sin(\xi_2 - \xi_1)}; \quad K_2 = \frac{\partial \xi_2}{\partial \theta_1} = -\frac{L_1 \sin(\theta_1 - \xi_1)}{L_{CD} \sin(\xi_2 - \xi_1)};$$
$$K_3 = \frac{\partial \xi_1}{\partial \theta_2} = -\frac{L_2 \sin(\xi_2 - \theta_2)}{L_4 \sin(\xi_2 - \xi_1)};$$

$$K_4 = \frac{\partial \xi_2}{\partial \theta_2} = -\frac{L_2 \sin(\theta_2 - \xi_1)}{L_{CD} \sin(\xi_2 - \xi_1)}; \quad K_5 = \frac{\partial \xi_1}{\partial \theta_3} = -\frac{L_3 \sin(\xi_2 - \theta_3)}{L_4 \sin(\xi_2 - \xi_1)};$$
$$K_6 = \frac{\partial \xi_2}{\partial \theta_3} = -\frac{L_3 \sin(\theta_3 - \xi_1)}{L_{CD} \sin(\xi_2 - \xi_1)}.$$

This means that $K_1 \ldots K_6$ will be equal to zero or undefined if one of the following conditions is met:

$$\sin(\xi_2 - \theta_1) = 0; \; \sin(\theta_1 - \xi_1) = 0; \; \sin(\xi_1 - \theta_2) = 0; \; \sin(\theta_2 - \xi_1) = 0;$$
$$\sin(\xi_2 - \theta_3) = 0; \; \sin(\theta_3 - \xi_1) = 0; \; \sin(\xi_2 - \xi_1) = 0. \quad (4)$$

Let us find the conditions for the appearance of singular positions by solving Eq. (3) taking into account each of the expressions (4), respectively (Table 1).

Table 1 Singular positions of the six-link mechanism

№	Solution	Singularity conditions	Singularity type
1	$\theta_1 - \theta_2 = k\pi$	Links L_1 and L_2 are collinear	The first type (Fig. 3a)
2	$\theta_3 - \xi_1 = k\pi$	Links L_3 and L_4 are collinear	The first type (Fig. 3b)
3	$\xi_2 - \xi_1 = k\pi$	Links L_4 and L_{CD} are collinear	Second type (Fig. 3c)
4	$\xi_2 - \theta_2 = k\pi$	Links L_2 and L_{CD} are collinear	Second type (Fig. 3d)
5	$\theta_1 - \xi_1 = k\pi$	Links L_1, L_4 and L_2, L_{CD} are collinear	The third type (Fig. 3e)
6	$\theta_2 - \xi_1 = k\pi$	Links L_2, L_{CD} and L_4 are collinear	The third type (Fig. 3f)
7	$\xi_2 - \theta_1 = k\pi$	Links L_1, L_2 and L_{CD} are collinear	The third type (Fig. 3g)
8	$\xi_2 - \theta_3 = k\pi$	Links L_3, L_4 and L_{CD} are collinear	The third type (Fig. 3h)

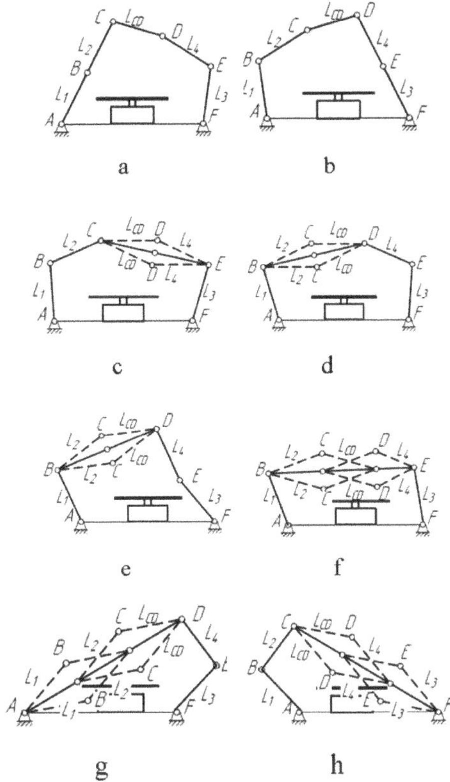

Fig. 3 Singular positions of the six-link mechanism

The lengths of the links of the six-link mechanism are not determined, which means that all singular positions from Table 1 may arise. The singular positions are shown in Fig. 3.

4 Determination of the Workspace Considering Singular Positions

The workspace is one of the main characteristics of any manipulation mechanism. To find the JRMM workspace, an iterative method was used, the essence of which is to enumerate the values of generalized coordinates with a given step within certain limits, followed by solving the direct problem of kinematics to find the coordinates of the working body of the mechanism.

Let us analyze the working space of the six-link mechanism JRMM with dimensions: $L = 250$ mm, $h = 60$ mm, $L_1 = 200$ mm, $L_2 = 200$ mm, $L_3 = 200$ mm, $L_4 = 200$ mm for the following criteria for enumerating variables: $\theta_1 = 0° : 5° : 120°, \theta_2 = 0° : 5° : 120°, \theta_3 = 0° : 5° : 120°$, $\xi_1 = 0° : 5° : 120°$. For this, an algorithm for finding the workspace with checking the bifurcation conditions was implemented (4) (Fig. 4).

Figure 5 shows the workspace with allowable positions (black dots) and singular positions (pink dots).

In a similar way, we will find the workspace of the entire mechanism of joint-relative manipulation. The following values

$\theta_1 = 0° : 15° : 120°, \theta_2 = 0° : 10° : 90°, \theta_3 = 0° : 10° : 90°, \xi_1 = 0° : 15° : 120°$ were used as constraints for the six-link mechanism and $\theta_5 = -100\text{ mm} :$

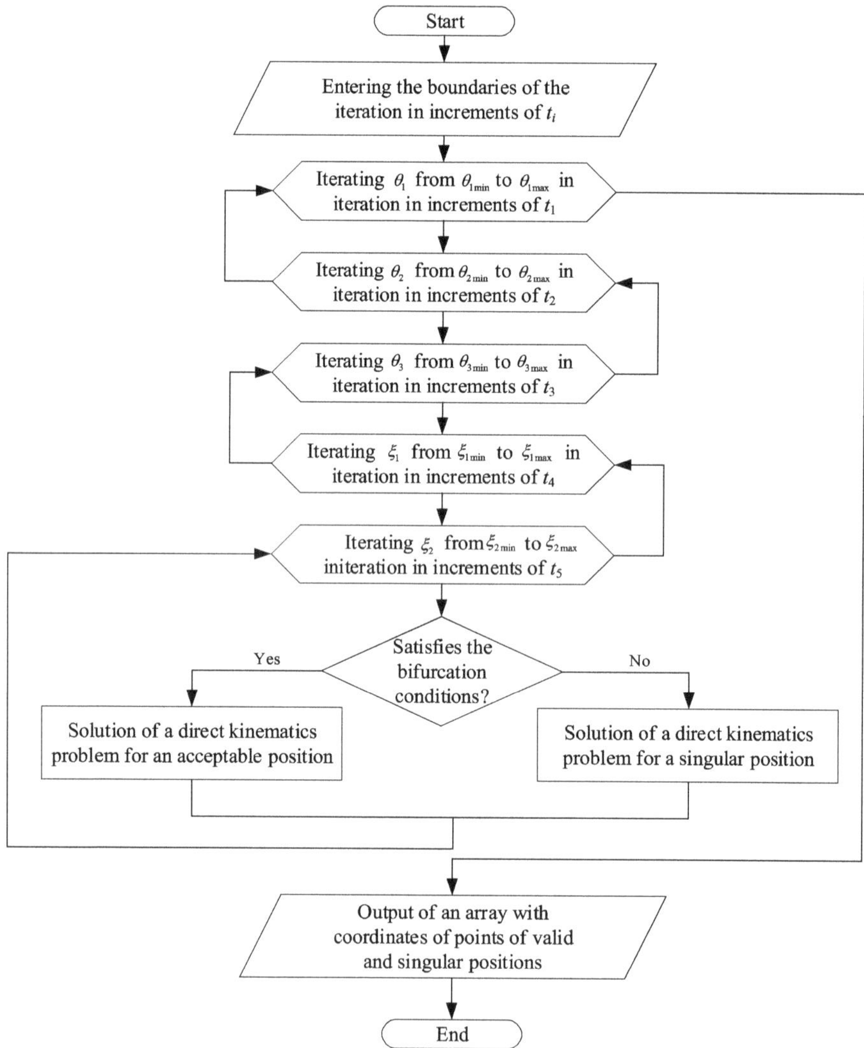

Fig. 4 Algorithm for finding the working space of a six-link mechanism, taking into account bifurcation conditions

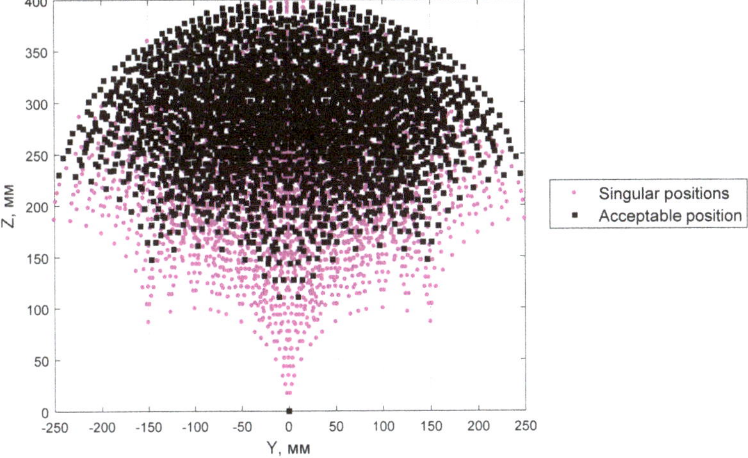

Fig. 5 Working space of the mechanism in the projection on the YZ plane

20 mm : 100 mm, $\theta_6 = 0° : 30° : 360°$ for the progressive one, for the rotary mechanism (Fig. 6).

Thus, the use of the translational-rotary mechanism in conjunction with the six-link mechanism within the JRMM allows significantly "expanding" the reach of the output link.

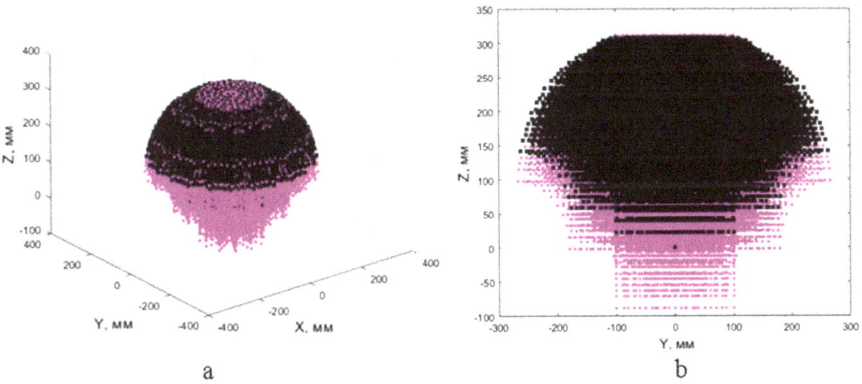

Fig. 6 Working space of the mechanism in the projection on the YZ plane

5 Solving the Problem of Forming a Program Trajectory in the Space of Generalized Coordinates Taking into Account Singularities

In the practical application of the mechanisms of joint-relative manipulation, it is necessary to use methods for reducing the number of potential singular points. Consider the problem of the trajectory passing by the output link located on the surface of a sphere centered at the origin with a radius of 125 mm. To do this, we will set the following points through which the planned trajectory passes:

$$s_1 = (0; -55.91; 111.41); \quad s_2 = (0; -39.47; 118.61); \quad s_3 = (0; -6.34; 124.84);$$
$$s_4 = (0; 23.17; 122.83); \quad s_5 = (0; 55.91; 111.41); \quad s_6 = (0; 94.52; 81.80).$$

Reaching points s_1, s_5 with a given orientation of the output link lead to the emergence of singularities of the third type.

The structure of the JRMM under consideration assumes the use of a translational-rotary mechanism, which allows avoiding some of the singular positions by increasing the number of different ways to achieve the same position of the output link on a given surface. The orientation angles of the mechanism under consideration are the angles, which determine the rotation of the output link around the axes and the base coordinate system, respectively.

The structure of the JRMM under consideration assumes the use of a translational-rotary mechanism, which allows avoiding some of the singular positions by increasing the number of different ways to achieve the same position of the output link on a given surface. The orientation angles of the mechanism under consideration are the angles α and β, which determine the rotation of the output link around the axes Z and X of the base coordinate system, respectively.

Consequently, the second-order surface under consideration can be rotated and moved in space by changing the position of the points reached relative to the coordinate systems of the mechanism, but not changing their position on the surface. Therefore, for a point S_1, you can choose the orientation angles $\alpha = -30°$, $\beta = 44.42°$ and the offset of the working table $\theta_5 = -27.96$ mm, and for the point: s_5:$\alpha = 30°$,$\beta = -44.42°$ and $\theta_5 = 27.96$ mm. As a result of applying these parameters, the joint-relative manipulation mechanism avoids singular positions and reaches all the necessary points on the surface.

6 Conclusions

The paper considers methods for determining special positions on the example of a flat six-link mechanism using the method for determining the conditions of bifurcations. The algorithm for finding the working area of the flat six-link mechanism, as well as the singularity area, is shown. It is shown that JRMM can significantly increase the volume of the working area, as well as reduce the volume of the singularity region. The paper considers the order of solving the problem of forming the program trajectory JRMM, taking into account the special positions of the mechanism.

References

1. Kalendarev, A.V., Glazunov, V.A.: Possible mechanisms of parallel structure in textile, light and other industries. Collect. Sci. Pap. Grad. Students **18**, 56–60 (2012)
2. Glazunov, V.A., Rashoyan, G.V.: Parallel structure mechanisms for air-dynamic tubes. In: International Conference on the Methods of Aero-106physical Research, pp. 106–107. Kazan, Russia (2012)
3. Artemenko, Yu.N., Glazunov, V.A., Demidov, S.M., Nyat, C.: Development and analysis of mechanisms of a parallel structure, designed to manipulate antennas of space telescopes. Ref. Eng. J. **5**, 30–34 (2012)
4. Glazunov, V.A.: Parallel structure mechanisms and their application: robotic, technological, medical, training systems. Inst. Comput. Res. Moscow–Izhevsk (2018)
5. Pashchenko, V.N., Meleshchenko, D.I., Rashoyan, G.V., Malyshev, D.I., Kuzmina, V.S.: Decision of the direct position problem of the joint relative manipulation mechanism with five degrees of freedom. Int. J. Appl. Mechan. Eng. **23**, 1025–1033 (2018)
6. Glazunov, V.A., Chunikhin, AYu.: Development of parallel structure mechanisms. Probl. Mechan. Eng. Mach. Reliab. **3**, 37–43 (2014)
7. Müller, A.: Geometric characterization of the configuration space of rigid body mechanisms in regular and singular points. In: ASME 2005 International Design Engineering Technical Conferences & Computers and Information in Engineering Conference. Long Beach, California USA (2005)
8. Pashchenko, V.N.: Construction of the working area of a six-degree manipulator of a parallel structure based on a crank mechanism. Internet J. Sci. **3**(34), 8, 135 (2016)
9. Merlet, J.P., Gosselin, C.M., Mouly, N.: Workspaces of planar parallel manipulators. Mech. Mach. Theor. **33**(1–2), 7–20 (1998)
10. Pashchenko, V.N., Sharapov, L.V., Rashoyan, G.V., Bykov, A.I.: Construction of a working area for the manipulation mechanism of simultaneous relative manipulation. J. Mach. Manuf. Reliab. **46**(3), 225–231 (2017)
11. Skvortsov, S.A., Demidov, S.M., Glazunov, V.A., Kalendarev, A.V.: On the analysis of special positions and dynamic properties of mechanisms of parallel structure. Ref. Eng. J. **5**, 23–29 (2015)
12. Rashoyan, G.V., Lastochkin, A.B., Glazunov, V.A.: Kinematic analysis of a spatial parallel structure mechanism with a circular guide. J. Mach. Manuf. Reliab. **44**(7), 54–60 (2015)
13. Glazunov, V.A., Rashoyan, G.V., Dubrovsky, V.A., Novikova, N.N.: The criterion of proximity to special provisions associated with the loss of the degree of freedom of mechanisms of a parallel structure. In: Problems of the Mechanics of Modern Machines. Materials of the V International Conference, pp. 32–36. VSGUTU, Ulan-Ude (2012)
14. Dimentberg, F.M.: On the special provisions of spatial mechanisms. Mach. Sci. **5**, 53–58 (1977)
15. Glazunov, V., Nosova, N., Kheylo, S., Tsarkov, A.: Design and analysis of the 6-DOF decoupled parallel kinematics mechanism. In: Dynamic Decoupling of Robot Manipulators, pp. 125–170. Springer, Cham (2018)
16. Laryushkin, P., Glazunov, V.: A new 3-DOF translational parallel manipulator: kinematics, dynamics and workspace analysis. Romansy 19—Robot design, dynamics and control. In: Proceedings of the Ninth CISM-IFToMM Symposium, pp. 11–18. Springer, Vienna (2012)
17. Pashchenko, V.N., Pashchenko, V.V., Lachikhin, A., Timoshenko, A., Shalyukhin, K., Skvortsov, S.: Positioning error calculation of the relative manipulation mechanism output

link. In: Proceedings of 14th International Conference on Electromechanics and Robotics "Zavalishin's Readings", pp. 197–208. Springer, Cham (2020)
18. Gosselin, C., Angeles, J.: Singularity analysis of closed-loop kinematic chains. IEEE Trans. Robot. Autom. **6**(3), 281–290 (1990)
19. Lu, Q.C.: Bifurcation and Singularity. Scientific and Technological Education Publishing, pp. 65–108. House, Shanghai, China (2004)
20. Tarnai, T.: Rigidity and kinematic bifurcation of structures. In: 40th Anniversary Congress of the International Association for Shell and Spatial Structures (IASS), Vol. 24, No. 1, p. B2. CEDEX, Madrid, Spain (1999)
21. Lengyel, A.: Analogy between equilibrium of structures and compatibility of mechanisms. Doctoral dissertation. Oxford University, UK (2002)
22. Yuan, X.F., Zhou, L., Duan, Y.F.: Singularity and kinematic bifurcation analysis of pin-bar mechanisms using analogous stiffness method. Int. J. Solids Struct. **49**(10), 1212–1226 (2012)
23. Kumar, P., Pellegrino, S.: Computation of kinematic paths and bifurcation points. Int. J. Solids Struct. **37**(46–47), 7003–7027 (2000)
24. Glazunov, V.A.: The structure of spatial mechanisms. Screw group and structural groups. Direct. Eng. J. **3**, 1–4 (2010)

Kinematic Modeling in Study of Manipulative Mechanism of Combined Movement

Sergey Orekhov, Nikita Zaychikov, Konstantin Petrukhin, Alexander Tsepurkin, and Nikolay Tsepurkin

Abstract The aim of the work is to develop a mechanism for sequential manipulation with combined movement. This article is relevant because the problems of kinematics and dynamics of mechanisms are widely and thoroughly studied, but current approaches, such as modeling and designing collaborative robots, and the development of executive systems for technical vision, require consideration of these problems in a complex. This article is devoted to particular issues of solving direct and inverse kinematics problems, determining the working area and solving the problem of trajectory control of the mechanism during combined movement.

1 Introduction

The development of the robotics industry over the past 10 years has shown that the main trends have "shifted" from the classical tasks of robotics, such as kinematics calculations, dynamics, control system synthesis to the tasks of creating collaborative robots, estimation work safety, developing control principles, and programming robots. At the same time, a number of tasks are solved: the speed of changeover of industrial robotic cells, the possibility of safe operation inside the cell directly in the production cycle, and increased flexibility of automated production. From the point of view of designing such systems, the tasks become more complicated, since additional sensors of sensitivity, spatial orientation sensors, technical vision systems,

S. Orekhov · N. Zaychikov (✉) · K. Petrukhin · A. Tsepurkin · N. Tsepurkin
Kaluga Branch of Bauman Moscow State Technical University, 4, Bazhenova str., Kaluga 248004, Russia
e-mail: zanik.2000@inbox.ru

S. Orekhov
e-mail: serg31057@mail.ru

K. Petrukhin
e-mail: farlamov.maxim@yandex.ru

A. Tsepurkin
e-mail: tsepurkin00@mail.ru

A. Ronzhin and V. Shishlakov (eds.), *Electromechanics and Robotics*, Smart Innovation, Systems and Technologies 232, https://doi.org/10.1007/978-981-16-2814-6_4

and other complex subsystems for estimation joint movement and positioning can be introduced. However, developers have to conduct kinematic analysis, investigate dynamics, controllability, and study the principles of trajectory movement.

1.1 Analysis of Combined Movement of the Manipulator

We introduce the main definitions and indicate the tasks to be solved in this research. Under the manipulation mechanism, we will understand a set of spatial lever mechanism that performs similar actions of the human hand [1].

Combined movement of the manipulator is a combination of a number of actions in one manipulation mechanism: analysis of the coordinates of the object of interest of the manipulation mechanism based on technical vision and movement of the manipulator links to the obtained position of the desired object.

Thus, the combined movement of the developed manipulation mechanism consists in changing the trajectory of the manipulator links when moving to the object of interest, taking into account features of the kinematics of the system under study.

2 Analysis Work of the Mechanism

The model is a sequential manipulation mechanism with links $L_1 = 140$ mm, $L_2 = 270$ mm, $L_3 = 260$ mm, $L_4 = 50$ mm, the block diagram of which is shown in Fig. 1.

The presented mechanism of sequential kinematics, on the one side, is quite simple and widespread, and on the other side, without losing the connected of the approach, it allows us to work out all the design features of such systems on a practical example.

The model has four degrees of freedom and four kinematic pairs.

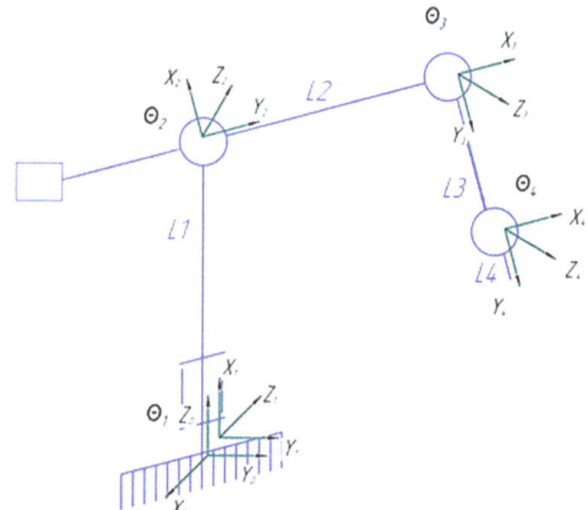

Fig. 1 Structural scheme

Table 1 Rotation limits of the generalized coordinates

Generalized coordinate	Rotation angle
Θ_1	(0; 360)
Θ_2	(−45;135)
Θ_3	(−45;135)
Θ_4	(0;180)

The rotation limits of the generalized coordinates are given in Table 1.

In our research, we took a sequential mechanism, because it has a wider working area than parallel mechanisms with comparable dimensions, which is the main advantage, despite the lower load capacity and lower design rigidity. For laboratory implementation in the form of a stand, it was decided to use a classical system with a counterweight.

3 Kinematic Analysis

The first stage of research on the manipulation mechanism is kinematic analysis: determining the position of an arbitrary point of the executive mechanism link. Based on the literature analysis [2–4], such calculations can be performed in different coordinate systems. The classical approach is to consider two types of tasks: direct and reverse. When solving a direct task, we determine the position of the tong relative to a fixed coordinate system for known generalized coordinates. In the inverse task, we determine the generalized coordinates and position of the tong at the known coordinates of the output link.

4 Solving the Direct Position Problem

The result of calculating the direct problem for the manipulation mechanism is the Cartesian coordinates of the position (x, y, z) of the output link of the manipulation mechanism for known generalized coordinates $(\Theta_1, \Theta_2, \Theta_3, \Theta_4)$.

Let us arrange the coordinate systems using the Denavit–Hartenberg method and make up the transition matrices. The structural scheme with the coordinate systems applied to it is shown in Fig. 1.

Let us create transition matrices. The matrix A_0 reflects a rotation by an angle Θ_1 around the z axis. The matrix A_1 reflects a shift by a height of 140 along the x-axis. The matrix A_2 reflects the rotation by an angle Θ_2 around the x-axis. The matrix A_3 reflects a shift by height 270 along the z-axis. The matrix A_4 reflects a rotation by an angle Θ_3 around the x-axis. The matrix A_5 reflects a shift by height 260 along the z-axis. The matrix A_6 reflects a rotation by an angle Θ_4 around the x-axis. The matrix A_7 reflects a shift by a height of 50 along the z-axis. The general transition matrix is obtained by multiplying the homogeneous transformation matrices. The matrices are shown below (1–8).

$$A0 = \begin{pmatrix} \cos(T1) & -\sin(T1) & 0 & 0 \\ \sin(T1) & \cos(T1) & 0 & 0 \\ 0 & 0 & 1 & 0 \\ 0 & 0 & 0 & 1 \end{pmatrix}; \tag{1}$$

$$A1 = \begin{pmatrix} 1 & 0 & 0 & 140 \\ 0 & 1 & 0 & 0 \\ 0 & 0 & 1 & 0 \\ 0 & 0 & 0 & 1 \end{pmatrix}; \tag{2}$$

$$A2 = \begin{pmatrix} 1 & 0 & 0 & 0 \\ 0 & \cos(T2) & -\sin(T2) & 0 \\ 0 & \sin(T2) & \cos(T2) & 0 \\ 0 & 0 & 0 & 1 \end{pmatrix} \tag{3}$$

$$A3 = \begin{pmatrix} 1 & 0 & 0 & 0 \\ 0 & 1 & 0 & 0 \\ 0 & 0 & 1 & 270 \\ 0 & 0 & 0 & 1 \end{pmatrix}; \tag{4}$$

$$A4 = \begin{pmatrix} 1 & 0 & 0 & 0 \\ 0 & \cos(T3) & -\sin(T3) & 0 \\ 0 & \sin(T3) & \cos(T3) & 0 \\ 0 & 0 & 0 & 1 \end{pmatrix}; \tag{5}$$

$$A5 = \begin{pmatrix} 1 & 0 & 0 & 0 \\ 0 & 1 & 0 & 0 \\ 0 & 0 & 1 & 260 \\ 0 & 0 & 0 & 1 \end{pmatrix}; \tag{6}$$

$$A6 = \begin{pmatrix} 1 & 0 & 0 & 0 \\ 0 & \cos(T4) & -\sin(T4) & 0 \\ 0 & \sin(T4) & \cos(T4) & 0 \\ 0 & 0 & 0 & 1 \end{pmatrix}; \tag{7}$$

$$A7 = \begin{pmatrix} 1 & 0 & 0 & 0 \\ 0 & 1 & 0 & 0 \\ 0 & 0 & 1 & 50 \\ 0 & 0 & 0 & 1 \end{pmatrix}, \tag{8}$$

where $T1-\theta_1$; $T2-\theta_2$; $T3-\theta_3$; $T4-\theta_4$.

Multiplying the transition matrices from the base coordinate system to the coordinate system of the grip, we get the matrix A_{com}. Thus, the calculation of Cartesian coordinates depending on the generalized coordinates is presented in (10–12):

$$A_{\text{com}} = \begin{pmatrix} a_1 & a_2 & a_3 & x \\ a_4 & a_5 & a_6 & y \\ a_7 & a_8 & a_9 & z \\ 0 & 0 & 0 & 1 \end{pmatrix}; \tag{9}$$

$$x = (\sin(T1) * (13000 * \sin(T2 + T3) + 13500 * \sin(T2) + 2500 * \sin(T2 + T3 + T4) - 6939))/50; \tag{10}$$

$$y = -(\cos(T1) * (13000 * \sin(T2 + T3) + 13500 * \sin(T2) + 2500 * \sin(T2 + T3 + T4) - 6939))/50; \tag{11}$$

$$z = 260 * \cos(T2 + T3) + 270 * \cos(T2) + 50 * \cos(T2 + T3 + T4) + 140 \tag{12}$$

$$a_1 = \cos(T1); \tag{13}$$

$$a_2 = \sin(T4) * (\cos(T2) * \sin(T1) * \sin(T3) + \cos(T3) * \sin(T1) * \sin(T2)) - \cos(T4) * (\cos(T2) * \cos(T3) * \sin(T1) - \sin(T1) * \sin(T2) * \sin(T3)); \tag{14}$$

$$a_3 = \cos(T4) * (\cos(T2) * \sin(T1) * \sin(T3) + \cos(T3) * \sin(T1) * \sin(T2)) + \sin(T4) * (\cos(T2) * \cos(T3) * \sin(T1) - \sin(T1) * \sin(T2) * \sin(T3)); \tag{15}$$

$$a_4 = \sin(T1); \tag{16}$$

$$a_5 = \cos(T4) * (\cos(T1) * \cos(T2) * \cos(T3) - \cos(T1) * \sin(T2) * \sin(T3)) - \sin(T4) * (\cos(T1) * \cos(T2) * \sin(T3) + \cos(T1) * \cos(T3) * \sin(T2)); \tag{17}$$

$$a_6 = -\cos(T4) * (\cos(T1) * \cos(T2) * \sin(T3) + \cos(T1) * \cos(T3) * \sin(T2)) - \sin(T4) * (\cos(T1) * \cos(T2) * \cos(T3) - \cos(T1) * \sin(T2) * \sin(T3)); \tag{18}$$

$$a_6 a_7 = 0; \tag{19}$$

$$a_8 = \cos(T4) * (\cos(T2) * \sin(T3) + \cos(T3) * \sin(T2)) + \sin(T4) * (\cos(T2) * \cos(T3) - \sin(T2) * \sin(T3)); \quad (20)$$

$$a_9 = \cos(T4) * (\cos(T2) * \cos(T3) - \sin(T2) * \sin(T3)) - \sin(T4) * (\cos(T2) * \sin(T3) + \cos(T3) * \sin(T2)); \quad (21)$$

where $T1 - \theta_1$; $T2 - \theta_2$; $T3 - \theta_3$; $T4 - \theta_4$.

5 Calculating the Robot Working Envelope

The working area is the space in which the working body can be located during the operation of the manipulator.

Based on the obtained solution of the direct position problem, we substitute all possible sets of generalized coordinates and get all possible positions of the output link of the mechanism. From the obtained coordinates, we will find the maximum and minimum values and construct the robot working envelope. The calculation principle is shown in Fig. 2.

The result of the program is the construction of the work area, which is shown in Figs. 3, 4, 5, and 6.

6 Calculation of the Mechanism Maneuverability

The maneuverability of the manipulating mechanism is understood as the number of degrees of freedom in a stationary grip. When comparing different manipulator schemes, it is assumed that maneuverability depends both on the number of degrees of freedom of the manipulator and on the location of the kinematic pairs.

Maneuverability shows the ability of the manipulator links to change their position, provided that the grip is connected to the object of manipulation.

Let us calculate the degree of freedom using the Chebyshev–Grabler–Kutzbach formula (22) for a given mechanism:

$$W = 3(n - 1) - 2p_5 = 3(5 - 1) - 2 \cdot 4 = 12 - 8 = 4, \quad (22)$$

where n—number of mechanism links (including one fixed link); p_5—the number of kinematic pairs of the fifth order (one degree of mobility).

Fix the input link of the lower part of the mechanism. Then, $n = 4$ and $p5 = 4$ in formula (22):

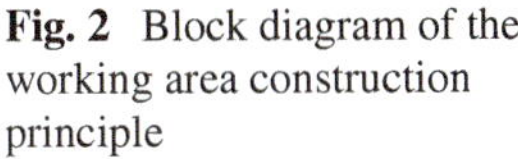

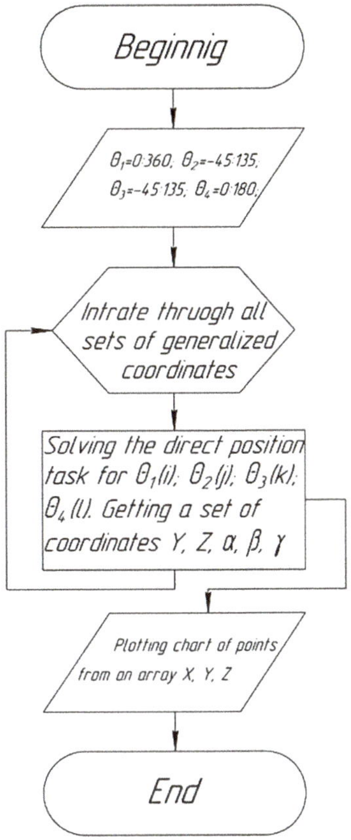

Fig. 2 Block diagram of the working area construction principle

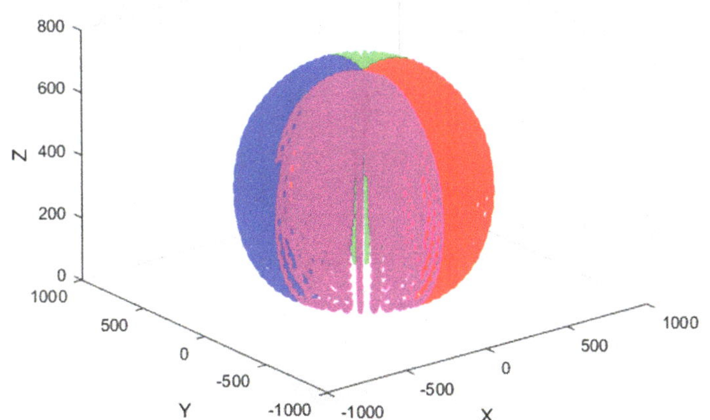

Fig. 3 Working area (axonometric view)

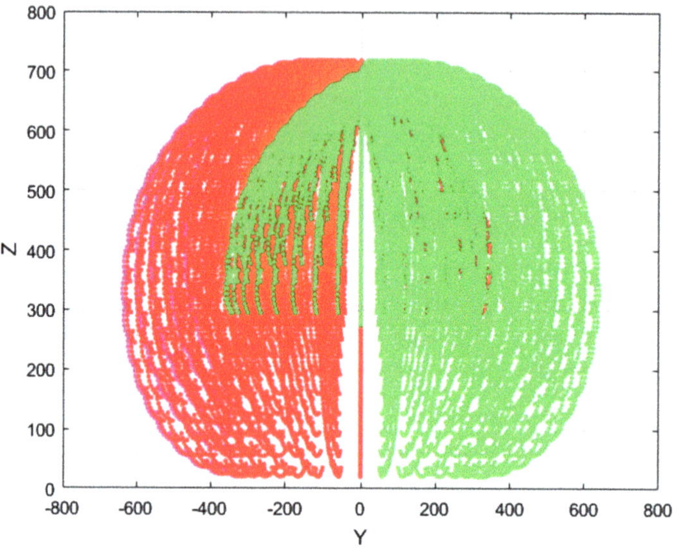

Fig. 4 Working area (front view)

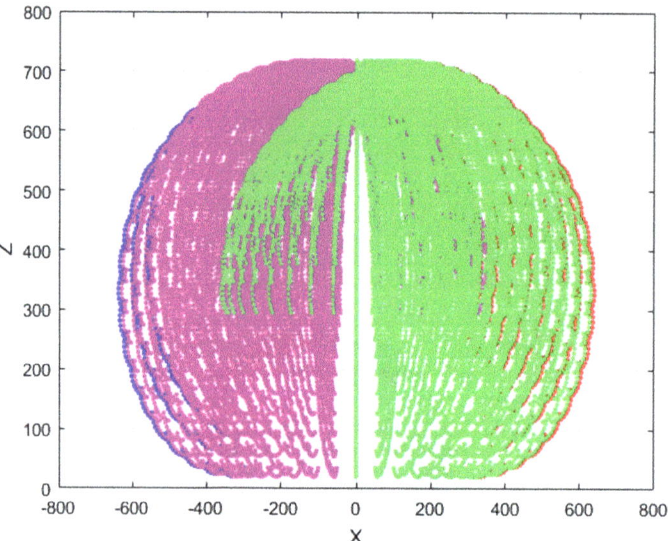

Fig. 5 Working area (side view)

$$M = 3(n - 1) - 2p_5 = 3(4 - 1) - 2 \cdot 4 = 9 - 8 = 1. \tag{23}$$

The studying mechanism has a maneuverability of 1. The studying mechanism has a maneuverability of 1.

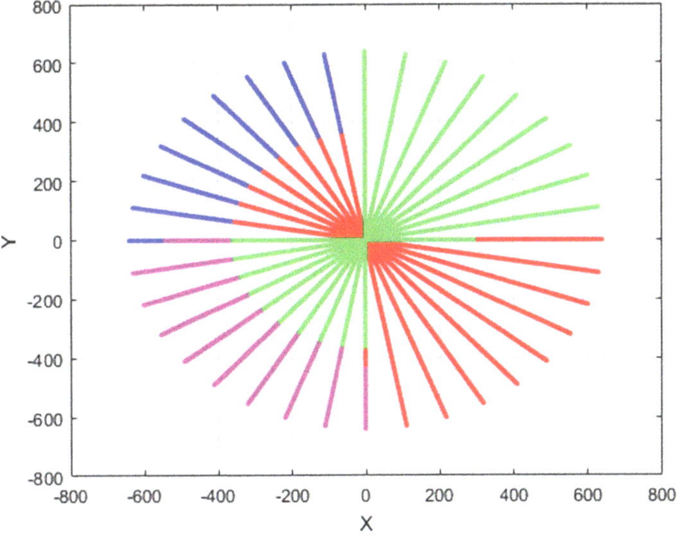

Fig. 6 Working area (top view)

7 Trajectory Problem

The solution of the trajectory problem can be carried out in different ways. The most universal method is the solution of the problem "key points" (Table 2). Each position of the grip will be described by a set of generalized coordinates (angles of rotation in the joints) (Q_1, Q_2, Q_3, Q_4), and the time that has passed from the beginning of the movement to the end point of the grip.

We solve the direct problem of the position of the mechanism, and find the generalized coordinates of each point (Table 3).

The trajectory of the output link in the coordinate system of the mechanism is shown in Fig. 7.

Table 2 Trajectory points

Point no	x	y	z	Point in time
1	0	138	720	0
2	−30	174	715	1
3	−20	160	710	2
4	−10	150	710	3
5	−15	140	700	4
6	−20	150	695	5
7	−25	165	690	6
8	−30	180	695	7

Table 3 Generalized coordinates of mechanism points

Point no	Q_1	Q_2	Q_3	Q_4	Point in time
1	0	34.0869	69.3428	35.2553	0
2	9.7824	33.2215	67.5884	34.2305	1
3	7.1250	32.2680	65.6550	33.3585	2
4	3.8141	31.8514	64.8101	32.9503	3
5	6.1155	28.4539	57.9161	29.2597	4
6	7.5946	21.4968	43.7817	21.9116	5
7	8.6156	23.0904	47.0214	23.8628	6
8	9.4623	24.2612	49.4007	25.1394	7

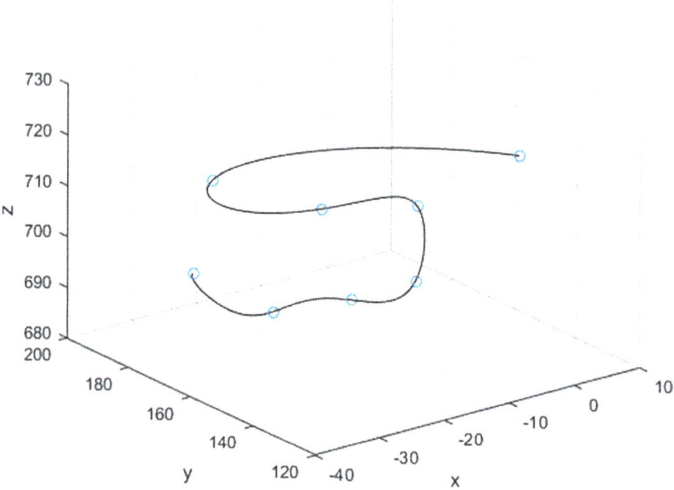

Fig. 7 Graphic of the trajectory of the output link of the mechanism

8 Conclusion

The main problems arising in the calculations is the exact definition of working and handling, because these parameters are "input" for the development of vision systems, synthesis of the joint management system, and assess the accuracy of joint positioning.

As a result of the first part of the study, the combined movement of the manipulator was analyzed, direct and inverse kinematics problems were solved, the working area of the mechanism was determined, and the problem of trajectory control of the mechanism was solved.

References

1. Pashchenko, V.N., Sharapov, I.V., Rashoyan, G.V., Bykov, A.I.: Construction of a working area for the manipulation mechanism of simultaneous relative manipulation. J. Mach. Manuf. Reliab. **46**(3), 225–231 (2017)
2. Frolov, K.V.: Mechanics of industrial robots. 1: Kinematics and dynamics. Higher School (1988)
3. Krainev, A.F., Glazunov, V.A.: New mechanisms of relative manipulation. Probl. Mechan. Eng. Mach. Reliab. **5**, 106–117 (1994)
4. Karris, S.T.: SimMechanics User's Guide. Orchad Publication, USA (2006)

Simulation of Static Walking in an Exoskeleton

Sergey Jatsun, Andrey Yatsun, Andrey Fedorov, and Ekaterina Saveleva

Abstract This article discusses aspects of the operators static walking in a human–machine exoskeletal system increasing the physical capabilities of a person. The development of BTWS models and algorithms, the control strategy implemented by the human–machine interface (HMI) and ensuring the human functionality expansion, is an urgent scientific and technical task. The most important element of BTWS, providing high precision control of the links of the exoskeleton system, is the HMI. In this article, BTWS is considered as a collection of elements, united by heterogeneous and multilevel types of links, as an integral object. The development of control algorithms for the BTWS exoskeleton, using a systematic approach, methods of decomposition, analysis and synthesis, the apparatus of the theory of automatic control and related modern methods of mathematical modeling, allows to explore a complex object as a whole. At the framework of this article, the following results are achieved. The mathematical model of a human exoskeleton during static walking is developed. A model of the system center of mass movement and also a model foot plane-parallel movement during walking on a horizontal rough surface has been developed, which ensures a stable vertical position during walking. Diagrams and trajectories of ankle joint and the center of mass movement during walking were built. The method for determining the force effect on the human–machine system at the interaction of the foot and lower leg is proposed.

1 Introduction

One of the ways of improving the working conditions quality is the exoskeletons usage. They significantly increase the person power capabilities, while the body functional tension level decreases [1–3]. The maximum effect is achieved with the usage of "active" exoskeletons in which the hinges are equipped with controlled electric drives. This effect is achieved when the operator and the exoskeleton form

S. Jatsun (✉) · A. Yatsun · A. Fedorov · E. Saveleva
Department of Mechanics, Mechatronics and Robotics, South-West State University, 94, 50 Let Oktyabrya Str., 305040 Kursk, Russia
e-mail: teormeh@inbox.ru

A. Ronzhin and V. Shishlakov (eds.), *Electromechanics and Robotics*, Smart Innovation, Systems and Technologies 232, https://doi.org/10.1007/978-981-16-2814-6_5

an integrated human–machine system (HMS) [4, 5]. The performance of the system is determined by the degree of its elements consistency (synchronicity) work [6–10]. The system elements include the person (operator) and the exoskeleton. To ensure this, the system is complemented by another important element—a human–machine interface (HMI), which is a group of technical means that provides the operator-exoskeleton interaction [11]. Such a system, consisting of an operator, an exoskeleton and an HMI, is called the biotechnical walking system (BTWS) [2, 12].

The development of control algorithms for the exoskeleton BTWS walking involves the use of a systematic approach, methods of decomposition, analysis and synthesis, the apparatus of the automatic control theory and related modern mathematical modeling methods, which makes it possible to study a complex object as a whole, as well as to consider a set of BTWS elements, united by diverse and multi-level types of links, as an integral object. For developing BTWS, it is important to ensure effective human-exoskeleton interaction. For this purpose, it is necessary to develop the HMI structure, solve the problem of synthesizing the exoskeleton control system parameters and implement control based on the processing and use of signals from angular measurement sensors and force sensors [1, 13–16]. To build the design tools for the BTWS, system links movement mathematical models are necessary. It must be taken into account the selected control strategy implemented by HMI, as well as the anatomical and physiological operator characteristics. However, in this formulation, the issues related to the HMI structure and mathematical models development, ensuring the HMS movement control, remain insufficiently developed. The nature of the HMI parameters influence on the HMS quality indicators has not been studied, which hinders the development of exoskeletal HMS. Thus, the development of BTWS models and algorithms, the control strategy implemented by the HMI and ensuring the human functionality expansion, is an urgent scientific and technical task.

2 BTWS Structural Scheme

The general BTWS view is shown in Fig. 1. The exoskeleton structure includes a power frame, a personified attachments system to a person, LGC, angular displacement and force sensors. The human thigh, lower leg, foot and back are fixed in the exoskeleton required position.

The exoskeleton consists of the following main assemblies and modules (Fig. 1): (1) shoulder protection module; (2) force–moment sensor of interaction between the operator's back and exoskeleton; (3) exoskeleton back tilt sensor; (4) operator's back tilt sensor; (5) exoskeleton power frame; (6) lumbar spine fasteners; (7) hip articulation with D43 relative angle sensors; (8) active hip joint; (9) force sensors on the fixing stops hips; (10) thigh cuffs; (11) clips for personalized thigh-length adjustment; (12) knee protection; (13) clips for personalized shin-length adjustment; (14) foot attaching adjustable clip; and (15) two-link foot.

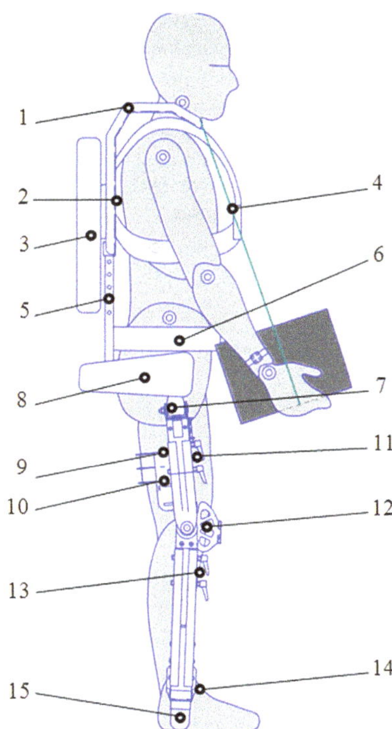

Fig. 1 BTWS general view

3 Static Walking BTWS Scheme

In Fig. 2, the diagram of a person's walking in an exoskeleton is shown in single-support phase. The following designations are accepted here: 1—right foot; 2—right shin; 3—right thigh; 4—human torso and exoskeleton body; 5—left thigh; 6—left lower leg; and 7—left foot.

Each human foot is connected to the lower leg by an ankle joint. Human feet are considered as two-link, consisting of the foot body and toes, which are interconnected by the metatarsophalangeal joints. The exoskeleton feet are connected to the lower leg by means of the ankle joint and are considered as a two-link mechanism. The mechanism links are connected by cylindrical hinges. All other joints and hinges are modeled in the same way. In the model, it is assumed that the hinges axes and joints coincide, and the links of the human limbs and the exoskeleton are considered as a rigid non-deformable body.

Absolute back link q_{40} and right foot angles q_{10} can be found as

$$q_{40} = \sum_{i=4}^{7} q_{i,i+1} + q_{80}, \quad q_{10} = \sum_{i=1}^{4} q_{i,i-1}.$$

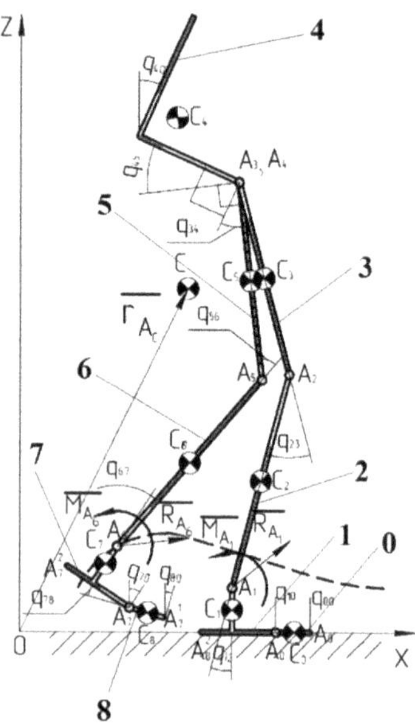

Fig. 2 Scheme of a person's walking in an exoskeleton in single-support phase. q_{ij}—rotation angles of link i relative to j, C_i—center of mass of the i-th link, A_i—hinges, M_{Ai}—moments of rotation, R_{Ai}—radius vector of the i-th hinge

4 BTWS Feet Movement Modeling

The links providing the BTWS interaction with the supporting surface are feet 1 and 7. Further, it is assumed that the supporting surface is horizontal, rough and non-deformable. BTWS walking occurs when the feet move according to a given law. We will assume that the left foot moves from the left ankle joint (AJ) (point A_6) according to the law:

$$\bar{r}_{A_6} = \bar{r}_{A_6}(t),$$
$$\bar{r}_{A_6}(t) = \left| X_{A_6}\ Y_{A_6}\ Z_{A_6} \right|^T.$$

Thus, for determination the AJ position in every moment of time, it is necessary to know the point A_6 coordinate dependence on time. To determine these laws, studies have been carried out on the movement of the human foot while walking [3, 17–20]. As a result of these studies, the dependences of the corresponding coordinates on time were obtained in the following form:

$$X_{A_6}(t);\ Y_{A_6}(t);\ Z_{A_6}(t).$$

In addition, the foot simultaneously rotates around the AJ at an angle $q_{70}(t)$.

After mathematical processing of these experimental data, equations describing the A_6 point motion were developed that (Fig. 3). In this case, the AJ movement is decomposed into two time intervals. On the first, the AJ moves in a circle, and on the second, according to the polynomial law:

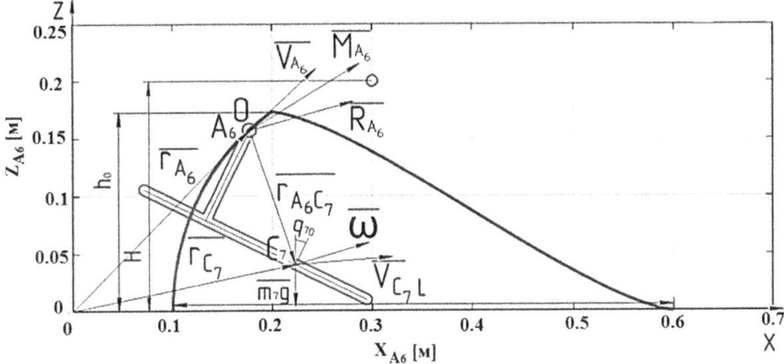

Fig. 3 Scheme of the BTWS left foot movement with a graph of the desired trajectory. h_0—variable parameter of the foot rising height; H—the maximum foot rising height, C_i—center of mass of the i-th link, A_6—ankle joint, M_{A6}—moment of ankle joint rotation, R_{Ai}—radius vector of the 6-th hinge, $\bar{\omega}$—angular velocity

$$t_0 + t_1 \geq t \geq t_0 \; : \; \begin{cases} X_{A_6} = H - H \cdot \cos(q_{70}) + 0.1 \\ Z_{A_6} = -H \cdot \sin(q_{70}) \end{cases},$$

$$T \geq t \geq t_0 + t_1 \; : \; \begin{cases} X_{A_6} = \sum\limits_{i=0}^{3} a_i \cdot t^i \\ Z_{A_6} = \sum\limits_{i=0}^{5} b_i \cdot t^i \end{cases}.$$

where $T \geq t \geq t_0 + t_1$: and $t_0 + t_1 \geq t \geq t_0$:.

Coefficients a_i and b_i can be found from boundary conditions (1), (2) as a solution of the following algebraic equations system (3):

$$\left| H - H \cdot \cos(q_{70}) + 0.1 \; 0.6 \; 0 \; 0 \right|^{\mathrm{T}}, \tag{1}$$

$$\left[-H \cdot \sin(q_{70}) \; 0 \; 0 \; 0 \; 0 \; 0 \right], \tag{2}$$

$$\begin{vmatrix} 1 & t_1 & t_1^2 & t_1^3 \\ 1 & t_T & t_T^2 & t_T^3 \\ 0 & 1_1 & 2 \cdot t_1 & 3 \cdot t_1^2 \\ 0 & 1 & 2 \cdot t_T & 3 \cdot t_T^2 \end{vmatrix}^{-1} \cdot \begin{vmatrix} X_0 \\ X_k \\ \dot{X}_0 \\ \dot{X}_k \end{vmatrix} = \begin{vmatrix} a_0 \\ a_1 \\ a_2 \\ a_3 \end{vmatrix}, \tag{3}$$

$$\begin{bmatrix} 1 & t_1 & t_1^2 & t_1^3 & t_1^4 & t_1^5 \\ 0 & 1 & 2\cdot t_1 & 3\cdot t_1^2 & 4\cdot t_1^3 & 5\cdot t_1^4 \\ 0 & 0 & 2 & 6\cdot t_1 & 12\cdot t_1^2 & 20\cdot t_1^3 \\ 1 & t_T & t_T^2 & t_T^3 & t_T^4 & t_T^5 \\ 0 & 1 & 2\cdot t_T & 3\cdot t_T^2 & 4\cdot t_T^3 & 5\cdot t_T^4 \\ 0 & 0 & 2 & 6\cdot t_T & 12\cdot t_T^2 & 20\cdot t_T^3 \end{bmatrix}^{-1} \cdot \begin{bmatrix} Y_0 \\ \dot{Y}_0 \\ \ddot{Y}_0 \\ Y_k \\ \dot{Y}_k \\ \ddot{Y}_k \end{bmatrix} = \begin{bmatrix} b_0 \\ b_1 \\ b_2 \\ b_3 \\ b_4 \\ b_5 \end{bmatrix}.$$

It is also important to determine the left foot center of mass coordinates C_7:

$$\bar{r}_7(t) = \left| X_7 \ Y_7 \ Z_7 \right|^T,$$
$$\bar{r}_{C_7} = \bar{r}_{A_6} + \bar{r}_{A_6C_7},$$
$$\bar{r}_{C_7} = \bar{r}_{A_6} + T_{70}\bar{r}^{(7)}_{A_6C_7}.$$

where T_{70}—foot rotation matrix in the absolute coordinate system, $\bar{r}^{(7)}_{A_6C_7}$—radius vector C_7 in the seventh coordinate system.

The change in the coordinates of point A_6 depends both on the length and height of the step and on the absolute angle q_{70}, which is approximately determined in the work using the formula:

$$q_{70} = \frac{\pi}{3} \cdot \sin(-2 \cdot \pi \cdot t).$$

5 Description of the Point C Center of Mass Motion

An important point for studying of walking is the BTWS center of mass. Radius vector defining the position of point C on a given trajectory can be represented as:

$$\bar{r}_C = \begin{vmatrix} X_C = \sum_{i=0}^{5} c_i \cdot t^i \\ Y_C = \sum_{i=0}^{6} d_i \cdot t^i \\ Z_C = \sum_{i=0}^{6} e_i \cdot t^i \end{vmatrix}. \quad (4)$$

The coefficients are found from the boundary conditions as a solution to a system of algebraic equations with boundary conditions:

The coefficients c_i, d_i, e_i can be found from the boundary conditions as an algebraic equations system solution with boundary conditions:

$$\begin{gathered}\left[C_X^0\ 0\ 0\ L/2\ 0\ 0\right],\\ \left[C_Z^0\ 0\ 0\ C_Z^0\ 0\ 0\ C_Z^{\max}-C_Z^0\right],\end{gathered} \tag{5}$$

$$\begin{bmatrix} 1 & t_0 & t_0^2 & t_0^3 & t_0^4 & t_0^5 \\ 0 & 1 & 2\cdot t_0 & 3\cdot t_0^2 & 4\cdot t_0^3 & 5\cdot t_0^4 \\ 0 & 0 & 2 & 6\cdot t_0 & 12\cdot t_0^2 & 20\cdot t_0^3 \\ 1 & t_1 & t_1^2 & t_1^3 & t_1^4 & t_1^5 \\ 0 & 1 & 2\cdot t_1 & 3\cdot t_1^2 & 4\cdot t_1^3 & 5\cdot t_1^4 \\ 0 & 0 & 2 & 6\cdot t_1 & 12\cdot t_1^2 & 20\cdot t_1^3 \end{bmatrix}^{-1} \cdot \begin{bmatrix} X_0 \\ \dot{X}_0 \\ \ddot{X}_0 \\ X_1 \\ \dot{X}_1 \\ \ddot{X}_1 \end{bmatrix} = \begin{bmatrix} b_0 \\ b_1 \\ b_2 \\ b_3 \\ b_4 \\ b_5 \end{bmatrix},$$

$$\begin{bmatrix} 1 & t_0 & t_0^2 & t_0^3 & t_0^4 & t_0^5 & t_0^6 \\ 0 & 1 & 2\cdot t_0 & 3\cdot t_0^2 & 4\cdot t_0^3 & 5\cdot t_0^4 & 6\cdot t_0^5 \\ 0 & 0 & 2 & 6\cdot t_0 & 12\cdot t_0^2 & 20\cdot t_0^3 & 30\cdot t_0^4 \\ 1 & t_1 & t_1^2 & t_1^3 & t_1^4 & t_1^5 & t_1^6 \\ 0 & 1 & 2\cdot t_1 & 3\cdot t_1^2 & 4\cdot t_1^3 & 5\cdot t_1^4 & 6\cdot t_1^5 \\ 0 & 0 & 2 & 6\cdot t_1 & 12\cdot t_1^2 & 20\cdot t_1^3 & 30\cdot t_1^4 \\ 1 & (\frac{t_1-t_0}{2})^6 & (\frac{t_1-t_0}{2})^2 & (\frac{t_1-t_0}{2})^3 & (\frac{t_1-t_0}{2})^4 & (\frac{t_1-t_0}{2})^5 & (\frac{t_1-t_0}{2})^6 \end{bmatrix}^{-1} \cdot \begin{bmatrix} Z_0 \\ \dot{Z}_0 \\ \ddot{Z}_0 \\ Z_1 \\ \dot{Z}_1 \\ \ddot{Z}_1 \\ Z_0+H \end{bmatrix} = \begin{bmatrix} c_0 \\ c_1 \\ c_2 \\ c_3 \\ c_4 \\ c_5 \\ c_6 \end{bmatrix},$$

where $\left[C_X^0\ C_Z^0\right]$—initial coordinates of point C, L—step length, $C_Z^{\max}$– maximum center of mass height.

6 Planning the Trajectory of the BTWS Feet During Walking

We will consider gait with a step equal $2L$, step time T.

$$\bar{r}_{K_6} = \begin{bmatrix} X_{K_6} = X_{K_6}^0 + k\cdot(2\cdot L_{X_{K_6}X_{D_6}}) \\ Y_{K_6} = Y_{K_6}^0 \end{bmatrix},$$

$$\bar{r}_{D_6} = \begin{bmatrix} X_{D_6} = X_{D_6}^0 + k\cdot(2\cdot L_{X_{K_6}X_{D_6}}) \\ Y_{D_6} = Y_{D_6}^0 \end{bmatrix},$$

$$\bar{r}_{K_2} = \begin{bmatrix} X_{K_2}^2 = X_{K_2} + k\cdot(2\cdot L_{X_{K_2}X_{D_2}}) \\ Y_{K_2} = Y_{K_2}^0 \end{bmatrix},$$

$$\bar{r}_{D_6} = \begin{bmatrix} X_{D_2}^2 = X_{D_2}^0 + k\cdot(2\cdot L_{X_{K_2}X_{D_2}}) \\ Y_{D_2} = Y_{D_2}^0 \end{bmatrix}.$$

where k—step number.

BTWS will be stable in the case of providing the center of mass projections on the horizontal plane inside the reference polygon. For this, the following conditions

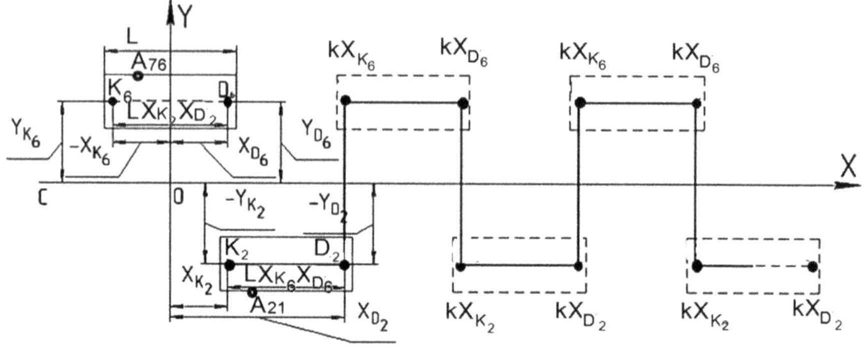

Fig. 4 Scheme of the feet movement and the BTWS center of mass projections when performing a static gait along the *Ox* axis

are met as follows:

$$\begin{cases} X_{K_6} > X_C > X_{D_6} \vee X_{K_2} > X_C > X_{D_2} \\ Y_{K_6} > Y_C > Y_{D_6} \vee Y_{K_2} > Y_C > Y_{D_2} \end{cases}.$$

Figure 4 shows the feet positions when walking at the moment of contact with the supporting surface and also shows the center of mass trajectory, at which conditions (5) are satisfied.

The work considers a piecewise center of mass motion linear trajectory as shown in Fig. 4. Point C moves along these trajectories according to Eq. (4).

7 Simulation of Feet and Center of Mass Movement

Input simulation parameters:

$$\bar{r}_{D_6} = \begin{bmatrix} X_{D_6} = 0.14 \\ Y_{D_6} = 0.153 \end{bmatrix}, \; \bar{r}_{K_6} = \begin{bmatrix} X_{K_6} = -0.14 \\ Y_{K_6} = 0.153 \end{bmatrix},$$

$$\bar{r}_{D_2} = \begin{bmatrix} X_{D_2} = 0.14 \\ Y_{D_2} = -0.153 \end{bmatrix}, \; \bar{r}_{K_2} = \begin{bmatrix} X_{K_2} = -0.14 \\ Y_{K_2} = -0.153 \end{bmatrix}.$$

Time intervals used in calculations

$$t = t_0 + t_1 + t_2 + t_3,$$

$$t = \begin{bmatrix} t_0 \\ t_1 \\ t_2 \\ t_3 \end{bmatrix} = \begin{bmatrix} 0 \\ 0.33\,c \\ 0.33\,c \\ 0.34\,c \end{bmatrix}.$$

Step length:

$$L = \left|X_{D_6} + \left|X_{K_6}\right|\right|.$$

Step initial conditions:

$$L = 0.5\,\text{m}, \quad H = 0.2\,\text{m}.$$

We divide the step into three stages, then $T = t_0 + t_1 + t_2$, where $t_0 = 0$ s, $t_1 = 0.25$ s, $t_2 = 0.75$ s.

8 Modeling Results

Figure 5a, b shows a graph of the AJ position change along the Ox axis in the form of $X_{A6}(t)$ and speed $\dot{X}_{A6}(t)$ and both the $Z_{A6}(t)$ and the speed $\dot{Z}_{A6}(t)$ along Oz axis. The foot starts moving at the lowest speed, and the AJ accelerates in the interval 0.25–0.55 s. At the point $t = 0.37$ s, the AL is in the highest position. The graph of the speed change along the Oz axis crosses the zero level. Further, the foot slows down, and the AJ position along the Oz axis decreases. The speed graph along the Ox axis with the completion of the step comes to zero. The graph of the speed changes along the Oz axis. After AJ passing the highest mark, it lies in the negative region and with the completion of the step also comes to zero.

In Fig. 5c, three trajectories are shown—center of mass movements, the AJ movement trajectory for right and left legs. From the graphs of the trajectories of the AJ movement, it can be seen that the projections of the center of mass during BTWS move transversely along the right and left supporting feet in single-supporting phases. In the phase in which both feet are on the surface, the center of mass projection moves within the support surface from the foot that is preparing to make a step to the foot for which the supporting phase begins.

9 Conclusions

Within the framework of this article, the following scientific and technical problems were solved:

1. A mathematical model of the BTWS with static walking has been developed, which makes it possible to determine the patterns of movement of the exoskeleton and person links.
2. The mathematical model of the system center of mass motion, which provides a stable vertical position when walking, is developed.

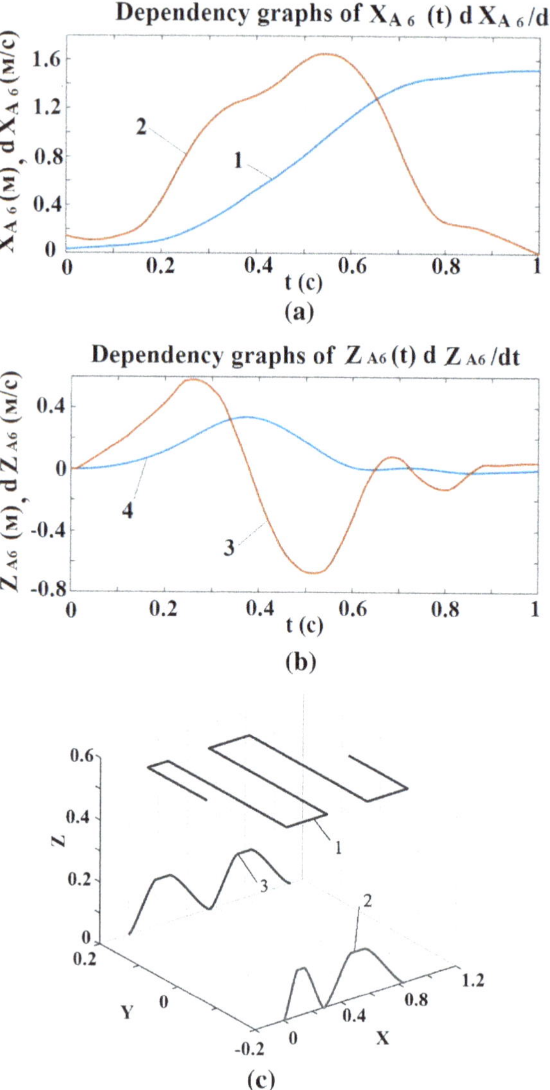

Fig. 5 **a, b** Graphs of parameters changes 1—X_{A6}, 2 –X_{A6}/dt, 3—Z_{A6}, 4—dZ_{A6}/dt; **c** trajectories: 1—center of mass, 2—A_1 point, 3—A_6 points

3. A plane-parallel movement of the foot model when walking on a horizontal rough surface is developed. The dependence of the foot rotation angle relative to the AJ was established. The AJ and the metatarsal joint movement trajectories were determined.
4. Diagrams and trajectories of AJ and the center of mass movement during walking were constructed. It is proposed to use piecewise linear trajectories of the center of mass motion. The conditions are formulated to ensure the vertical position of the system stability in the frontal and sagittal planes.

5. A method for the calculated determination of the force impact on the AJ during the interaction of the foot and lower leg is proposed. To create design tools for BTWS, mathematical models of the system links movement were created. The anatomical and physiological characteristics of the operator must be considered.

Acknowledgements The work was supported by Russian Federation President grant for young scientists, candidates of sciences MK-901.2020.8

References

1. Jatsun, S., Malchikov, A., Loktionova, O.: Modeling of human-machine interaction in an industrial exoskeleton control system. In: International Conference on Interactive Collaborative Robotics, pp. 116–125. Springer, Cham (2020)
2. Chernikova, L.A., Demidova, A.E., Domashenko, M.A., Truhanov, A.I.: The effect of using robotic devices ("Erigo" and "Lokomat") in the early stages after ischemic stroke. Bull. Restor. Med. **5**, 73–75 (2008)
3. Jatsun, S., Malchikov, A., Yatsun, A.: Simulation of a walking robot-exoskeleton movement on a movable base. In: CLAWAR 2020 Proceedings, pp. 15–22. CLAWAR Association Ltd, UK (2020)
4. Al-, M.S., Elamvazuthi, I., Daud, S.A., Parasuraman, S., Borboni, A.: EEG-based control for upper and lower limb exoskeletons and prostheses: a systematic review. Sensors **18**(10), 3342 (2018)
5. Kawamoto, H., Lee, S., Kanbe, S., Sankai, Y.: Power assist method for HAL-3 using EMG-based feedback controller. In: SMC'03 Conference Proceedings. 2003 IEEE International Conference on Systems, Man and Cybernetics. Conference Theme-System Security and Assurance vol. 2, pp. 1648–1653. IEEE (2003)
6. Bhagat, N.A., Venkatakrishnan, A., Abibullaev, B., Artz, E.J., Yozbatiran, N., Blank, A.A., French, J., Karmonik, C., Grossman, R., O'Malley, M.K., Francisco, G.E., Contreras-Vidal, J.L.: Design and optimization of an EEG-based brain machine interface (BMI) to an upper-limb exoskeleton for stroke survivors. Front. Neurosci. **10**, 122 (2016)
7. Rosen, J., Brand, M., Fuchs, M.B., Arcan, M.: A myosignal-based powered exoskeleton system. IEEE Trans. Syst. Man Cybern. Part A: Syst. Hum. **31**(3), 210–222 (2001)
8. Aguirre-Ollinger, G., Colgate, J.E., Peshkin, M.A., Goswami, A.: Active-impedance control of a lower-limb assistive exoskeleton. In: 2007 IEEE 10th International Conference on Rehabilitation Robotics, pp. 188–195. IEEE (2007)
9. Anam, K., Al-, A.A.: Active exoskeleton control systems: state of the art. Procedia Eng. **41**, 988–994 (2012)
10. Kazerooni, H., Steger, R., Huang, L.: Hybrid control of the Berkeley lower extremity exoskeleton (BLEEX). Int. J. Robot. Res. **25**(5–6), 561–573 (2006)
11. Jatsun, S., Malchikov, A., Yatsun, A.: Comparative analysis of the industrial exoskeleton control systems. In: Proceedings of 14th International Conference on Electromechanics and Robotics "Zavalishin's Readings", pp. 63–74. Springer, Singapore (2020)
12. Sheean, G.: The pathophysiology of spasticity. Eur. J. Neurol. **9**, 3–9 (2002)
13. Stampacchia, G., Rustici, A., Bigazzi, S., Gerini, A., Tombini, T., Mazzoleni, S.: Walking with a powered robotic exoskeleton: subjective experience, spasticity and pain in spinal cord injured persons. Neuro Rehabil. **39**(2), 277–283 (2016)
14. Hill, A.V.: First and Last Experiments in Muscle Mechanics. Cambridge University Press, New York (1970)

15. Heo, P., Gu, G.M., Lee, Sj., Rhee, K., Kim, J.: Current hand exoskeleton technologies for rehabilitation and assistive engineering. Int. J. Precis. Eng. Manuf. **13**(5), 807–824 (2012)
16. Veneman, J.F., Kruidhof, R., Hekman, E.E., Ekkelenkamp, R., Van Asseldonk, E.H., Van Der Kooij, H.: Design and evaluation of the LOPES exoskeleton robot for interactive gait rehabilitation. IEEE Trans. Neural Syst. Rehabil. Eng. **15**(3), 379–386 (2007)
17. Pratt, G.A., Williamson, M.M.: Series elastic actuators. In: Proceedings 1995 IEEE/RSJ International Conference on Intelligent Robots and Systems. Human Robot Interaction and Cooperative Robots vol. 1, pp. 399–406. IEEE (1995)
18. Lu, R., Li, Z., Su, C.Y., Xue, A.: Development and learning control of a human limb with a rehabilitation exoskeleton. IEEE Trans. Industr. Electron. **61**(7), 3776–3785 (2013)
19. Rajasekaran, V., Aranda, J., Casals, A., Pons, J.L.: An adaptive control strategy for postural stability using a wearable robot. Robot. Auton. Syst. **73**, 16–23 (2015)
20. Brahmi, B., Saad, M., Ochoa-Luna, C., Rahman, M.H., Brahmi, A.: Adaptive tracking control of an exoskeleton robot with uncertain dynamics based on estimated time-delay control. IEEE/ASME Trans. Mechatron. **23**(2), 575–585 (2018)

Algorithm for Controlling Manipulator with Combined Array of Pressure and Proximity Sensors in Gripper

Aleksei Erashov and Konstantin Krestovnikov

Abstract This paper considers the development of gripping algorithm to grasp different objects by a two-finger manipulator of an anthropomorphic robot using a composite array of pressure and proximity capacitance sensors. The developed algorithm of link motion utilizes data from the composite array of pressure and proximity sensors, controlling the process of grip approach in the small neighborhood of gripping point through setting of position and orientation, required for grip, with specified accuracy levels. Based on the two-finger manipulator grip and on the data from the composite sensor arrays, embedded into robot fingers, several experiments were performed, using the developed algorithm of object grip. Dependencies for successful object grip were obtained, derived from orientation-related accuracy parameters and from finger configuration during grasp. It was revealed in process of testing of the developed algorithm that by optimal values of the specified parameters, the probability of successful grip was 97%. The proposed algorithm with certain modifications, which comply with the problem in question, is applicable for gait control of walking and anthropomorphic robots, as well for secure human–machine interactions, involving multi-link robotic systems.

1 Introduction

The current course of research in the domain of force-torque sensing of manipulators [1, 2] and pedipulators [3, 4] in robots is a vibrant topic. Many papers are published regarding this domain, which detail, particularly, development of new pressure sensor designs [5, 6], as well of combined sensors, intended for measurement of various quantities—such as external applied pressure and object proximity [7, 8]. Development of algorithms for object gripping, which relies on signaling from such sensors, is a relevant problem. Provided it could be solved, robotic manipulators

A. Erashov (✉) · K. Krestovnikov
St. Petersburg Institute for Informatics and Automation of the Russian Academy of Sciences, St. Petersburg Federal Research Center of the Russian Academy of Sciences (SPC RAS), 39, 14th Line, St. Petersburg 199178, Russia
e-mail: erashov.a@iias.spb.su

A. Ronzhin and V. Shishlakov (eds.), *Electromechanics and Robotics*, Smart Innovation, Systems and Technologies 232, https://doi.org/10.1007/978-981-16-2814-6_6

could securely grasp and hold objects, move them, interact with other robots in a group, and human–machine interactions would be facilitated as well.

Utilization of combined pressure and proximity sensors, embedded in the manipulator grip, provides more comprehensive data on object, than utilization of separate sensors each for a certain quantity. Data, obtained from the combined sensors allow to improve positioning control and grip orientation when approaching and object, considering the allowable error margin, as well set the grip fingers equally spaced from the object and adjust the pressure, exerted by the fingers on the object.

Pressure and proximity sensors can be combined through their close positioning to each other [9], setting one of the sensors into the hole [10, 11] or in a gap within the another sensor [12].

Using the solutions [7, 9], it is possible to measure pressure and proximity via the same operational approach, based on the measurement of intensity of the reflected light. A disadvantage of the proximity sensor consists in the dependency of the signal value from the color of the object. On the contrary, utilization of such solutions can simplify the design and implementation of the combined pressure and proximity sensor.

Using the capacitive connection, being established between the object and the combined sensor [8], object approach process can be measured, as well the pressure, exerted on the one of the electrodes. The advantages of such sensors are unified structural design of the sensor and a relatively low number of outbound wires.

Gripping algorithm, which relies on pressure and proximity sensors, can be divided into two categories: algorithms, that describe the gripping act at the moment, when the grip position and orientation are close to target point, and the algorithms, showing, which tasks are performed by the manipulator control system by arbitrary initial conditions—position and orientation of each link can take any value from the operational range of the manipulator.

Algorithms from the first group usually come into play, when the grip is positioned closely to the target point, relatively to which the object is grasped [10]. In this paper, authors employed the following workflow to solve the grasp-related problems: approaching, control of positioning of gripping fingers and grasp with adjustment of pressure. In [12], based on the readings of the proximity sensor, the control system ensures such positioning of links and grip that the object would be situated within the allowable error margin between the grip fingers; then, the grasp occurs. To grasp the target object, in [11], a three-stage algorithm was developed; having received the grip instruction, the grip fingers approach each other, until the touch with the object will be detected. Further, the manipulator control system adjusts the grip position, until every finger acquires touch with the object surface. Then, the object is grasped, and if it slips between the fingers, the grip should fasten. Otherwise, the finger positions remain unchanged. An optical-based algorithm, which prevents object slip, is proposed in [13]. Object grasp depends on adjustment of force, exerted by four-finger manipulator grip. The robot grasps the object with fingertips, thereby exerting the default force, and reduces the force value until the slip event is detected. Then, the grasp fastens for more reliable object gripping. Should a more significant slip be detected, the pressure force increases even more. The gripping algorithm in

[14] consists of the following stages: mutual approaching of all the fingers in grip, until they reach approximately equal distance to object each. After grasp, the contact of fingers with object surface is retained with a minor delay, and then, the controller determines the maximum force value within this timeframe and multiplies it by the specified ratio of mechanical stress factor to the velocity factor of the finger close-in. Further, the grasped object is lifted via adjustment of grip force.

To the first subset of gripping algorithms, the approach can be attributed that is proposed in [8], where the object grasp routine is defined by the conductivity of the object; if the object is made of conductive material, then the proximity signal suffices for grip fingers to close in; otherwise, the touch event on surfaces with combined surfaces is required.

The algorithms from the second group often employ a path planner for manipulator link motions to the target point. With such planner and image-based object recognition, in [15] the gripper is approached to the grasp point, and then, the search module iteratively infers positions and orientation of grip fingers, to grasp the object. The paper presents a touch detection algorithm, which takes the data of finger positions as reference and compares this data with a threshold value. A similar design has the object detection algorithm, which bears on proximity sensor data. The control algorithm in [16] is implemented with an open-source software robot operating system (ROS). After the grasp point has been reached, the algorithm proceeds to bear on proximity sensor data. Should object touch at one finger be detected, so only the finger proceeds moving, which has not touched the object surface yet. Before the touch event the actuator could be repositioned, if object motion would be detected by the sensor, embedded into the robotic fingertip. Grip positioning relative to the object is performed iteratively via finger opening and close-in; thereby, distance from every finger to the object is determined. In [17], an algorithm for operation of a three-finger gripper is presented. This algorithm is based on data, obtained via image processing by an artificial neural network. The algorithm is divided as follows: perception, grip planning and execution, holding of the grasped object and manual adjustment of the grip. Having inferred the position and orientation of the object using a depth map and point cloud, grip parameters from the respective database are generated. Further, the OpenRave planner generates a path for collision-free motion of the gripper to the target point. Then, the object is grasped. The developed classifier evaluates the grip, interpreting its reliability upon images, obtained by means of technical vision.

For the purpose of this paper, it is more preferable to develop a control algorithm that could be attributed to the second group; hence, it would enable link repositioning into any point of the operational space and control the grasp workflow, bearing on data from arrays of combined sensors. In this context, it also remains feasible to extend the algorithm for sensor-aided obstacle avoidance, as well for human–machine interactions and to ensure safety during operational workflow.

2 Algorithm for Object Grasping

The development of an algorithm for grasping and further manipulations with the target object (TO) requires solving the following subproblems:

1. Defining of grasp point for the TO.
2. Actuator path construction to reach the grasp point with specified accuracy level.
3. Solution of the forward and inverse kinematics problem to provide the system of manipulator link control with necessary data about position and orientation of the manipulator grip.
4. Reposition of the manipulator grip to reach the vicinity of the grasp point with accuracy γ. Then, correct the orientation of the manipulator grip, based on signals from the cells of the combined pressure and proximity sensors embedded in each finger of the grip.

The grasp point for the TO can be found with neural network-related methods and technical vision [18].

This point determines the required target position and orientation of the actuator relative to the frame of reference of the manipulator after moving its links. This information is used as input for the trajectory generation unit of the developed object gripping algorithm using combined array pressure and proximity sensors, which is shown in Fig. 1.

The algorithm uses the following notation:

1. ε is deviation between the signals of the cells of the array, embedded in one of the gripper fingers. This value corresponds to the accuracy of the orientation of the gripper relative to the operated object.
2. μ is deviation between the signals of the cells of the array, embedded in different gripper fingers. This value corresponds to the accuracy of the orientation of the gripper relative to the operated object.
3. λ is deviation of signals from the cells of array sensors at the moment of object being touched by the manipulator grip.
4. σ is deviation of signals from the cells of the array sensors, corresponding to the specified pressure value that must be reached by the grip of the manipulator before starting to manipulate the object.

The initial position and orientation of the manipulator actuator and the grasp point are involved in path generation for the links. After setting the preliminary configuration of the gripper, the through-point of the path is computed. Then, the inverse kinematics problem is solved to calculate the necessary angles of rotation for moving the links and the actuator of the manipulator to the intermediate (junction) points of the generated path.

In the process of moving the links along the path, it is ensured that the actuator of the manipulator reaches the vicinity of the capture point. The check is performed according to the accuracy metric γ, which depends on the tasks performed and on the approaches to their solution.

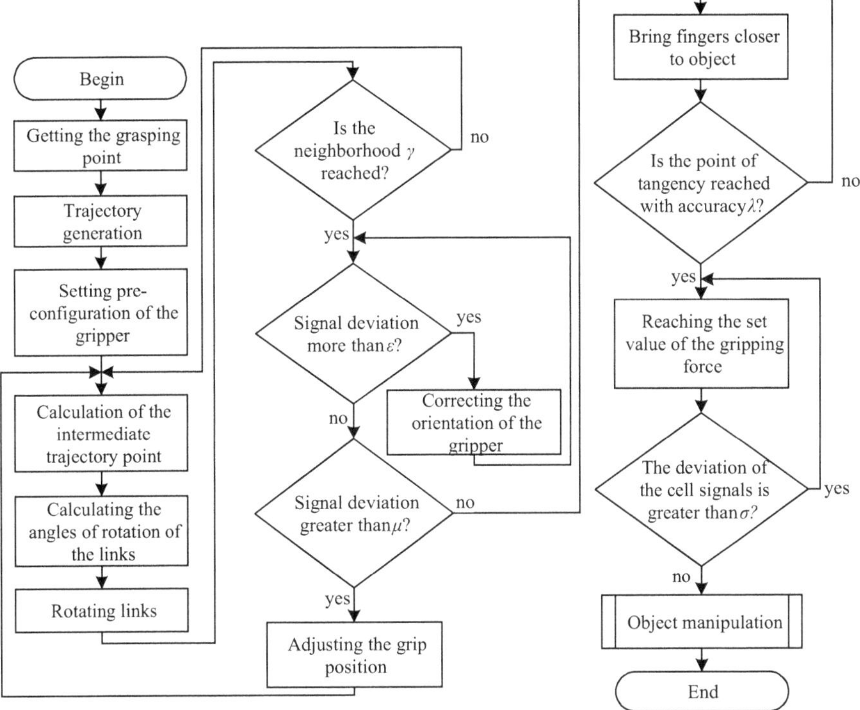

Fig. 1 Algorithm of manipulator operation with arrays of pressure and proximity sensors on the gripper

Further steps are intended to achieve the specified deviation values ε and μ between the signals of the cells of each sensor array.

After the actuator reaches the TO grip point, the orientation and position of the gripper are adjusted according to signals from the cells of the array sensors. In this case, the following cases are possible:

1. The deviation between the signals of the cells of the array of one finger is greater than ε, which indicates an incorrect orientation of the gripper relative to the TO. Then, the alignment of the orientation of the manipulator grip is performed according to the signals from each sensor array until the deviation between the signals of the array cells becomes less than ε:

$$\sqrt{\frac{\sum_{i=1}^{N}(\overline{s}-s_i)^2}{N}} \leq \varepsilon, \quad (1)$$

where N is the number of array cells, $\overline{s} = \frac{\sum_{i=1}^{N} s_i}{N}$ is the average value of a sensor array, and s_i is the signal from each array cell.

2. The deviation between the signals of the array cells located on different fingers of the manipulator grip is greater than μ. This indicates that one of the grasping fingers is closer to the object than the other. In this case, the position of the actuator of the manipulator is adjusted until the deviation of signals between array cells is less than μ:

$$\sqrt{\frac{\sum_{i=1}^{M}(\overline{m}-\overline{s}_i)^2}{M}} \leq \mu, \tag{2}$$

where M is the number of arrays of combined sensors, $\overline{m} = \frac{\sum_{i=1}^{M}\overline{s}_i}{M}$ is the average value among the all sensor arrays, and $\overline{s}_i$ is the average value of the signal from each array.

The developed algorithm, which involves utilization of array pressure and proximity sensors, assumes the orientation of the gripper relative to the grasp point so that the point associated with its geometric center is in the middle between the fingers with an accuracy of ε and μ. After reaching the specified accuracy in position and orientation, the manipulator fingers are iteratively closed in, while the correspondence to the range of the touchpoint of the levels of the output signals of the array sensors cells is checked with an accuracy λ. The check is performed according to the similar (2) expression:

$$\sqrt{\frac{\sum_{i=1}^{M}(\overline{m}-\overline{s}_i)^2}{M}} \leq \lambda \tag{3}$$

When the signal levels fall into this range, the control algorithm increases the gripping force until a certain value of the pressure force is reached with the deviation of signals from the cells of the array sensors σ:

$$\sqrt{\frac{\sum_{i=1}^{M}(\overline{m}-\overline{s}_i)^2}{M}} \leq \sigma \tag{4}$$

Hence, the algorithm does not depend on the number of combined sensor arrays in use and on the number of cells in each array. The values, given in (1)–(4), are subject to variations and are set at the implementation step.

3 Experiments

Based on previous research, an array with cells of a combined pressure and proximity sensor was developed [19]. The array design corresponds to the structure, given in [20]. For utilization in a gripper of an anthropomorphic robot [21, 22], based on the developed circuits and design solutions, experimental prototypes of the sensors were assembled in the layout 2×2 (Fig. 2a). The four-cell array of combined sensors is embedded into each finger of the two-finger gripper of the anthropomorphic robot (Fig. 2b).

By installing an array sensor in each finger, it is possible to determine the desired orientation and position of the gripper relative to the target when approaching it.

To specify the required positions and orientations of the manipulator links, it is necessary to solve the forward and inverse kinematics problems taking into account the relevant parameters. To derive kinematic equations, a modified Denavit-Hartenberg technique was used [23]. Kinematic scheme of the robotic arm is given in Fig. 3. The point α is the geometric center of the manipulator grasp.

The inverse kinematics problem has to be solved to determine the rotation angles in the joints, where the actuator ensures the matching of the α point with the target point for object grasp. However, in this case, intermediate values of the angles should be controlled, i.e., choose the type of control routine, according to which the generalized coordinates will be changed. To do this, path planner RRT-Connect is used [24], which, depending on the relevant motion constraints, performs path planning for angle changes in joints. The path being planned is laid along the through-points, which allow to compute intermediate angle values according to the kinematic equations. Reaching the grasp point with γ accuracy is based on markers and technical vision systems. The developed algorithm for object grip using an array of capacitive combined sensors was tested when capturing conductive and non-conductive objects [25].

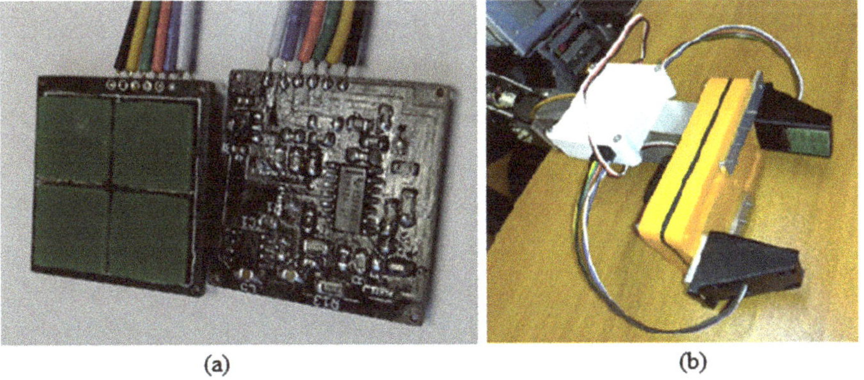

(a) (b)

Fig. 2 The four-cell array of combined pressure and proximity sensors (**a**), embedded into each finger of the two-finger gripper of the anthropomorphic robot (**b**)

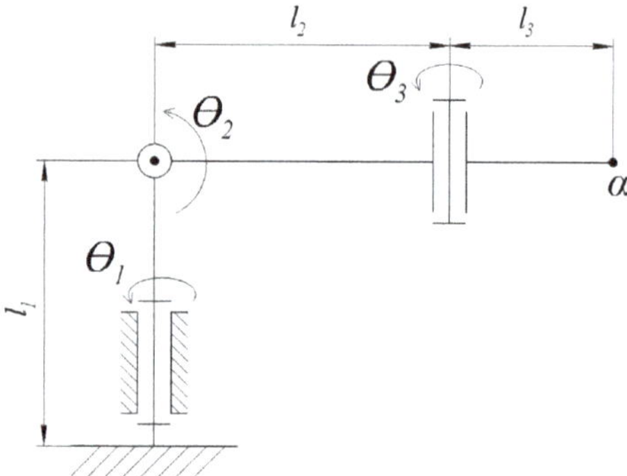

Fig. 3 Kinematic scheme of the robotic manipulator arm

Experiments were performed with gripping of cylindrical, spherical and parallelepiped-like objects. The objects were made of different materials: conditionally non-deformable with a hardness of more than 50 Shore and easily deformable with a material hardness of less than 10 Shore. For each object, the values of λ and σ metrics were empirically determined. Then, by setting the orientation necessary for the grasp, the value of ε was determined, with which, for different values of μ metric, the characteristic of the probability of successful grasp was obtained. The probability of successful gripping is understood as the value of the ratio of the number of grips of an object without deformations and retention without slipping out to the total number of tests. Finally, taking the experimentally value of μ metric corresponding to the highest probability of successful grasp, the dependency of the probability of successful grasp on the ε metric was obtained. The curves of the obtained dependencies are given in Fig. 4.

According to the characteristics given above, it is evident that the more stringent requirements for the accuracy of setting the orientation and position will be imposed

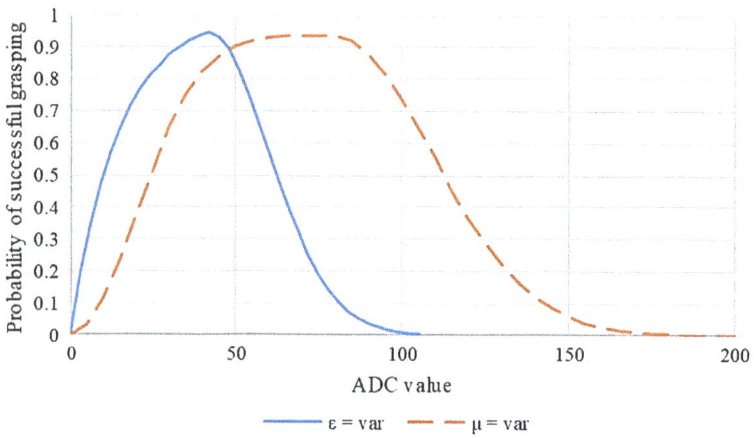

Fig. 4 Probabilistic dependencies of successful grip by variations of ε and μ

on the grip, the less likely the grip can be performed due to the influence of uncompensated errors, which can force the algorithm to loop into infinite search for the position without reaching it due to the small value of ε or μ. Less stringent accuracy requirements can increase the likelihood of capture; however, at high values of ε and μ, the probability of successful object grip decreases again. At high predefined values of ε and μ, the object may be displaced by the finger that is the closest to the object. The object also can slip from the fingers if the orientation of the fingers or the grip as a whole is incorrectly set.

The sensor arrays embedded into the manipulator grip of the manipulator enable to determine the touchpoints on the object and measure the contact forces in each of them. Feedback enables to grip easily deformable objects without damaging their shape when the allowable gripping force is preset. The data on the force of the gripping pressure allows to minimize the number of the slipping instances in manipulation with objects to 0.89%. Proximity data regarding the grasping fingers and their pressure enabled to successfully grip the object in 97% of cases.

4 Conclusion

In this paper, an algorithm was proposed, which enables object gripping by robotic manipulators, based on data, obtained from arrays of combined pressure and proximity sensors. This algorithm is intended to control the position and orientation of the grasp, as well to adjust the pressure of the fingers, thereby ensuring reliable grasping of objects without displacement, deformations and slipping. The proposed solution requires no structural changes in sensor arrays when the number of sensor or cells in array changes. Probabilistic dependencies were obtained for grip success depending on the accuracy of gripper orientation relative to the object being manipulated ε and accuracy from gripper position μ. It was experimentally confirmed that successful gripping without deformations, displacements and slipping with optimal values of accuracy parameters was achieved in 97% of cases. The proposed solution with some adjustments in the implementation of the algorithm structure is applicable for the implementation of the gait of walking and anthropomorphic robots, as well as to determine the control operations of pedipulators [26].

In the following research, it is intended to improve the algorithm to implement obstacle detection capabilities in it when manipulator links are moving along the planned path.

References

1. Yang, J., et al.: Flexible, tunable, and ultrasensitive capacitive pressure sensor with microconformal graphene electrodes. ACS Appl. Mater. Interfaces. **11**(16), 14997–15006 (2019)

2. Jamone, L., Natale, L., Metta, G., Sandini, G.: Highly sensitive soft tactile sensors for an anthropomorphic robotic hand. IEEE Sens. J. **15**(8), 4226–4233 (2015)
3. Yasin, A., Xu, Q., Chen, B., Lu, Q., Khan, M. W.: Design of a 23-DoF small humanoid robot with ZMP force sensors. In: Informatics in Control, Automation and Robotics, pp. 31–38. Springer, Heidelberg (2011)
4. Krestovnikov, K., Saveliev, A., Cherskikh, E.: Development of a circuit design for a capacitive pressure sensor, applied in walking robot foot. In: 2020 IEEE 20th Mediterranean Electrotechnical Conference (MELECON), pp. 243–247. IEEE (2020)
5. Ramalingame, R., et al.: Highly sensitive capacitive pressure sensors for robotic applications based on carbon nanotubes and PDMS polymer nanocomposite. J. Sens. Sens. Syst. **8**(1), 87–94 (2019)
6. Yussof, H., Wada, J., Ohka, M.: Sensorization of robotic hand using optical three-axis tactile sensor: Evaluation with grasping and twisting motions (2010)
7. Araki, R., Abe, T., Noma, H., Sohgawa, M.: Miniaturization and high-density arrangement of microcantilevers in proximity and tactile sensor for dexterous gripping control. Micromachines **9**(6), 301 (2018)
8. Tavakoli, M., et al.: Autonomous selection of closing posture of a robotic hand through embodied soft matter capacitive sensors. IEEE Sens. J. **17**(17), 5669–5677 (2017)
9. Konstantinova, J., Stilli, A., Althoefer, K.: Force and proximity fingertip sensor to enhance grasping perception. In: 2015 IEEE/RSJ International Conference on Intelligent Robots and Systems (IROS), pp. 2118–2123. IEEE (2015)
10. Hasegawa, H., Mizoguchi, Y., Tadakuma, K., Ming, A., Ishikawa, M., Shimojo, M.: Development of intelligent robot hand using proximity, contact and slip sensing. In: 2010 IEEE International Conference on Robotics and Automation, pp. 777–784. IEEE (2010)
11. Saen, M., Ito, K., Osada, K.: Action-intention-based grasp control with fine finger-force adjustment using combined optical-mechanical tactile sensor. IEEE Sens. J. **14**(11), 4026–4033 (2014)
12. Patel, R., Correll, N.: Integrated force and distance sensing using elastomer-embedded commodity proximity sensors. In: Robot. Sci. Syst. (2016)
13. Maldonado, A., Alvarez, H., Beetz, M.: Improving robot manipulation through fingertip perception. In: 2012 IEEE/RSJ International Conference on Intelligent Robots and Systems, pp. 2947–2954. IEEE (2012)
14. Romano, J.M., Hsiao, K., Niemeyer, G., Chitta, S., Kuchenbecker, K.J.: Human-inspired robotic grasp control with tactile sensing. IEEE Trans. Rob. **27**(6), 1067–1079 (2011)
15. Patel, R., Curtis, R., Romero, B., Correll, N.: Improving grasp performance using in-hand proximity and contact sensing. In: Robotic Grasping and Manipulation Challenge, pp. 146–160. Springer, Cham (2016)
16. Jiang, L. T., Smith, J. R.: Seashell effect pretouch sensing for robotic grasping. In: 2012 IEEE International Conference on Robotics and Automation, pp. 2851–2858. IEEE (2012)
17. Dang, H., Allen, P.K.: Stable grasping under pose uncertainty using tactile feedback. Auton. Robot. **36**(4), 309–330 (2014)
18. Yakovlev, R., Rubtsova, Yu., Erashov, A.: Comparative assessment of approaches to determining the points of capture of objects by robotic mean. Mechatron. Autom. Control **22**(5), 83–93 (2021) (In Russ.)
19. Krestovnikov, K., Cherskikh, E., Zimuldinov, E.: Combined capacitive pressure and proximity sensor for using in robotic systems. In: Proceedings of 15th International Conference on Electromechanics and Robotics "Zavalishin's Readings", pp. 513–523. Springer, Singapore (2020)
20. Krestovnikov, K., Erashov, A., Bykov, A.: Development of circuit solution and design of capacitive pressure sensor array for applied robotics. Robot. Tech. Cybern. **8**(4), 296–307 (2020). (In Russ.)
21. Pavlyuk, N.A., Bizin, M.M.: Constructive solutions for the anthropomorphic robot ANTARES. In: Mathematical Methods in Technique and Technologies—MMTT, vol. 9(91), pp. 138–141 (2016). (In Russ.)

22. Kodyakov, A.S., Pavlyuk, N.A., Budkov, V.Y., Prakapovich, R.A.: Stability study of anthropomorphic robot antares under external load action. J. Phys. Conf Ser. **803**(1), 012074 (2017)
23. Craig, J.J.: Introduction to Robotics: Mechanics and Control, 3rd edn. Pearson Education India (2009)
24. Kuffner, J.J., LaValle, S.M.: RRT-connect: an efficient approach to single-query path planning. In: Proceedings 2000 ICRA. Millennium Conference. In: IEEE International Conference on Robotics and Automation. Symposia Proceedings, vol. 2, pp. 995–1001. IEEE (2000)
25. Laboratory of Autonomous Robotic Systems. https://www.youtube.com/watch?v=Ltn5oReLZtU. Last accessed 11 Jan 2021
26. Gorobtsov, A.S., et al.: Features of solving the inverse dynamic method equations for the synthesis of stable walking robots controlled motion. SPIIRAS Proc. **18**(1), 85–122 (2019) (In Russ.)

Simulation of Underwater Robot Autonomous Motion Along Predetermined Straight Path

Jamil Safarov, Sergey Jatsun, Andrey Yatsun, and Sergey Knyazev

Abstract The development of systems for automated water monitoring for quality control in reservoirs (SAWM) allows the prompt real-time assessment of the environmental situation at various points in the monitoring area. One of the main steps preceding the study of the water quality is sampling. The stationary character of such pollution analyzers makes it practically impossible to provide control at various sites of the monitoring object. In this article, there is considered a problem of specified rectilinear motion separation, in definition of the miniature unmanned underwater vehicles (MUUV) "specified point" desired position and perturbed motion resulting from an external disturbance. The research objectives are the development of mathematical models describing the process of MUUV trajectory generation, as well as the development of a mathematical model of the MUUV system. The development of a motion control strategy in the study of water quality is carried out to organize the sampling process. A mathematical model which describes perturbed motion is proposed, and a method for the synthesis of the parameters of control system regulator based on the decomposition of the MUUV motion equations and transient process is developed. Mathematical models of the force factors acting on the MUUV were developed in view of external pulse disturbances. Mathematical model of autonomous space motion of miniature unmanned underwater vehicles (MUUV) is presented. The method of regulator parameters definition by use of roots approach is considered.

1 Introduction

The system of control of environment involves carrying out measures to monitor the state of water in almost every reservoir. For automated monitoring of water quality in reservoirs, (SAWM) allows for the prompt real-time assessment of the environmental situation at various points in the monitoring area [1, 2]. One of the

J. Safarov · S. Jatsun (✉) · A. Yatsun · S. Knyazev
Department of Mechanics, Mechatronics and Robotics, South-West State University, 94, 50 Let Oktyabrya Str., 305040 Kursk, Russia
e-mail: teormeh@inbox.ru

A. Ronzhin and V. Shishlakov (eds.), *Electromechanics and Robotics*, Smart Innovation, Systems and Technologies 232, https://doi.org/10.1007/978-981-16-2814-6_7

main steps preceding the study of the water quality is sampling [3–6]. One of the main stages for investigation of water quality is sampling. Under current conditions, water sampling is carried out by an operator at stationary monitoring posts fitted out with required equipment. However, the stationary character of such pollution analyzers makes it practically impossible to provide control at various sites of the monitoring object.

In these cases, the operators use floating platforms and underwater equipment. This is a time-consuming and unsafe process. Therefore, there is a problem of prompt and safe information acquisition relevant and sufficient for the assessment of the quality of the reservoir and the appropriate response. The solution of this problem could be found with application of underwater robots [7, 8]. Such kind approach allows providing sampling process automatically without person [9–10].

2 The MUUV Autonomous Locomotion Mathematical Model

Consider mathematical model of the miniature unmanned underwater vehicles (MUUV) autonomous locomotion. The scheme of reservoir by use of MUUV monitoring process is presented on Fig. 1.

In this paper, an approach which is based on idea of motion on specified rectilinear motion separation is considered, in definition of the MUUV "specified point" desired

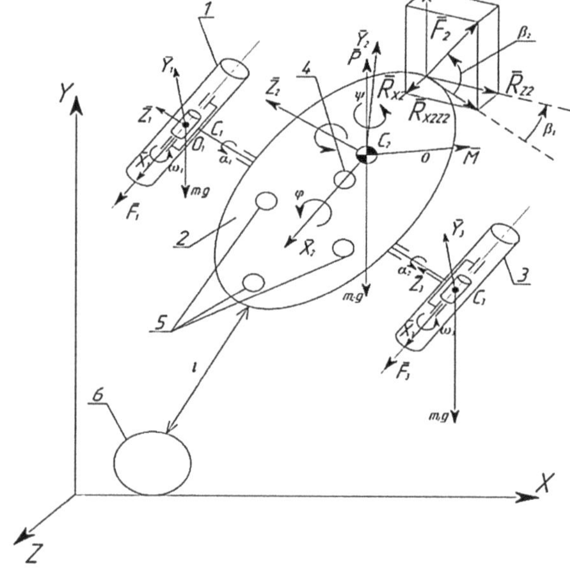

Fig. 1 Scheme of monitoring process of reservoir by use of MUUV: 1—right electric screw drive; 2—the body of MUUV; 3—left electric screw drive; 4—on-board computer; 5—sensors; 6—point of sampling

position of and perturbed motion resulting from an external disturbance. So, it means that we investigate the system motion by use of the principle of "follow by specified point" (FSP) [11, 12].

Therefore, radius vector, which defining real position of MUUV center of mass, is presented at the next form:

$$\bar{r}_C = \bar{r}_M + \bar{e},$$
$$\bar{r}_M = (xM_M, y_{M,}z_M)^T,$$

where $\bar{r}_M$—is the radius vector, defining the MUUV center of mass desired position; $\bar{e}$—vector, defining deviation of MUUV real position relatively to desired position.

Respectively, the absolute speed and acceleration are defined by formulas:

$$\begin{aligned} \dot{\bar{r}}_C &= \dot{\bar{r}}_M + \dot{\bar{e}}, \\ \ddot{\bar{r}}_C &= \ddot{\bar{r}}_M + \ddot{\bar{e}}, \end{aligned} \tag{1}$$

The control forces created by screw thrusters and rudders of depth and direction at first and second coordinate system are presented in (2):

$$\bar{F}_1^{(1)} = \begin{vmatrix} F_1 \\ 0 \\ 0 \end{vmatrix}; \bar{F}_3^{(3)} = \begin{vmatrix} F_3 \\ 0 \\ 0 \end{vmatrix} \tag{2}$$

Control forces $\bar{F}_1^{(1)}$, $\bar{F}_3^{(3)}$ depend on the angular velocity of the screws. The mathematical model of second control force shows influence of angles of direction and depth rudders β_1, β_2.

$$F_2^{(2)} = (R_0 \cos\beta_2 \sin\beta_1, R_0 \sin\beta_{21}, R_0 \cos\beta_2 \cos\beta_1)^T,$$

where $R_0 = r_0\bar{V}|\bar{V}|$

Differential equation describing locomotion of MUUV could be presented in following form (3):

$$m\ddot{\bar{r}}_C = -\bar{G} + \bar{H} + \bar{P} + \bar{F} + \bar{W}, \tag{3}$$

where $\overline{G}$, $\overline{P}$ forces—are the weight and the power of Archimedes; the resistance force $\overline{H}$, acting in the point C, $\overline{W}$—disturbance $G = P$. Rotation motion of MUUV is described by the differential Eq. (4):

$$\bar{M}_C^e = \frac{\mathrm{d}\bar{L}}{\mathrm{d}t} + \bar{L} \times \bar{\omega}, \tag{4}$$

where $\bar{L}$—kinetic moment MUUV; $\bar{M}_C^e$—the main moment of external forces.

The kinetic moment of MUUV $\bar{L} = I\bar{\omega}$, where I—inertia tensor; $\bar{\omega}$—angular velocity of MUUV.

We decompose the motion of the MUUV into a given and perturbed one:

$$\bar{F} = \bar{F}_1 + \bar{F}_2, \tag{5}$$

where $\overline{F_1}$—the control forces, providing desired locomotion of MUUV and $\overline{F_2}$–correcting forces of the perturbed motion.

The resistance force is presented in a form:

$$\bar{H} = \bar{H}_1 + \bar{H}_2.$$

From (1) is defining:

$$\bar{F} = m\ddot{\bar{r}}_C - \bar{H} + \bar{W}.$$

For definition of the control forces providing the MUUV, desired locomotion direct dynamics problem should be solved. For correcting perturbed motion forces, the inverse dynamic problem should be solved.

Substitute these expressions into (5), we will get (6):

$$\bar{F}_1 + \bar{F}_2 = m(\ddot{\bar{r}}_M + \ddot{\bar{e}}) - \bar{H}_1 - \bar{H}_2. \tag{6}$$

From here, it could be found:

$$\bar{F}_1 = m\ddot{\bar{r}}_M + \bar{H}_1,$$
$$\bar{F}_2 = m\ddot{\bar{e}} + \bar{H}_2 + \bar{W}.$$

For definition of the main control moment, the following equation is used:

$$\bar{M}_C^e = \frac{\mathrm{d}\bar{L}}{\mathrm{d}t} + \bar{L} \times \bar{\omega},$$

where $\bar{\omega} = (\omega_x + \omega_y + \omega_z)^T$, $\bar{\omega} = \bar{\omega}_1 + \Delta\bar{\omega}$, $\overline{\omega_1}$—desired angular velocity; $\Delta\bar{\omega}$—deviation of real angular velocity relatively desired.

We decompose the motion of the MUUV into a given and perturbed one.

$$\bar{M}_C^e = \bar{M}_{C1}^e + \bar{M}_{C2}^e,$$

where $\bar{M}_{C1}^e$—the control moment, providing desired direction of MUUV locomotion and $\bar{M}_{C2}^e$—correcting moment of the perturbed motion direction.

$$\bar{M}^e_{C1} + \bar{M}^e_{C2} = \frac{\mathrm{d}I(\bar{\omega}_1 + \Delta\bar{\omega})}{\mathrm{d}t} + I(\bar{\omega}_1 + \Delta\bar{\omega}) \times (\bar{\omega}_1 + \Delta\bar{\omega}).$$

For case, when $\bar{\omega}_1 << \Delta\bar{\omega}$:

$$\bar{M}^e_{C1} = \frac{\mathrm{d}I\bar{\omega}_1}{\mathrm{d}t} + I\bar{\omega}_1 \times \bar{\omega}_1,$$
$$\bar{M}^e_{C2} = \frac{\mathrm{d}I\Delta\bar{\omega}}{\mathrm{d}t} + I\Delta\bar{\omega} \times \Delta\bar{\omega}.$$

Denote the vector of control actions:

$$\bar{U} = (\bar{F}, \bar{M}_C(\bar{F}))^T,$$
$$\bar{U} = \bar{U}_1 + \bar{U}_2.$$

where $\bar{U}_1$—control actions that ensure movement along a given trajectory; $\bar{U}_2$—corrective control actions with disturbed movement.

3 Mathematical Model of the Specified Rectilinear Motion

Consider the diagram shown in Fig. 2. The desired position of the MC of MUUV is set by the point M. The movement of the MUUV occurs in a horizontal plane in a straight line, at the depth specified by the coordinate $y_m = 0$.

Point M moves along the M_iX_i axis. The coordinates of point M are defined as follows (7):

$$x_M(t) = \sum_{i=0}^{n} d_i \cdot t^i, \quad (7)$$

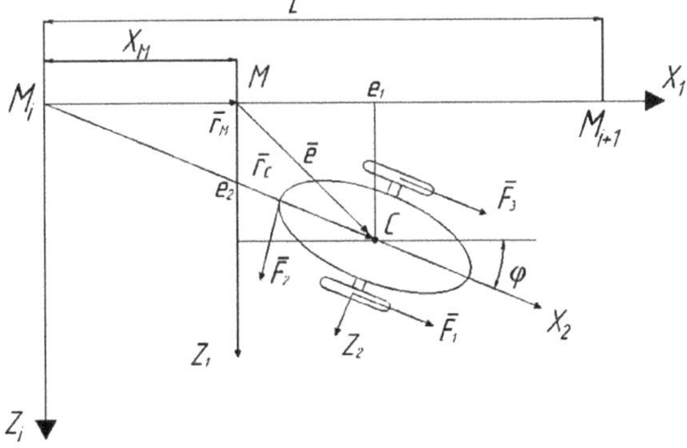

Fig. 2 Scheme of MUUV perturbed motion

where the d_i constants are determined from the boundary conditions.

In Fig. 2, $M_iX_iZ_i$—fixed coordinate system; MX_IZ_I—a mobile coordinate system, associated with a given point M; CX_2Z_2—a mobile coordinate system associated with MUUV; $0 < x_M < L$, $z_M = 0$; L—the length of a straight-line segment.

The boundary conditions on this section at the beginning of the trajectory and at the end are represented as:

$$\begin{aligned} &\mathrm{t} = \mathrm{t}_1,\ y = y_1,\ x = x_1, \dot{y} = 0,\ \dot{x} = 0,\ \phi = \phi_0,\ \dot{\phi} = 0, \\ &\mathrm{t} = \mathrm{t}_2, y = y_1,\ x = x_1, \dot{y} = 0,\ \dot{x} = 0, \phi = \phi_{01},\ \dot{\phi}_{01} = 0. \end{aligned} \tag{8}$$

$$t = t_2, \tag{9}$$

The coefficients are found from (8) and (9). The system of algebraic equations comes from these equations:

$$A_2 D = B_2. \tag{10}$$

The solution of (10) is presented in the form:

$$D = A_2^{-1} B_2,$$

where $A_2 B_2$—the coefficients matrix; D—a coefficients vector. Substituting previous equations, we define $\bar{F}_1$—the vector of control forces that ensure movement along a given trajectory is represented as: $\bar{F}_1 = (\bar{F}_{1x} \bar{F}1_y \bar{F}_{1z})^T$.

Since the movement is set by the coordinate: $x_M(t) = \sum_{i=0}^{n} d_i \cdot t^i$, then the time derivatives will take the form (11):

$$\dot{x}_M(t) = \sum_{i=1}^{n} d_i \cdot i t^{i-1}, \tag{11}$$

$$\ddot{x}_M(t) = \sum_{i=2}^{n} d_i \cdot i(i-1) t^{i-2}. \tag{12}$$

Then:

$$\begin{aligned} F_{1x} &= m\ddot{x}_M + \mu_x \dot{x}_M, \\ F_{1x} &= m \sum_{i=2}^{n} d_i \cdot i(i-1) t^{i-2} + \mu_x \sum_{i=1}^{n} d_i \cdot i t^{i-1}. \end{aligned}$$

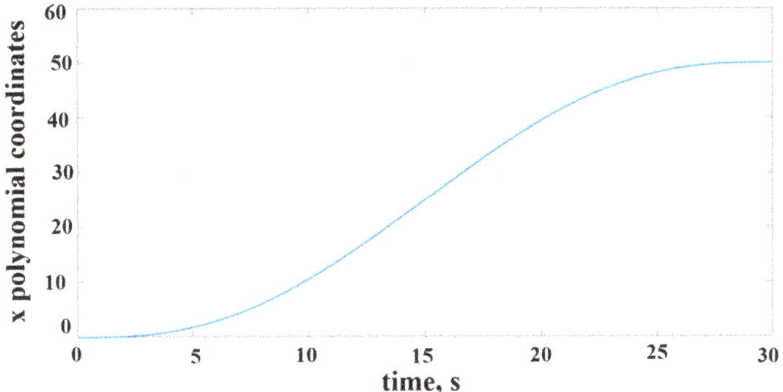

Fig. 3 Diagram of desired function of coordinate center of mass of MUUV $x_m(t)$

Diagram of specified rectilinear motion $x(t)$ is presented on Fig. 3. This diagram shows that initial and final position of center of mass has zero velocity and acceleration.

4 The Perturbed MUUV Motion Mathematical Modeling

To find the $\overline{F_2}$—vector of forces correcting the perturbed motion, consider the MUUV perturbed motion in the horizontal plane Oxz along a straight-line segment [13]. It is possible to use the scheme in Fig. 2. To study the perturbed motion of an object relative to a given one, the motion is considered in a mobile Fresnel coordinate system associated with a point M. That point determines the desired position of the object on a given trajectory. In this case, the system of differential equations describing the perturbed motion can be presented in the form (13):

$$\frac{d\bar{\xi}}{dt} = A\bar{\xi} + B\bar{U} + f(t), \tag{13}$$

where $\bar{e} = (e_1, e_2, e_3)^T$; $\bar{\xi} = (v_1, v_2, v_3)^T$;$\bar{U} \in R^3$; A—positive definite matrix 3×3; B—positive definite matrix 3×3 (13).

$$B = \frac{1}{m}\begin{vmatrix} 1 & 0 & 1 \\ 0 & 0 & 1 \\ l_1 & l_2 & -l_1 \end{vmatrix}; A = -\frac{1}{m}\begin{vmatrix} \mu_x & 0 & 0 \\ 0 & \mu_z & 0 \\ 0 & 0 & \mu_\phi \end{vmatrix}; \bar{f}(t) = \begin{vmatrix} -\frac{\mu_x}{m}\dot{x}_M + \ddot{x}_M(t) \\ 0 \\ 0 \end{vmatrix}. \tag{14}$$

The vector of correcting the disturbing motion of the effects is represented in the form (15):

$$\bar{U}_1 = \begin{vmatrix} F \\ F_2 \\ M_C(F) \end{vmatrix}. \tag{15}$$

Let us assume that the regulator forms a control action according to the following law (16):

$$\begin{gathered} \bar{U} =_1 \bar{\xi} + B_1\bar{e}, \\ A_1 = \begin{vmatrix} a_{11} & 0 & 0 \\ 0 & a_{22} & 0 \\ 0 & 0 & a_{33} \end{vmatrix}; B_1 = \begin{vmatrix} b_{11} & 0 & 0 \\ 0 & b_{22} & 0 \\ 0 & 0 & b_{33} \end{vmatrix}. \end{gathered} \tag{16}$$

Then (16), taking into account the previous expression, will take the form:

$$\bar{U} = \begin{vmatrix} {}_{11} & 0 & 0 \\ 0 & {}_{22} & 0 \\ 0 & 0 & {}_{33} \end{vmatrix} \begin{vmatrix} \nu_1 \\ \nu_2 \\ \nu_3 \end{vmatrix} + \begin{vmatrix} b_{11} & 0 & 0 \\ 0 & b_{22} & 0 \\ 0 & 0 & b_{33} \end{vmatrix} \begin{vmatrix} e_1 \\ e_2 \\ e_3 \end{vmatrix}.$$

Let us represent the components of the vector $\bar{U} \in R^3$ in the form:

$$\begin{aligned} F &= F_1 + F_3 = b_{11}e_1 + a_{11}\nu_1, \\ F_2 &= b_{22}e_2 + a_{11}\nu_2 \end{aligned} \tag{17}$$

$$M_C(F) = F_1 l_1 + F_2 l_2 - F_3 l_1 = b_{33}e_3 + a_{33}\nu_3. \tag{18}$$

After rearrangements of Eq. (13), we obtain three independent differential equations describing the perturbed position of the MUUV center of mass and orientation:

$$\begin{aligned} &\ddot{e}_1 + (n_x +_{11})\dot{e}_1 + b_{11}e_1 = -n_x\dot{x}_M - \ddot{x}_M, \\ &\ddot{e}_2 + (n_z +_{22})\dot{e}_2 + b_{22}e_2 = 0, \\ &\ddot{e}_3 + (n_\phi +_{33})\dot{e}_3 + b_{33}e_3 = 0. \end{aligned} \tag{19}$$

where $\dot{x}_M$, $\ddot{x}_M$—the desired speed and acceleration of the MUUV center of mass. n_x, n_z, n_φ—reduced drag coefficients for the corresponding generalized coordinates.

Definition of the coefficients of regulator is very important problem because coefficients of matrix A1 and B1 seriously influence on the quality of control system of MUUV. Precision, quick actions and stability strongly depends on those parameters. Equations (17), (18), (19) are linear, and it allows to consider transient process separately independent from a given movement of MUUV, which was described at Sect. 2 of this paper.

In this paper, definition of the regulator coefficients is used an approach bases on the analysis of the transient process [14].

To do this, it is offered to convert the formulas of roots definition in to next form:

$$\lambda_{1,2} = -\frac{n_x + a_{11}}{2} \pm \sqrt{\frac{(n_x + a_{11})^2}{4}(1 - \beta)},$$
$$\beta = \frac{4b_{11}}{(n_x + a_{11})^2},$$

where β—is parameter determining quality of transient process. If $\beta > 1$, we have oscillatory process, and in case when $\beta < 1$, we get a periodical process.

After that we define parameters $a_{11}b_{11}$:

$$_{11} = \frac{(\lambda_1 + \lambda_2)^2}{2} - (\lambda_1 - \lambda_2)^2,$$
$$b_{11} = \lambda_1 + \lambda_2 + n_x.$$

Parameters $a_{22}, a_{33}, b_{22}, b_{33}$, also, can be defined from consideration of Eqs. (18), (19).

Diagram of deviation between real position of the MUUV‘s center of mass and desired position during the starting velocity is presented on Fig. 4. In the result of action of initial impulse, center of mass has initial velocity $\dot{e}_0(0) = 0, 1$. In the beginning, MUUV has disturbance velocity near 0,1. After transient process, MUUV has got small additional velocity, which do not influence on the general motion process.

Diagram of control force allowing to compensate deviation of MUUV‘s real position relatively desired during impulse action is presented on Fig. 5. It is very oscillatory function, which is difficult to realize by control system.

Analysis of these diagrams shows that for these parameters of regulator ($a_{11} = 20$ and $b_{11} = 20$), transient process has oscillatory character, and this is unacceptable for practical operation of MUUV. Time of transient process of MUUV is long around 10 s. It is very long.

Therefore, further, the case ($a_{11} = 40$ and $b_{11} = 40$) when $\beta < 1$ is considered.

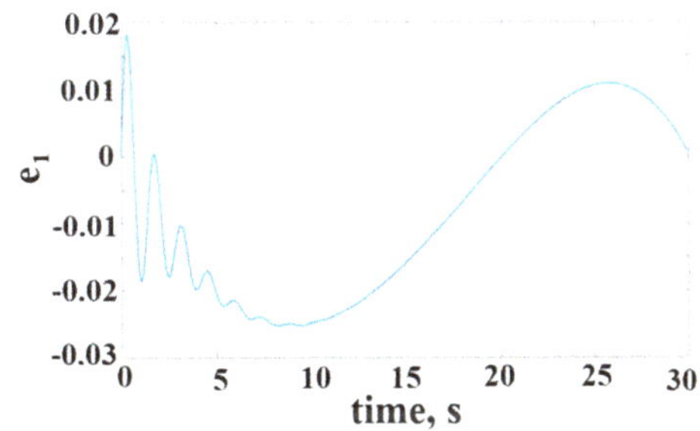

Fig. 4 Diagram of deviation of real position of the MUUV‘s center of mass relatively desired position during the starting velocity $\dot{e}_0(0) = 0, 1, \beta > 1$

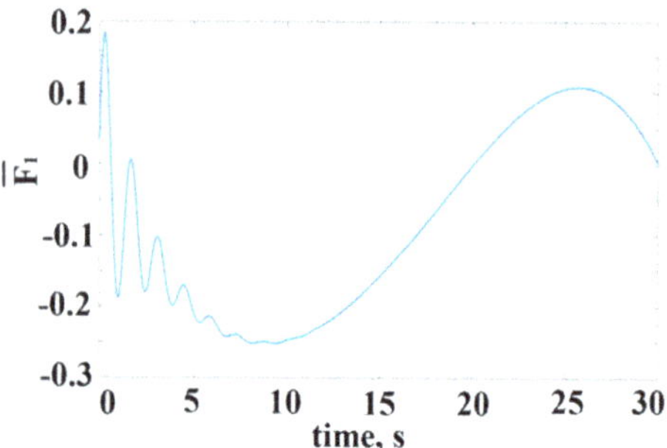

Fig. 5 Diagram of control force allowing to compensate deviation of real position of the MUUV's relatively desired during impulse action $\dot{e}_0(0) = 0, 1, \beta > 1$

Diagram of deviation real position of the MUUV center of mass relatively desired position during the starting offset $\beta < 1$ is presented on Fig. 10. It is very smooth function, which quite similar to get by control system. Deviation of MUUV real position is very small on all the trajectory of MUUV motion. It changes inside in interval from 0,07 till 0,03. It is very smooth function, and it is good result if we take into account that MUUV goes 50 m and relative mistake is less than 1%.

5 Conclusions

1. An approach based on idea of separation motion on specified rectilinear motion is considered in definition of the desired position of MUUV "specified point" and perturbed motion resulting from an external disturbance. So, it means that we investigate the motion of this system by use of the principle of "follow by specified point" (FSP).
2. Mathematical model of autonomous space motion of miniature unmanned underwater vehicles (MUUV) is presented. The model includes a description of the spatial motion of the MUUV hull under the action of controlled influences created by two screw drives with a variable thrust vector and rudders of motion and direction. Random and impulse external disturbances are also taken into account.
3. The monitoring of water quality in the points M_j where it is necessary to take samples in accordance with information about the reservoir, temperature, and pollution is settled by operator. The planning of points that need to be visited by the MUUV in the autonomous mode of movement is carried out using different approaches, including the method of evenly distributed sequences or planning an experiment.
4. The method of regulator parameters definition by use of roots approach is considered. Definition of the coefficients of regulator is very important problem because they seriously influence on the quality of control system of MUUV. Precision, quick actions and stability strongly depends on those parameters.

The linear equations given in this work allow to consider transient process separately independent from a given movement of MUUV.

Acknowledgements The work was supported by RFBR, research project № 19-08-00440.

References

1. Inzartsev, A.V., Pavin, A.M., Yeliseyenko, G.D., Rod'kin, D.N., Sidorenko, A.V., Lebedko, O.A., Panin, M.A.: Reconfigurable cross-platform environment for modeling the behavior of an unmanned underwater vehicle. Underwater Res. Robot. **2**(20), 28–34 (2015). (in Russ.)
2. Inzartsev, A.V., Pavin, A.M., Lebedko, O.A., Panin, M.A.: Recognition and inspection of small underwater objects using autonomous unmanned underwater vehicles. Underwater Res. 36–43 (2016) (in Russ.)
3. Levin, V.I.: Modeling of optimization problems in conditions of interval uncertainty. Izvestiya of the Penza State University. Phys. Math. Tech. Sci. **26**, 589–595 (2011) (in Russ.)
4. Ageev, M.D., Naumov, L.A., Illarionov, G.Y.: Unmanned underwater vehicles for military purposes. Dal'nauka, Vladivostok (2005). (in Russ.)
5. Fish robot for diagnosing water pollution. http://aquavitro.org/2010/12/24/ryba-robot-dlya-diagnostiki-zagryazneniya-vody/. Last accessed 15 Jan 2020 (in Russ.)
6. Inzartsev, A.V., Matviyenko, Y.V., Pavin, A.M., Rylov, N.I.: Monitoring of the seabed with the use of technologies for intelligent processing of data from search devices on board an autonomous unmanned underwater vehicle. Underwater Res Robot **2**(20), 20–27 (2015). (in Russ.)
7. Inzartsev, A.V., Pavin, A.M., Bagnitskiy, A.V.: Planning and implementation of actions of a survey underwater robot based on behavioral methods. Underwater Res Robot **1**(15), 4–16 (2013). (in Russ.)
8. Yatsun, S.F., Knyazev, S.I., Yatsun, A.S.: Controlled movement of a small-sized underwater complex (SSUC). Proc. Southwest State University **23**(5), 185–196 (2019). (in Russ.)
9. Knyazev, S.I.: Methodological aspects of creating seaglider-type underwater robots. In: Vibration Technologies, Mechatronics and Controlled Machines, pp. 138–142. Kursk (2016) (in Russ.)
10. Knyazev, S.I., Yatsun, A.S., Jatsun, S.F.: Automated terminal control system for monitoring ponds with the aid of a miniature unmanned underwater vehicle (MUUV). RusAutoCon, Russia, Sochi (2020)
11. Bondyrev, V.V., Knyazev, S.I., Korolev, V.I., Yatsun, S.F.: Control algorithm for an unmanned underwater complex. Proc. Southwest State University **23**(7), 185–196 (2020). (in Russ.)
12. Yefimov, S.V., Knyazev, S.I., Yatsun, S.F.: Study of the controlled movement of a small-sized underwater complex—water area pollution analyzer. J. Cloud Sci. **7**(3), 488–497 (2020). (in Russ.)
13. Zoteyev, V.Ye., Zausayev, A.F.: Application of perturbation methods in mathematical modeling: lab. workshop and textbook. method. Decree. Samar. Gos. Tekhn. Un-t, Samara (2010) (in Russ.)
14. Efimov, S.V., Knyazev, S.I., Yatsun, A.S., Jatsun, S.F.: Simulation of the automated control system for monitoring ponds with the aid of a miniature unmanned underwater vehicle. FarEastCon, pp. 813–817. Vladivostok (2020) (in Russ.)

15. Knyazev, S.I., Yatsun, A.S., Yatsun, S.F.: Controlled movement of a small-sized underwater robotic complex (MBPK). In the collection. Baltic Maritime Forum—materials of the VII International Baltic Maritime Forum: in 6 volumes, pp. 40–45 (2019) (in Russ.)
16. Knyazev, S.I., Maslov, A.A., Yatsun, S.F.: Mathematical model of a small-sized unmanned underwater complex for ecological monitoring of water bodies. In: XXXI International Innovative Conference of Young Scientists and Students on Mechanical Engineering Problems (MIKMUS-2019), pp. 480–483 (2020) (in Russ.)

In-Pipe Modular Robot: Configuration, Displacement Principles, Standard Patterns and Modeling

Ildar Nasibullayev, Oleg Darintsev, and Dinar Bogdanov

Abstract The article shows the relevance of the topic, the results of search studies and computational experiments performed in solving the problem of synthesis of a modular in-pipe robot intended for pipe inspection with complex topology. Based on the analysis of structures and applications of snake-like robots, methods of moving along the inner surface of the pipe, algorithms for switching between typical movements in the presence of various branches and fittings on the trajectory are proposed. A mathematical model of kinematics is implemented, taking into account the specifics of the robot's motion, which allowed us to form requirements for the design of modules, connecting joints with geometric constraints, specify the requirements for the control and information system. Testing of the mathematical model was carried out using a script written in the Python programming language with visualization of the robot's displacement in the Blender program. In the course of computational experiments, the parameters of the three-dimensional configuration of the modular robot were determined as it moves along the following trajectory patterns: Along a plane with transitions to a circular trajectory inside the pipe, to a helical line, followed by a rectilinear movement along the pipe. The features of the trajectories of the driven modules are analyzed. The proposed model is the basis for the development of a dynamic robot model and the subsequent synthesis of the design of a real robot prototype.

1 Introduction

Pipelines are an integral part of modern production, as well as a necessary means for implementing the smooth functioning of various systems in industry and urban infrastructure. The total length of pipelines significantly exceeds the size of other transport systems: Railways, highways. Most of the products necessary for the production or

---

I. Nasibullayev (✉) · O. Darintsev · D. Bogdanov
Mavlyutov Institute of Mechanics UFRC RAS, prosp. Oktyabrya 71, 450054 Ufa, Russia

O. Darintsev
USATU, K. Marksa str., 12, 450008 Ufa, Russia

A. Ronzhin and V. Shishlakov (eds.), *Electromechanics and Robotics*, Smart Innovation, Systems and Technologies 232, https://doi.org/10.1007/978-981-16-2814-6_8

the population activity are clean air, drinking and industrial water, heat transfer agents, gas, oil products and others are delivered to consumers through pipelines. Therefore, maintenance pipelines, monitoring the condition, timely diagnostics and repairs are the most urgent task [1]. The current standards define the maintenance of pipelines as the following: External inspection of pipelines to detect faults in welded joints and/or flanges, determination of thermal insulation, and/or coating, external inspection and minor repair of pipe fittings in the position of use, repair or replacement of emergency sections of pipe or fittings. The analysis of the damage causes to the pipelines, as well as the frequent use of their hidden installation, points to the need for an inspection of the internal surfaces, which is currently carried out only for trunk line pipes of a large diameter, for which the values of gradients and rotational radii «threads» are strictly regulated. Urban and intra-plant pipeline systems are characterized not only by smaller pipe diameters, but also compared to trunk pipes, and they have more complex topology, a large number of different branches and locking and regulating fittings (Fig. 1).

Currently, penetration or mobile tele-inspection systems with limits of up to 400 m in length and a minimum diameter of 120 mm in the test pipe are used for the diagnosis of intra-pipe space, required cable and operator permanently present [2, 3]. Studies are conducted mainly on straight or slightly curved sections, so for the inspection of real pipelines, it is necessary to develop new designs of inspection robots that have high autonomy, the ability to move in any direction, pass through holes in fittings, etc.

One of the promising types of robots for pipelines are snake-like robots, which are multi-link mechanisms using specific principles of displacement. For a wheelless variant of the structure, the movement is realized on the basis of the «mode of locomotion» change—the change of the robot shape [4], for the wheeled one—coordinated movement of the modules according to a given trajectory, with simultaneous

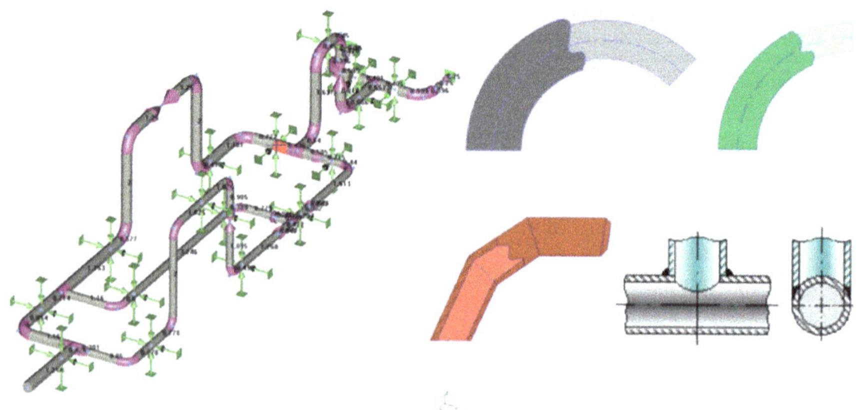

Fig. 1 The characteristic topology of the pipeline with a diameter of 200 mm (left) and the typical branches of the pipelines (bends and tee) (right)

change of the modules orientation relative to each other [5]. For such robots, the possibility of their reconfiguration [5–7] shall be indicated as follows:

- the implementation of a given form to perform a specific function;
- the splitting a chain of modules into separate segments that perform agreed tasks;
- the possibility of changing the number of modules in the chain without reworking the information control circuits.

The possibility of reconfiguration is provided by universal structural and information interfaces of robot modules, algorithms of their interaction, as demonstrated in the work [8–10]. One of the goals of robot reconfiguration—is to change the way the robot moves (serpentine movement, "walking", rolling), which is necessary to adapt to the environment of movement. When moving in pipelines on different sections of it, different ways of moving may be required. The slow movement of snake-like robots reduces the efficiency of their use in most pipeline maintenance tasks, however, the redundancy of the degrees of freedom makes it possible to form special configurations of robots inside the pipeline and realize the required movements at a higher speed.

The problems of choosing the configuration of snake-like robots and the ways of their implementation, taking into account the design features, are considered in sufficient detail in [11, 12]. The peculiarities of the functioning of modular robots in closed spaces require the development of other approaches due to some specific problems. One of them is the need to provide the specified reaction forces of the supports at the points of contact of the wheels with the pipeline inner surface for the selected robot configuration; the other is related to the functional purpose of the robot—diagnostics, and requires maintaining the specified orientation of the robot's working modules in the inner pipe space.

The simulation of robot configurations makes it possible to determine the required number of degrees of freedom, as well as other basic parameters for the synthesis of the structure, information and control systems, types and number of sensors. The sensor system of snake-like robots [13] comprises a wide range of sensors: Accelerometers, gyroscopes, internal state parameter sensors, joint angle position, speed, and wheel position sensors. Infrared, laser, and ultrasonic sensors, as well as video cameras, are used to perform operations, monitor the environment, and make measurements. If the surfaces on which the movement takes place may contain obstacles, tactile sensors or contact force measurements shall be used to detect them. For example, the work [4] shows the use of tactile sensors when the robot moves along an obstacle, and the [14] uses a system of contact force sensors to determine the point of contact of the modular robot with the surface.

In order to correctly choose the design of the modules of the robot, the method of connecting them to each other and the selection of its optimal configurations, and it is necessary to develop a specialized mathematical model and to carry out several computational experiments. For example, in the work [15], a model of a modular wheeled robot moving in the plane along a line, circle or spiral was proposed and the deviation of the tail from the trajectory of the drive module was evaluated.

The snake-like and wheeled modular robots discussed above are designed to move in the plane to overcome obstacles, the configurations are made to adapt to the terrain, but the chosen mode of movement of such robots significantly limits their applicability as a pipeline robot inspector.

The development of mathematical models of modular systems is greatly simplified by using specialized software with simultaneous visualization of the configuration of the systems under study. One of the widely used tools in recent years is the 3D modeling program Blender [16]. Thus, in the work [17], a kinematic manipulator model consisting of four segments with the same degree of freedom between the segments is proposed; in [18] a multi-segment manipulator model is presented, which makes it possible to monitor the movement of a person with the aid of a video camera and sensors; in [19] a robotic arm motion is modeled in Blender and signals are sent simultaneously via USB port to replicate the movements in the real model.

This article proposes a kinematic model of a wheeled modular robot capable of moving both in the plane, along the pipe, and along of a helical line trajectories, on the inner surface of the pipe, which will allow scanning of the surface of the pipes in a pipeline with a complex topology. A mathematical model is presented that describes the 3D configuration of a modular robot when it moves inside a pipe, based on the use of a set of patterns of movement trajectories. The proposed model allows the expansion of possible robot configurations by adding new trajectory patterns and new algorithms for transitions between patterns. For the subsequent testing of the mathematical model, Python code was developed, and the results were visualized in Blender.

2 Mathematical Model

To build a modular robot motion model, it is necessary to consider the real topology of pipelines using standard types of fittings. The topology analysis allows to form a set of modular robot configuration patterns and transition algorithms between them. Figure 2 shows the geometry of the module in three plane sections. Two coordinate

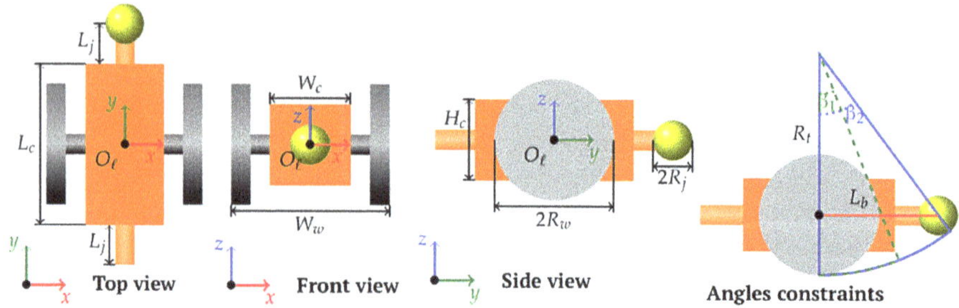

Fig. 2 Module geometry and structural angles constraints

systems were used in the simulation: The global coordinate system (GCS) with the Ox, axis coinciding with the pipe axis of radius R_t, the vertical axis Oz and the horizontal axis Oy perpendicular to the Ox, and the local coordinate system (LCS) with the origin O_ℓ, located in the geometric center (GC) of the module $O_\ell x$, the horizontal axis $O_\ell y$ in the direction of motion, and the vertical axis $O_\ell z$. We define the geometric constraints due to the ratio between the module size and the pipe radius. When the robot moves around the circumference, the only points of contact between the robot and the pipe should be the points of contact between the wheels and the surface of the pipe. If the radius of the pipe is not sufficient, it is possible to touch the lower part of the central module (Fig. 2). Module geometry and structural angles constraints, determined by the design angle β_1 or the ball joint (determined by the design angle β_2). For a robot consisting of several modules to fit into a pipe, the following conditions must be met: $R_t > \max(L_c/(2\sin\beta_1), L_b/(2\sin\beta_2))$.

To form standard patterns, let us assume that the robot modules are connected by ball joints. When the robot is moved, the first module is the lead one, and the rest are driven, meaning that the spatial configuration of the entire robot is determined by the required trajectory of the first module, the relative position of the other modules depends on the design features of the robot, the geometric characteristics of the pipe and the implemented pattern.

The simulation of geometric center position of the j module (x_j, y_j, z_j) and the local rotation angles $(\gamma_j^x, \gamma_j^y, \gamma_j^z)$ where the superscript means the axis of rotation, was carried out as follows. At the first stage, a mathematical model of the robot's motion is built within a single pattern. Then, a pattern for the next trajectory is selected, and an intermediate model of transition from one pattern to the other is constructed. It was considered that when moving segments along the trajectories of different patterns it is necessary to recalculate the movement parameters of modules from the first module of the robot to the last.

The main goal of the simulation is the correct synthesis of the robot configuration, while the physical parameters (gravity, friction forces, centrifugal acceleration) are not considered at the current stage of study.

As an example, the robot's model of motion follows a sequence of patterns: The robot moves in a horizontal plane (Pattern Line); perpendicular to the hole in the tube (Pattern Line-to-Circle); the robot moves in the tube along the circumference (Pattern Circle); the modules are sequentially injected into the pipe (Pattern Circle-to-Helix); the robot moves inside the pipe along the helical line (Pattern Helix); and from the helical line moved on the path along the pipe (Helix-to-Tube).

Within the Line template, the robot moves along the Oy axis at a velocity of v. The position of the GC will change over time τ: $\delta x = 0$, $\delta y = \delta\ell$, $\delta z = 0$, where $\delta\ell = v\tau$—is the value of the linear displacement. At time t the position of the j module is determined by

$$x_j = x_0,\ y_j = y_0 + vt + (j-1)L_g,\ z_j = z_0, \tag{1}$$

where the vector (x_0, y_0, z_0) is the initial position of the GC of the first module; $L_g = 2L_b = L_c + 2L_j$—the distance between the GC of the adjacent modules, and L_b—the length of the base line from the GC to the center of the hinge joint.

Within the Line-to-Circle pattern, part of the modules moves around the circle and some along the plane. At the time points t_k the module with the number k enters the pipe through the hole and begins to move around the circle. In LCS, the $O_\ell z$ axis for the k module entering the pipe is directed to the center of the Oyz section of the pipe and forms the angle γ_k^x associated with the radius of the pipe R_t and the current moment of time t as follows:

$$\gamma_k^x = \beta_0(t - t_k),\ \beta_0 = \delta\ell / R_c,\ R_c = R_t - R_w, \tag{2}$$

where the distance from the pipe axis to the GC module R_c when moving around the circle is a constant value. For convenience, we choose GCS so that the origin is located on the pipe axis Ox. Accordingly, in Eq. (1): $x_0 = 0,\ y_0 = -vt,\ z_0 = -R_c$. The motion along the circle is described by the parametric equation:

$$x_k = 0,\ y_k = R_c \sin \gamma_k^x,\ z_0 = -R_c \cos \gamma_k^x. \tag{3}$$

Since the ball joint has only rotational degrees of freedom, when the k module moves, the coordinates of its rear part $z_{b,k}$ should coincide with the coordinates of the front $z_{f,j}$ of the j module moving in the plane, i.e., the j module rotates at an angle α_1:

$$\alpha_1 = \arcsin\left((z_{f,j} - z_{b,k})/L_b\right),\ z_{f,j} = -R_c,\ z_{b,k} = z_k - L_b \sin \gamma_k^x. \tag{4}$$

The rotation of the j module will lead to a rotation of $j + 1$, by the formula:

$$\gamma_j^x = (-1)^{-j-k}\alpha_1,\ j > k. \tag{5}$$

When moving along the circle of the k module, the arc of length $\delta\ell$ will be passed. The movement along the Oy for the j and subsequent modules are determined by the angle α_1 and coordinate GC y_k:

$$y_{k-1} = y_k - L_b(\cos\alpha_1 + \cos\gamma_k^x),\ y_j = y_{k-1} - 2(j - k + 1)L_b \cos\alpha_1. \tag{6}$$

When the j module at time t_k starts moving around the circle, the angles $\alpha_1 = 0$ and γ_k^x are determined by (2) and the coordinates of GC (3). The formulas (2)–(6) are applied to the modules until all the modules reach a circular trajectory. The results of modeling the robot configuration when moving along the intermediate Line-to-Circle pattern are shown in Fig. 3. Robot configuration from a linear pattern to a circle motion inside the pipe. This pattern describes a situation when a robot enters a pipe through a process hole on the side of the pipe.

Within the circle pattern, the motion of modules in a pipe along a circle is described by Eqs. (2) and (3).

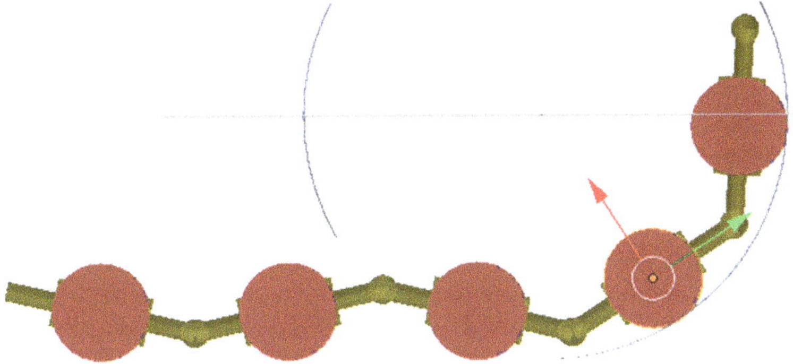

Fig. 3 Robot configuration from a linear pattern to a circle motion inside the pipe

Within the Circle-to-Helix pattern, the modules are turned sequentially around the local $O_\ell z$ axis by an angle γ^z. The configuration is constructed relative to the module with index k, which is currently turning. We denote by t_k the time points of the beginning k module turning. When rotating the module by an angle γ_k^z, the GC displacement must be considered, since, unlike the previous patterns, the R_c distance is not a constant value. Figure 4 shows the limiting case of the rotation $\gamma_k^z = \pi/2$. In the direction of the local axis $O_\ell z$, the distance between the GC and the pipe axis Ox will change by the value δH determined by the design angle β_2, depending on the angle γ_k^z.

From geometric considerations, we get $R_{c,k} = \sqrt{R_t^2 - \left(W_w \sin \gamma_k^z/2\right)^2} - R_w$, where the values γ_k^z change linearly from zero to the set value γ^z; $R_{c,k}$—the new distance from the pipe axis Ox to the GC of the k module.

The equation of the GC module motion is described by the parametric helix equation [20]. The parameters of this equation are determined in such a way that

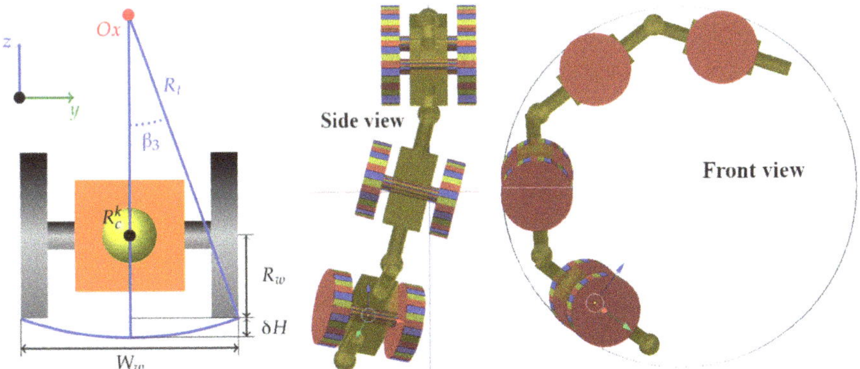

Fig. 4 Changing the position of the GC when turning by angle γ_k^z (left). Projections of the robot configuration within Circle-to-Helix pattern (right)

there is no gap in the joint when moving. The x_k coordinate is determined by the geometric parameters of the module and the angle of rotation γ_k^z. The module moves along a circle of a smaller radius and a displacement along the Ox axis is added, therefore, the angular velocity must be less in order to maintain a constant value of the velocity modulus v of the first module. The rotation angle γ_k^z with the scale factor s_k also changes accordingly. Changes in the linear velocity (along the GCS Ox axis) and angular velocity (around the LCS $O_\ell x$ axis) are nonlinear, and to improve the accuracy of the model, it is necessary to calculate the position of the module with a small time step at time points t_i with absolute displacement per time step $\delta\ell$. First, the robot configuration is calculated with the coefficient $s_k = 1$. For the k module:

$$\begin{aligned}\gamma_k^x(t_i) &= \gamma_k^x(t_{i-1}) + \delta\ell\cos\gamma_k^z\tau_i s_k/R_{c,k},\ \tau_i = t_i - t_{i-1},\\ x_k &= -L_b\sin\gamma_k^z s_k,\ y_k = R_{c,k}\sin\gamma_k^x,\ z_k = R_{c,k}\cos\gamma_k^x.\end{aligned} \tag{7}$$

For the preceding modules $j < k$, the rotation angle and coordinates of the j module are calculated using the formulas:

$$\begin{aligned}\gamma_j^x(t_i) &= \gamma_j^x(t_{i-1}) + (\cos\gamma_k^x(t_i)\\ &\quad + [2(k-j)-1]\cos\gamma_j^x)\tau_i s_k\cos\gamma_k^z\delta\ell/R_{c,k},\\ x_j &= -[2\sin\gamma_k^z + (2[k-j]-1)\sin\gamma^z]L_b s_k,\\ y_j &= R_{c,k}\sin\gamma_j^x,\ z_j = R_{c,k}\cos\gamma_j^x.\end{aligned} \tag{8}$$

For modules $j > k$, when the position of the GC of the k module is changed to preserve the adhesion in the hinges, an additional rotation around the Ox axis is realized in subsequent modules by an amount:

$$\begin{aligned}\alpha_2 &= -\arcsin\big((z_{f,k} - z_{b,k-1})/L_b\big),\\ z_{f,k} &= z_k + L_b\sin\gamma_k^x\cos\gamma_k^z,\ z_{b,k-1}\\ &= z_{k-1} - L_b\sin\gamma_{k-1}^x\cos\gamma_{k-1}^z.\end{aligned}$$

Accordingly, the angles of rotating and coordinates will take the form:

$$\begin{aligned}\gamma_j^x(t_i) &= \gamma_j^x(t_{i-1}) + \cos\gamma_k^x\cos\gamma_k^z\tau_i s_k\delta\ell/R_{c,k} + (-1)^{j-k}\alpha_2 s_k,\\ x_j &= 0,\ y_j = R_{c,j}\sin\gamma_j^x,\ z_j = -R_{c,j}\cos\gamma_j^x.\end{aligned} \tag{9}$$

The results of the simulation of the robot configuration when moving according to the Circle-to-Helix pattern are shown in Fig. 4.

If it is necessary to calculate the motion of the robot with a scale coefficient s_k, which ensures the constancy of the module velocity v for the k module, the coefficient s_k is determined by the formula:

$$s_k = v/\sqrt{\delta x_k^2 + (R_{c,k}\delta\gamma_k^x)^2},\ x_k = x_k(t_i) - x_k(t_{i-1}),\ \delta\gamma_k^x = \gamma_k^x(t_i) - \gamma_k^x(t_{i-1}).$$

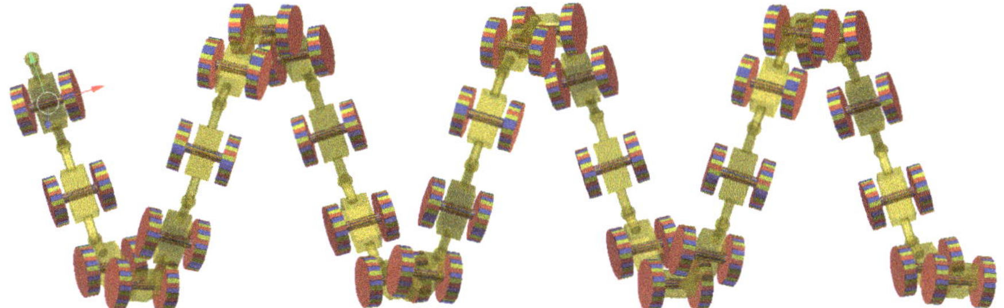

Fig. 5 Robot configuration within the Helix pattern

The robot configuration is then recalculated according to the formulas (7), (8), and (9) for the known value s_k.

Note that since rotations are not commutative, the order of rotations must be considered: At first, there is a rotation around the axis $O_\ell X$, and then—around the axis $O_\ell Z$.

Within the Helix pattern, all robot modules move along a helix. The motion is described by the following equations:

$$x_k = x_{k-1} - L_b \sin\gamma^z,\ y_k = R_{c,k} \sin\gamma^x,\ z_k = -R_{c,k} \cos\gamma^x,\ \gamma^x = \cos\gamma^z \tau_k \delta\ell / R_{c,k}$$

and the simulation results are shown in Fig. 5. This pattern describes the operating mode of the robot, in which the robot moves along a helical line and inspecting the pipe inner surface.

Within the Helix-to-Tube pattern, the robot moves from a helical trajectory to a linear one along the tube. The algorithm for constructing the configuration is like the Circle-to-Helix pattern, only the initial configuration of the robot does not correspond to a circle, but to a helix. If the movement along the helix goes with a turn by angle γ^z, then an additional turn by angle $\pi/2 - \gamma^z$ is necessary to get the movement along the pipe. This pattern describes the situation when the robot exits the operating mode and moves to another work area.

Note that in the transition patterns, it is possible to change the direction of movement to the opposite. Then, the pattern Line-to-Circle will describe the transition of the robot from the circle to the line (for example, in the transition to perpendicular to the pipe bend), Circle-to-Helix—the transition from the helix to the circle, and Helix-to-Pipe—the transition from the linear motion along the pipe to the helical line (the transition to the operating mode inside the pipe).

When driving straight ahead, for a circle inside the tube and the motion along the pipe trajectory the GC of the driven modules coincides with the trajectory of the GC lead module. In the Circle-to-Helix transition pattern, the trajectories of the GC and the points of the wheels contact with the pipe surface in the lead module and the driven modules differ. This is because the rotation of the master module around the $O_\ell Z$ axis causes the other modules to rotate. Let us demonstrate this by using the example of turning the robot on the plane. The results of modeling the configuration

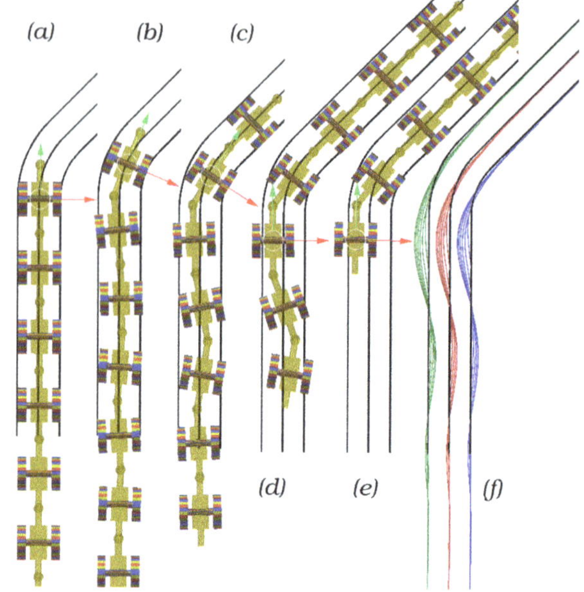

Fig. 6 Configuration of the robot when turning on the plane. Initial configuration (**a**); rotation of the first (**b**) and second modules (**c**); maximum deviation from the trajectory of the fifth (**d**) and seventh (**e**) modules; trajectory of the GC and wheel contact points (**f**)

of the robot, consisting of seven modules, on different sections of the trajectory are shown in Fig. 6. It is seen that when the drive module is rotated, the movement of the GC takes place in the positive direction of Ox and the movement of the hinge joint takes place in the negative direction, which leads to the wave-shaped trajectory of the other modules (the same pattern is observed in motion of the robotic train [15] or the road-train [21]). The maximum deviation of the last module from the trajectory depends on the number of modules.

When all the robot modules enter a trajectory that is implemented according to a single pattern, the deviations of the driven modules' trajectories become negligible in a region of order $2L_b$. It should be noted that the rotation on the plane can be considered as a simplified pattern of the robot's transition from the pipe to the pipe branch.

3 Conclusions

A mathematical model is proposed for finding parameters of a 3D configuration of a modular robot as it moves along patterns: In the plane with transitions to a circular trajectory inside the pipe, on the helical line with subsequent exit to the rectilinear motion along the pipe.

The resulting mathematical model makes it possible to correctly synthesize the 3D configuration of the robot modules, which enables movement along a complex trajectory while preserving contact between the wheels of the modules and the inner surface of the pipe. The basic patterns are formed based on real pipeline topologies, and the transition algorithms between the resulting configurations are shown in sufficient detail, the performance of which is confirmed by computational experiments.

The mathematical model was tested using a Python script with visualization in the Blender program. When the entire robot moved within a single pattern, the configuration accuracy was ensured automatically. In transition patterns, where part of the modules moves according to one pattern, and the other part—in a different way, the accumulation of error is possible. In one step over time, the modules moved along the trajectory by a distance $L_b/100$, i.e., the error in the configuration construction did not exceed 1%. The simulation time of the robot's movement along the chain of all the considered patterns was about 10 min.

Further work is expected to separate the computational part from the visualization program, which will improve both the accuracy of the configuration calculation and the calculation speed. Adding new patterns will allow to expand the scope of the robot's applicability (for example, overcoming obstacles). On the basis of the proposed model, a dynamic robot motion model will be constructed, taking into account inertia, gravity, and centrifugal force when moving along a curved trajectory, the friction force between the wheels and the surface taking into account the deviation of the trajectory of the driven modules from the trajectory of the leading module. The dynamic model will allow the definition of the design and hardware of the modules, which will prevent slipping of the wheels, as well as reduce the load on the actuators and configuration changes. Estimated weight and size of the module, it is possible to synthesize dynamic model of the robot, allowing to choose the onboard mechanisms, actuators and sensors with the subsequent synthesis of the real robot prototype design.

Acknowledgements This work was supported by the state assignment No. 0246-2018-007.

References

1. Isayev, V.N., Khurgin, RYu.: Truboprovodnyye kommunal'nyye sistemy. Santekhnika **3**, 52–59 (2006). (in Russian)
2. Gamzayev, B.A.: Sostoyaniye i osobennosti razvitiya truboprovodnogo transporta Rossii na sovremennom etape. Molodoy uchenyy **3**, 155–159 (2019). (in Russian)
3. Poyezzhayeva, Y.V., Fedotov, A.G., Zaglyadov, P.V.: Razrabotka robota dlya kontrolya truboprovodov. Molodoy uchenyy **16**, 218–222 (2015) (in Russian)
4. Hirose, S.: Biologically Inspired Robots. In: Snake-like Locomotors and Manipulators, Oxford University Press (1993)
5. Pfotzer, L., Staehler, M., Hermann, A., Roennau, A., Dillmann R.: KAIRO 3: Moving over stairs & unknown obstacles with reconfigurable snake-like robots. In: 2015 European Conference on Mobile Robots (ECMR), pp. 1–6 (2015)
6. Maxim, P., Spears, W., Spears, D.: Robotic chain formations. IFAC Proc. Vol. (IFAC-PapersOnline) **42**(22), 19–24 (2009)

7. Yim, M., Duff, D.G., Roufas, K.D.: PolyBot: a modular reconfigurable robot. In: Proceedings 2000 ICRA, Millennium Conference, IEEE International Conference on Robotics and Automation, Symposia Proceedings (Cat. No.00CH37065), vol. 1, pp. 514–520 (2000)
8. Murata, S., Yoshida, E., Kamimura, A., Kurokawa, H., Tomita, K., Kokaji, S.: M-TRAN: self-reconfigurable modular robotic system. IEEE/ASME Trans. Mechatron. **7**(4), 431–441 (2002)
9. Hauser, S., Mutlu, M., Léziart, P.-A., Khodr, H., Bernardino, A., Ijspeert, A.J.: Roombots extended: challenges in the next generation of self-reconfigurable modular robots and their application in adaptive and assistive furniture. Robot. Autonom. Syst. **127**, 103467 (2020)
10. The new MIT "milli-motein" reconfigurable robot. https://www.wired.com/2012/11/the-new-mit-milli-motein-reconfigurable-robot. Last accessed 2021/01/26
11. Lipkin, K., Brown, I., Peck, A., Choset, H., Rembisz, J., Gianfortoni, P., Naaktgeboren, A.: Differentiable and piecewise differentiable gaits for snake robots, pp. 1864–1869 (2007)
12. Liu, C., Whitzer, M., Yim, M.: A distributed reconfiguration planning algorithm for modular robots. IEEE Robot. Autom. Lett. **4**(4), 4231–4238 (2019)
13. Wright, C., et al.: Design and architecture of the unified modular snake robot. In: 2012 IEEE International Conference on Robotics and Automation, Saint Paul, MN, pp. 4347–4354 (2012)
14. Liljebäck, P., Fjerdingen, S., Pettersen, K.Y., Stavdahl, Ø: A snake robot joint mechanism with a contact force measurement system. In: 2009 IEEE International Conference on Robotics and Automation, pp. 3815–3820 (2009)
15. Evgrafov, V.V., Pavlovskiy, V.Y., Petrovskaya, N.V.: Issledovaniye dinamiki dvizhe-niya tsepochki «Robopoyezd», Upravlyayemoye dvizheniye, Preprinty IPM im. M. V. Keldysha, vol. 120, p. 31 (2005) (in Russian)
16. Blender: 3D Content Creation Noob to Pro. Wikibooks. https://upload.wikimedia.org/wikipedia/commons/b/b4/BlenderDocumentation4.pdf. Last accessed 2021/01/26
17. Bruyninckx, H.: Blender for robotics and robotics for Blender. Department of Mechanical Engineering, KU Leuven, Belgium (2004)
18. Buys, K., De Laet, T., Smits, R., Bruyninckx, H.: Blender for robotics: integration into the Leuven paradigm for robot task specification and human motion estimation. In: International Conference on Simulation, Modeling, and Programming for Autonomous Robots, pp. 15–25 (2010)
19. Díaz-Andrade, A., Álvarez-Cedillo, J., Herrera-Lozada, J., Rivera-Zarate, I.: Robotic arm control with blender. J. Emerg. Trends Comput. Inf. Sci. **4**(4), 382–386 (2013)
20. Sochi, T.: Principles of Differential Geometry. arXiv:1609.02868 (2016)
21. Ayupov, V.V.: Issledovaniye manevrennykh svoystv avtopoyezdov na osnove sistemnogo podkhoda. In: Perm': Izd-vo FGBOU VPO Permskaya GSKHA (2012) (in Russian)

Multi-criteria Optimization of the Mobile Robot Group Strategy Using the Ant Algorithm

Darintsev Oleg and Migranov Airat

Abstract One of the problems of group control, the distribution, of tasks in a group of mobile robots, is considered. The basic solution algorithm is proposed to use the "swarm intelligence" implemented on the basis of ant algorithms. This article is a development of the previous authors' work, which reviewed the known approaches to the distribution of problems using swarm algorithms, described the ant algorithm in detail, and analyzed the resulting solution to the problem of single-criteria optimization. The problem statement, the working space model are presented, the robot function objectives and parameters characterizing their work are formalized. To solve the multi-criteria problem, the following quality criteria were selected: The amount of energy expended, the time it took to complete the tasks, and the number of robots involved. For multi-criteria optimization, the resulting vector optimality criterion is linearly converged by introducing additional parameters characterizing group control: The total efficiency of a group of robots, the specific amount of energy for the operation of the support group and the energy for moving one robot to the given coordinates. To implement the ant algorithm, the problem was presented in the form of a set of undirected weighted graphs on which the "ants" will build solutions. An example is given that shows a detailed transition from the initial problem setting to graphs. The experiments were carried out on the basis of a software-implemented algorithm for finding the optimal group behavior strategy for a multi-criteria objective function with weight coefficients. For a group of three robots, experimental data were obtained and the results analysed, and a solution was modeled for several groups of robots with different weights.

D. Oleg (✉) · M. Airat
Mavlyutov Institute of Mechanics, Ufa Investigation Center, R.A.S., Prosp. Oktyabrya 71, 450054 Ufa, Russia

D. Oleg
Ufa State Aviation Technical University, Ul. K. Marxa 12, 450008 Ufa, Russia

A. Ronzhin and V. Shishlakov (eds.), *Electromechanics and Robotics*, Smart Innovation, Systems and Technologies 232, https://doi.org/10.1007/978-981-16-2814-6_9

1 Introduction

In recent years, the development of robotics has been characterized by two main directions: The expansion of the scope of robotics (apart from industry, there are real examples of robotization in the domestic, social and military fields) and the adoption of robotic complexes and groups of robots. Furthermore, the tendency to replace a single-multifunctional robot by groups (collectives) of simpler robotic agents is maintained, due to the greater functional saturation, flexibility, and reliability of the group of robots (multi-agent robotic complex). Actual experience in the use, applicability and specificity of such robotic complexes are given in the following works: For the Ministry of Emergency Situations [1, 2], military use [3], social robotics [4].

The group of robots is a complex technical system, but the performance of the tasks is characterized by certain parameters, the number of which depends on the specifics of the application, design, and program features. Two main techno-economic indicators are most often identified: The speed at which the task is accomplished and the resources spent on it due to the quality of team control. Synthesis of mobile robot grouping control algorithms that take into account at the same time several parameters of performance of a given task or a list of tasks is a current and actively researched direction of robotics [5–7].

Solving the problem of planning and allocating goals among robots leads to a whole range of additional tasks: Simulation and analysis of the maximum realistic interaction of robots with environment [8, 9], interaction analysis with limited communication [10], analysis of the information security of communication algorithms (protocols) in groups [11].

The article considers the problem of constructing an optimal algorithm (strategy) for the behavior of a group of robots on a working field containing several tasks, therefore, the objective function is constructed in the class of multi-criteria, and one of the modifications of the ant algorithm was used for synthesis. The authors have previously solved a similar problem using neural network computing [12] and genetic algorithms [13], therefore, in order to continue the research and compare the different approaches, it was decided to investigate the quality of multi-criteria optimization of the behavior strategy of a group of mobile robots using the ant algorithm. The works [14, 15] show the applicability of ant algorithms related to swarm algorithms [16] for solving this class of problems when optimizing one efficiency criterion.

Currently, the following paradigms of group interaction among robots are used in the research carried out by various scientific groups [17]:

(1) Multi-Robot System (MRS) [18, 19];
(2) Multi-Agent System (MAS) [20, 21];
(3) Swarm Robotics System (SRS) or Robotic Swarm (RS) [22, 23].

In this article, a group of robots is defined as robots that perform coordinated operations and interact with each other in accordance with the Multi-Robot System (MRS) principles. In this work, the number of criteria is extended and an ant algorithm is used to optimize the distribution problem, and due to the application of the ant algorithm, the problem will be reduced to a single-criterion using scalarization. The results of the selected robot behavior strategies simulation confirmed the effectiveness of the proposed algorithm.

2 Statement of the Problem and Objectives of the Study

The source data is a discrete $N \times M$ sized working field is used, on which n robots are located with coordinates (x_i, y_i) and m tasks—$(x_j{*}, y_j{*})$. The purpose of a group of robots is to perform tasks located in the working field, with only one robot and/or one task in each field cell.

The aim of the study is to develop an algorithm for optimizing the distribution of tasks among available robots, based on the ant algorithm and implementing the optimization of a multi-criteria objective function.

3 Multi-criteria Optimization Problem

Quality criteria are selected and described to define a multi-criteria problem.

(1) The first criterion is to use the amount of energy spent on the implementation of the proposed task distribution plan.

Each robot has unique characteristics (parameters):

- energy consumption by the robot in operation w_i^{FN},
- energy consumption by the robot when moving w_i^{MV},
- movement speed v_i.

In this case, the consumed energy can be calculated:

$$W_\Sigma = \sum_{i=1\ldots n} \left(S_i w_i^{\mathrm{MV}} + \tau_i w_i^{\mathrm{MV}}\right),$$

where S_i is the distance covered by the i-robot; τ_i is the time spent by the i-robot to complete a group of tasks.

The distance shall be calculated as Euclidean between the points of the route:

$$S_i = \sum_m S_m^i = \sum \sqrt{(x_m - x_{m-1})^2 + (y_m - y_{m-1})^2}.$$

(2) The second criterion is the time $t_{\max}$ required to complete all tasks by a group of mobile robots.

This criterion depends not only on the movement speed and the distance traveled, but also on the shape of the path of each robot (the number and angles of rotation of the robot).

The calculated time of each robot determines the maximum time that will determine the time of the robot team's tasks:

$$t_{\max} = \max_{m \in N_W}(t_m).$$

(3) The third criterion is the number of robots involved in the implementation of tasks N_w, and since when the quantitative and functional composition of a group of robots is changed some tasks can be postponed even if «free» agents, it is necessary to keep monitoring of the number of outstanding tasks.

In the final version, the multi-criteria optimization problem in the given setting and under the accepted assumptions is formulated as follows: The values of variable parameters $\bar{z}$ which minimize all conditions (criteria) 1–3 should be defined. The vector optimality criterion $(\phi_1(\bar{z}), \phi_2(\bar{z}), \phi_3(\bar{z}))$ is defined on the set **Z**, and the operator's task is reduced to minimizing the values of each of the particular optimality criteria $\phi_i(\bar{z})$, $i = \overline{1, n}$:

$$J\colon \min_{\bar{z} \in Z}(\phi_1(\bar{z}), \phi_2(\bar{z}), \phi_3(\bar{z})). \tag{1}$$

Here $\bar{z}$ is a vector of task numbers, each element of which z_i, $i = \overline{1, n}$ defines the problem for the i-robot for the next step. The area of valid vector values $\bar{z}$ forms a set **Z**. Optimality criteria:

(1) total energy consumption—$\phi_1(\bar{z}) = W_\Sigma(\bar{z}) \to \min_{\bar{z} \in Z}$;
(2) maximum implementation time—$\phi_2(\bar{z}) = t_{\max}(\bar{z}) \to \min_{\bar{z} \in Z}$;
(3) the number of robots involved—$\phi_3(\bar{z}) = N_w(\bar{z}) \to \min_{\bar{z} \in Z}$.

4 Scalarization of the Objective Function

Multi-criteria optimization can be performed automatically without the involvement of an operator if linear convolution (scalarization) of the vector criterion is performed.

For the task under consideration, all decisions made ultimately affect the energy costs of the mobile robots group and the operation of the support group (delivery of the group to the initial location, monitoring).

Taking this approach into account, we introduce the following parameters for linear convolution:

- efficiency of robots operation: (η);
- specific energy required for the functioning of the support group during the entire time of the operation (w^τ);
- the energy spent on placing one robot on the field (W^{ST}).

Entering additional parameters (in fact, weighting factors) allows you to reduce conditions 1–3 to the integral condition:

$$\Phi(\bar{z}) = \frac{1}{\eta} \cdot W_{\Sigma}(\bar{z}) + w^{\tau} \cdot t_{\max}(\bar{z}) + W^{\mathrm{ST}} \cdot N_{w}(\bar{z}), \tag{2}$$

and the multi-criteria optimization function (1) to the following scalar optimal search function:

$$J\colon \min_{\bar{z}\in Z}(\Phi(\bar{z})). \tag{3}$$

All objective functions are reduced to a common dimension—energy units, so operator participation is minimized and no additional decision-making methods are required.

5 Choosing an Ant Algorithm

To select the type of swarm algorithm, we used the results of the comparative analysis of the algorithms of the bee swarm, ant swarm and swarm of particles from work [14]. The problem considered in the work is close to the MD-H-VRP problem; therefore, the ant multi-colony optimization (AMCO) method was chosen as the basic ant algorithm [24], for which the problem must be represented as a set of components and transitions (a set of undirected weighted graphs) on which ants can construct solutions.

In the first stage, a group of two robots was considered, with four tasks to be performed, using a working field of size 10 * 10. The initial data for the considered example are shown in Table 1.

Table 1 Parameters of robots and tasks

Robots		
№.	1	2
Coordinates	{4, 10}	{1, 7}
Speed v_i, unit distance/unit time	30	20
Energy for commissioning, W_i^{ST}, unit of energy	60	40
Energy consumption for moving w_i^{MV}, unit energy/unit distance	30	40
Energy consumption for operation w_i^{FN}, unit energy/unit time	120	100

Tasks				
№.	1	2	3	4
Coordinates	{5, 8}	{1, 6}	{5, 4}	{2, 7}

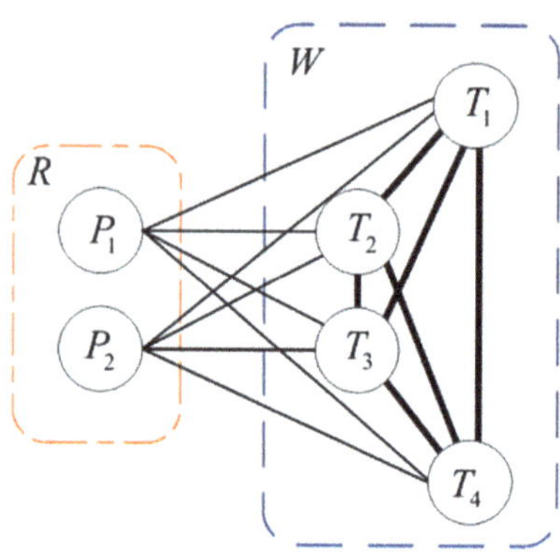

Fig. 1 The original graph G

In order to move to an undirected graph, it is necessary and sufficient to represent the whole group of robots and tasks on the field as vertices of the graph $G = \{V, A\}$, where A is the set of undirected edges; $V = \{R \cup W\}$—union of disjoint sets of vertices R, corresponding to robots, and vertices W—tasks. Moreover, in the graph G, the subgraph W is complete; any pair of vertices from the subset R is not adjacent; each edge is weighted. The weight ${w_{ij}}^k$ of the edge determines the energy costs of the corresponding transition, which depend on the direction of the transition (from vertex W_i to vertex W_j) and the robot class (k) performing the transition.

The graph built for this set is shown in Fig. 1. To accomplish the task, it is necessary to define the order of the tasks for each robot in such a way that all tasks are completed and the total energy consumption is minimized. On a graph, this corresponds to the construction of paths from each vertex of a subset R that together must cover all vertices of G, have no adjacent vertices, and provide a minimum sum of the rib weights. Four options (out of a possible 120) for constructing a set of routes for this example are shown in Fig. 2.

After presenting the task as a graph, it is obvious that the given problem belongs to the class of generalized VRP problems (Vehicle Routing Problem). The classical VRP is a combinatorial optimization problem in which a pair of identical vehicles (robots) are required to define an optimal set of closed routes from a single depot (the starting point of the robot's position) to multiple remote clients (tasks).

To use the previously selected AMCO method, values must be set for the following set of parameters:

- t is the number of iterations;
- g is the number of intercolonial groups of ants;
- α is the weight of the arcs pheromone concentration;
- β is the weight of the heuristic attractiveness of the arcs;
- p is the pheromone evaporation coefficient.

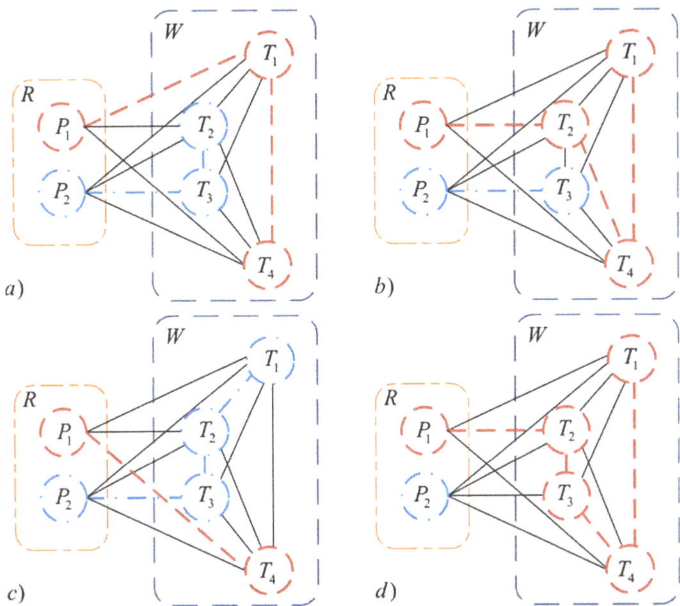

Fig. 2 Examples of solution routes on graph G

6 Solving the Task Distribution Problem in a Group of Robots with Different Weights for Each of the Criteria Using the Ant Algorithm

For the computational experiment in the MATLAB environment, an algorithm for finding the optimal strategy of the group of mobile robots using the AMCO method for the multi-criteria target function with weights has been implemented.

The study examined a group of three robots, who were assigned 10 tasks, located on working field 10 * 10. The source data for this example is given in Table 2, and the location of the robots and tasks is shown in Fig. 3.

The following initial values of the parameters of the multi-colonial ant algorithm were chosen:

$t = 50$ is the number of iterations;
$g = 40$ is the number of intercolonial groups of ants;
$\alpha = 0.4$ is the weight of the arcs pheromone concentration;
$\beta = 3$ is the weight of the heuristic attractiveness of the arcs;
$p = 0.3$ is the pheromone evaporation coefficient.

During the experiment, the set of weights was changed, and the baselines of the robots and field tasks remained constant. The experimental plan with indicating variable criteria weights is presented in Table 3.

Table 2 Parameters of robots and tasks

Robots			
№.	1	2	3
Coordinates	{2, 10}	{5, 10}	{8, 10}
Speed v_i, unit distance/unit time	30	20	25
Energy consumption for moving w_i^{MV}, unit energy/unit distance	30	40	35
Energy for commissioning, W_i^{ST}, unit of energy	120	100	110

Tasks										
№.	1	2	3	4	5	6	7	8	9	10
Coordinates	{5, 8}	{10, 8}	{2, 7}	{1, 6}	{7, 6}	{5, 4}	{9, 4}	{3, 3}	{7, 2}	{1, 1}

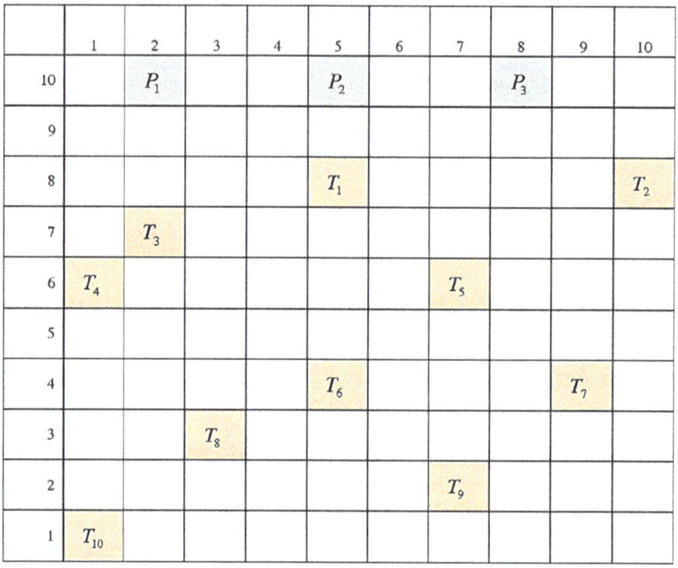

	1	2	3	4	5	6	7	8	9	10
10		P_1			P_2			P_3		
9										
8					T_1					T_2
7		T_3								
6	T_4						T_5			
5										
4					T_6				T_7	
3			T_8							
2							T_9			
1	T_{10}									

Fig. 3 Location of robots and tasks on the work field

The results of the algorithm are strategies for the behavior of robots in the form of a chain of tasks performed, quantitative indicators of the chosen strategy are also calculated: The number of tasks completed, the route duration, the route time and the energy costs of each robot. The total energy costs of the group, taking into account the costs of locating and maintaining the robot team.

Results for different strategies are shown in Table 4. Robot behavior strategies for four groups of experiments are shown in Fig. 4.

Table 3 Experiment plan

№.	Efficiency	specific energy w^{τ}, unit energy/unit time	placement energy, W^{ST}, unit energy	Note
1	1.0	0	0	Minimizing the energy consumption of a group of robots without taking into account external energy costs
2	1.0	200	0	Minimizing the total processing time of all tasks
3	1.0	0	200	Minimizing the number of robots used
4	0.9	100	50	Minimizing the integral criterion

Table 4 Results of strategy modeling for different input data by weight criteria

№.	Robot	Number of tasks	The length of the route, unit of distance	Route time, unit time	Power consumption of robots, units energy		Total cost, units energy
1	1	9	25.08	8.26	1755.4	1978.9	1978.9
	2	0	–	–	–		
	3	1	2.83	1.13	223.5		
2	1	4	10.85	3.62	759.4	2221.1	3003.5
	2	3	7.66	3.83	689.1		
	3	3	9.78	3.91	772.6		
3	1	10	30.16	10.05	2111.5	2111.5	2311.5
	2	0	–	–	–		
	3	0	–	–	–		
4	1	5	13.95	4.65	1085.1	2346.4	2980.0
	2	1	2.00	1.00	200.0		
	3	4	12.09	4.83	1061.3		

7 Analysis of the Obtained Results

The set of parameters No. 1 (Fig. 4). When optimizing without taking into account external energy costs, the algorithm «recommends» to use robots #1 and #3 for the first step. The result of the algorithm is due to the fact that the first robot has minimal operating costs, and the delegation to the third robot of one of the tasks is determined by its proximity to robot #3 and a significant distance from other robots.

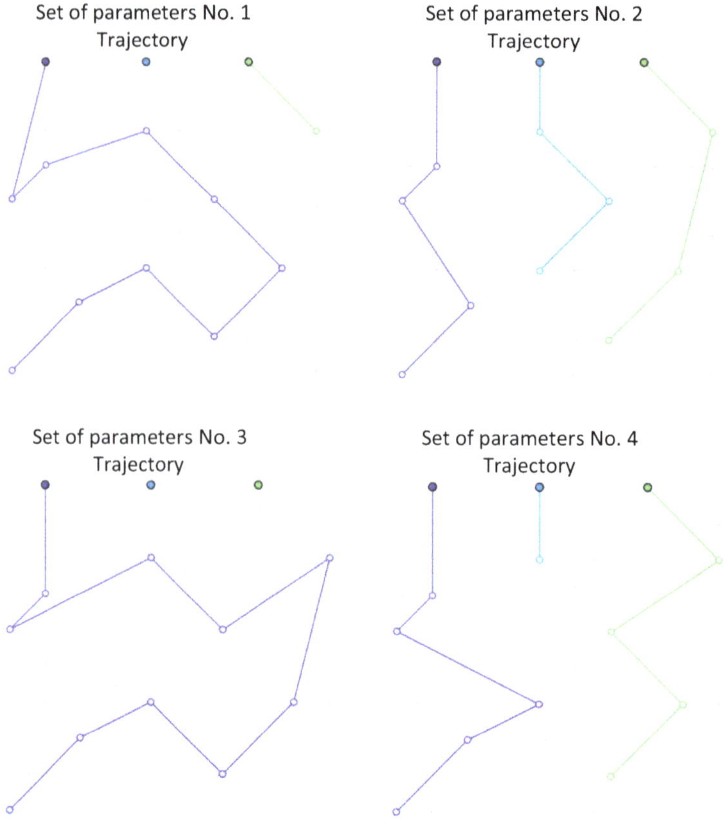

Fig. 4 Strategies for robot behavior on the field for different sets of weight criteria values

Set of parameters No. 2. When considering the energy costs proportional to the total time of the tasks, the algorithm redistributes the tasks among all the robots, taking into account the minimization of the total time for tasks implementation. In this case, the local times of task implementation are close for each of the robots, which indicates the effectiveness of the proposed strategy.

Set of parameters No. 3. When minimizing the number of robots, the algorithm offers the full load for robot #1, which has minimal the energy consumption movement and functions among the considered robots in the group. At the same time, the algorithm constructs an optimal distance path among the proposed vertices, i.e., solves the traveling salesman problem.

Set of parameters No. 4. For the proposed set of criteria weights, a strategy of behavior involving all robots is obtained, which indicates the effectiveness of AMCO in solving the problem under consideration.

8 Conclusion

The article considers the problem of constructing an optimal control strategy for an arbitrary group of mobile robots for performing an arbitrary number of tasks, taking into account several efficiency criteria. An ant algorithm (AMCO) is used to solve the problem of finding a group of open routes, the movement of which is carried out with minimal energy costs. A multi-criteria optimization function has been formulated and scalarized for use with AMCO. The algorithm is implemented in MATLAB. The application of the selected solution algorithm to the source problem is shown with different weights in the linear efficiency function. The effectiveness of the algorithm is confirmed by the results of modeling the selected robot behavior strategies obtained during four groups of experiments with different initial data.

As part of further research, it is planned to:

- analyze the impact of AMCO input parameters on the efficiency and speed of the solution;
- extend the proposed algorithm for the problem with the presence of obstacles in the field of the group of robots, which will require calculating the real path of robots to the tasks;
- compare the results of the ant algorithm with other heuristic algorithms.

Acknowledgements This work was supported within the framework of state assignment No. 0246-2018-007.

References

1. Casbeer, D.W., Beard, R.W., McLain, T.W., Li, S.M., Mehra, R.K.: Forest fire monitoring with multiple small UAVs. In: Proceedings of the 2005, American Control Conference, IEEE, pp. 3530–3535 (2005)
2. Sujit, P.B., Kingston, D., Beard, R.: Cooperative forest fire monitoring using multiple UAVs. In: 2007 46th IEEE Conference on Decision and Control, pp. 4875–4880 (2007)
3. Appel'ganc, A.V., Pyatakova, O.I., Solov'ev, A.A.: Group control of military robots. In: Povyshenie kachestva obrazovaniya, sovremennye innovacii v nauke i proizvodstve, pp. 562–568 (2019) (in Russian)
4. Nekrasov, K.V.: Innovative methods of warehouse logistics at the enterprises of agroindustrial complex. In: Agrarnoe obrazovanie i nauka, vol. 4, p. 22 (2019) (in Russian)
5. Khamis, A., Hussein, A., Elmogy, A.: Multi-robot task allocation: a review of the state-of-the-art. Cooper. Robot. Sensor Netw. **2015**, 31–51 (2015)
6. Liao, Y.L., Su, K.L.: Multi-robot-based intelligent security system. Artif. Life Robot. **16**(2), 137 (2011)
7. Marino, A., Parker, L.E., Antonelli, G., Caccavale, F.: A decentralized architecture for multi-robot systems based on the null-space-behavioral control with application to multi-robot border patrolling. J. Intell. Rob. Syst. **71**(3), 423–444 (2013)
8. Shcherbakov, V.S., Korytov, M.S.: On a modification of the ant colony algorithm for planning the trajectory of cargo movement in space with obstacles, taking into account the angular

orientation. Izvestiya vysshih uchebnyh zavedenij. Povolzhskij region. Tekhnicheskie nauki **3**(15), 142–150 (2010) (in Russian)
9. Motorin, D.E.: Investigation of a polymodel complex of a motion planning system for a heterogeneous group of autonomous robots under conditions of spatial and situational uncertainty. Robototekhnika i tekhnicheskaya kibernetika **7**(4), 291–299 (2019). (in Russian)
10. Ivanov, D.Y.: Distribution of roles in robot coalitions with limited communications based on role-based interaction. Upravlenie bol'shimi sistemami. Institut problem upravleniya im. VA Trapeznikova RAN **78**, 23–45 (2019) (in Russian)
11. Basan, E.S., Veselov, E.S, Semikoz, E.S.: Modeling of elements of the group control system for mobile robots in order to analyze their security. In: INFOBEZOPASNOST'-2019, pp. 100–107 (2019) (in Russian)
12. Darincev, O.V., Migranov, A.B.: Intelligent distribution of tasks in a group of mobile robots. Sovremennye naukoemkie tekhnologii **5**, 30–34 (2019). (in Russian)
13. Migranov, A.B.: Cloud-based task distribution system infrastructure for group of mobile robots. In: Proceedings of 15th International Conference on Electromechanics and Robotics "Zavalishin's Readings", ER (ZR) 2020, Ufa, Russia, pp. 409–420 (2020)
14. Migranov, A.B., Darintsev, O.V.: Choosing a swarm algorithm to synthesis an optimal mobile robot team control strategy. In: 2020 International Multi-Conference on Industrial Engineering and Modern Technologies (FarEastCon), Vladivostok, pp. 1–5 (2020)
15. Jian-Ping, W., Yuesheng, G., Xiao-Min, L.: Multi-robot task allocation based on ant colony algorithm. J. Comput. **7**(9), 2160–2167 (2012)
16. Zajcev, A.A., Kurejchik, V.V., Polupanov, A.A.: Review of evolutionary optimization methods based on swarm intelligence. Izvestiya YUFU **12**(113), 7–12 (2010). (in Russian)
17. Zakiev, A., Esoy, T., Magid, E.: Swarm robotics: remarks on terminology and classification. In: International Conference on Interactive Collaborative Robotics, pp. 291–300 (2018)
18. Cornejo, A., Nagpal, R.: Distributed range-based relative localization of robot swarms. Algorithmic Found. Robot. **XI**, 91–107 (2015)
19. Serebrenny, V., Shereuzhev, M.: Dependence of dynamics of multi-robot system on control architecture. In: Robotics: Industry 4.0 Issues and New Intelligent Control Paradigms, pp. 125–132 (2020)
20. Kang, S.M., Ahn, H.S.: Design and realization of distributed adaptive formation control law for multi-agent systems with moving leader. IEEE Trans. Ind. Electron. **63**(2), 1268–1279 (2016)
21. Chen, D., Dong, S.: Study on teamwork in robot football game based on multi-agent system (MAS). Int. J. Control Autom. **6**(2), 283–292 (2013)
22. Kim, J.H., Kwon, J.W., Seo, J.: Mapping and path planning using communication graph of unlocalized and randomly deployed robotic swarm. In: 2016 16th International Conference on Control, Automation and Systems (ICCAS), pp. 865–868 (2016)
23. Zhiguo, S., Jun, T., Qiao, Z., Lei, L., Junming, W.: A survey of swarm robotics system, pp. 564–572. Advances in Swarm Intelligence, Springer, Berlin Heidelberg (2012)
24. Kubil, V.N.: The research and development of methods for solving multi-criteria vehicle routing problem based on ant colony optimization algorithm: dissertation for the degree of candidate of technical sciences. In: YUzhno-Rossijskij gos. politekhnicheskij universitet imeni M.I. Platova, Novocherkassk (2019) (in Russian)

Analysis of the Allocation and Implementation of Tasks in the Heterogeneous Team of the Collaborative Robotic System

Rinat Galin, Mark Mamchenko, and Roman Meshcheryakov

Abstract The paper discusses the functioning of a collaborative robotic system that includes both collaborative robots (cobots) and humans (operators). Group control (management) methods and algorithms to solve the problem of increasing the efficiency of human–robot interaction are proposed to be used. The article deals with the principles of the formation and functioning of heterogeneous teams within a collaborative robotic system (CRS), as well as the methods of collective tasks allocation among the participants of these teams. The members of the CRS heterogeneous team perform jointly their specific sub-tasks in order to achieve a single global goal of the system (or accomplish its single common mission). The methods and algorithms of task allocation for the CRS members are considered in the context of a hybrid management strategy. The authors also propose a structural scheme of the modular control system of the CRS (based on a distributed hierarchical principle) that takes into account the indeterminacy of the external environment and the lack of posterior information provided.

1 Introduction

Industrial production (as any type of business activity) aims at increasing the revenue, while minimizing the expenses. This is mainly achieved by improving the quality of the products being manufactured, based on the introduction of advanced technologies and innovations into industrial processes.

The Fourth Industrial Revolution, which is currently in process, implies the technological transformation of the industry; the overall result is the creation of a network of "smart production facilities" (or "smart factories"). Among the key technologies of these "smart facilities" are virtual and augmented reality (AR), industrial and collaborative robots, simulation tools, digital models and digital twins, artificial intelligence, additive technologies, Internet of Things, cloud services, big

R. Galin (✉) · M. Mamchenko · R. Meshcheryakov
V.A. Trapeznikov Institute of Control Sciences of Russian Academy of Sciences, 65, Profsoyuznaya str., Moscow 117342, Russia

A. Ronzhin and V. Shishlakov (eds.), *Electromechanics and Robotics*, Smart Innovation, Systems and Technologies 232, https://doi.org/10.1007/978-981-16-2814-6_10

data, and integration of heterogeneous technological elements into cyber-physical systems (CPS) [1, 2]. A single production line of a "smart factory" can be used for the simultaneous manufacturing of several types of products through dynamic routing of resources and logistics, self-organizing capabilities of the entire system or its individual parts, as well as high-speed data exchange between the elements of the CPS [2].

Comprehensive digital models of all production processes and agents used gives the most complete awareness of the current production state, provides feedback from the agent of the CPS, and allows to identify and take measures to increase the overall efficiency of the "smart factory" in real time via its "reconfiguration".

Collaborative robots (or cobots) designed to perform production operations together with humans are now becoming increasingly common in the industry. According to "BIS Research", the market of cobots is expected to reach the revenue of $9.13 billion by 2024 [3]. Handling complex and high-precision operations, where manual labor cannot yet be completely replaced, cobots have several advantages over conventional industrial robots, including their flexible relocation, frequent task changes, simplicity of programming, smaller mass-dimensional characteristics, as well as their ability to work directly and safely with human operators.

The use of cobots in "smart production facilities" is often reduced to the solution of the problem of effective and safe human–robot interaction (HRI). While HRI safety issues are already described in the relevant standards [4, 5], the problem of the effectiveness of such interactions is complicated by the need to define one or several performance criteria, which are specific to each production operation or process. In other words, the main indicator of the efficiency of HRI between the human worker and the cobot can be the quantity of products manufactured per unit of time, the minimization of the number of needed elementary operations, energy costs, time to perform a single operation, or even a set of criteria (i.e., a combination of several of the above-mentioned efficiency indicators).

In this case, the problem of safe and effective HRI in large industrial enterprises will be hampered by the following factors:

- the use of several heterogeneous cobots with different characteristics and capabilities;
- the impossibility for a cobot (a group of cobots) to predict human actions, especially during complex, multi-stage operations;
- further sophistication of the cobot's behaviors and HRI models when using more sensors and systems (e.g., the AR [1]);
- the need to deal with possible "conflict situations" (e.g., when a cobot and a person simultaneously take the same tool or part);
- increasing complexity of the operations' planning and control processes with the growth of the number of operators and cobots involved.

Thus, the need to engage multiple operators and cobots (both homogeneous and heterogeneous) to perform complex production operations necessitates considering them as the elements of a multi-agent robotic system with a common purpose (e.g., manufacturing of products) and appropriate restrictions and limitations (both HRI safety principles and performance criteria).

2 Statement of the CRS Functioning Problem

The actions of heterogeneous group participants within the CRS should be coordinated in order to achieve a global objective in the most effective way. The article discusses the CRS, the overall performance of which is measured by the effectiveness of formation and functioning of a heterogeneous team. The state of the CRS may vary depending on the influence of external forces, and this means that the system is not stationary. Control of the heterogeneous team implies the identification of effective operation of its participants, while the system passes from its initial state to the final one.

Given that the members of the team perform different types of work of a non-negative volume, and the cost function is additive, the problem of effectiveness of forming a heterogeneous team is reduced to solving the optimization problem.

Let E be some medium, where a heterogeneous CRS team $\Re$ interacts. The team consisting of N participants $U_j \in \Re$ $(j = \overline{1, N})$, represented by both humans and robots (cobots). Then, the status of each member of the group will be described by the following vector-valued function (or vector function):

$$U_j(t) = \left[u_{1,j}(t), u_{2,j}(t), \ldots, u_{k,j}(t)\right]^T,$$

where $u_{i,j}(t)$ $\left(i = \overline{1, k}\right)$ are k variable states of j team participant.

Then, the general state of the group $\Re$ at t time will be as follows:

$$\Re(t) = U_1(t), U_2(t), \ldots, U_N(t).$$

We then set the state of the environment around the working space of j member of CRS at t moment of time as $e_{i,j}(t)$ $\left(i = \overline{1, w}\right)$, and this state will be a part of the set of all w acceptable similar states relative to the j member of CRS in the accepted workspace E (at t moment of time):

$$E_j(t) = \left[e_{1,j}(t), e_{2,j}(t), \ldots, e_{w,j}(t)\right]^T.$$

The values of the variables describing the state of each member of the group (e.g., taken as their positions in the environment, actions, speed of the movement, angles of rotation, energy resource of the cobots, etc.) will be given at t moment of time.

In this case, the set of possible states of the heterogeneous group of CRS participants will be described by the values (points) of the $N \cdot (k + w)$-dimensional space of all states of the group $\{S_{\text{CRS}}\}$. The initial and final states of the system shall be defined as the set of the overall state of the group and the state of the surrounding environment as ($S^0_{\text{CRS}} = \Re^0, E^0$ and $S^f_{\text{CRS}} = \Re^f, E^f$), respectively. Given the vector functions (1) and (2), the overall state of the group and the state of the surrounding environment at the initial t_0 and the final t_f moments of time will be as follows:

$$\Re^0 = \Re(t_0), E^0 = E(t_0), \tag{1}$$

$$\Re^f = \Re(t_f), E^f = E(t_f). \tag{2}$$

The CRS (i.e., all elements of the system) must perform q different works (actions) from the set $A = \{a_n(t)\}_{n=1}^{q}$. Then, let the actions of the heterogeneous CRS group be described as follows i.e., as a system of differential equations (3):

$$S_{\text{CRS}} = f_{\text{CRS}}(S_{\text{CRS}}(t), A_{\text{CRS}}(t), g(t), t) \tag{3}$$

where g(t) are external forces that change the state of the CRS (given that the environment is non-stationary), S_{CRS} is the state of the medium, $A(t)$ is the action vector, and f_{CRS} is a function that describes the evolution of the state of CRS participants from the initial to the final moments of time, taking into account the changes in external forces.

The workspaces of each group member of the CRS should be separated to avoid both the conflicts and reducing the overall efficiency of tasks implementation. Furthermore, there are logical limitations related to the impossibility of location of more than one participant in the same place, performing the same actions at the same time. In this regard, the appropriate general limitations of the system's states should be introduced as follows:

$S_{\text{CRS}}(t) \in \{S^p_{\text{CRS}}(t)\} \subset \{S_{\text{CRS}}\}$,

where $\{S^p_{\text{CRS}}(t)\}$ is a set of valid states of the system at t moment of time.

Then, the limitations on the actions performed by the participants will be as follows:

$A_{\text{CRS}}(t) \in \{A^p_{\text{CRS}}(t)\} \subset \{A_{\text{CRS}}\}$,

where $\{A^p_{\text{CRS}}(t)\}$ is a set of valid states of actions of the participants at t moment of time.

The problem of collective control (management) for a heterogeneous group of CRS participants implies the determination of $\overline{A_j}(t)$ actions for each participant $U_j \in \Re$ that provide the functional extremum. This allows setting the objective (global task) for the functioning of the heterogeneous group, as well as assessing the quality of the management (control) processes as (4):

$$Y_{CRS} = \Phi\left(S^f_{\text{CRS}}, t_f\right) + \int_{t_0}^{t_f} F(S_{\text{CRS}}(t), A_{\text{CRS}}(t), g(t), t)\text{d}t. \tag{4}$$

3 Principles of CRS Formation

Group control (management) of CRS participants requires solving a number of problems that are broadly divided into four types:

- the choice of the control strategy (i.e., the general principles of management/control of the CRS participants);
- global planning—decomposing the global objective into simple tasks and allocating (distributing) them between all participants of the CRS;
- local planning—the implementation of allocated/assigned tasks by individual CRS participants;
- processing data received from the CRS participants, while performing local tasks.

The first two problems should be solved only once (in the initial stages of group management), while other operations (control and dispatch of the participants carrying out local tasks, as well as getting feedback and processing the incoming data) should be executed continuously (i.e., iteratively and repetitively over a fixed time interval).

The strategy of group control and management directly affects the composition and the structure of the CRS participants [6–8]. As a whole, group management strategies are divided into centralized and decentralized, depending on the operational characteristics of the participants of the collaborative robotic system. Both the global task (objective) of the CRS, and the number of the participants should be taken into account, while choosing an optimal group control strategy. In particular, in terms of few participants in the group the use of a centralized strategy is most effectively. However, the growth of the number of CRS elements results in the linear increase in time needed to solve the problem of group control [9, 10], and a decentralized strategy would be the best choice in this situation.

However, in some instances, mixed (hybrid) management and control strategies (a combination of centralized and decentralized approaches) may be the most efficient ones, depending on the system's performance. For the CRS, which consists of a group of cobots interacting with a group of human operators, the hybrid structure of group control will be as shown in Fig. 1.

This article proposes a modular CRS system based on the hierarchical principle of a distributed system. The advantage of modularity of the control system is the possibility of modifying (replacing) its individual parts without changing the other elements. The structure of the proposed modular CRS management (control) system is shown in Fig. 2.

4 Collective Task Allocation in a CRS Heterogeneous Team

Once the overall structure of the CRS management and control system has been selected and established, and it is necessary to proceed to the choice of methods and

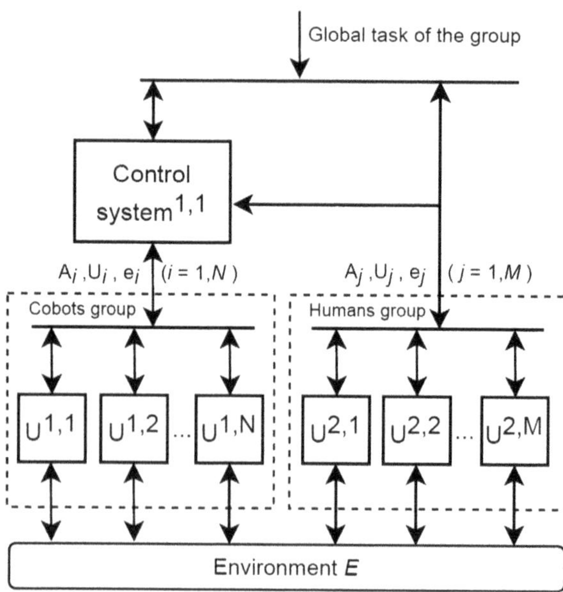

Fig. 1 CRS hybrid structure of group control

algorithms of tasks allocation in a heterogeneous team. The elements of the team perform joint local actions in order to achieve a single global objective of the system.

The CRS is composed of heterogeneous participants, resulting in the need to apply collective management (control) approaches, algorithms and methods to the elements of a heterogeneous team [8, 11–23]. According to [8, 24], two stages of the life cycle of a heterogeneous team are usually considered—its formation and functioning.

The stage of forming a heterogeneous team with the formal problem statement (in relation to the CRS) was considered in article [11]. Participants of the CRS have to perform q different tasks (types of work) with respective volumes given as a vector $V = \left(V^1, \ldots, V^q\right)$. Suppose that the participants' performance of non-negative volumes of work results in costs, depending on the volumes x_{ik} of work of each k type for i participant, and corresponding performance $r_{ik} \in [0,\ 1]$ (efficiency of i participant, performing k type of work). Then, the dependence of x_{ik} volume for k type of work and the i participant efficiency for this type of work will be described as a $c_{ik}(x_{ik}, r_{ik})$ function. In addition, we set the initial value of the function to zero $c_{ik}(0,\ \cdot) = 0$, as it is obvious that in terms of the absence of the assigned work of k type to be performed, the cost value will be zero.

Provided that each element of the system can perform a non-negative volume of work $V^k \geq 0, k \in [1, q]$ of each q type, and assume that the cost functions are additive for different types of work, the problem of forming a heterogeneous team is reduced to solving the optimization problem. The above-mentioned problem was considered on the example of the performance of collective actions of human teams with the cost functions presented as the generalized Cobb–Douglas production

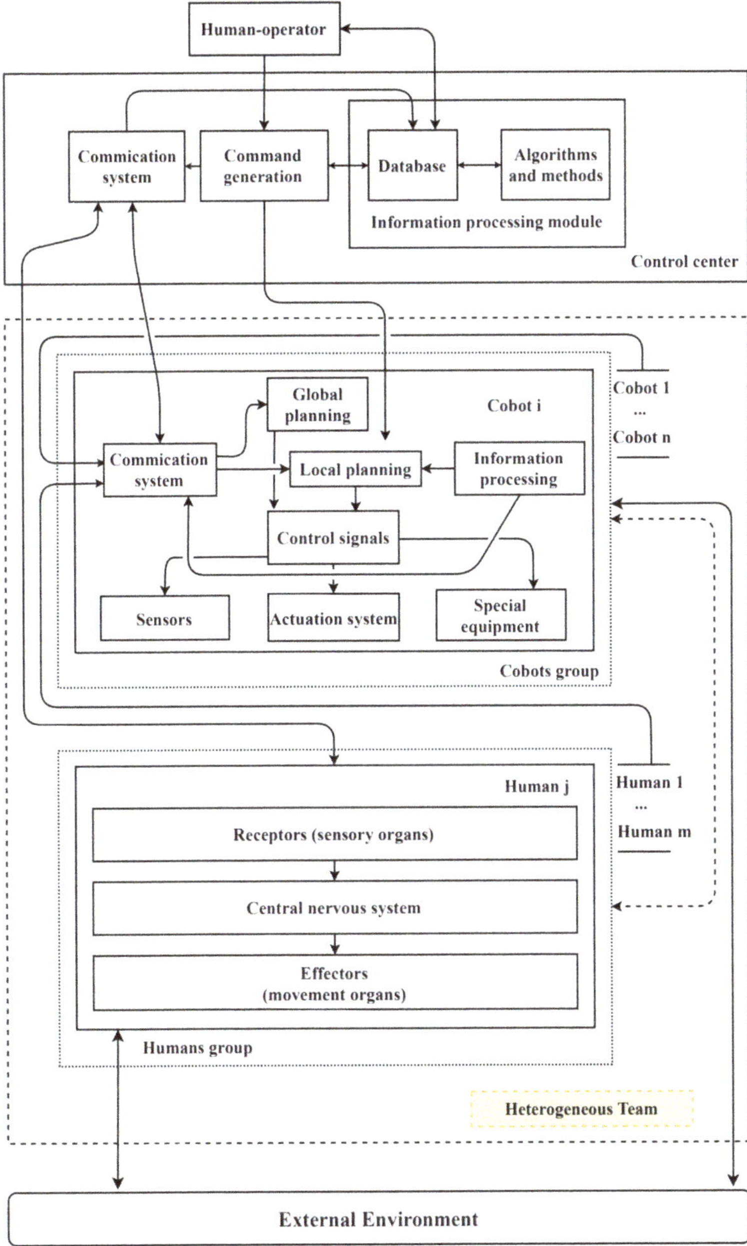

Fig. 2 The structure of the CRS modular management (control) system

functions [8, 11]:

$$c_i(x_i, r_i) = \sum_{k=1}^{q} r_{ik}\varphi\left(\frac{x_{ik}}{r_{ik}}\right), \ i \in [1, m],$$

where $\varphi\left(\frac{x_{ik}}{r_{ik}}\right) = \left(\frac{x_{ik}}{r_{ik}}\right)^2$, while $\varphi(0) = 0$.

Each participant in a heterogeneous team will act in accordance with the following principles of collective management and control:

- the participant determines own actions independently on the basis of the current situation at each moment of time;
- the selection of the optimal actions is based on the current data on the status of the team's global task (objective);
- actions of the participants are considered to be most effective when maximizing the CRS objective functional during the shift of the system's state in time (from the initial to the final one);
- an alternative action (differing from the one recognized as the most effective) is possible, if it has the least impact on the overall performance of the team's achievement of the system's global objective.

Compliance with the above-mentioned principles characterizes the collective management, and the heterogeneous team control method is used as the method of collective tasks allocation within the CRS. According to [6], methods of collective control are focused on the distributed and hierarchical distributed systems, including the heterogeneous team of the CRS.

In general, the solution of the problem of heterogeneous group control lies in the identification of the most effective actions of the CRS participants, when the system passes from its initial to the final state (i.e., finding the best value of the continuous vector function of the actions $\overline{A_{\text{CRS}}(t)}$. Given that both sensory data (generated by the cobots) and cycling of computational processes are discrete, the vector function of actions $\overline{A_{\text{CRS}}(t)}$ will also be discrete, i.e., discontinuous (depending on the Δt time interval considered). Since the influence on both team members and the environment cannot be determined in advance, the solution of the problem of collective tasks allocation lies in the identification of the actions that maximize the system's functional (provided that the functional value exceeds some threshold $Y_{\min}$):

$$\max \ Y_{\text{CRS}} \geq Y_{\min}.$$

Thus, the system and its processes are considered to be discrete. Such problems can be solved using high a priori and low posterior information algorithms [24]. In this case, the objective function can be presented as follows:

$$Y_{\text{CRS}} = \sum_{t=t_0}^{t_f-1} F(\Re(t), E(t), A_{\text{CRS}}(t), g(t))\Delta t.$$

Assume that the external influence on the CRS participants and the environment are negligible. Then, the problem of maximization of the objective functional can be described as follows:

$$Y_{\text{CRS}} = Y_{\text{CRS}}(t) \rightarrow Y_{\text{CRS}}^{\max}. \tag{5}$$

The problem of finding a maximum functional value (5) can be solved using dynamic programming or variational calculus methods. However, due to the non-linearity and lack of a priori information for tasks allocation, it is proposed to use an iterative approach in case of the heterogeneous team of the CRS.

5 Conclusion

The rapid development of collaborative robotics and the continuing increase in the number of cobots used in both manufacturing and everyday life necessitate the consideration of a CRS as a complex multi-agent socio-cyber-physical systems, in which people interact with the cobots in order to achieve common objectives as effectively as possible.

The article proposes to use a modular structure of the CRS control system, based on the hierarchical principle of distributed systems, as well as the hybrid strategy of heterogeneous teams' management and control. It is proposed to implement efficient allocation of tasks among the participants of heterogeneous teams using iterative methods, which allow solving the optimization problems in relation to teams' control within the CRS. The solution of the problem of group control of CRS elements operating in a dynamic non-deterministic environment lies in defining the actions of each element to maximize the growth of the system's objective functional. This problem is proposed to be solved using algorithms with high a priori and low posterior information, dynamic programming or variational calculus methods, and iterative approach.

Acknowledgements The reported study was partially funded by RFBR according to the research project No. 19-08-00331.

References

1. Blaga, A., Tamas L.: Augmented reality for digital manufacturing. In: 2018 26th Mediterranean Conference on Control and Automation (MED), pp. 173–178. Zadar (2018)

2. Sadikova, A.M.: Prospects for the introduction of smart production. Sci. Pract. Electron. J. Alley Sci. **3**(30), 1, 434–443 (2019)
3. Global Collaborative Robot (Cobot) Market to Reach to $9.13 Billion by 2024. https://www.prnewswire.com/news-releases/global-collaborative-robot-cobot-market-to-reach-to-9-13-billion-by-2024--300914808.html. Last accessed 2021/02/10
4. ISO/TC 299 Robotics: ISO/TS 15066:2016 Robots and Robotic Devices—Collaborative Robots. https://www.iso.org/standard/62996.html. Last accessed 2021/02/10
5. ISO 10218-1, 2:2011: Robots and Robotic Devices—Safety Requirements for Industrial Robots—Part 1, 2: Robots. https://www.iso.org/ru/standard/51330.html. Last accessed 2021/02/10
6. Kalyaev, I.A., Gaiduk, A.R., Kapustyan, S.G.: Models and Algorithms of Collective Control in Groups of Robots, 1st edn. Fizmatlit, Moscow (2009)
7. Vorotnikov, S., Ermishin, K., Nazarova, A., Yuschenko, A.: Multi-agent robotic systems in collaborative robotics. In: International Conference on Interactive Collaborative Robotics, vol. 11097, pp. 270–279. Springer, Cham (2018)
8. Novikov, D.A.: Mathematical Models of the Formation and Functioning of Teams, 1st edn. Physical and Mathematical Literature Publishing House, Moscow (2008)
9. Gaiduk, A.R., Kalyaev, I.A., Kapustyan, S.G.: Team management of intellectual objects based on flocking principles. Proc. Southern Sci. Center **1**(2), 20–27 (2005)
10. Urevich, E.I.: Foundations of robotics, 4th edn. BHV-Peterburg, Saint Petersburg (2018)
11. Galin, R.R., Serebrennyj, V.V., Tevyashov, G.K., Shiroky, A.A.: Human-robot interaction in collaborative robotic systems. Proc. Southwest State Univ. **24**(4), 180–199 (2020)
12. Kulinich, A.A.: A decision-making support model for coalition formation under uncertainty. Artif. Intell. Dec. Making **2**, 95–106 (2012)
13. Bruce, J., Bowling, M., Browning, B., Veloso, M.: Multi-robot team response to a multi-robot opponent team. IEEE Int. Conf. Robot. Autom. **2**, 2281–2286 (2003)
14. Noh, S., Park, J.: System design for automation in multi-agent-based manufacturing systems. In: 2020 20th International Conference on Control, Automation and Systems (ICCAS), pp. 986–990. Busan (2020)
15. Nugroho, A., Yuniarno, E.M., Purnomo, M.H.: Cooperative multi-agent for the end-effector position of robotic arm based on consensus and PID controller. In: 2019 IEEE International Conference on Computational Intelligence and Virtual Environments for Measurement Systems and Applications (CIVEMSA), pp. 1–6. Tianjin (2019)
16. Ghassemi, P., DePauw, D., Chowdhury, S.: Decentralized dynamic task allocation in swarm robotic systems for disaster response: extended abstract. In: 2019 International Symposium on Multi-Robot and Multi-Agent Systems (MRS), pp. 83–85. New Brunswick (2019)
17. AL-Buraiki, O., Payeur, P.: Probabilistic task assignment for specialized multi-agent robotic systems. In: 2019 IEEE International Symposium on Robotic and Sensors Environments (ROSE), pp. 1–7. Ottawa (2019)
18. Ruch, C., Gächter, J., Hakenberg, J., Frazzoli, E.: The +1 method: model-free adaptive repositioning policies for robotic multi-agent systems. In: Jianwei, H. et al. (eds.) IEEE Transactions on Network Science and Engineering, vol. 7(4), pp. 3171–3184. IEEE, New York (2020)
19. Sini, J., Violante, M., Dessì, R.: Computer-aided design of multi-agent cyber-physical systems. In: 2018 IEEE 23rd International Conference on Emerging Technologies and Factory Automation (ETFA), pp. 677–684. Turin (2018)
20. Ka Hong, S.G., Ahmad, A., Yee Chong, A.T.: Formulation of mandatory task algorithm using task prioritization technique. In: 2018 International Symposium on Agent, Multi-Agent Systems and Robotics (ISAMSR), pp. 1–4. Putrajaya (2018)
21. Esmaeili, A., Ghorrati, Z., Matson, E.: Multi-agent cooperation using snow-drift evolutionary game model: case study in foraging task. In: 2018 Second IEEE International Conference on Robotic Computing (IRC), pp. 308–312. Laguna Hills (2018)
22. Karimadini, M., Karimoddini, A., Lin, H.: Modular cooperative tasking for multi-agent systems. In: 2018 IEEE 14th International Conference on Control and Automation (ICCA), pp. 618–623. Anchorage (2018)

23. Jia, H., Ding, B., Wang, H., Gong, X., Zhou, X.: Fast adaptation via Meta learning in multi-agent cooperative tasks. In: 2019 IEEE SmartWorld, Ubiquitous Intelligence and Computing, Advanced and Trusted Computing, Scalable Computing and Communications, Cloud and Big Data Computing, Internet of People and Smart City Innovation (SmartWorld/SCALCOM/UIC/ATC/CBDCom/IOP/SCI), pp. 707–714. Leicester (2019)
24. Beer, M.: Organization Change and Development: A System View, 1st edn. Goodyear Publishing Company, Santa Monica (1980)

Algorithm of Target Point Assignment for Robot Path Planning Based on Costmap Data

Lev Kuznetsov, Polina Kozyr, and Dmitriy Levonevskiy

Abstract Most mobile robots which use the operating systems (ROS) employ costmap-based path planner for navigation in environment. Besides the systems, in which robotic devices function in completely autonomous manner, also such systems exist, where user specifies in a self-contained interface the target point for robot motion, without using any costmap-related data. In such settings, the robot sometimes cannot reach the user-specified point because it is treated as an obstacle on the costmap. To solve this problem an algorithm was developed, which accepts the coordinates of the target point, specified through user interface, and finds possible target points in vicinity of the specified one, regarding the predefined distance constraints. Conceptually, the search of possible points consists in looping through costmap cells of the global cost planner in orthogonally related directions from the obtained point and in defining of the preferable target points, upon reaching which the task is considered to be completed. In this paper, the algorithm was tested as part of the navigational module of the robot. Experiments were performed, particularly, in the Gazebo simulation environment; robot model TURTLEBOT3 Waffle Pi was also utilized. The experimental results showed that this module can be combined with any planner. When the value of validation step was set at the minimum value of 0.01 m, it was required no more than 55 ms to find a target point. The minimum duration of such search was 1 ms.

1 Introduction

Currently, graphical user interfaces are ubiquitous in interactions with mobile robots and robotic platforms [1, 2]. Among the most common interfaces of such kind is RViz, embedded in the robot operating system (ROS) [3], being utilized in the most robotic devices. Such graphical interfaces commonly visualize the indoor layout, charted with the SLAM (simultaneous localization and mapping) algorithm [4], or

L. Kuznetsov (✉) · P. Kozyr · D. Levonevskiy
St. Petersburg Institute for Informatics and Automation of the Russian Academy of Sciences, St. Petersburg Federal Research Center of the Russian Academy of Sciences (SPC RAS), 39, 14th Line, St. Petersburg 199178, Russia

A. Ronzhin and V. Shishlakov (eds.), *Electromechanics and Robotics*, Smart Innovation, Systems and Technologies 232, https://doi.org/10.1007/978-981-16-2814-6_11

by other means [5]. The layout is used for specifying and editing of mobile robot target points, but operator has no information, whether certain point is accessible for the robot.

The navigational package of ROS also contains global planners, which build paths to the target points [6]. Exactly, these planners ensure definition of the specific point as potential target for a robot. In real-world setting, the target point of the robot can be close to a certain physical object, such as wall, table, stack, etc. The possible practical instances of such kind are delivery of payload to the specified object, rescue activities, maintenance-related tasks, associated with some physical object. In such cases because of possible collisions of robots with obstacles, the global planner denies planning a path to the specified point. The essence of the task is to find an alternative point, having reached that the robot virtually completes its mission. Thereby it is assumed that possible minor deflections from the initial target point does not influence the overall success of the operation: The task is considered to be completed, if the robot reaches the initial target point in the range of predefined distance margin.

In this paper, a possible solution for target point search is considered, for the cases, when the planner is unable to build a path to the user-specified destination. This solution is a costmap-based algorithm for target point search, where the pool of target points is contained in the global costmap. The algorithm is implemented and tested as a ROS module.

2 Related Work

Besides of the integrated navigation package, the ROS operating system is equipped with many planners that can be utilized together with machine learning methods [7–9]. The principle tooling of the global planners includes visibility graphs, based on image skeleton establishment, probabilistic roadmap planners, rapid exploration of random trees, state lattices and navigation functions. The navfn planner utilizes the Dijkstra's algorithm and A* algorithm to discover an optimized path between the origin and target positions. It assumes utilization of a spherical robot and uses the costmap to ensure the navigation of this robot [10]. The navfn planner is based on the NF1 approach [11] and included into the default navigation package of ROS [12].

Another planner, included into the default navigational package of ROS, is global_planner. It is also based on the NF1 approach and is a fast interpolated global planner, designed as a more flexible alternative to navfn.

The asr_navfn planner builds upon the navfn. A crucial feature of the planner consists in its ability to find the nearest point, available to reach, if the target point belongs to an obstacle. Having revealed such alternative destination point, the planner follows along the established shortest path from the origin point according to the Dijkstra'a algorithm and searches for the nearest unoccupied point. A disadvantage of this approach is the impossibility to set the distance from the origin point to the new discovered one. This planner is also not supported in the latest ROS versions [10].

Sbpl_lattice_planner employs the state lattice approach to navigate from the origin. The planner employs the self-paced balance learning (SPBL) method, developed by Maxim Likhachev [13]. The planner utilizes the kinematic robot model and generates the path using combinations of motion primitives. Discovery and building of global robot path are performed using ARA* [14] or AD* [15] algorithms.

The carrot_planner planner establishes a vector from the origin to the obtained target point. If it reveals that the target point belongs to an obstacle, it proceeds to the search of another target points available in the direction from target to origin along the established vector. The disadvantage is that the planner does not respect the global area map [10].

Along with the widely used global planners, some solutions also exist, intended for more specific applications, or insufficiently studied. So, in [7] a novel planner is described, developed for complicated and narrow environments. To search for the target point, the authors proposed to use the A* algorithm. During robot recovery in the instance of collision with physical obstacle, the target point is substituted with a random one in the vicinity of the current position of the robot. The case, when a possible collision point is set as a target point, is not detailed in this paper.

In [8], a novel path planning approach for autonomous vehicles is described, which is based on the time elastic band (TEB) approach. Authors assume that any collisions in target point are excluded, and that the local planner, developed by the authors, also ensures for the vehicle a collision-free path to the target point.

In [9], a novel path planner is described, which is based on a modified spline-algorithm. To do this, authors use a precomputed Voronoi graph, based on an area map. The paper describes only the cases of such path planning, where the target point does not belong to any obstacle.

Besides of classic algorithms, some systems involve machine learning methods or neural networks. So, in [16] it is proposed to leverage neural networks to model situations with people, occurring in the operational area of the robot. To do this, authors propose a global planner, based on the Dijkstra's algorithm to find the shortest path. The neural network is employed only in the local planner to compute linear and angular velocities.

Machine learning methods can be used in global planners as well. Thus, the authors in [17] use the «hallucination learning» (LfH) approach, as well as their hallucinated learning and sober deployment (HLSD) paradigm, to train the robot to avoid obstacles in the environment. The novel method eliminates the need for a high-resolution global planner map, which allows to boost the system performance.

The paper [18] presents a new approach to constructing a lattice-based cost map. It is proposed to build a separate costmap layer for each type of obstacle, which is subsequently superimposed on other layers, and a basic map is received as output, that embraces every kind of obstacles. In a separate layer, it is to arrange the so-called "blowing" of obstacles so that the robot would prefer to move as far as possible from them. With this technique, the authors create initial collision-free conditions.

In [19], the authors describe a new global planner that allows rebuilding the path of the mobile robot to the target point if it is in motion. This planner is based on the D* algorithm, which searches for the initial path. The data from the so-called search trees of this algorithm is then used to quickly recalculate the grid of the cost map, as well as composition a new path. With any movement of the target, the algorithm recalculates the grid of the costmap of the grid cells nearest to the target to find the optimal and best path. When calculating a new path for each cell, its cost is recalculated depending on the costs of movement through this cell and the cost of costs of neighboring cells, provided that this cell is not an obstacle. Thus, when constructing search trees for each cell, it is considered whether there is an obstacle in it or not.

In addition, there is the problem of exploring an unknown environment. In [20], the authors describe a new method for investigating an unknown environment using a method based on the borderline approach. Movement planning is based on data obtained from sensors. When the map is updated based on sensor data, a new costmap grid is composition. Possible target points are selected from the newly obtained grid.

In general, global planners, which consider whether the target point is an obstacle, are either not supported or do not consider the space surrounding the robot. Therefore, a component is needed that connects the task of user to send the robot to a target point inaccessible to him and the robot, which must execute the command of user, avoiding obstacles and, if possible, minimizing the distance to the target point. To solve this problem, a target point search algorithm is proposed, which can be implemented as a separate module and can be used with any planner that uses a global costmap.

3 Proposed Solution

A costmap is a two-dimensional array, whose dimensions correspond to the dimensions of the operational area map of the robot. The values of the array points are in the range [0, 100], where 0—unoccupied point, 100—occupied point. Generally, the search for a possible point in terms of the developed algorithm is performed through iterating over points in positive or negative direction from the user-specified target point.

It is assumed that when a point is received from the user through the interface, the module obtains the costmap of the global planner. After that the search for the specified value is performed in the costmap. In Fig. 1, the high-level logic of execution of the developed algorithm is given. After the target point has been specified, the function is passed the relevant costmap of the global planner from ROS. This map is utilized only to compute points at the moment, when a new point is obtained through user interface.

Algorithm 1 describes the search for a possible target point in the positive y-axis direction. This algorithm also respects the verification of the threshold value of the distance from the target point, based on the threshold parameter. Algorithm 1 also contains the data of the dimensions of the costmap array and relevant search

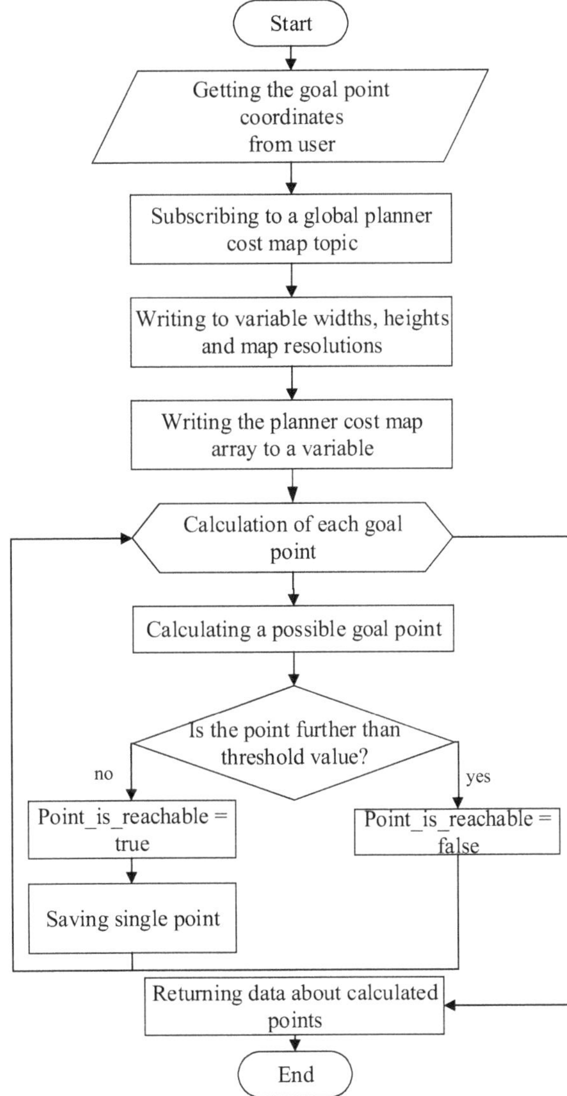

Fig. 1 High-level algorithm for search of a possible target point

parameters. To clarify the interaction of the program module with the interface, in lines 4 and 5, the variables calculate_pixel_y and calculate_pixel_ are declared, which are used to convert the coordinates of a point from meters to pixels. After all the transformations, a search loop to find the required value in the costmap array is initiated. For clarity, in line 9 of the pseudocode, the cell value is compared with 0, which indicates the most likely absence of an obstacle. The search continues with incrementing the pixel value of the cost map with step_of_search. After finding the desired value, the verification begins with the allowable distance, which is shown in line 15. As a result, the value of the required coordinate and values about whether the point is reachable or not are returned. For clarity, this algorithm is described by the block diagram presented in Fig. 2.

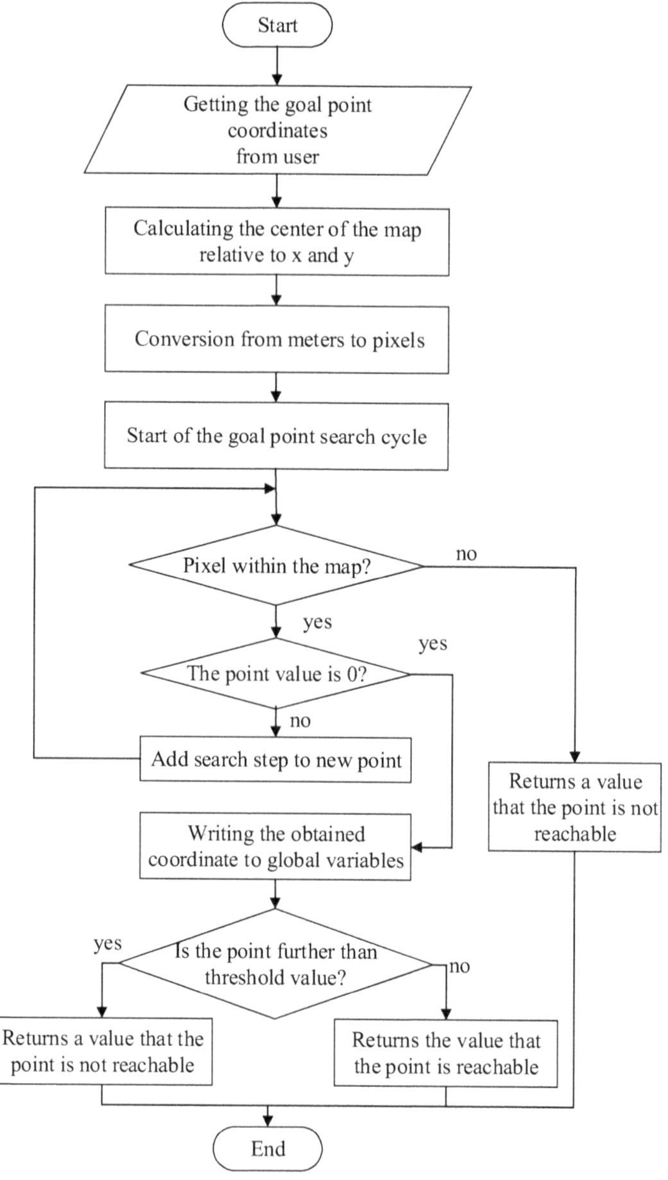

Fig. 2 Algorithm of the function to compute one of four points

Algorithm 1

```
1: n_point_is_reachable, y_n_point = calculate_n_point(y_point_from_interface,
   x_point_from_interface, planner_points_width, planner_points_height, dis-
   tance_from_point, threshold, step_of_search, grid_global_planner)
2:   int center_x = planner_points_height/2
3:   int center_y = planner_points_width/2
4:   int calculated_pixel_y = center_y + y_point_from_interface / map_resolution
5:   int calculated_pixel_x = center_x + x_point_from_interface / map_resolution
6:   bool point_is_reachable
7:    while true
8:      calculated_pixel_y = center_y + y_point_from_interface / map_resolution
9:      if grid_global_planner [calculated_pixel_x][ calculated_pixel_y] != 0
10:        y_point_from_interface += step_of_search
11:     else
12:        break
13:     end if
14:   end while
15:   if calculated_pixel_y > distance_from_point + threshold
16:     point_is_reachable = false
17:   else
18:     point_is_reachable = true
19:   end if
20: return point_is_reachable, y_point_from_interface
```

4 Experiments

As the reference data about the delivery area, either the coordinates of the center of mass of the polygon, or, in the case of a rectangle, the coordinates of the vertices of its diagonal, were transmitted from the user interface to the software package to search for the target point. Therefore, the problem was reduced to two special cases. In the first case, no additional calculations are performed, as shown in Algorithm 1. In the second case, the center point and size of the polygon are calculated from the coordinates of points on the diagonal. After that target points are searched based on the size of the polygon relative to its center point. In this case, only lines 4 and 5 of the pseudocode will be modified.

The software developed on the basis of the obtained algorithm was tested in a three-dimensional Gazebo simulation environment. It modeled a real-world environment, in which the robot operates. Then the gmapping algorithm, integrated into ROS SLAM was applied [4] and an area map was composed. Further, the model of the robot was led along the path using a specific navigation package [6, 12]. For this environment, the size of the costmap was 384 $\times$ 384 pixels (see in Fig. 3) with a scale of 0.05 m/pix. A three-dimensional model of the TURTLEBOT3 Waffle Pi robot was used, on which LiDARs, encoders for obtaining odometry, and a gyro stabilizer

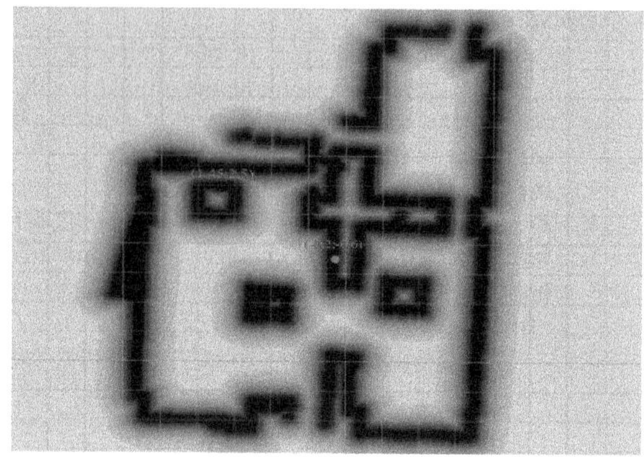

Fig. 3 Costmap for the operational area of the robot. The area map was modeled in Gazebo simulation environment

(IMU) are installed. During testing, two points, inaccessible for robot, were selected, which is evident from Fig. 3.

Testing was performed with different parameters to find target points, specifically the acceptable distance from the target point to the new point (in meters) and the step for finding a point (in meters). Thus, the experiments were performed regarding the values of the distance from the origin point to the possible target point equal to 1, 2, 5 m. The search step was taken equal to 0.1 and 0.01 m.

It was experimentally revealed that the search time for an available target point decreases with increase in step value, because of the reduction in the amount of calculations required to find an unoccupied cell of the costmap. The average value of the required time over 6 experiments to find an available point is 867 ms for a verification step of 0.0001 m, for a verification step of 0.01 m, the average search time is 12.8 ms, for a verification step of 0.1 m, the average time was 4.7 ms.

The experiments involved three planners, integrated into the ROS navigation package: navfn, carrot_planner, global_planner. One area map and two points were tested with each planner. It was found that the average time spent on finding a point with a search step of 0.1 m for the navfn planner is 7.75 ms, for the global_planner planner 25.85 ms, for the carrot_planner planner 13.13 ms. With a step of 0.005 m, the average time for the navfn planner was 5.63 ms, for the global_planner, 4.625 ms, and for the carrot_planner, 4.375 ms. For all planners, the same coordinate values of the discovered points were obtained. As is evident from the results, the time of execution of the algorithm and the obtained coordinate values do not significantly depend on any specific planner. The experiments were performed in two different scales of the costmap, which are presented in Fig. 4 for clarity. Thus, it can be concluded, that the developed algorithm can be employed with any planner.

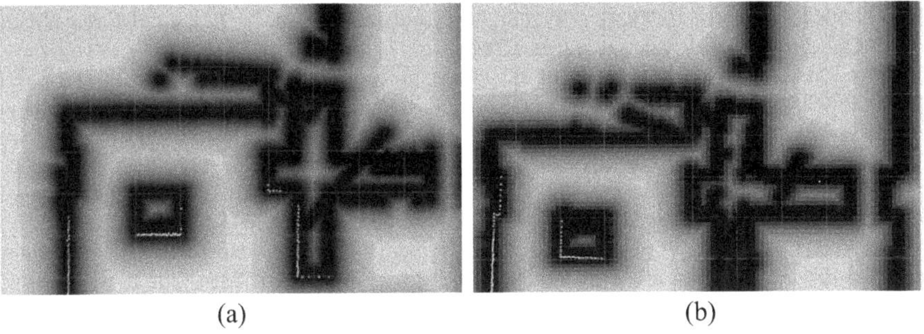

(a) (b)

Fig. 4 Costmaps, obtained in Gazebo in scales of 0.005 m/px (**a**) and 0.1 m/px (**b**)

5 Conclusion

The paper presents an algorithm for finding a target point in the case, when the initially specified target point belongs to an obstacle. The principal prerequisite for using of this algorithm is the availability of a costmap in the planner. The efficiency of the algorithm does not significantly depend on the specific planner in use, but only on the parameters of the cost map, which has been demonstrated experimentally. The results of the experiments showed that with a minimum value of the verification step of 0.01 m, the search time for the target point did not exceed 55 ms. The least search time was 1 ms.

The proposed solution can be used within robot navigation software to solve such problems as the delivery of payload to various real-world objects, which are obstacles for the global planner. It is anticipated to implement the calculation of rotation angles of rotation of the robot in the coordinates of the received area map, since the map built using SLAM can be rotated relative to the coordinate system used in ROS. This feature allows to optimize the robot positioning and enables user control to set the desired position when approaching a target object.

References

1. Collett, T.H.J., MacDonald, B.A.: Developer oriented visualisation of a robot program. In: Proceedings of the 1st ACM SIGCHI/SIGART Conference on Human-Robot Interaction, pp. 49–56 (2006)
2. Ily, M., Roman, L., Magid, E.: Development of a graphical user interface for a crawler mobile robot Servosila engineer. In: 2018 11th International Conference on Developments in eSystems Engineering (DeSE), pp. 192–197. IEEE (2018)
3. Kam, H.R., Lee, S.H., Park, T., Kim, C.H.: Rviz: a toolkit for real domain data visualization. Telecommun. Syst. **60**(2), 337–345 (2015)
4. Balasuriya, B.L.E.A., et al.: Outdoor robot navigation using Gmapping based SLAM algorithm. In: 2016 Moratuwa Engineering Research Conference (MERCon), pp. 403–408. IEEE (2016)
5. Yoshisada, H., et al.: Indoor map generation from multiple LiDAR point clouds. In: 2018 IEEE International Conference on Smart Computing (SMARTCOMP), pp. 73–80. IEEE (2018)

6. Guimarães, R.L., et al.: ROS navigation: concepts and tutorial. In: Robot Operating System (ROS), pp. 121–160. Springer, Cham (2016)
7. Watanabe, A., Endo, D., Yamauchi, G., Nagatani, K.: Neonavigation meta-package: 2-D/3-DOF seamless global-local planner for ROS—development and field test on the representative offshore oil plant. In: 2016 IEEE International Symposium on Safety, Security, and Rescue Robotics (SSRR), pp. 86–91. IEEE (2016)
8. Marin-Plaza, P., Hussein, A., Martin, D., Escalera, A.D.L.: Global and local path planning study in a ROS-based research platform for autonomous vehicles. J. Adv. Transp. **2018** (2018)
9. Lavrenov, R.O., et al.: Development and implementation of spline-based path planning algorithm in ROS/GAZEBO environment. SPIIRAS Proc. **18**(1), 57–84 (2019)
10. Filotheou, A., et al.: Quantitative and qualitative evaluation of ROS-enabled local and global planners in 2D static environments. J. Intell. Robot. Syst. 1–35 (2019)
11. Brock, O., Khatib, O.: High-speed navigation using the global dynamic window approach. In: Proceedings 1999 IEEE International Conference on Robotics and Automation (Cat. No. 99CH36288C) 1, pp. 341–346. IEEE (1999)
12. Zheng, K.: ROS Navigation Tuning Guide. arXiv:1706.09068 (2017)
13. Likhachev, M.: Search-Based Planning with Motion Primitives. https://wiki.ros.org/Events/CoTeSys-ROS-School?action=AttachFile&do=get&target=robschooltutorial_oct10.pdf. Last accessed 2021/01/10
14. Likhachev, M., Gordon, G.J., Thrun, S.: ARA*: Anytime A* with provable bounds on sub-optimality. Adv. Neural. Inf. Process. Syst. **16**, 767–774 (2003)
15. Likhachev, M., Ferguson, D.I., Gordon, G.J., Stentz, A., Thrun, S.: Anytime dynamic A*: an anytime, replanning algorithm. ICAPS **5**, 262–271 (2005)
16. Pokle, A., et al.: Deep local trajectory replanning and control for robot navigation. In: 2019 International Conference on Robotics and Automation (ICRA), pp. 5815–5822. IEEE (2019)
17. Xiao, X., Liu, B., Stone, P.: Agile Robot Navigation Through Hallucinated Learning and Sober Deployment. arXiv:2010.08098 (2020)
18. Lu, D.V., Hershberger, D., Smart W.D.: Layered costmaps for context-sensitive navigation. In: 2014 IEEE/RSJ International Conference on Intelligent Robots and Systems, pp. 709–715. IEEE (2014)
19. Drake, D., Koziol, S., Chabot, E.: Mobile robot path planning with a moving goal. IEEE Access **6**, 12800–12814 (2018)
20. Kulich, M., Miranda-Bront, J.J., Přeučil, L.: A meta-heuristic based goal-selection strategy for mobile robot search in an unknown environment. Comput. Oper. Res. **84**, 178–187 (2017)

Feasibility of Synthesized Optimal Control Approach on Model of Robotic System with Uncertainties

Elizaveta Shmalko

Abstract The paper is focused on practical feasibility of numerical solutions of the optimal control problem for complex robotic systems. The question of the loss of optimality of the computational programmed trajectories during the implementation of stabilization systems and the need to take into account possible uncertainties of models of control objects are discussed. The problem statement of the optimal control problem for a plant with uncertainties is presented. The approach of synthesized optimal control is proposed for its solution. It consists in two stages. At the first stage, the object is stabilized at some point in the state space. This stage of numerical synthesis of stabilization system makes it possible to embed control into the object in such a way that the system of differential equations has the necessary property of feasibility. In this case, the equilibrium point can be changed after some time, but the object maintains equilibrium at every moment of time. Then, we control the position of the equilibrium point. A computational example of the optimal control problem for a mobile robot moving in a space with phase constraints is presented. Study of the model behavior in the presence of uncertainties shows effectiveness of the proposed approach.

1 Introduction

Calculation of the optimal control for complex robotic systems is made on the basis of their mathematical models. The model is a simplified version of the object itself, covering only its main essential properties. On one hand, detailed dynamic models that accurately describe the behavior of robots are too complex and may not be available at all. On the other hand, the most common control algorithms have been developed using simplified linear models. But the performance of these models can be poor as the robot dynamics can be far from linear. As a result, when transferring the obtained calculated optimal data to a real object, deviations of the real object from the

E. Shmalko (✉)
Federal Research Center "Computer Science and Control" of Russian Academy of Sciences, 44/2, Vavilova str., Moscow 119333, Russia

A. Ronzhin and V. Shishlakov (eds.), *Electromechanics and Robotics*, Smart Innovation, Systems and Technologies 232, https://doi.org/10.1007/978-981-16-2814-6_12

calculated model occur. For this, additional stabilization systems and control loops are introduced into real systems.

However, the introduction of additional control loops changes the model of the control object. As a result, the calculation of optimality was performed for one model, but in fact it is used for another closed system. Consequently, optimality in this case is not guaranteed.

Thus, in view of the objectively existing differences of the model from the control object, or in other words some uncertainties, in practice it is necessary to build a feedback stabilization system. But since the construction of this feedback system changes the object model, then when solving the optimal control problem, it is necessary to first build a stabilization system, introduce it into the control object model, and only then for the stabilized object to calculate the optimal control and construct optimal trajectories.

The construction of the stabilization system can be carried out by various existing analytical, technical, or numerical methods. Methods of modal control [1] can be applied for linear systems, as well as other analytical methods such as backstepping [2], analytical design of aggregated regulators [3], or synthesis based on the application of the Lyapunov function [4]. Note that all known analytical synthesis methods for nonlinear systems, when implemented, are associated with a specific type of model, therefore, they cannot be considered universal. In practice, linear controllers, such as PID or PI controllers, are often used to ensure stability. Their use is also associated with specifics of the model.

The wide and rapid spread of automation and robotization requires the development of universal approaches both to solving the problem of synthesizing the object stabilization system, and then to solving the problem of optimal control for the stabilized object.

This work is devoted to the study of the method of synthesized optimal control [5] for the practical implementation of optimal control for robotic systems. According to this approach, the optimal control is calculated for a stabilized object. The stabilization system is synthesized numerically using modern symbolic regression methods. These methods belong to the class of machine learning control. They allow to find a mathematical expression for the control function in the form of a special code with the help of an evolutionary genetic algorithm. The found mathematical expression describing the stabilization system ensures the presence of an equilibrium point in the state space. According to the approach, the synthesized function has a set of parameters that affect the position of the equilibrium point. Optimal control is realized by changing the position of the equilibrium point.

This approach has several advantages. The main strength of this approach is its versatility and applicability to various dynamic models of objects. Another advantage is the creation of optimal control systems with the feasibility property, which appears due to stabilization of the control object at the first stage. The stabilization system provides the object with an equilibrium point in the state space. This means that even with small deviations, the system will always be attracted to a given equilibrium point, which means that the system will function stably even in the presence of discrepancies between the real object and the mathematical model.

The paper presents a mathematical formulation of the optimal control problem with uncertainty. A brief description of the synthesized optimal control method is proposed. And a numerical example of solving the optimal control problem for a mobile robot moving in a space with phase constraints is presented, with the study of the model behavior in the presence of uncertainties.

2 Problem Statement

Let us consider the formulation of the optimal control problem for a model of a robotic system with uncertainty.

Commonly, the object model is described by a system of ordinary differential equations:

$$\dot{\mathbf{x}} = \mathbf{f}(\mathbf{x}, \mathbf{u}), \tag{1}$$

where $\mathbf{x}$ is a state space vector, $\mathbf{x} \in \mathbb{R}^n$, $\mathbf{u}$ is a control vector $\mathbf{u} \in \mathrm{U} \subseteq \mathbb{R}^m$, U is a compact set, $m \leq n$.

For the system (1) initial conditions are set:

$$\mathbf{x}(0) = \mathbf{x}^0 \in \mathbb{R}^n. \tag{2}$$

Given a terminal condition:

$$\mathbf{x}(t_f) = \mathbf{x}^f \in \mathbb{R}^n, \tag{3}$$

where t_f is a time of reaching the terminal condition (3), t_f is limited but is not given,

$$t_f = \begin{cases} t, & \text{if } t < t^+ \text{ and } \|\mathbf{x}^f - \mathbf{x}(t)\| \leq \varepsilon \\ t^+, & \text{otherwise} \end{cases},$$

ε is a given small positive value, t^+ is a given limit time of control.

There are phase constraints in the workspace of object movement:

$$\varphi_i(\mathbf{x}) \leq 0, \quad i = 1, \ldots, S, \tag{4}$$

where S is the number of static phase constraints.

Given a quality criterion in the form of integral functional with addition of a penalty for violation of the phase constraints (4):

$$J = \int_0^{t_f} f_0(\mathbf{x}, \mathbf{u})\mathrm{d}t + \alpha \int_0^{t_f} \sum_{i=1}^{S} \vartheta(\varphi_i(\mathbf{x}))\mathrm{d}t \to \min, \quad (5)$$

where α is a penalty coefficient, $\vartheta(a)$ is a Heaviside function:

$$\vartheta(a) = \begin{cases} 1, & \text{if } a > 0 \\ 0, & \text{otherwise} \end{cases}.$$

If we consider the problem (1)–(5) and search for the control as a function of time:

$$\mathbf{u} = \mathbf{v}(t), \quad (6)$$

which for any moment of time satisfies the control constraints $\mathbf{u} \in \mathrm{U}$, and after its substitution into the right-hand side of the system (1) changes this system of differential equations so that its particular solution from the given initial conditions (2) reaches terminal conditions (3) with optimal value of the quality criterion (5) while all the conditions of phase constraints (4) are satisfied, we obtain the known problem of optimal control.

The existing methods for solving the optimal control problem can be divided into two classes, direct and indirect. Both classes transform the optimal control problem from an optimization problem in an infinite-dimensional space into a nonlinear programming problem in a finite-dimensional space. Direct methods [6] directly look for control function as a function of time. The time axis is divided into a finite number of intervals, and in each interval the control function is approximated by a certain function determined with an accuracy to parameters, for example, a low-order polynomial. Indirect methods solve the optimal control problem based on the Pontryagin maximum principle [7, 8]. In this case, the problem is transformed to a boundary value problem, in which, in the classical setting, it is necessary to find the initial conditions for the vector of conjugate variables.

One of the main problems of all existing approaches to solving the optimal control problem is the feasibility of the obtained solution. Obviously, the time function obtained as a result of solving the optimal control problem cannot be directly used in a real object. This is due not only to the accuracy of the mathematical model of the control object, but also to the requirement to implement state control based on feedback. The generally accepted assumption on the implementation of the solution to the optimal control problem is that on the basis of the obtained solution, it is possible to construct the optimal trajectory of the object in the state space. Further, it is supposed to design on a real object a system to stabilize motion along a programmed trajectory. Two significant circumstances should be noted here. First, the motion of the object in the neighborhood of the programmed trajectory may differ significantly from the optimal one in terms of the value of the functional [9]. Secondly, the mathematical model of the object with stabilization system differs from the model without the stabilization system, which was used to solve the problem of optimal control, therefore, the trajectory obtained on the basis of solving the problem of optimal control is not optimal for real object with the stabilization system. There is also a third circumstance that complicates the implementation of the solution to the optimal control problem. In the classical formulation of the optimal control problem (1)–(6) no additional requirements are put forward for the mathematical model of the control object. It follows that the problem is solved for any object, including not stable or possessing special properties, bifurcations, cycles, poles. In practical implementation, the inaccuracies of the mathematical model behave differently depending on the qualitative characteristics of the system of differential equations of the model.

Thus, let us consider the formulation of the problem with a model of a control object, adding an additive uncertainty function to it.

$$\dot{\mathbf{x}} = \mathbf{f}(\mathbf{x}, \mathbf{u}) + \psi(t), \tag{7}$$

where $\psi(t)$ is an unknown bounded function,

$$\psi^- \leq \psi(t) \leq \psi^+, \tag{8}$$

ψ^-, ψ^+ are known vectors of uncertainty constraints.

Then the optimal control problem, taking into account the uncertainty of the model of the control object, can be formulated as follows.

It is necessary to find a control in the form:

$$\mathbf{u} = \mathbf{h}(\mathbf{x}, t), \tag{9}$$

such that after substituting this control function (9) into the right-hand sides of the system (7), we obtain the following system of differential equations:

$$\dot{\mathbf{x}} = \mathbf{f}(\mathbf{x}, \mathbf{h}(\mathbf{x}, t)) + \psi(t), \tag{10}$$

a particular solution $\mathbf{x}(t, \mathbf{x}^0)$ of which from the initial state (2) in the absence of uncertainty $\psi(t) \equiv 0$ provides a minimum to functional (5), and in the presence of uncertainty $\psi(t) \neq 0$ satisfying constraints (8), the value of the functional changes in no more than a given amount:

$$J(\mathbf{h}(\mathbf{x}, t)) = J_{\text{opt}} + \Delta_\psi, \tag{11}$$

where Δ_ψ is a given value of the functional change, J_{opt} is the optimal value of functional (5), obtained in the absence of uncertainty.

In practice, engineers have long understood the difficulties of controlling objects with uncertainties, so they initially make the object stable, and then solve the problems of control. It is known that objects possess good properties for control, when their mathematical models in the phase space have a stable equilibrium point. Based on the analysis of practically implemented control systems, a new numerical approach to solving problems of optimal control, a method of synthesized optimal control, was developed. Following the principle of practical feasibility of optimal control, in the proposed approach, the optimal control problem is solved after ensuring the stability of the object in the state space. At the first stage, the system without perturbations is stabilized at some point in the state space. This stage of the synthesis of the stabilization system makes it possible to embed control into the object in such a way that the system of differential equations has the necessary property of feasibility. In this case, the equilibrium point can be changed after some time, but the object maintains equilibrium at every moment of time. Then for solving the optimal control problem, we control the position of the stable equilibrium point.

3 Description of the Approach

The approach is called synthesized optimal control, since it requires, firstly, a solution of stabilization system synthesis. The approach consists in finding such a control function that will change system (1) without uncertainty so that this system will always have a stable equilibrium point in some region in which there is a solution to the optimal control problem. In the desired control function, a parametric vector is introduced. It affects the position of a stable equilibrium point in a closed control system:

$$\mathbf{u} = \mathbf{g}(\mathbf{x}, \mathbf{q}^*), \tag{12}$$

where $\mathbf{q}^*$ is a parametric vector of the same dimension as the state vector.

The control function (12) provides for the system (1) without uncertainty:

$$\dot{\mathbf{x}} = \mathbf{f}(\mathbf{x}, \mathbf{g}(\mathbf{x}, \mathbf{q}^*)), \tag{13}$$

existence of a stable equilibrium point $\mathbf{x}^*(\mathbf{q}^*)$:

$$\mathbf{f}(\mathbf{x}^*(\mathbf{q}^*), \mathbf{g}(\mathbf{x}^*(\mathbf{q}^*), \mathbf{q}^*)) = 0, \tag{14}$$

where $\mathbf{x}^*(\mathbf{q}^*)$ is a vector of coordinates of the position of a stable equilibrium point.

Solution of the synthesized optimal control includes two stages.

1st stage: It is necessary to solve the problem of control synthesis to ensure the existence of a stable equilibrium point in the state space. To solve the synthesis problem in the considered mathematical formulation, it is necessary to find the control function (12). Most of the known methods specify the control function with an accuracy to parameters' values, for example, methods associated with the solution of the Bellman equation, as well as the use of various controllers, including controllers based on popular now artificial neural networks [10]. But there are now general numerical approaches to solve the synthesis problem based on the application of symbolic regression methods [11]. They can be applied to find a solution without reference to specific model equations. In the result of the first stage, the structure and parameters of the function **g** are obtained.

2nd stage: Once the function (12) is found, the optimization problem of positioning of equilibrium point is solved. It is necessary to find a parametric vector $\mathbf{q}^*$ that when switching from one equilibrium point $\mathbf{x}^*(\mathbf{q}^*)$ to another after a certain time interval Δt, a partial solution of the system (13) from the initial conditions (2) achieves the terminal conditions (3) with optimal value of the quality criterion (5).

Thus, in the synthesized optimal control approach the uncertainty of model (7) is compensated by the stability of the system relative to a point in the state space. Near the equilibrium point, all solutions converge and feasibility principle is satisfied. This first step of stabilization system synthesis is a key idea of the approach, and it provides achievement of better results in the tasks with complex environment and uncertainties.

4 Computational Example

As an example, consider the problem of optimal control for one mobile robot moving from some initial state to the terminal position in the complex environment with multiple state constraints. The mathematical model of a robot has the following form.

$$\begin{aligned}\dot{x} &= 0.5(u_1 + u_2)\cos(\theta),\\ \dot{y} &= 0.5(u_1 + u_2)\sin(\theta),\\ \dot{\theta} &= 0.5(u_1 - u_2),\end{aligned} \tag{15}$$

where $\mathbf{x} = [x\ y\ \theta]^T$ is a vector of state space, $\mathbf{u} = [u_1\ u_2]^T$ is a control vector.

Control boundaries are given:

$$u_i^- \leq u_i \leq u_i^+, i = 1, 2. \tag{16}$$

Initial conditions are set:

$$x(0) = x^0, y(0) = y^0, \theta(0) = \theta^0. \tag{17}$$

Terminal conditions are set:

$$x(t_f) = x^f, y(t_f) = y^f, \theta(t_f) = \theta^f. \tag{18}$$

Static phase constraints are given:

$$\varphi_i(x, y) = r_i^2 - (x_i^C - x)^2 - (y_i^C - y)^2 \le 0, \tag{19}$$

where r_i, x_i^C, y_i^C are given parameters (radius and center coordinates) of the static phase constraints, $i = 1, \ldots, S$, S is the number of phase constraints.

A quality criterion is given that minimizes the time to reach the goal and includes penalty functions for deviations from the terminal state and for violation of phase constraints:

$$\tilde{J} = t_f + \alpha_1 \sqrt{(x^f - x(t_f))^2 + (y^f - y(t_f))^2 + (\theta^f - \theta(t_f))^2} + \alpha_2 \int_0^{t_f} \sum_{i=1}^{S} \vartheta(\varphi_i(x)) \mathrm{d}t \to \min \tag{20}$$

where α_1, α_2 are penalty coefficients, $\vartheta(a)$ is a Heaviside function.

We have received the optimal control by two different approaches (the synthesized optimal control and the direct approach of piece-wise linear approximation of control) and then we studied the behavior of the model with the obtained controls in the presence of uncertainties in the model, which were added in the form of noise.

4.1 Synthesized Optimal Control Approach

On the first step, in order to provide a steady state of the object, the problem of stabilization system synthesis was solved.

To solve the synthesis problem numerically, the set of initial conditions was set:

$$\mathbf{X}_0 = \{\mathbf{x}^{0,1}, \ldots, \mathbf{x}^{0,K}\}. \tag{21}$$

One terminal state was set:

$$\mathbf{x}^* = [x^* y^* \theta^*]^T. \tag{22}$$

The functional was:

$$J_{\mathrm{st}} = \sum_{i=1}^{K} \left(t_{f,i} + \alpha_1 \left\|\mathbf{x}^* - \mathbf{x}\left(t_{f,i}, \mathbf{x}^{0,i}\right)\right\|\right) \to \min, \tag{23}$$

where $t_{f,i}$ is time of reaching terminal condition (22) from the initial state $\mathbf{x}^{0,i}$, $i = 1, \ldots, K$.

The synthesis of the stabilization system implies finding one control function (24):

$$\mathbf{u} = \mathbf{h}(\mathbf{x}^* - \mathbf{x}), \tag{24}$$

which ensures the minimum of functional (23) for all given initial conditions (21). Eqt (24) is a particular case of (12), where $\mathbf{q}^* = \mathbf{x}^*$.

To solve the control synthesis problem, we use one of the numerical methods of symbolic regression—the network operator [11]. In calculations, the following parameters' values were set: $X_0 = (\mathbf{x}^{0,1} = [-5\ -5\ -\pi/2]^T$, $\mathbf{x}^{0,2} = [-5\ -5\ \pi/2]^T$, $\mathbf{x}^{0,3} = [-5\ 5\ -\pi/2]^T$, $\mathbf{x}^{0,4} = [-5\ 5\ \pi/2]^T$, $\mathbf{x}^{0,5} = [5\ -5\ -\pi/2]^T$, $\mathbf{x}^{0,6} = [5\ -5\ \pi/2]^T$, $\mathbf{x}^{0,7} = [5\ 5\ -\pi/2]^T$, $\mathbf{x}^{0,8} = [5\ 5\ \pi/2]^T)$, $u^- = -10$, $u^+ = 10$, $x^f = 0$, $y^f = 0$, $\theta^f = 0$.

As a result, the following control function was obtained:

$$u_i = \begin{cases} u^+, & \text{if } u_i \geq u^+ \\ u^-, & \text{if } u_i \leq u^- \\ \tilde{u}_i, & \text{otherwise} \end{cases}, \quad i = 1, 2 \tag{25}$$

where

$$\begin{aligned}
\tilde{u}_1 &= A^{-1} + \sqrt[3]{A} + \operatorname{sgn}(q_3(\theta^* - \theta)) \exp(-|q_3(\theta^* - \theta)|) + \operatorname{sgn}(\theta^* - \theta) + \mu(B), \\
\tilde{u}_2 &= \tilde{u}_1 + \sin(\tilde{u}_1) + \arctan(H) + \mu(B) + C - C^3, \\
A &= \tanh(0.5D) + \left(B + \sqrt[3]{x^* - x}\right)^3 + C + \sin(q_3(\theta^* - \theta)), \\
B &= C + \tanh(0.5G) + x^* - x, \\
C &= G + \operatorname{sgn}(\operatorname{sgn}(x^* - x)q_2(y^* - y)) \exp(-|\operatorname{sgn}(x^* - x)q_2(y^* - y)|) \\
&\quad + \sin(x^* - x), \\
D &= H + C - C^3 + \operatorname{sgn}(q_1(x^* - x)) + \arctan(q_1) + \vartheta(\theta^* - \theta), \\
G &= \operatorname{sgn}(x^* - x)q_2(y^* - y) + q_3(\theta^* - \theta) + \tanh(0.5q_1(x^* - x)), \\
H &= \arctan(q_1(x^* - x)) + \operatorname{sgn}(W)\sqrt{|W|} + W + V \\
&\quad + 2\operatorname{sgn}(W + \tanh(0.5V)) \\
&\quad + \sqrt[3]{W + \tanh(0.5V) + \sqrt[3]{x^* - x}} \\
&\quad + \operatorname{sgn}\left(x^* - x)\sqrt{|x^* - x|} + \sqrt[3]{x^* - x} + \tanh(0.5V\right), \\
W &= \operatorname{sgn}(x^* - x) + \operatorname{sgn}(q_2(y^* - y))\operatorname{sgn}(x^* - x)Q, \\
V &= q_3(\theta^* - \theta) + \operatorname{sgn}(x^* - x)q_2(y^* - y) + Q, \\
Q &= \tanh(0.5(x^* - x)),
\end{aligned}$$

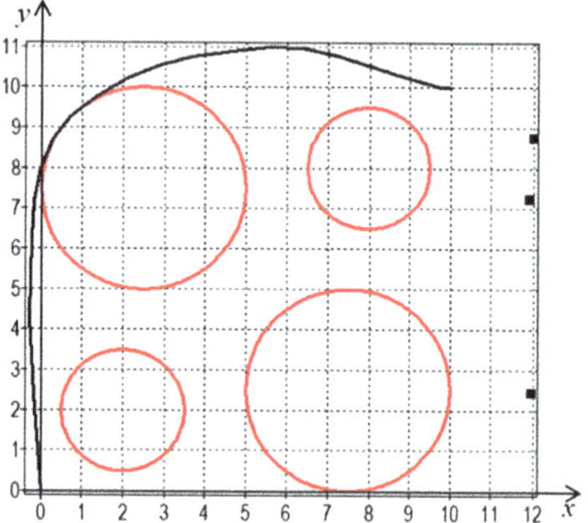

Fig. 1 Optimal trajectory of the robot by the synthesized approach

$$\mu(\alpha) = \max\{0, \alpha\}, \tanh(\alpha) = \frac{1 - \exp(-2\alpha)}{1 + \exp(-2\alpha)},$$
$$q_1 = 11.72876, q_2 = 2.02710, q_3 = 4.02222.$$

The obtained control functions (25), which ensure the stabilization of the object, are substituted into the equations of the model (15). In the result of the solution of control synthesis problem a stable equilibrium point in the state space is appeared. Position of the equilibrium point depends on the terminal vector (22).

On the second step of the approach, to obtain an optimal trajectory of the robot, we search for the optimal positions of the equilibrium point. The set of $P = 3$ points (22) was searched for the robot on criterion (20) including the given phase constraints (19). Parameters for phase constraints were set as following: $x_1^C = 8$, $y_1^C = 8$, $r_1 = 1.5$, $x_2^C = 2$, $y_2^C = 2$, $r_2 = 1.5$, $x_3^C = 2.5$, $y_3^C = 7.5$, $r_3 = 2.5$, $x_4^C = 7.5$, $y_4^C = 2.5$, $r_4 = 2.5$. Boundaries for values of point positions were set $-2 \le x^* \le 12$, $-2 \le y^* \le 12$, $-\pi/2 \le \theta^* \le \pi/2$. To find the optimal locations of the equilibrium point, particle swarm optimization algorithm [12] was used.

The following optimal solution was found $\mathbf{x}^{*,1} = [11.8264\ \ 4.3850\ \ 0.0280]^T$, $\mathbf{x}^{*,2} = [11.1745\ \ 6.6811\ \ 0.3700]^T$, $\mathbf{x}^{*,3} = [12.0000\ \ 8.8290\ \ 0.1496]^T$. These points were switching in some time interval $\Delta t = 0.625$ sec for control function (25). The functional value was $\tilde{J} = 2.4735$.

Figure 1 shows the moving trajectory of the robot with the obtained synthesized optimal control. The robot trajectory is a black line, black squares are positions of the equilibrium point, and red circles are given phase constraints.

4.2 Direct Approach of Piece-Wise Linear Approximation

For comparison, the same optimal control problem was solved by the direct approach. The control was approximated by a piece-wise linear function in each interval $(k -$

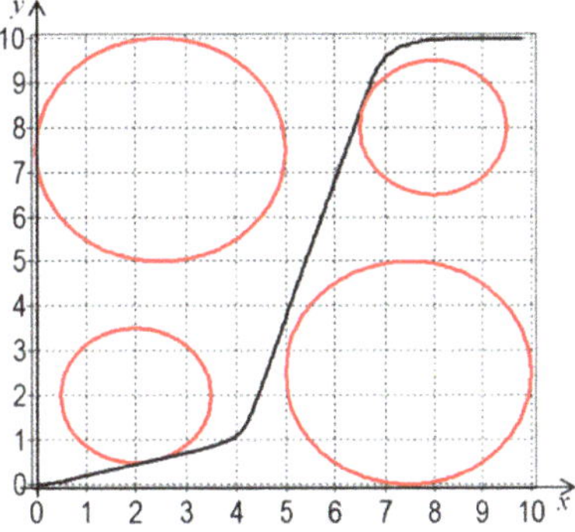

Fig. 2 Optimal trajectory of the robot by the direct approach

$1)\delta t \leq t < k\delta t$:

$$\begin{array}{c}\tilde{u}_1(t) = q_j + \frac{q_{j+1}-q_j}{\delta t}(t-(j-1)\delta t)\\ \tilde{u}_2(t) = q_{j+L} + \frac{q_{j+L+1}-q_{j+L}}{\delta t}(t-(j-1)\delta t)\end{array}, \tag{26}$$

where $j = 1, \ldots, L$, δt is a value of time interval, L is a number of intervals $L = \lfloor t^+/\delta t \rfloor$.

As a solution of the optimal control problem by the direct method values of parameters are to be found $\mathbf{q} = [q_1 \ldots q_{2(L+1)}]^T$. For this purpose, PSO algorithm [12] was used too. In the experiment $\delta t = 0.25$, $L = 10$. Values of parameters were bounded $-20 \leq q_i \leq 20$, $i = 1, \ldots, 2(L+1)$. As a result, the following vector of parameters was obtained:

$$\begin{aligned}\mathbf{q} = [&11.802\ \ 16.617\ \ 19.537\ \ 19.448\ \ 18.873\ \ 13.517\ \ -1.994\ \ 12.411\\ &10.234\ \ -17.480\ \ -15.859\ \ 5.592\ \ 9.182\ \ 9.182\ \ 1.403\ \ 17.248\\ &19.408\ \ 19.782\ \ 15.421\ \ 18.457\ \ 6.950\ \ 19.851\ \ -4,101]^T.\end{aligned}$$

Figure 2 shows the moving trajectory of the robot with the optimal control obtained by the direct approach. The functional value was $\tilde{J} = 1.9191$.

4.3 *Study of Sensitivity to Uncertainties*

The obtained controls were studied in the presence of uncertainties in model. For this purpose, perturbations were added to the mathematical model of the robot (15).

$$\begin{aligned}\dot{x} &= 0.5(u_1+u_2)\cos(\theta) + \beta\xi(t),\\ \dot{y} &= 0.5(u_1+u_2)\sin(\theta) + \beta\xi(t),\\ \dot{\theta} &= 0.5(u_1-u_2) + \beta\xi(t),\end{aligned} \tag{27}$$

Table 1 Values of the quality criterion $\tilde{J}$ in the presence of perturbations in model

β	Synthesized approach	Direct approach
0	2.4865	2.5835
5	2.6742	6.7678
10	4.3449	7.4849

where β is a constant positive parameter, $\xi(t)$ is a random function that takes values from -1 to 1.

The results of comparative experiments are presented in Table 1. Values of the functional were calculated as mean values of 10 experiments. Trajectories of the robot with the obtained optimal control at different noise level are shown in Figs. 3 and 4. As can be seen, the synthesized control occurs less sensitive to the uncertainties of model.

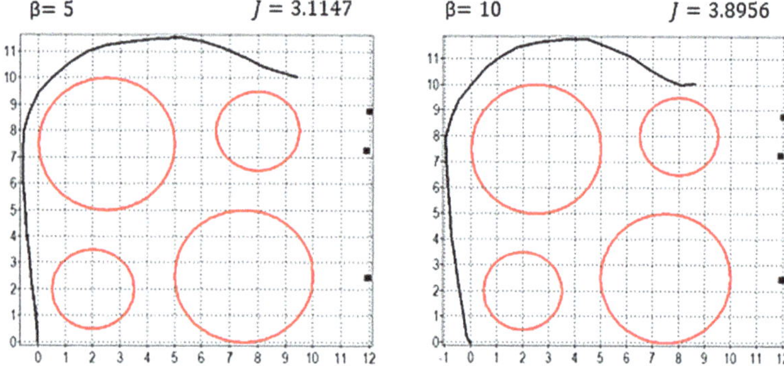

Fig. 3 Optimal trajectory of the robot by the synthesized approach with perturbations

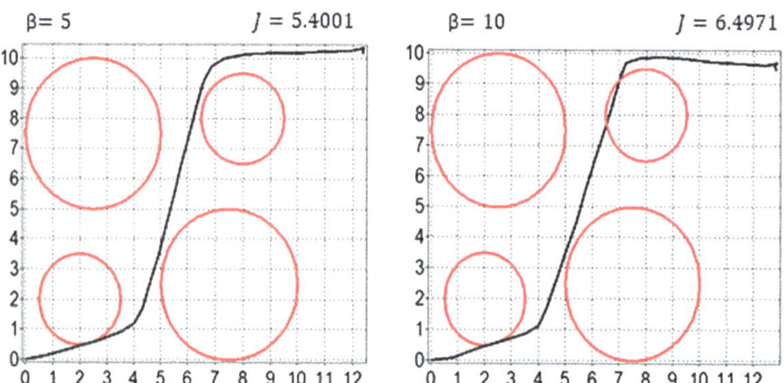

Fig. 4 Optimal trajectory of the robot by the direct approach with perturbations

5 Conclusion

Studies of the synthesized optimal control have shown that this approach allows to obtain a control that is insensitive to model inaccuracies, and the obtained optimal control can be directly implemented on the plant without creating an additional system for stabilizing the object's motion along a programmed trajectory. The only difficulty of this approach is the solution of a rather complex problem of general control synthesis at the first stage. Here, it is proposed to use the methods of symbolic regression, the use of which has recently expanded in the field of control.

Acknowledgements This work was supported by the Russian Science Foundation under project 19-11-00258.

References

1. Simon, J.D., Mitter, S.K.: A theory of modal control. Inf. Control **13**(4), 316–353 (1968)
2. Wang, J.: Speed-assigned position tracking control of SRM with adaptive backstepping control. IEEE/CAA J. Autom. Sin. **5**(6), 1128–1135 (2018)
3. Kolesnikov, A.A., Kuz'menko, A.A.: Backstepping and ADAR method in the problems of synthesis of the nonlinear control systems. Mekhatronika, Avtomatizatsiya, Upravlenie **17**(7), 435–445 (2016)
4. Agarwal, R., O'Regan, D., Hristova, S.: Stability by Lyapunov like functions of nonlinear differential equations with non-instantaneous impulses. J. Appl. Math. Comput. **53**, 147–168 (2017)
5. Diveev, A., Shmalko, E., Serebrenny, V., Zentay, P.: Fundamentals of synthesized optimal control. Mathematics **9**(1), 21 (2021)
6. Gill, P.E., Murray, W., Wright, M.H.: Practical Optimization. Academic Press (1981)
7. Arutyunov, A., Karamzin, D.: A survey on regularity conditions for state-constrained optimal control problems and the non-degenerate maximum principle. J. Optim. Theory Appl. **184**(3), 697–723 (2020)
8. Pontryagin, L.S., Boltyanskii, V.G., Gamkrelidze, R.V., Mishchenko, E.F.: Pontryagin Selected Works: The Mathematical Theory of Optimal Process, vol. 4. Gordon and Breach Science Publishers, New York (1985)
9. Young, L.C.: Lecture on the calculus of variations and optimal control theory. AMS Chelsea Publishing, Providence Rhode Island (2000)
10. Yang, J., Lu, W., Liu, W.: PID controller based on the artificial neural network. In: Yin, F.L., Wang, J., Guo, C., (eds.) Advances in Neural Networks, Lecture Notes in Computer Science, vol. 3174. Springer: Berlin/Heidelberg, Germany (2004)
11. Diveev, A.I.: Numerical method for network operator for synthesis of a control system with uncertain initial values. J. Comput. Syst. Sci. Int. **51**(2), 228–243 (2012)
12. Kennedy, J.; Eberhart, R.: Particle swarm optimization. Proc. ICNN'95 Int. Conf. Neural Netw. **4**, 1942–1948 (1995)

Algorithm for Calculating Coordinates of Repeaters for Combining Stationary and Mobile Devices into Common Cyber-Physical System

Alexander Denisov and Irina Vatamaniuk

Abstract When configuring a cyber-physical system (CPS) in large areas and connecting distributed autonomous devices to it via wireless communication channels, it is necessary to take into account a number of factors that negatively affect the quality of data transfer among CPS elements: Radio signal attenuation, interference, data transfer rate and others, which can lead to the appearance of zones without coverage. The growth in the number of heterogeneous devices with different properties and actual purposes, interacting over a wireless network, leads to the emergence of the problem of providing a speed-satisfactory method for configuring a wireless communication network between heterogeneous devices in a distributed CPS. To solve this problem, algorithms for the repeater locating were developed, which provide configuration of a wireless data transmission network for CPS deployment. The proposed solution allows to increase the speed of configuration with a minimum number of repeaters utilized in a given service area, with the presence of obstacles in the form of surface features of the ground and green spaces.

1 Introduction

One of the main scientific problems, concerning the usage of cyber-physical systems (CPS) in outdoor areas with unprepared environments is the need of establishment of a comprehensive engineering infrastructure, ensuring the CPS operability, what incurs significant time and resource costs and thereby makes it impossible to quickly deploy CPS in open areas [1]. In addition, these areas are subject to dynamic changes, which often requires reorganization and modernization of the system infrastructure in order to ensure fault-tolerant operation of the CPS [2]. Thus, today there are practically no CPS, suitable for quick deployment and adjustment of their own infrastructure to changing environmental conditions. Such situation not only limits the applicability of such systems in dynamically changing outdoor environments, but also almost

A. Denisov (✉) · I. Vatamaniuk
St. Petersburg Federal Research Center of the Russian Academy of Sciences (SPC RAS), St. Petersburg Institute for Informatics and Automation of the Russian Academy of Sciences, 39, 14th Line, 199178 St. Petersburg, Russia

A. Ronzhin and V. Shishlakov (eds.), *Electromechanics and Robotics*, Smart Innovation, Systems and Technologies 232, https://doi.org/10.1007/978-981-16-2814-6_13

completely eliminates the possibility of ad-hoc short-term use of CPS, since in such cases, the duration of CPS deployment significantly exceeds the time of their target functioning. Thus, there is an urgent need today to ensure radio communication of stationary and mobile devices over extended areas. The problem is nontrivial and assumes many conditions to be met that complicate the development of a suitable solution method. The main prerequisites are the low cost of configuring and maintaining a data transmission network, providing communication with randomly moving mobile objects, providing communication over extended areas with different surface features and landscaping.

To solve this problem, many different solutions have been proposed, each with its own advantages and disadvantages. There are methods that consist in connecting to the gateway (central element of the network) one device of the system, depending on the location and distance from the gateway using repeaters, the distance between which is calculated based on the required data transfer rate [3–6]. To ensure communication with mobile devices moving through a known area, methods of repeater positioning by «tiling» have been proposed [7–9]. There are also methods using so-called passive repeaters for broadcasting digital television programs into the radio-hot zone in mountainous areas. The principle of a passive repeater is that it has a high reflectance, where the reflected wave is directed toward the radio shade. In open areas with a direct line of sight, the Vvedensky formula [10] can be used to predict the signal level [11]. Other methods focus on using different types of antennas. On a damp and well-conductive surface, antennas of the asymmetrical vertical dipole type are used. The use of a whip antenna on a rocky surface reduces its efficiency by 2–2.5 times, therefore, counterweights and directional antennas must be used on dry or stony soil. In winter (frosty weather), antennas of the ABV type are used, during thaw and wet snow, λ-shaped antennas are used. When operating in mountains with steep slopes and sharp breaks, VHF radio communication can be provided for a much longer range along the natural waveguide due to multiple reflections of waves from the slopes. With an increase in operating wavelength, the influence of mountain obstacles on the propagation of radio waves decreases. When the radio channels are located close to each other along the frequency grid step, each repeater is equipped with double band-notch filters using separate transmit and receive antennas. Filters are tuned to skip a certain frequency and simultaneously cut out a different frequency. Band-notch filters are indispensable in the presence of nearby frequencies or interference [12]. There are methods for placing repeaters to ensure peer-to-peer (P2P) communication in areas with difficult terrain [13–16]. The general equation for calculating the Fresnel zone radius at any point between end nodes is used to calculate the height of the repeater mounting points. This approach allows you to calculate suitable height for repeater locations, to ensure connecting devices to the network and yields meaningful results when calculating the location of repeaters in areas with difficult terrain. Also, methods based on models can be used, which take into account: total propagation losses in conditions of difficult terrain, wooded areas and uneven buildings, the number of segments into which the propagation path is divided depending on the type of segment (forest, urban development, complex terrain), distance from the transmitting antenna, propagation loss open space, additional propagation loss within a segment, defined as the sum of losses in the environment and losses due to uneven terrain [17–20].

These methods are geometrical and allow the use of a minimum number of repeaters to establish a network from the elements of the system. However, the first methods do not take into account the relief, buildings, as well as green spaces, the influence of which on the transmission of the radio signal can take on a critical nature, and the latter are intended only for combining stationary devices into a network, which is not enough for the correct functioning of the CPS. This paper presents an algorithm for calculating the coordinates of the location of repeaters to connect to a common network of stationary and mobile devices CPS in areas with difficult terrain and landscaping.

2 Algorithm for Calculating Location Coordinates of Repeaters for Territories with Variable Surface Features

Thus, it is required to provide communication with N devices on a terrain of the size $A \times B$ and area S with various obstacles. To do this, it is required to calculate the coordinates of repeaters location between the devices and the gateway, thereby establishing the proper configuration of a data transmission network.

Landscape, trees, and various buildings impair the radio signal, as well as cause the appearance of zones without adequate coverage. To solve this problem, an algorithm was developed for determining the coordinates of the location of repeaters for territories with variable surface features and in the presence of obstacles (Fig. 1). To ensure the operation of this algorithm, it is necessary to compile tables of surface altitudes and obstacles to the passage of a radio signal, as well as attenuation of radio waves when passing through obstacles (for example, trees, surface features, etc.). The attenuation of radio waves when passing through obstacles requires a closer spacing of repeaters to ensure data transmission at the required speed [21]. The total attenuation (dB) of radio waves in areas with green spaces was determined by the expression [22]:

$$W_n = V_\partial + M_1 + M_2,$$

where V_∂ is the diffraction component of the attenuation, $M_{1,2}$ are the interference factors of attenuation on the sections of the emitter-obstacle path, the obstacle-receiver.

$$\left|M_{1,2}\right| = \sqrt{1 + \Phi_{1,2}^2 - 2\Phi_{1,2}\cos\delta_{1,2}},$$

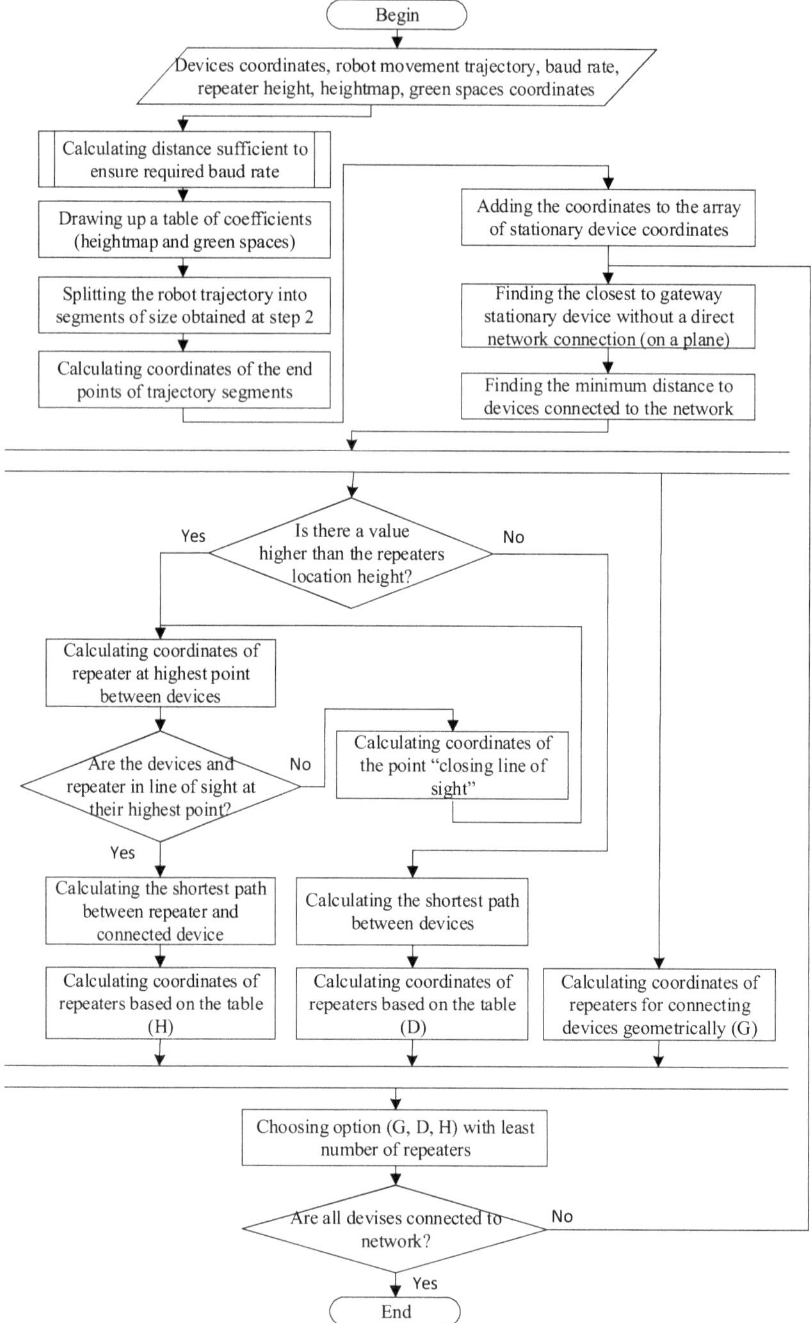

Fig. 1 Algorithm for determining the coordinates of the repeater locations for areas with variable terrain features and the presence of obstacles

where $\delta_{1,2} = 2kh_{1,2}\sin\Psi_{1,2}$, $k = 2\pi/\lambda$, λ is the length of the radio wave, $h_{1,2}$ are the heights of the transmitter and receiver antennas, $\Psi_{1,2}$ are the grazing angles in the sections of the diffraction paths, $\Phi_{1,2}$ are the reflection coefficients. According to [22], in a mixed forest with an average density of trees, the average seasonal linear attenuation of the radio signal is 0.1 dB/m for frequencies of 330 MHz.

In this work, to calculate the attenuation of propagation of radio waves in open space and the effect of the surface features on the propagation of radio waves, we use the classical calculation formulas [23]. According to the Rayleigh criterion, you can get the acceptable height of the irregularities, i.e., the height of the irregularities at which the reflection can be considered mirror-like:

$$h < \frac{\lambda}{8\cos\theta},$$

where h is the permissible irregularity height, λ is the wavelength, θ is the angle of incidence of the wave.

Attenuation coefficient, calculated as follows:

$$|W| = \frac{1}{2p_1(r_1 + r_2)}\sqrt{1 + \frac{4}{\pi}p_1(r_1 + r_2)\frac{r_2}{r_1}},$$

where

$$p = \frac{\pi r}{\lambda\sqrt{\varepsilon_2^2 + (60\gamma_2\lambda)^2}},$$

r_1,r_2 are the path lengths, $\varepsilon_2{}^2$ is the dielectric constant, γ_2 is determined by the averaged electrical parameters of the soil on the path and the terrain.

This algorithm computes the connection of devices to the network by using repeaters in three possible ways: (1) geometric; (2) based on the highest elevation point between the connected devices; (3) using the path of least resistance based on the coefficient tables and Dijkstra's algorithm. It can be seen from the figure that data collection and the compilation of coefficient tables are performed first. This algorithm was developed to calculate the location of the repeaters along the known path of robotic vehicles; therefore, the coordinates of the repeaters are calculated geometrically, based on the coefficient tables. Next comes the calculation of the coordinates of the repeaters for connecting to a network of stationary devices. As described above, the calculation is performed in three different ways. The method of calculation based on the highest terrain point between the connected devices allows you to select points for repeaters on the vertices (terrain features or objects) to provide a line of sight to the devices. This solution is more suitable for areas with very uneven terrain or a large number of buildings that are poorly penetrated by the radio signal, where it is not possible to mount devices high above the surface or objects. The next method considers the option of avoiding high and poorly transmitting radio signal obstacles

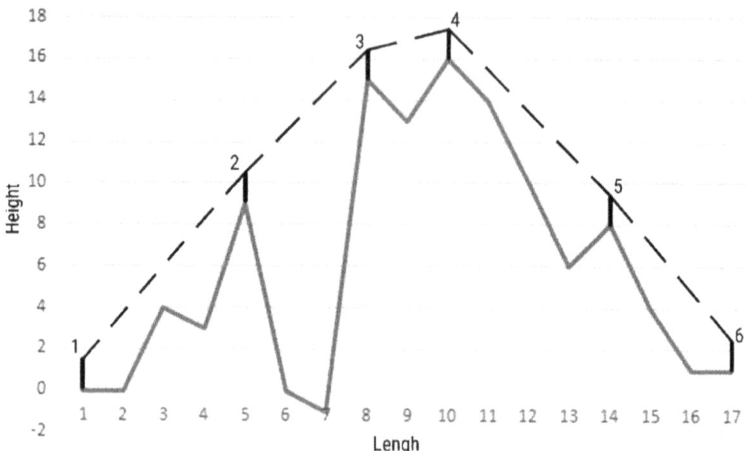

Fig. 2 The result of the algorithm operation based on the highest terrain point between the connected devices (bypass over the obstacle)

by the side through the use of Dijkstra's algorithm and tables of coefficients. At the end, a comparison is made of the number of repeaters required to connect the devices, and the option with the least number of devices involved is selected.

The result of the algorithm execution based on the highest point of the terrain between the connected devices (obstacle avoidance from above) is shown in Fig. 2.

The figure shows a height map in a straight line between the connected devices. The gray line shows the change in heights between devices, black strokes show the recommended positions of repeaters, the black dotted path indicates line of sight between devices and repeaters. The figure shows that it is required to provide communication between devices with positions 1 and 6. To do this, according to the table of heights in a straight line between the devices, the highest terrain point is searched (heights below the height of the repeater tower are not regarded), where the repeater is installed (position 4). Further, if it is impossible to provide direct communication at the required rate between the installed repeater and the devices, the steps of the algorithm are repeated. Thus, the result obtained requires a minimum number of repeaters to connect devices into a network while avoiding obstacles from above. But during the operation of the algorithm, a variant of bypassing from the side of large changes in the relief using Dijkstra's algorithm is also considered. This version of the algorithm enables to calculate coordinates for the location of repeaters in areas with a smaller change in altitude (avoiding obstacles from the side). The values of the table of heights are used to construct a graph connecting two devices of the system with a minimum weight. The weights of the graph are the elevations of the terrain of the selected area with the distance to the nearest radio module already connected to the network. Further, depending on the selected radio modules, the coordinates of the repeater positions [10] are calculated along the edges of the constructed graph.

In the proposed approach aimed to find a solution for providing communication between devices using the least number of repeaters, both methods are coupled. Thus, the output is a combined solution (since the algorithm in Fig. 1 uses both methods to calculate the position of each of the repeaters and in the subsequent iteration uses the option with the least number of repeaters), containing both the coordinates of the repeaters for avoiding the obstacle from above, and for bypassing from the outside, to reduce the number of repeaters used.

For areas with vegetation, this algorithm is used in a similar way, but instead of a table (map) of heights, a table of radio signal attenuation when passing through vegetation is built. Thus, to obtain the best result, the algorithm combines two advantages: The distance between repeaters when passing through green spaces is reduced and it is bypassed from the side of zones with vegetation.

For the experiment, a height map was used, shown in Fig. 3. The height of the radio beacon rack was equal to 2 conventional units, the maximum relief height was 16 units, and the minimum was equal to 0 units. The table of coefficients in this

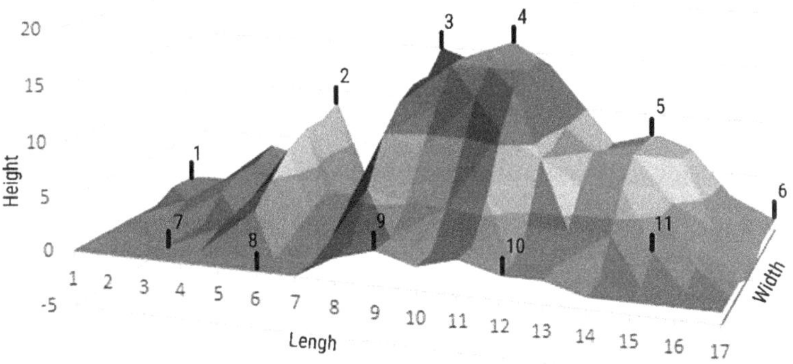

Fig. 3 The result of the algorithm for calculating the location of repeaters for connection of devices

Table 1 Table of coefficients for this problem

Coors	1	2	3	4	5	6	7	8	9	10	11	12	13	14	15	16	17
1	0	0	0	0	0	0	0	2	3	2	3	2	2	1	1	1	1
2	0	0	0	0	0	0	0	2	3	2	3	2	2	1	1	1	1
3	0	0	0	1	3	1	1	6	5	5	6	4	4	4	3	2	1
4	0	0	0	1	5	2	2	10	9	9	11	9	7	7	5	2	0
5	0	1	2	2	7	3	3	12	12	13	14	9	9	9	5	3	0
6	1	1	3	3	8	1	0	13	15	15	15	10	8	9	8	4	1
7	1	1	3	3	8	1	0	13	15	15	15	10	8	9	8	4	1
8	0	0	4	3	9	0	0	15	13	16	14	10	6	8	4	1	1

example is the height map (Table 1). The devices that need to be networked are located on opposite sides of the elevation of the terrain.

As evident from Fig. 3, the height map provides two possible ways to connect two devices (positions 1 and 6) into a network. The strokes in the upper part of the figure indicate the way of avoiding the obstacle from above, the strokes in the lower part—the way of avoiding the obstacle from the side. When traversing from above, the algorithm recommends using 4 repeaters to connect devices, when walking from the side for a given height map—5 repeaters. For a given terrain and location of devices, the most suitable way to place repeaters is bypassing the obstacle from above.

3 Conclusion

This paper represents possible solutions for the problem of configuring a wireless communication network between CPS devices, distributed over vast territories. An algorithm is proposed for calculating the coordinates of the location of repeaters for connecting to a common network of stationary and mobile devices of the CPS in areas with difficult terrain and landscaping. The distinguishing feature of the proposed algorithm is combination of two approaches to solve the problem and to provide communication between two devices: Directly through a variable terrain (buildings, green spaces) and indirectly, bypassing the obstacles.

Further research will be aimed at improving the proposed algorithm for automating workflows and data exchange between robotic devices and distributed sensor networks of CPS. Additional noise immunity of radio modules is also among the key objectives of the study, as well as work on eliminating collisions when relaying data packages.

Acknowledgements The research was performed with the support of RSF №. 20-79-10325.

References

1. Vatamaniuk, I.V., Iakovlev, R.N.: Generalized theoretical models of cyberphysical systems. Izvestiya Yugo-Zapadnogo gosudarstvennogo universiteta. Proc. Southwest State Univ. **23**(6), 161–175 (2019) (in Russian). https://doi.org/10.21869/2223-1560-2019-23-6-161-175
2. Iakovlev, R., Saveliev, A.: Approach to implementation of local navigation of mobile robotic systems in agriculture with the aid of radio modules. Telfor J. **12**(2), 92–97 (2020)
3. Denisov, A.V.: Development of a recommender system for parameter calculation in wireless network of sensor devices. Modeling Optimiz. Inf. Technol. **7**(4) (2019) (in Russian)
4. Denisov, A., Sivchenko, O.: Modeling wireless information exchange between sensors and robotic devices. In: Proceedings of 15th International Conference on Electromechanics and Robotics "Zavalishin's Readings". Springer, Singapore, pp. 317–326 (2021). https://doi.org/10.1007/978-981-15-5580-0_26

5. Permyakov, V.A., Mikhailov, M.S., Malevich, E.S.: Analysis of propagation of electromagnetic waves in difficult conditions by the parabolic equation method. IEEE Trans. Antennas Propag. **67**(4), 2167–2175 (2019)
6. Tikhomirov, A., Omelyanchuk, E., Semenova, A.: Radio wave propagation impact on signal parameters of local positioning systems. In: 2016 IEEE NW Russia Young Researchers in Electrical and Electronic Engineering Conference (EIConRusNW), IEEE, pp. 456–460 (2016)
7. Denisov, A.V., Usina, E.E., Iakovlev, R.N., Strutz, T., Narandzic, M., Guzey, M., Jokisch, O.: Algorithms for radio beacon mesh network establishment for navigation of robotic systems in agriculture. Bullet. MSTU "Stankin" **3**, 57–65 (2019) (in Russian)
8. Denisov, A., Iakovlev, R., Lebedev, I.: Mathematical and algorithmic model for local navigation of mobile platform and UAV using radio beacons. Proc. ICR-2019 **11659**, 53–62 (2019)
9. Denisov, A., Shabanova, A., Sivchenko, O.: Data exchange method for wireless UAV-aided communication in sensor systems and robotic devices. In: International Conference on Interactive Collaborative Robotics, pp. 45–54. Springer (2020)
10. Recommendation ITU-R P.1546. Method for point-to-area predictions for terrestrial services in the frequency range 30–3 000 MHz (in Russian)
11. Ojimamadov, I.T., Daminov, Sh.R.: Distribution radio waves of the digital television broadcasting on the example of the Dushanbe city and nearby areas. Bullet. Tajik Nat. Univ. **2**(42), 14–16 (2018). (in Russian)
12. Nazarov, E.A., Plaksitskiy, A.B.: Construction of a sustainable communication system of the head office of the emergency ministry of Russia in the republic of Adhyege in different geographical conditions. Probl. Ensuring Safety Elimination Conseq. Emerg. Situat. **1**, 137–143 (2017). (in Russian)
13. Ortega, A.J.: Finding repeater placement for P2P wireless links with NLOS in extremely mountainous regions (2018)
14. Leushin, D.A.: Possibility of building a radio communication network in a high-mountainous area of localities. Innov. Sci. **5** (2020) (in Russian)
15. Katerynchuk, I., Rachok, R., Mul, D., Balender, A.: Modelling of radio waves propagation and creation of radio networks using geoinformation systems. In: 2016 13th International Conference on Modern Problems of Radio Engineering, Telecommunications and Computer Science (TCSET). IEEE, 2016, pp. 677–681
16. Malevich, E.S.: Modeling the propagation of radio waves over the earth's surface taking into account the complex relief of the area. In: Radioelectronics, Electrical Engineering and Power Engineering, pp. 58–58 (2018) (in Russian)
17. Tikhomirov, A.V., Omelyanchuk, E.V., Semenova, A.Y., Mikhailov, V.Y.: Hybrid model of radio wave propagation in conditions of difficult terrain, wooded areas and uneven building. Eng. Bullet. Don **4**(51) (2018) (in Russian)
18. Grishko A.K.: Geospatial analysis of electromagnetic fields in radio electronic systems taking into account reflections. Models Syst. Netw. Econ. Technol. Nature Soc. **2**(22) (2017) (in Russian)
19. Malevich, E.S., Mikhailov, M.S., Volkova, A.A., Permyakov, V.A.: The research of radio wave propagation mechanisms in the forest vegetation with the complex landscape. In: 2019 Antennas Design and Measurement International Conference (ADMInC), IEEE, pp. 87–89 (2019)
20. Zheng, X., Wang, H., Wen, H., Guo, J.: Propagation characteristics of electromagnetic waves in mine roadways. J. Appl. Geophys. **173**, 103922 (2020)
21. Kofnov, O.V., Lebedev, E.L., Mikhailenko, A.V.: Computer simulation of the diffraction of millimeter electromagnetic waves to detect internal defects of products made using additive technology. SPIIRAS Proc. **1**(56), 76–94 (2018). (in Russian)
22. Abarykov, V.N., Batoroev A.S.: The influence of vegetation on the scattering and diffraction of radio waves at small grazing angles. Collect. Rep. XXIII All-Russian Sci. Conf. Radio Wave Propag. Yoshkar-Ola **2**, 9–12 (2011) (in Russian)
23. Grudinskaya, G.P.: Propagation of Radio Waves, p. 280. Moscow, Higher School (1975) (in Russian)

Simulation of Trainable Control System for Quadruped Robot

Dmitry Dobrynin

Abstract In this article, we consider a trainable control system for a four-legged robot with 12 degrees of freedom. The control system consists of a training part built using mathematical methods of inverse kinematics and an intelligent learning JSM system. The training part of the system uses a virtual model of a four-legged robot to synthesize movements and works in simulation mode. The trained part of the system is built using the intelligent JSM decision-making method. The trained JSM system, thanks to its high speed, can control the walking robot in real time. A model of the control system is constructed, which is used to generate the gaits of a virtual robot model. The article describes the operation of the control system in training mode and in working mode. Simulation of the robot control system showed a high learning rate. The article considers the variants of static and dynamic stability of the robot. The calculated parameters of the diagonal gait of a four-legged robot are given.

1 Introduction

Currently, there is an increased interest in the creation of walking robots in the world. Interest in this topic has increased due to the development and first steps in the commercialization of walking robots by Boston Dynamics and their analogues, the creation of a number of exoskeletons for medical and industrial purposes.

One of the main reasons to develop a four-legged walking robot is to overcome the lack of mobility of wheeled vehicles on irregular terrains. The ability to traverse uneven or varying terrain at high speeds, turn sharply, and start or stop suddenly are all ordinary aspects of four-legged locomotion for a variety of cursorial mammals. Researches on four-legged robots have been widely carried out [1–3]. A quadruped robot with multi-degrees of freedoms is a high nonlinear system. Firstly, sometimes, the uncertain parameters and the external disturbance apply a very adverse effect on the control performance. Secondly, motion of a quadruped robot under the four legs

D. Dobrynin (✉)
Russian State University for the Humanities, Moscow, Russia

Federal Research Center for Computer Science and Control RAS, Moscow, Russia

A. Ronzhin and V. Shishlakov (eds.), *Electromechanics and Robotics*, Smart Innovation, Systems and Technologies 232, https://doi.org/10.1007/978-981-16-2814-6_14

full support can be regarded as a dynamic system under holonomic constraint, and it makes the control more complicated.

Various methods are used to control walking robots. In [4], a robust control method for generating stable gaits of a four-legged robot is considered. Paper [5] is devoted to the use of fuzzy logic for controlling a walking robot. In paper [6], the application of neural networks for robot control is considered. Papers [7–9] are devoted to the management of walking systems during the implementation of different gaits.

This article discusses the trainable control system of a four-legged robot. A model of the control system is constructed, which is used to generate the gaits of a virtual robot model. The variants of static and dynamic stability are considered. The calculated parameters of the diagonal gait of a four-legged robot are given.

2 Robot Model

We will consider the following mechanism as a robot model (see Fig. 1).

The robot is built like an animal with four legs. The robot's legs protrude downward. This design allows you to reduce the necessary moments of the drives in comparison with the insect-like arrangement of the legs. Note that this design reduces the area of the reference polygon and thereby reduces the stability. The robot's knees look forward.

Each leg of the robot has three degrees of freedom. The M_1 drive moves the robot's leg perpendicular to its body. This drive is necessary for making turns. The M_2 drive lifts the entire leg. The M_3 drive bends the leg at the knee. The support sole of the foot rests on the surface on which the robot moves. The surface acts on the sole with a force of F_y. This is the reaction force of the support. The sum of all the reaction forces of the foot support is equal to the weight of the robot.

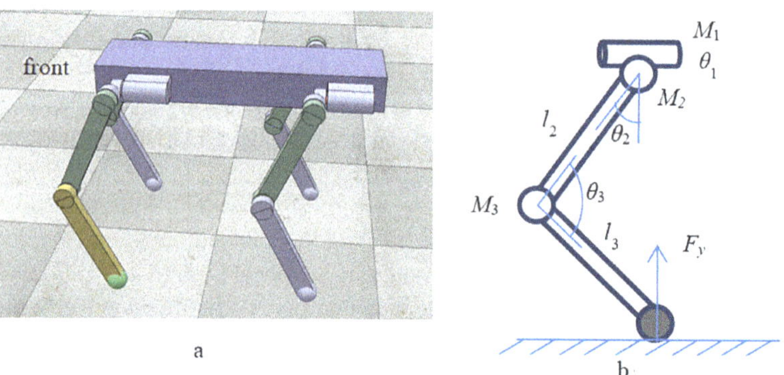

Fig. 1 **a** Model of robot; **b** model of robot leg

3 Trainable Control System

To control the robot, a control system is used, the block diagram of which is shown in Fig. 2. The system consists of two large parts—a training system that uses mathematical methods to solve inverse kinematics problems and a learning intelligent system.

In the training system, there are blocks for calculating the center of mass, calculating the reference polygon, the reaction forces of the support, and the leg movement shaper. The training system cannot work in real time, because it uses methods that require large computational resources and long calculations. The calculated drive control parameters are used to train the second system. It is constructed using the intelligent JSM method [10].The JSM-trained system has high performance and can control the robot in real time.

The trained JSM system contains blocks for analyzing the reaction forces of the support, stability analysis, and the block for forming the trajectories of the robot's legs. The operating modes of the entire system are set by the phase generator, which determines the sequence of movement of the robot's legs.

JSM-the method of automatic generation of hypotheses is a theory of automated reasoning and a way of presenting knowledge for solving forecasting problems in conditions of incomplete information. The JSM method was developed by Professor V. K. Finn in the 70s of the twentieth century. The JSM method has been successfully applied to extract knowledge in such poorly structured areas as chemical compound toxicity prediction, medical diagnostics, technical diagnostics, and others.

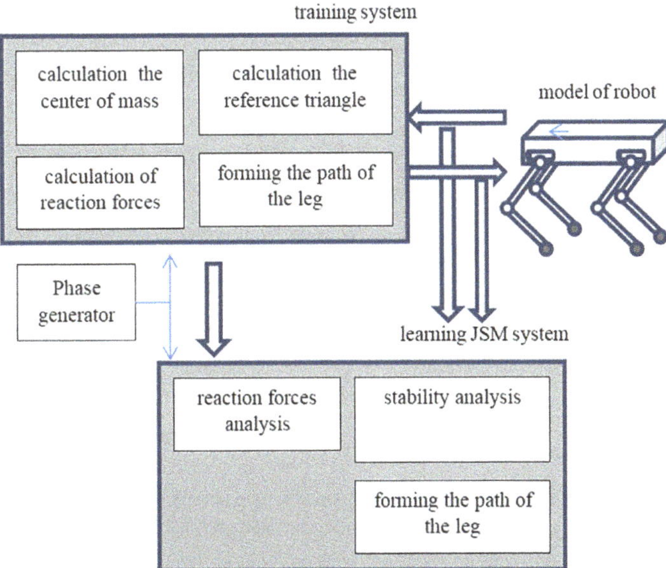

Fig. 2 Trainable robot control system

For robotics, the JSM method was not previously used due to a number of reasons. The JSM method refers to the methods of artificial intelligence, along with such well-known methods as neural networks, decision trees, and rule-based systems. This work is an attempt by the author to apply this method to build a trainable control system for a walking robot.

The classical JSM method [11] works with a closed set of source examples, which is formed by an expert and forms a knowledge base. Each example is described by a set of elementary features and the presence (or absence) of a target property. Using special logical procedures from this knowledge base, the JSM system obtains hypotheses that explain the properties of the original examples due to the presence or, conversely, the absence of a certain set of features in the structure of the examples. Thus, the JSM system selects essential sets of features from the initial information in the knowledge base, i.e., it performs automatic classification. The JSM method is successfully applied in those areas of knowledge where the example can be represented as a set (or tuple) of elementary features.

In contrast to the classical JSM method, which works with a closed set of source examples and their predefined properties, the dynamic JSM method allows you to work in an open environment with an unknown number of examples in advance [10].

Dynamic JSM operates in two modes:

- learning mode (see Fig. 3), when the fact base is filled in (a lot of training examples) and hypotheses are generated that make up the knowledge base;
- operating mode (see Fig. 4), when previously obtained hypotheses are used to generate control signals.

A set of training examples is a set of pairs of the form

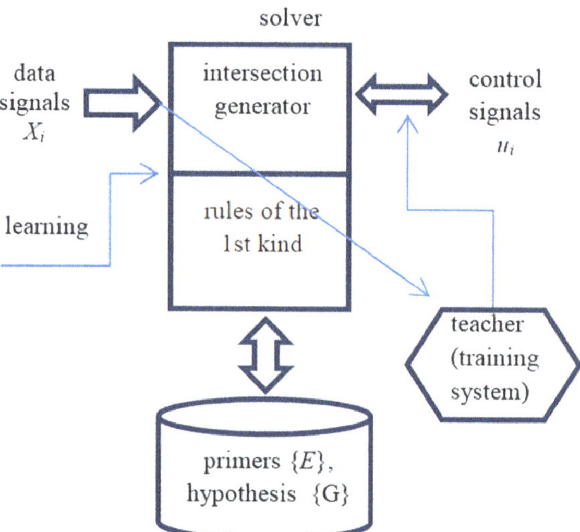

Fig. 3 Learning mode of JSM system

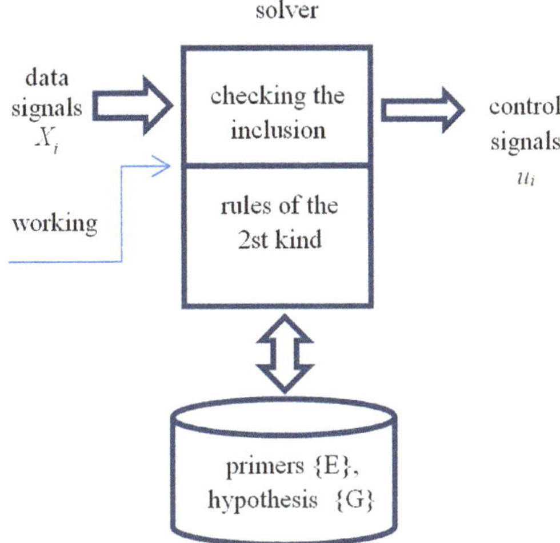

Fig. 4 Working mode of JSM system

$$E = \{e_i\} = \{(X_i, u_i)\},$$

where X_i is the coordinate vector of the links,

u_i—control vector (state of actuators, such as servos).

Hypotheses are represented as a set of pairs of the form:

$$G = \{g_i\} = \{x_i, y_i\},$$

where x_i is part of the coordinate vector of the links,

y_i—the required control vector (control of actuators).

In the training mode (see Fig. 3), an external control system—the so-called «teacher»—is used to generate training examples. This system receives input information from position sensors and generates control signals necessary for the correct movement of the robot's foot drives. The set of coordinates of the links and the control actions developed for them determines one training example. This example is checked for uniqueness and entered by the JSM system in the fact database. After entering each new example in the set of training examples, a hypothesis search is performed. Additional restrictions may be imposed on the obtained hypotheses, for example, a ban on counterexamples, when a positive hypothesis should not be embedded in negative examples and vice versa. These limitations are determined by the JSM method used [12].

The resulting set of hypotheses will contain all possible intersections (common parts) of the training examples. Further, among them, the minimum hypotheses are selected, that is, those that are embedded in the rest. Thus, the number of "useful" hypotheses is sharply reduced. The obtained minimum hypotheses are checked for uniqueness and entered into the knowledge base.

The learning process of the JSM system itself is as follows:

- the training system receives information about the position of the links of the robot's legs and develops control actions-controls the drives. The calculation is carried out in simulation mode, since the models are multiparametric. After the calculations are completed, this information is fed to the JSM system at each elementary step. The JSM system forms a "training example," which it enters into the fact base. If there is already such an example in this fact base, then the solution was found in the previous steps;
- if a new training example appears that has not previously been found in the fact base, then it is passed to the JSM solver, which uses it to form a new hypothesis. If the resulting hypothesis meets the consistency criteria, it is added to the hypothesis database;
- updating the database of facts and obtaining new hypotheses is carried out as long as the training mode works.

After the end of the training mode, the JSM system has a set of hypotheses that are later used for the operation of the trained control system. This means that no new information is received at the input of the training system.

To form a JSM hypothesis, the system requires at least two examples. As a result, the learning rate is high. In the course of testing the system, it was enough to go through two or three step cycles for the system to learn and be ready for operation.

In the operating mode (see Fig. 4), the JSM system receives the input coordinates of the links in the current position, from which the test vector is formed. The decision is made by checking the embedding of hypotheses in this vector. If a hypothesis is embedded in the test vector of link coordinates, then the robot must act in accordance with it. If no hypothesis is found, then this is an unknown state for which you need to generate a control stop signal.

4 Stability of the Robot

The main requirement for ensuring static stability of the robot is that the projection of the center of mass should lie inside the polygon formed by the points of contact of the robot's legs with the support surface. Let us consider various variants of static stability (see Fig. 5).

In the case where the robot rests on the surface with four legs, the reference polygon is a quadrilateral (see Fig. 5a). Let the projection of the robot's center of mass M_g be located in the center of this quadrilateral. This position of the robot's center of mass is statically stable and the robot does not tip over. This case corresponds to a stationary robot.

If the robot lifts one leg to move it, the situation changes. As can be seen from Fig. 5b, the reference polygon turns into a triangle. The projection of the center of mass M_g turns out to be located on the long side of this triangle. The upper case of Fig. 5b corresponds to raised leg 1, the lower case corresponds to raised leg 3. It is

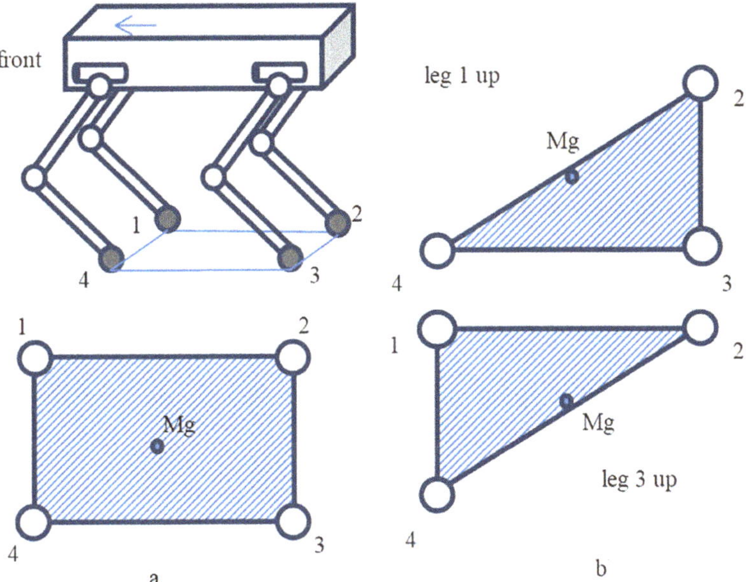

Fig. 5 **a** Static stability of the robot with four points of support; **b** static stability of the robot with three points of support

obvious that the position of the robot is unstable. A small external disturbance can disrupt the stability of the robot.

5 Performing a Step by the Robot

When the robot moves on the surface, its legs are alternately detached from it to transfer the fulcrum (Fig. 6a).

Let the step length be L and the step height be h (see Fig. 6a). The robot body is located at a height H above the surface. In the section of the trajectory 1-2, the leg rests on the surface. The reaction force of the support F_y must remain constant. To do this, the torques of the drives M_1, M_2 and M_3 must be changed according to the corresponding law:

$$\begin{aligned} F_y &= \text{const}, \\ M_1 &= f_1(\theta_1, \theta_2, \theta_3, H, l_2, l_3), \\ M_2 &= f_2(\theta_1, \theta_2, \theta_3, H, l_2, l_3), \\ M_3 &= f_3(\theta_1, \theta_2, \theta_3, H, l_2, l_3) \end{aligned}$$

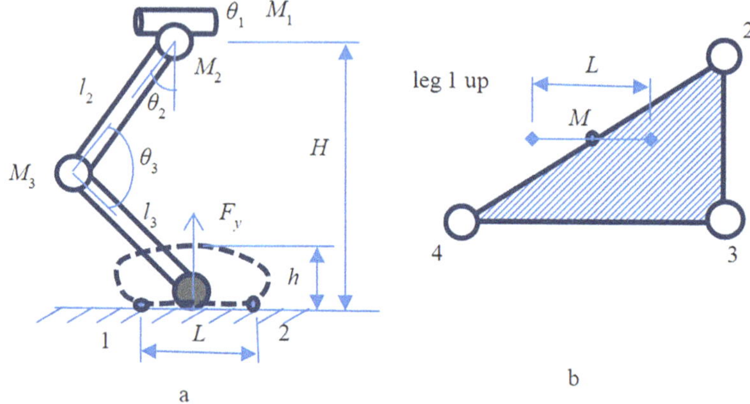

Fig. 6 **a** Step by leg; **b** dynamical stability

In section 2-1, the leg rises above the support surface and is carried forward. The height of the leg lift above the surface is h. The path of movement of the legs can be chosen arbitrarily. The leg transfer time should be no more than one quarter of the step time. Under this condition, the robot will always rest on three legs.

As shown in Fig. 6b, when the robot moves, the projection of the center of mass moves inside the reference triangle and can extend beyond it. This is because the robot body is moving forward at a certain speed. In this case, the robot rests on three legs and the support triangle remains stationary relative to the support. The case in Fig. 6b corresponds to the transfer of leg 1.

When the point of projection of the center of mass goes beyond the reference triangle, the robot begins to roll over. However, due to the large mass of the robots body, the rollover process is much slower than the leg transfer. After transferring leg 1, the reference triangle becomes different, as shown in Fig. 5b. At the same time, stability is restored. When moving the other legs of the robot, stability is restored in the same way.

One of the possible robot gaits is the following gait:

- transfer of the front right leg 1 (phase 2-1), the remaining legs are in the support phase 1-2;
- transfer of the rear left leg 3;
- transfer of the front left leg 4;
- transfer of the rear right leg 2.

This gait is named as diagonal. The transfer of the robot's legs takes place first on one diagonal, then on the other. During the leg transfer, the robot temporarily loses stability for a short time. Stability is restored after the end of the leg transfer.

Figure 7a shows the dependence of the drive angles M_2 and M_3 on the position of the touch point for the following parameters:

$$l_2 = 0.3\,\text{m}, l_3 = 0.3\,\text{m}, L = 0.2\,\text{m},$$

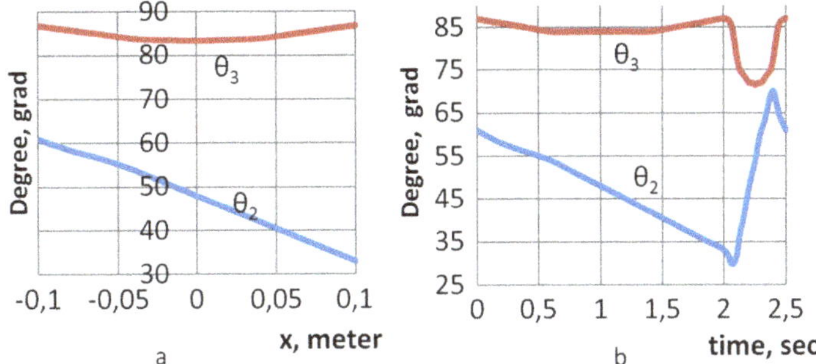

Fig. 7 **a** Dependence of the angles on the position of the touch point; **b** the dependence of the angles on the time when making a step

$$H = 0.4\,\mathrm{m},\ h = 0.05\,\mathrm{m};$$
$$x = [-0.1 \ldots 0.1]\mathrm{m}.$$

At $x = 0$, the tangent point is exactly under the axis of the drive M_2. From Fig. 7a, it can be seen that the angle θ_2 changes almost linearly, and the angle θ_3 changes slightly.

The time dependence of the angles for the full step is shown in Fig. 7b. The segment $t = [0, 2.0]$ s corresponds to the support phase of leg 1-2. The segment $t = [2.0, 2.5]$ s corresponds to the transfer of the foot to the front point of the reference segment 2-1. Figure 7b clearly shows that the greatest speed of movement of the leg corresponds to the transfer phase 2-1.

Cubic splines are used to approximate the dependence of angle changes in the JSM system. In the support phase, the drive angles change quite smoothly, so the cubic spline describes the initial dependence with high accuracy. Sharp changes in angles (see Fig. 7a) occur in the transition zone from the support phase to the leg transfer phase. In such transitions, the JSM system uses cubic splines with other coefficients to approximate the original dependence as accurately as possible. Note that the trajectory of the leg transfer is not critical for the nature of the gait, so high accuracy of approximation is not required here.

Simulation of the robot's gait shows that the robot's body sways when walking. The amplitude of the swaying is small, so it does not interfere with the movement of the robot.

6 Conclusions

This article presents a trainable control system for a four-legged walking robot. The training part of the system is built using mathematical methods of inverse kinematics. It uses complex equations to solve problems of generating the trajectories of the robot's legs. The speed of calculating motion parameters is low, so the training part works in simulation mode and controls the virtual robot. The calculated motion parameters are used to train the intelligent JSM system. The rained JSM system can control the robot in real time due to its high speed. Simulation of the control system showed a high learning rate of such a system.

The gait of a four-legged animal-type robot contains zones of instability. In static mode, the robot is stable if it rests on all four legs. With a diagonal gait, such a robot must quickly pass through areas of instability in order to maintain balance.

Modeling of the time dependence of the drive angles during the step shows that in the reference phase, a high accuracy of approximation of the initial trajectory of the foot can be achieved. High speed movement of the actuators is required in the leg transfer phase.

References

1. Shin, J.H., Park, K.B., Kim, S.W., Lee, J.J.: Robust adaptive control for robot manipulators using regressor-based form. Proc. IEEE Int. Conf. Syst. Man Cybern. IEEE **3**, 2063–2068 (1994)
2. Hirose, S., Kikuchi, H., Umetani, Y.: Standard circular gait of a quadruped walking vehicle. Adv. Robot. **1**(2), 143–164 (1986)
3. Jindrich, D.L., Full, R.J.: Many-legged maneuverability: dynamics of turning in hexapods. J. Exp. Biol. **202**(12), 1603–1623 (1999)
4. Li, K., Wen, R.: Robust control of a walking robot system and controller design. Proc. Eng. **174**, 947–955 (2017)
5. Stepanenko, Y., Su, C.Y.: Variable structure control of robot manipulators with nonlinear sliding manifolds. Int. J. Control **58**(2), 285–300 (1993)
6. Miller, W.T.: Real-time application of neural networks for sensor-based control of robots with vision. IEEE Trans. Syst. Man Cybern. **19**(4), 825–831 (1989)
7. Beranek, R., Ahmadi, M.: A Learning behavior based controller for maintaining balance in robotic locomotion. J. Intell. Rob. Syst. **82**(2), 189–205 (2016)
8. Jatsun, S.F. et al.: Control the movement of the exoskeleton of lower limbs when walking. In: Yugo-Zapadnyy universitet: monographiya. Kursk, p. 185 (2016)
9. Jatsun, S., Savin, S., Yatsun, A., Gaponov, I.: Study on a two-staged control of a lower-limb exoskeleton performing standing-up motion from a chair. Robot Intell. Technol. Appl. **4**, 113–122 (2016)
10. Dobrynin, D.A.: Dynamic JSM-method in the problem of intelligent robot control. In: Tenth National Conference on Artificial Intelligence CII-2006, Obninsk, Proceedings of the Conference, vol. 2. Fizmatlit (2006)
11. Finn, V.K.: Automatic generation of hypotheses in intelligent systems, p. 528. Book House "LIBROKOM", Moscow (2009)
12. Finn V.K.: Plausible reasoning in JSM type intelligent systems. Res. Sci. Technol. Inform. **15** (1991)

Hardware/Software Architecture for Research of Control Algorithms of a Quadcopter in the Presence of External Wind Loads

Sergey Jatsun, Oksana Emelyanova, Petr Bezmen, Andres Santiago Martinez Leon, and Luis Miguel Mosquera Morocho

Abstract The paper presents a hardware/software architecture for the study of algorithms for controlling an unmanned aerial vehicle-type quadcopter in the presence of wind loads; an algorithm for the stabilization of the position of the center of mass of the quadcopter has been implemented and tested on an experimental model. The purpose of the presented hardware/software architecture is a study of the UAV control system parameters influence on the accuracy of the positioning of the vehicle in laboratory conditions under the influence of wind loads, as well as the movement of the vehicle along the specified points of the trajectory based on inertial sensors. The article discusses about the necessary hardware and software elements for reaching the stabilization of the position of the quadcopter in the hover mode based on data obtained from sensors. As the result of the study, the theoretical and experimental data are compared and the permissible range of the positioning error of the device is determined.

1 Introduction

Expert assessment of the current state of robotics shows that there are problems in the issues of scientific and technical reserve in relation to complexes with unmanned aerial vehicles (UAVs), one of which is the development of methods and algorithms for autonomous control of UAVs [1–5]. There are problematic aspects of designing artificial intelligence systems as a task of synthesizing the behavior of an autonomous agent, for example, in the tasks of aerial monitoring of endemic animals in the mountainous area of the Andean Eastern Cordillera (Ecuador), where the slopes are windward throughout (affected by precipitation and wind loads) and deeply dissected by rivers. In conditions that are difficult for land vehicles and humans to access, it is

S. Jatsun · O. Emelyanova · P. Bezmen · A. S. Martinez Leon (✉) · L. M. Mosquera Morocho
Department of Mechanics, Mechatronics and Robotics, South-West State University, 94, 50 Let Oktyabrya Str., 305040 Kursk, Russia

S. Jatsun
e-mail: newteormeh@inbox.ru

A. Ronzhin and V. Shishlakov (eds.), *Electromechanics and Robotics*, Smart Innovation, Systems and Technologies 232, https://doi.org/10.1007/978-981-16-2814-6_15

important to obtain accurate information about the habitat and population of animals such as mountain tapir, vicuna, deer, and other inhabitants of the high Andes to preserve the unique biological diversity in Ecuador [6, 7]. This is a time-consuming process for observers, so it becomes important to collect static information for a long time in the conditions of unpredictable behavior of the air environment and the periodic absence of visual contact between the operator and the aircraft. Therefore, one of the main tasks is to analyze and design a control system to suppress the impact of such influences, since this negatively affects the static and dynamic stability of the UAV.

The best approach for this purpose is a multirotor vehicle of the quadcopter type, which allows to implement vertical take-off and landing from almost any surface, hovering over objects for a better survey of the site, and the speed of horizontal movement due to the maneuverable capabilities of this UAV configuration type.

Known models of quadcopters allow the device to stay in the air for about 20–30 min [7–9]. However, when performing aerial monitoring, there is a need for a longer stay in the air for a thorough examination. In this regard, it became necessary to conduct research on the creation of a promising model of a quadcopter, which provides a longer autonomous stay in the air and increases the accuracy of the binding of observed objects in the presence of wind loads and humidity.

The peculiarity of this quadcopter experimental model is that it is a hardware and software architecture performed and assembled in the Laboratory of Mechanics, Mechatronics, and Robotics of the SWSU. The model is equipped with various navigation systems that combine the use of inertial, correlation-extreme, correlation-type methods of remote, and autonomous piloting to perform a flight task—autonomous movement along an established route points, stabilization of the device under the action of wind disturbances, remote control of the payload by one operator [4, 10–12].

The purpose of this article is to study the parameters of the control system (CS) to ensure the accuracy of the positioning of the vehicle in laboratory conditions under the influence of wind loads, as well as the movement of the vehicle along the specified points of the trajectory based on inertial sensors.

This paper is organized as follows: Sect. 2 presents the mathematical modeling of the dynamics of the UAV; Sect. 3 presents a brief description of the hardware/software architecture and shows the main results obtained during indoor experiments; Sect. 4 shows the structure of the implemented control system; Sect. 5 concludes this purpose presenting the main validation results obtained during outdoor test performance.

2 Mathematical Modeling of the Dynamics of the UAV

The position of the center of mass of the convertiplane-quadcopter in space is characterized by the X_0, Y_0,Z_0 coordinates in a global Cartesian coordinate system and the three angles of rotation φ, θ, ψ—roll, pitch, and yaw angles around the axes of the local coordinate system $CX_1Y_1Z_1$ [3, 13–15] (Fig. 1). The body of the device is modeled as a solid body with six degrees of freedom, the motion of which can be

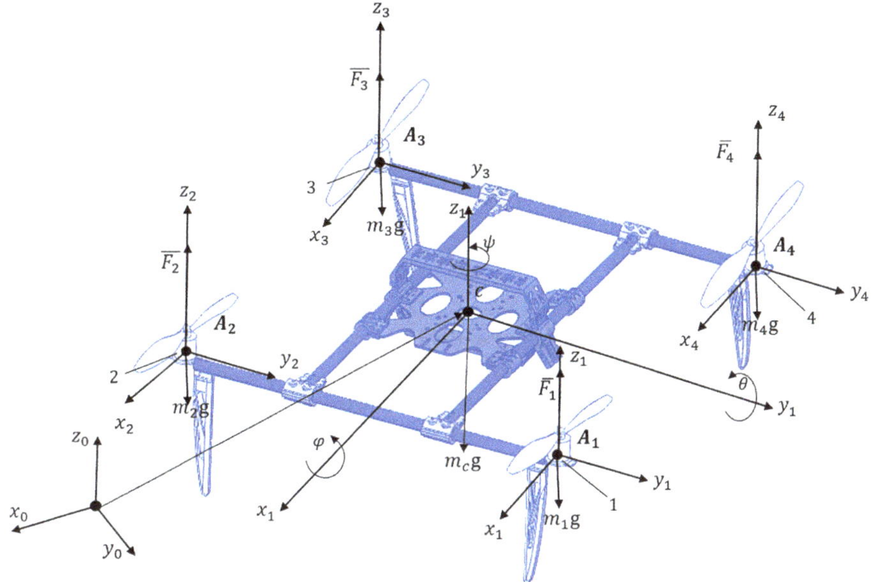

Fig. 1 Design scheme of the aerial vehicle taking into account external forces: $OX_0Y_0Z_0$—global coordinate system; $CX_1Y_1Z_1$—local coordinate system; F_i—traction forces of the rotors; m_cg—gravity force of the quadcopter

seen as a set of translational motion, determined by the movement of an arbitrary point of the body (pole), and the movement of the body around that point as a fixed one. As a pole, the point C—center of mass is taken. Since the research takes place indoors, the influence of the external environment will be extremely small, which does not show strong disturbance during flight.

During the flight, the following forces are applied to the aerial vehicle: the gravity of the body $m_c\overline{g}$ and the rotors' screws $m_i\overline{g}$ attached at points C and A_i, respectively, the aerodynamic forces of the rotors $\overline{F_i}$, attached to the mass centers of the rotors A_i, $(i = 1 - 4)$ [3].

The main vector of the aerodynamic forces of the rotor, due to the smallness of the coefficients of the longitudinal and transverse forces, is equal to the thrust force [14, 15]:

$$\begin{cases} F_i = \begin{bmatrix} F_{iX_1} & F_{iY_1} & F_{iZ_1} \end{bmatrix}^T = k_T\begin{bmatrix} 0 \; 0 \; \omega_{iZ_1}^2 \end{bmatrix}^T \\ k_T = \frac{c_T \rho F_H R^2}{2} \end{cases},$$

where

F_i—projection of the traction force on the axis of the mobile coordinate system;
k_T—aerodynamic component;
C_T—lift coefficient;

F_H—surface area swept by the propeller blades;
R—rotor radius;
ω_i—angular velocity of rotation of the i-th rotor.
The sum of all external forces on the axis $^{(0)}$ of the coordinate system:

$$\sum \overline{F_i}^{(0)} = T_{10} \sum_{i=1}^{4} \overline{F_i}^{(1)},$$

where
T_{10}—matrix of rotation of the local coordinate system [3, 12].
When the quadcopter propellers rotate, there is a torque moment M_{KP}, created by the traction forces F_i of the main propellers. This moment is directed in the side opposite to the rotation of the propellers and is proportional to the angular velocity of the propeller [1, 3, 13]:

$$M_{\mathrm{KP}} = \frac{m_{\mathrm{KP}}\rho}{2} F_H (\omega R)^2,$$

where
$m_{\mathrm{Kp}} = m_{\mathrm{P}} + m_{\mathrm{u}}$—torque coefficient;
m_{P}—torque coefficient, dependent on the profile resistance;
m_{u}—torque coefficient, dependent on the inductive resistance.
When constructing a complete system of equations, it is necessary to take into account the gravity force:

$$G = \begin{bmatrix} 0 \; 0 \; -G \end{bmatrix}^T$$

and the resistance force of the air medium, the vector of which is defined as:

$$\bar{R}_c^{(0)} = \begin{bmatrix} R_{C_x}^{(0)} \; R_{C_y}^{(0)} \end{bmatrix}^T, \; \bar{R}_c^{(0)} = -(\mu\bar{\upsilon} + \bar{\upsilon}^T k \bar{\upsilon}),$$

where μ, k—matrices of empirical resistance coefficients, depending on the geometric characteristics of the attack angle of the main propellers, $\bar{\upsilon}$—relative velocity of the center of mass of the device, which is the difference between the velocity vectors of the center of mass $\bar{\upsilon}_{\mathrm{C}}$ of the coordinate system OXZ and the velocity vector of the air flow $\bar{\upsilon}_B$:

$$\bar{\upsilon} = \bar{\upsilon}_{\mathrm{C}} - \bar{\upsilon}_B.$$

The law of change of the air flow is a random function of time, which can be represented as:

$$\upsilon_B(t) = \upsilon_0 + \upsilon_B \mathbf{sin}\Omega t,$$

where υ_B, Ω—random parameters of the amplitude and frequency of the wind load.

Taking into account the symmetry of the quadcopter and on the basis of the issues of mathematical modeling and research of the spatial motion of the convertiplane along the trajectory previously discussed in the articles [3, 4, 13, 16], the system of nonlinear differential equations will have the form:

$$\begin{aligned}
\dot{\upsilon}_C^x &= \left[(\sin\psi\sin\phi + \cos\psi\cos\phi\sin\theta)\cdot\sum F_i + R_x\right]/m;\\
\dot{\upsilon}_C^y &= \left[(\cos\phi\sin\psi\sin\theta - \cos\psi\sin\phi)\cdot\sum F_i + R_y\right]/m;\\
\dot{\upsilon}_C^z &= \left[(\cos\phi\cos\theta)\cdot\sum F_i + R_z\right]/m - g;\\
\dot{\omega}_{X_1} &= \left[\omega_{Y_1}\omega_{Z_1}\left(J_i^{Z_1} - J^{Y_1}\right) + \omega_{Y_1}\sum J_i^z\omega_i + +M_{X1}^C\right]/J^{X_1};\\
\dot{\omega}_{Y_1} &= \left[\omega_{X_1}\omega_{Z_1}\left(J^{X_1} - J_i^{Z_1}\right) - \omega_{X_1}\sum J_i^z\omega_i + M_{Y1}^C\right]/J^{Y_1};\\
\dot{\omega}_{Z_1} &= \left[J_i^z\dot{\omega}_i + \omega_{X_1}\omega_{Y_1}\left(J^{Y_1} - J^{X_1}\right) + M_{Z1}^C\right]/J^{Z_1}.
\end{aligned} \tag{1}$$

3 Description of the Hardware and Software Architecture

Structurally, the hardware and software architecture consist of a UAV-type quadcopter, which receives the control signals form a control unit. This control unit is the core of the stabilization and control position algorithm. The algorithm is processed and computed by the use of an Ardupilot microcontroller. The sensors onboard read the information coming from the measuring unit and determine the state of the system during the flight. The control unit block uses this data to compute and send the required PWM control signals to the quadcopter. The communication between the quadcopter and the user is carried out by the use of MavLink communication protocol [5, 9] which works at 915 MHz. Mission Planner software allows to the user to create a desired mission for the quadcopter by creating a trajectory waypoint array (Fig. 2).

At the initial stage, the hardware and software architecture laboratory experiments were conducted in a test bench platform. Thus, the quadcopter is fixed in the space, and the first tests can be carried out in a safe environment. The experiment is divided in three stages: I—take-off, II—trajectory, III—hovering, IV—take-off. According to the dimensions and geometrical characteristics of the test bench, the maximum distance traveled by the quadcopter is around 2 m during the indoor test. The prototype performs the experiment in a cycle of 60 s. During the validation process, the cycle is carried out two times (Fig. 3).

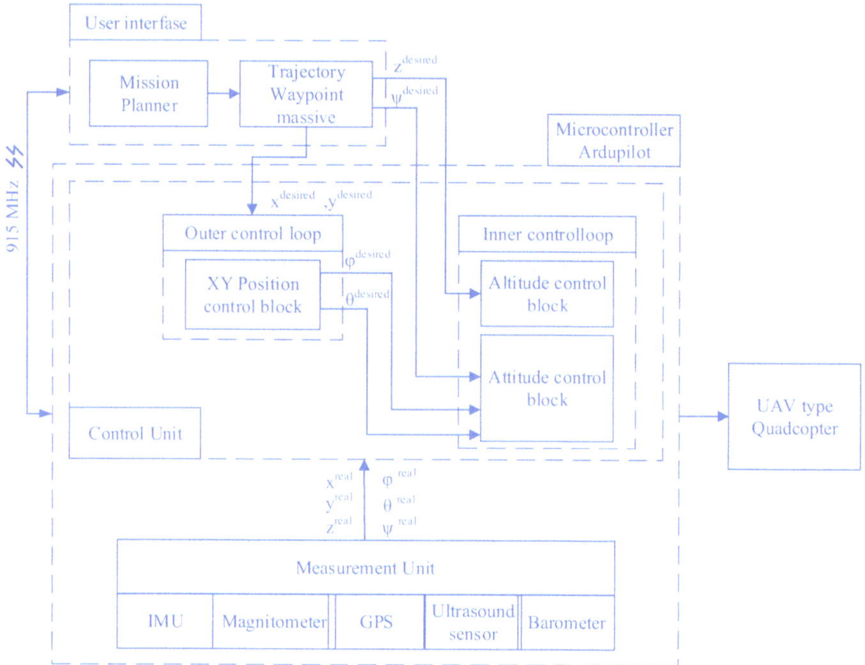

Fig. 2 Structure of the hardware and software architecture

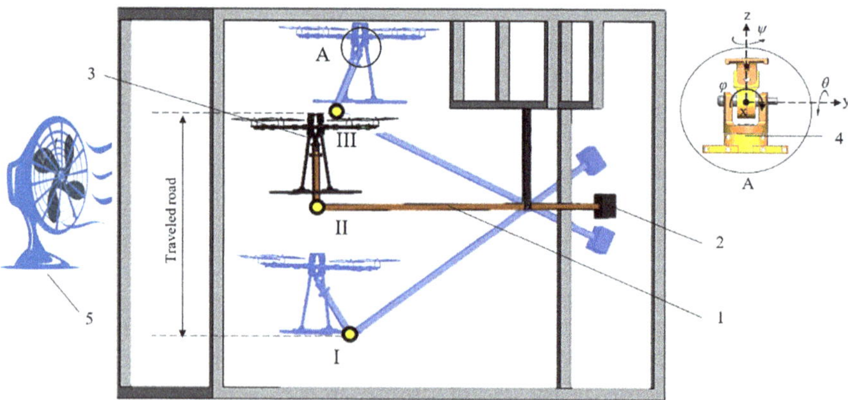

Fig. 3 Experimental test bench: 1 support, 2 load, 3 prototype, 4 universal joint (3DOF), 5 fan

During the realization of the experiment, the following assumptions have been considered: 1—The real start position of the quadcopter is 0.85 m and the final position is 1.75 m relative to the ground; 2—The maximum roll (φ) and pitch (θ) angles range is limited to $\pm$ 30°; 3—The desired orientation of the quadcopter in the space relative to the local coordinate system during the laboratory experiments

Fig. 4 Quadcopter prototype

is zero degrees; 4—External disturbances are created by the use of a fan to recreate the presence of wind load conditions.

The quadcopter we are using for this experiment is a self-developed prototype created in the laboratory of the Department of Mechanics, Mechatronics, and Robotics at Southwest State University (Fig. 4). The airframe of the prototype is completed in a symmetrical X configuration, its mass is equal to 1200 kg, the length of the arms (from the center of mass of the UAV to the center of mass of the rotors) is equivalent to 0.55 m. The lift force generated by each of the motors corresponds to 0.8 g. The autonomy of the prototype is approximately 35 min.

4 Control System Design

With the aim to achieve the control of the quadcopter, the orientation and the altitude of the system must be tracked and controlled. There are different control techniques; however, for practical implementation, performing a robust PID controller is a suitable alternative. The results carried out with the use of a PID algorithm are competitive over other advanced methods and satisfy the requirements. In Fig. 5, a diagram block of the proposed control strategy is presented.

When drawing up a block diagram of the UAV hardware and software architecture, two categories of control signals should be distinguished: input—information obtained from measuring devices and/or from the dynamic model of the control object; output—angular velocities generated by propeller pairs in the form of pulse-width modulation signals (PWM).

Therefore, there are two UAV control loops. The outer control loop analyzes the information obtained from the measuring instruments as well as the data array containing the required coordinates of the trajectory set by the operator. The output data of the outer control loop are the desired values of the roll φ and pitch θ angles, which in turn allow to indirectly control the position of the UAV (X and Y coordinates).

The inner control loop analyzes information from the measuring instruments, in particular information about the roll, pitch, and yaw angles, as well as the Z

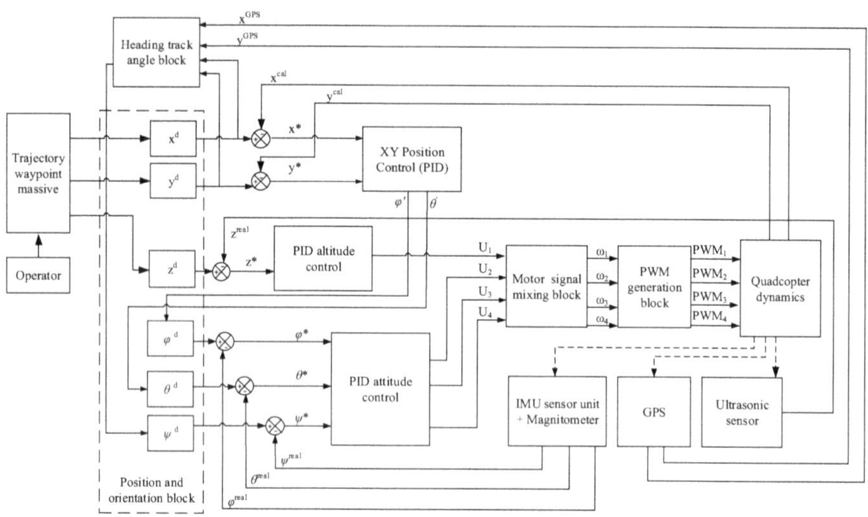

Fig. 5 Quadcopter control block diagram

coordinate. The output data of the inner control loop are the desired altitude and attitude values that stabilize the position of the UAV in space [7, 17, 18].

The quadcopter includes two pairs of motors rotating in a counterclockwise direction CCW (motors 1 and 2) and the remaining two motors rotating clockwise CW (motors 3 and 4). The four control variables U_i are expressed as [7, 13, 17]:

$$\bar{U} = \begin{bmatrix} F_1 + F_2 + F_3 + F_4 \\ (F_2 - F_4) \cdot L \\ (F_3 - F_1) \cdot L \\ -M_1 + M_2 - M_3 + M_4 \end{bmatrix} = \begin{bmatrix} k_T \cdot (\omega_1^2 + \omega_2^2 + \omega_3^2 + \omega_4^2 \\ (\omega_2^2 - \omega_4^2 \cdot k_T \cdot L) \\ (\omega_3^2 - \omega_1^2 \cdot k_T \cdot L) \\ k_M(-\omega_1^2 + \omega_2^2 - \omega_3^2 + \omega_4^2) \end{bmatrix} = \begin{bmatrix} U_1 \\ U_2 \\ U_3 \\ U_4 \end{bmatrix}$$

where l is the distance between the quadcopters center of gravity and propulsors axis, k_T is the thrust coefficient, k_M is the drag coefficient, ω_i is angular velocity of the propeller i.

Then, the vector of desired rotor speeds can be found as follows [7, 13, 17, 18]:

$$\omega_i^2 = \begin{bmatrix} \omega_1^2 \\ \omega_2^2 \\ \omega_3^2 \\ \omega_4^2 \end{bmatrix} = \begin{bmatrix} k_T & k_T & k_T & k_T \\ 0 & k_T \cdot L & 0 & k_T \cdot L \\ -k_T \cdot L & 0 & k_T \cdot L & 0 \\ -l_M & k_m & -k_M & k_M \end{bmatrix}^{-1} \cdot \begin{bmatrix} U_1 \\ U_2 \\ U_3 \\ U_4 \end{bmatrix} = \begin{bmatrix} \frac{1}{4k_T}U_1 - \frac{1}{2k_T \cdot L}U_3 - \frac{1}{4k_M}U_4 \\ \frac{1}{4k_T}U_1 - \frac{1}{2k_T \cdot L}U_2 - \frac{1}{4k_M}U_4 \\ \frac{1}{4k_T}U_1 - \frac{1}{2k_T \cdot L}U_3 - \frac{1}{4k_M}U_4 \\ \frac{1}{4k_T}U_1 - \frac{1}{2k_T \cdot L}U_2 - \frac{1}{4k_M}U_4 \end{bmatrix}.$$

In order to control the XY position of the quadcopter, from Eq. (1) can be singled out the equations related to the linear accelerations of the system on the axes X and Y [7, 18]:

$$\dot{\upsilon}_C^x = U_1(\sin\psi\phi + \cos\psi\theta)/m + R_x/m,$$
$$\dot{\upsilon}_C^y = U_1(\sin\psi\theta - \cos\psi\phi)/m + R_y/m.$$

If assume that the system is near the hover state, the thrust force U_1 has to be equal to the gravitational force $m_c g$. In this case, the component U_1/m can be replaced by the gravitational acceleration g [18, 19]. Also, the components of the resistance forces R_x and R_y can be ignored near hover state. Hence, the matrix notation of the expression, which describes the desired roll φ and pitch θ angles, can be presented as shown:

$$\begin{bmatrix} \varphi^{\text{des}} \\ \theta^{\text{des}} \end{bmatrix} = \frac{1}{g} \begin{bmatrix} \sin\psi & -\cos\psi \\ \cos\psi & \sin\psi \end{bmatrix} \begin{bmatrix} \dot{\upsilon}_C^x \\ \dot{\upsilon}_C^y \end{bmatrix}.$$

With the aim to control the position of the quadcopter, including its heading orientation while performing some trajectory, it becomes necessary to control the bearing angle, i.e., the required yaw angle ψ [11, 20]. The inaccuracy of heading angle orientation ψ_i determination can be denoted by $\Delta\psi$. Thus, during robot movement to a destination point, the robot executes the process of rotation around z-axis, so that the required heading angle ψ^{required} could be calculated as:

$$\psi^{\text{required}} = \psi^{\text{trajectory}} \pm \Delta\psi.$$

The actual heading angle orientation ψ^{actual} is calculated on the basis of the current coordinates of the robot (x_{C} and y_{C}), determined by the GPS, and the coordinates (x_i and y_i) of a destination point, coming from the trajectory waypoint array [11, 20]:

$$\psi^{trajectory} = \begin{cases} \left[\frac{\pi}{2} - \left[\arccos\left[\frac{x_i - x_C}{\left[\sqrt{(x_i - x_C)^2 + (y_i - y_C)^2}\right]}\right]\right]\right], & \text{if } (y_i - y_C) \geq 0, \\ \left[\pi \cdot \text{sign}(x_i - x_C) + \left[\frac{\pi}{2} - \left[\arccos\left[\frac{x_i - x_C}{\left[\sqrt{(x_i - x_C)^2 + (y_i - y_C)^2}\right] \cdot \text{sign}(x_i - x_C)}\right]\right]\right] \cdot (-\text{sign}(x_i - x_C))\right], \\ if\, (y_i - y_C) < 0 \end{cases}$$

where sign is the signum function:

$$\text{sign}(x) = \begin{cases} 1, & \text{if } x \geq 0, \\ -1, & \text{if } x < 0. \end{cases}$$

The criterion of robot movement completion can be denoted by following equations:

$$x_i - x_{\text{C}}| = \Delta x,\ |y_i - y_{\text{C}}| = \Delta y,\ |\varphi_{\text{trajectory}} - \varphi_{\text{real}}| = \Delta\varphi,$$

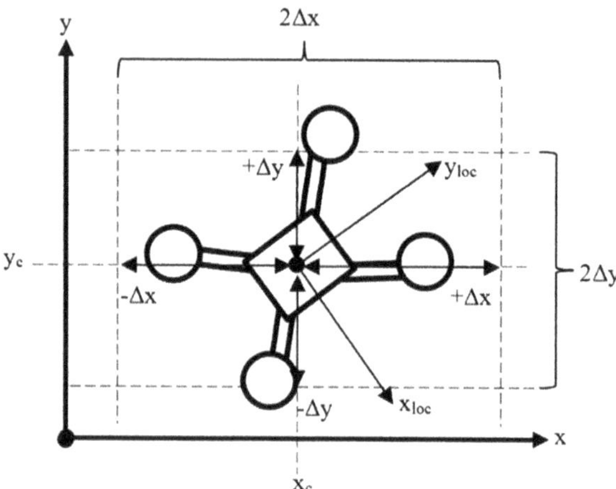

Fig. 6 Representation of the inaccuracies Δx and Δy: OXY—global coordinate system, x_C, y_C—robot coordinates in the global coordinate system, x_{loc} y_{loc}—local coordinate system, C—robot mass center

where x_i, y_i, and $\psi_{\text{trajectory}}$ are destination point coordinates and specified heading angle; x_C, y_C, and φ_{real} are current robot body coordinates and current heading angle; Δx, Δy, and $\Delta\psi$ are determination inaccuracies of coordinates x, y (induced by GPS receiver bias and obstacles for GPS signals), and heading angle ψ (induced by magnetometer and intense magnetic fields), respectively (Fig. 6).

5 Outdoor Test Validation

After working out the processes of stabilization and orientation of the quadcopter on the experimental test bench, an outdoor test for validating the results of the implemented hardware and software architecture has been carried out. When moving the device in space, continuous planning and construction of the trajectory of movement in the form of piecewise linear attacks are carried out (Fig. 7).

The main results obtained from the performed practical experiment are presented in Figs. 7, 8, 9 and 10.

6 Conclusions and Further Work

In this paper, a UAV hardwarel/software architecture for the study of UAV control algorithms with presence wind loads has been proposed, as well as a mathematical model of UAV dynamics, which take into account the presence of an air flow with randomize amplitude and frequency. In the first stage of the study, a stabilization and hovering algorithm has been proposed. As part of the implementation process, an indoor (laboratory conditions) test has been carried out fixing the quadcopter to an experimental test bench platform. In the second part of the study, an outdoor test

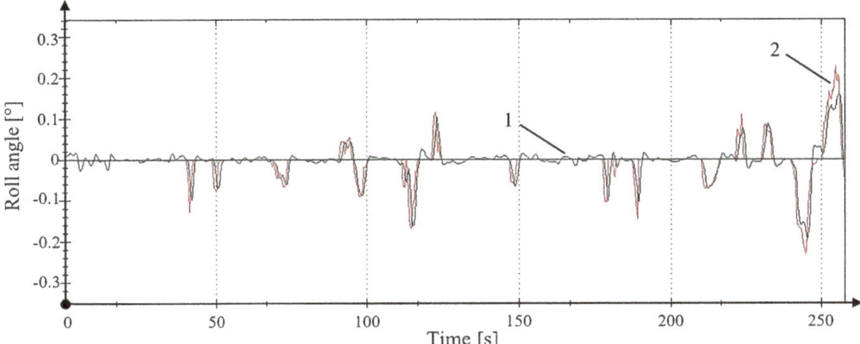

Fig. 7 Quadcopter outdoor test validation, roll angle control: 1 desired roll angle values, 2 real roll angle values

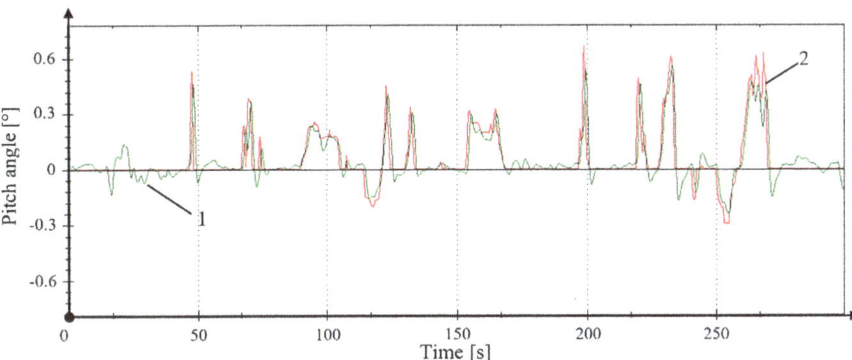

Fig. 8 Quadcopter outdoor test validation, pitch angle control: 1 desired pitch angle values, 2 real pitch angle values

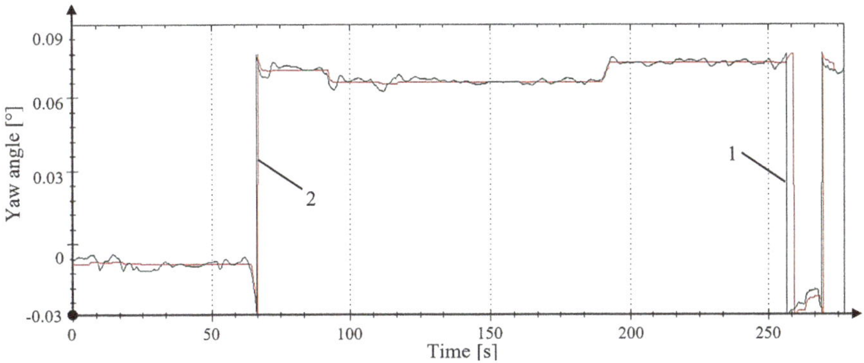

Fig. 9 Quadcopter outdoor test validation, yaw angle control: 1 desired yaw angle values, 2 real yaw angle values

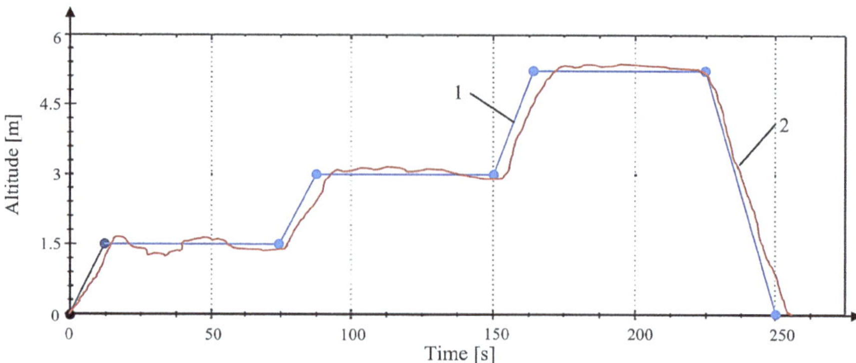

Fig. 10 Quadcopter outdoor test validation, altitude control: 1 desired altitude values, 2 real altitude values

has been performed. The aim of the outdoor test is related to validate the proposed control algorithm experimenting the robustness of the system to external conditions such as wind load influences while moving along a given by the operator piecewise linear trajectory. The deviation of the quadcopter from the given trajectory was about 0.15 m with a systematic shift proportional to the value of the desired speed. The achieved positioning error of the quadcopter relative to the desired trajectory was 0.1–0.2 m.

References

1. Mellinger, D., Kumar, V.: Control and planning for vehicles with uncertainty in dynamics. In: Proceedings of the IEEE International Conference on Robotics and Automation (ICRA), pp. 960–965 (2010)
2. Alves, A.N., Ferreira, M.A.S., Colombini, E.L., da Silva Simões, A.: An evolutionary algorithm for quadcopter trajectory optimization in aerial challenges. In: Proceedings of the IEEE 2020 Latin American Robotics Symposium (LARS), pp. 1–6. IEEE (2020)
3. Emelyanova, O.V., Kazaryan, G.K., Leon, A.M., Jatsun, S.F.: The synthesis of electric drives characteristics of the UAV of "convertiplane-tricopter" type. MATEC Web Conf. **99**, 249–263 (2017)
4. Jatsun, S., Morocho, L.M.M., Emelyanova, O., Leon, A.S.M.: Controlled adaptive flight of a convertiplane type tricopter in conditions of uncertainty for monitoring water areas. In: 2020 International Multi-Conference on Industrial Engineering and Modern Technologies (FarEastCon), pp. 1–7. IEEE (2020)
5. Sa, I., Corke, P.: System identification, estimation and control for a cost effective open-source quadcopter. In: Internal Conference on Robotics and Automation, pp. 2202–2209. IEEE, Minnesota (2012)
6. Flores, A., Scipión, D., Saito, C., Apaza, J., Milla, M.: Unmanned aircraft system for Andean Volcano monitoring and surveillance. In: Proceedings of the 2019 IEEE International Symposium on Safety, Security, and Rescue Robotics (SSRR), pp. 297–302 (2019)

7. Valencia, E.A., Palma, K.A., Changoluisa, I.D., Hidalgo, V.H., Cruz, P.J., Cevallos, C.E., Ayala, P.J., Quisi, D.F., Jara, N.G.: Wetland monitoring through the deployment of an autonomous aerial platform. IOP Conf. Ser. Earth Environ. Sci. **432**(1), 012002 (2019)
8. Doukhi O., Fayjie A.R., Lee D.J.: Intelligent controller design for quad-rotor stabilization in presence of parameter variations. J. Adv. Transp. (2017)
9. Casado R., Bermúdez A.A: Simulation framework for developing autonomous drone navigation systems. J. Electron. **1**(10), 7 (2021)
10. Zenkin, A., Berman, I., Pachkouski, K., Pantiukhin, I., Rzhevskiy, V.: Quadcopter simulation model for research of monitoring tasks. In: Proceedings of the 26th Conference of Open Innovations Association (FRUCT), pp. 449–457. IEEE (2020)
11. Krzysztofik, I., Koruba, Z.: Analysis of quadcopter dynamics during programmed movement under external disturbance. J. Nonlinear Dyn. Control, 177–185 (2020)
12. Jatsun S. et al.: Modeling and control architecture of an autonomous mobile aerial platform for environmental monitoring. In: 2019 International Conference on Information Systems and Computer Science (INCISCOS), pp. 177–182 IEEE. Quito, Ecuador (2019)
13. Jatsun S. et al.: Hovering control algorithm validation for a mobile platform using an experimental test bench. IOP Conf. Ser. Mater. Sci. Eng. **1**(1027), 012008 (2021)
14. Martinez Leon, A.S., Jatsun, S.F., Emelyanova, O.V.: Control of the electric drives of a multirotor system type convertertiplane. J. Fundamental Appl. Probl. Tech. Technol. J. **1**, 83–93 (2020)
15. Jatsun S. et al.: Synthesis of simmechanics model of quadcopter using solidworks CAD translator function. In: Proceedings of 15th International Conference on Electromechanics and Robotics" Zavalishin's Readings", pp. 125–137. Springer, Singapore (2020)
16. Martins, L., et al.: Feedback linearization with zero dynamics stabilization for quadrotor control. J. Intell. Rob. Syst. **1**(101), 1–17 (2021)
17. Bao, N. et al.: Research on attitude controller of quadcopter based on cascade PID control algorithm. In: 2017 IEEE 2nd Information Technology, Networking, Electronic and Automation Control Conference (ITNEC), pp. 1493–1497. IEEE (2017)
18. Suresh, H., Sulficar, A., Desai, V.: Hovering control of a quadcopter using linear and nonlinear techniques. Int. J. Mechatron. Autom. **6**, 120–129 (2018)
19. Dim C. et al.: Novel experiment design for unmanned aerial vehicle controller performance testing. IOP Conf. Ser. Mater. Sci. Eng. **1**(533), 012026 (2019)
20. Jatsun, S.F., Emelyanova, O.V., Martinez Leon, A.S.: Design of an experimental test bench for a UAV type convertiplane. IOP Conf. Ser. Mater. Sci. Eng. **714**(1), 012009 (2020)

Method for Inspecting High-voltage Power Lines Using UAV Based on the RRT Algorithm

Igor Lebedev and Valeriia Izhboldina

Abstract Diagnostics of high-voltage power lines using unmanned aerial vehicles enables to identify problems at the early stages. This method of inspection of power transmission lines allows for aerial imaging in hard-to-reach places in order to further analyze the images obtained. The construction of the optimal route in terms of avoiding obstacles and the time spent on calculating the trajectory itself and flying along it to study power lines directly affects the quality, speed, and results of examining the towers and elements of power lines. The paper proposes a method for examining power lines by detection of damage to the structure and elements of power transmission lines. The method consists of two algorithms: an algorithm for detecting operational altitudes and an algorithm for UAV movement along the power transmission line. In our proposed solution, the developed method of inspection of power lines is not limited to one straight section of power lines. The peculiarity of the method is that the path planning using the RRT algorithm along the power transmission line occurs in two-dimensional space at a given working height. To test the efficiency of the proposed survey method, simulation was performed in the Gazebo software environment. The path search time in our proposed survey method does not exceed 0.3 μs for two sections of power transmission lines with an angular tower and a total path length of 200 m. Moreover, this algorithm allows the UAV to fly around the towers at a safe distance, subject to a predetermined terrain map.

1 Introduction

Ongoing monitoring of high-voltage power lines (PL) can ensure the prevention of emergencies. Power transmission lines are diagnosed in various ways: walking tours; geodetic measurements; aerial photography; airborne laser scanning from manned aircraft; imaging from space [1]. Today, there exist some approaches that can reduce the cost and increase the speed of obtaining information about the condition of PL.

I. Lebedev · V. Izhboldina (✉)
St. Petersburg Federal Research Center of the Russian Academy of Sciences (SPC RAS), St. Petersburg Institute for Informatics and Automation of the Russian Academy of Sciences, 39, 14th Line, 199178 St. Petersburg, Russia

A. Ronzhin and V. Shishlakov (eds.), *Electromechanics and Robotics*, Smart Innovation, Systems and Technologies 232, https://doi.org/10.1007/978-981-16-2814-6_16

One approach is to use unmanned aerial vehicles (UAVs). Partial autonomy is the advantage of employing a UAV in this approach to monitoring PL.

The peculiarities of using UAVs when carrying out flaw detection of PL are presented in [2], where it is assumed that the UAV autonomously flies along pre-calculated feature points and performs multispectral survey. The resulting images are sent to the ground base station via a wireless link for further analysis to draw up a maintenance plan to eliminate the damage.

Finding the optimal route for the study of PL directly affects the quality, speed, and results of the survey of towers and elements of overhead power lines. Many UAVs already have in their onboard computers software and hardware-implemented tools for automating maneuvers and flight modes, as well as a software interface for accessing them. The UAV operator can transmit trivial commands for their execution by radio channel, for example—«fly forward», «gain altitude», «hover at a point». Thus, a relevant problem is the development of software for performing high-level tasks, such as mapping (composition and refinement of terrain maps in real time), localization, and trajectory planning. This paper considers a developed new approach to trajectory planning on the available map or terrain model.

2 Problem of Path Planning for an UAV to Monitor Power Lines

Often, the issue of planning a trajectory for a robotic agent is considered as a problem of finding a path on a graph. In this case, the agent itself is considered a material point. The vertices of the graph correspond to different positions of the agent in environment, the edges correspond to elementary trajectories, i.e., such trajectories, directing the agent along which is considered a trivial task. These can be line segments or curves of a certain radius. Thus, the problem of planning a trajectory is reduced to two subproblems: building a graph that simulates the agent's environment and finding a path on this graph. As graph models of the environment, graphs of regular decomposition (GRD) are often used, in English terminology—grids [3], similar models: metric and topological graphs (MTGraphs) [4], visibility graphs [5], probability terrain schemes [6], etc. A complete and informative overview of graph models used to solve the problem of planning a trajectory on a plane is given in [7]. Graph pathfinding methods are often based on iterative traversal of graph vertices—a principle described back in 1959 by Dijkstra in [8]. All sorts of heuristics are often used (usually based on the geometric properties of the work area and having a simple geometric interpretation) to reduce the search space [9]. Algorithms that implement heuristic search are modifications of the A* algorithm known in the literature on artificial intelligence [10]. There also exist decomposition algorithms for finding a path on a graph, based on dividing the original problem into several potentially easily solvable subproblems (local problems) with the subsequent decomposition of the resulting local solutions into the desired one. Examples of such algorithms are, particularly, R* [11], Hierarchical A* algorithm for MTGraphs (HGA*) [12], etc. Decomposition algorithms are advantageous compared to iterative algorithms in terms of the efficiency of using computing resources, but inferior in the quality of the solutions obtained.

Planning the path of a UAV near PL can also be based on the detection of power line elements [13]. In this case, the onboard computer calculates the further flight trajectory based on the data received from the camera and processed by the neural network. The Faster Region-based Convolutional Neural Network (Faster R-CNN) convolutional neural network is used to detect wires and towers. On the images, the nearest and the next tower is registered, after which the coordinates of a point near the next tower are calculated and the trajectory of movement is constructed. This method ensures the flight of the UAV at a safe distance from the current-carrying elements. But the authors of [13] do not say anything about the completeness of the recorded data on the state of the power transmission line: the state of insulators, fittings, fixtures, and other hard-to-reach places of the power transmission line.

The statement of the problem in this paper is as follows: the current and target positions of the UAV in environment are known, tied to the existing map of the area. The desired flight speed is set. The task of planning the trajectory is to describe a two-dimensional curve, associated with the existing map of the terrain, connecting the starting air and final position of the UAV. Any point on the curve must be passable for the UAV, i.e., the UAV can be located at this point, being at a safe distance from any obstacles. Thus, it is necessary to consider the limitations associated with the physical dimensions of the object under control and its spatial orientation. The proposed method of inspection of power transmission lines provides complete information about hard-to-reach elements of power transmission lines, by tracking components such as towers, wires, and insulators.

3 UAV Trajectory Planning Using the RRT Algorithm

In the problem of UAV path planning, the objective is to find a reliable and optimal path in terms of the criteria described below under given conditions from the starting aerial position to the target point. The criteria for the reliability and optimality of the path are: continuity of the proposed path; absence of excessive curvature of the path; path length; required time to find a path. To develop an algorithm for examining PL with the aid of UAV, the rapidly exploring random tree (RRT) algorithm was chosen, since it does not require large computational resources [14], flexibly configurable for any task [15], and easy to implement [16]. The original RRT algorithm was developed by Stephen Lavelli [17]. RRT is a type of algorithms based on probabilistic roadmap methods (PRM). When working with PRM algorithms, initial and target positions as well as the obstacle map are taken as input. The output data is the trajectory found for the UAV on this map. Using the RRT algorithm, the entire operational space can be divided into two areas: an area occupied by obstacles and an area that does not contain obstacles. The RRT algorithm constructs the trajectory so that all the vertices of the graph through which the UAV passes belong to an obstacle-free region.

One of the disadvantages of this algorithm is its low calculation time performance. The time for finding the trajectory should be short enough to enable the UAV to dynamically rebuild the trajectory if necessary. The second drawback is the curvature of the trajectory and, as a result, the longer distance to be traveled.

To solve the first problem, the endpoint orientation algorithm [18] is used. This algorithm is based on the fact that at each iteration of the tree construction, instead of a random state, with a certain probability either the target point or a random state from the vicinity of the target point can be taken. This solution aims to ensure that the tree mostly "grows" toward the end point [19]. This method also allows you to determine the moment when, to complete the construction of the path, it remains only to connect the top of the tree to the end point.

To eliminate the second drawback, the smoothing algorithm is used to optimize the obtained path [19]. The planned trajectory using the RRT algorithm contains redundant vertices, which cause excessive curvature of the trajectory, which, in turn, causes an increase in the path length. To smooth the trajectory, it is possible to apply interpolation algorithms for the computed trajectory, which enables to get rid of excessive tortuosity. To obtain a smoother trajectory, we will use the Bezier method [20, 21].

In order to develop an algorithm for examining PL, a classic RRT algorithm was chosen, supplemented by an endpoint orientation algorithm in order to reduce the trajectory construction time. And to smooth the found path and reduce its length, the Bezier method was chosen.

4 Development of a Method for UAV-Aided Power Line Examination

The tasks of PL examination are conventionally divided into two types: detecting changes in the geometric parameters of PL and detecting damage to the integrity of power lines. The first type of tasks includes determination of sag arrows, distances to intersected objects, distances between wires of different phases and comparing these values with acceptable values. Solving problems of the second type is aimed at detecting corrosion of wires, cables, tower structure elements, damage to insulators, malfunctions in fasteners and connections of wires and cables, as well as detecting burnt, split elements of towers and insulators. The solution of problems of the first type is given in [2]. To solve problems of the second type, it is necessary to establish key survey points above the PL and on one of the sides of the power line parallel to the wire.

The key points for UAV flight over PL can be the geometric centers of the towers. In this case, the survey takes place at a height that exceeds the height of the PL tower. The survey consists in the movement of the UAV to the next designated key point using the basic algorithms embedded in the flight controller. To carry out such a survey, it suffices to know the starting point, the coordinates of the geometric centers of the towers, and the operating flight height, which must be above the level of the transmission line tower. After the specified parameters are set, the examination is performed automatically.

In the case of a flight from one of the sides of a PL along a wire, it is necessary to plan a route in advance so that the UAV would be at a safe distance from the towers and wires. Here, the problem arises of choosing the optimal algorithm for planning a path, which ensures the construction of a trajectory for the movement of the UAV at such a distance that noise from electromagnetic radiation and the elements of the transmission line do not affect the operation of the UAV. Also, the images obtained from the UAV camera should provide the initial data for detecting the crucial defects. It is assumed that to solve problems of the second type in flight, the operator determines the key points of the flight, and after that, the task is performed without human intervention.

The proposed method for examining power lines consists of two algorithms: an algorithm for determining the number of operational heights and an algorithm, which ensures the motion of UAVs along the power lines. The operational height is the level at which the insulators, fittings, and other important components of the power line subject to examination are located. An algorithm for determining the number of operational heights is necessary in the case when the insulators are located on the tower at different heights. The result of this algorithm will be an array of operational heights.

The starting point $(X_0, Y_0, 0)$ of the UAV and the height of the power line tower H are used as input data for the algorithm for determining the number of operational heights (Fig. 1a), which are set manually by the personnel of the operational field team. Further, the UAV takes off to a height of 1 m, at which a search for a tower is performed on the image obtained from the UAV camera. This height is sufficient to locate the tower in the image obtained from the UAV camera. If the tower is not detected in the image, the operator receives a notification about this so that he can make further decisions. At the next step, the UAV is gaining flight altitude to detect insulators that are principal components in the survey of power lines. When the insulators are found, the operational height counter is incremented by one. The height at which the insulator is found is stored in an array of operational heights. The UAV gains altitude until all the insulators, i.e., all operational height levels will not be detected. This is achieved by comparing the current height of the UAV and the specified height of the transmission line tower H.

The input data for the algorithm of movement along the power line (Fig. 1b) are the coordinates of the end point (X_t, Y_t), the operational height taken from the array of operational heights, and a two-dimensional map of the location of the towers in order to build a path for the UAV to move without collisions using RRT algorithm. First, the UAV takes off to the launch air position at the operating altitude, and the search for an insulator is carried out there. After detecting the insulator, the UAV position is adjusted so that the insulator is located in the center of the image. If the isolator is not found, then the UAV rotates the camera in order to detect the isolator, or the

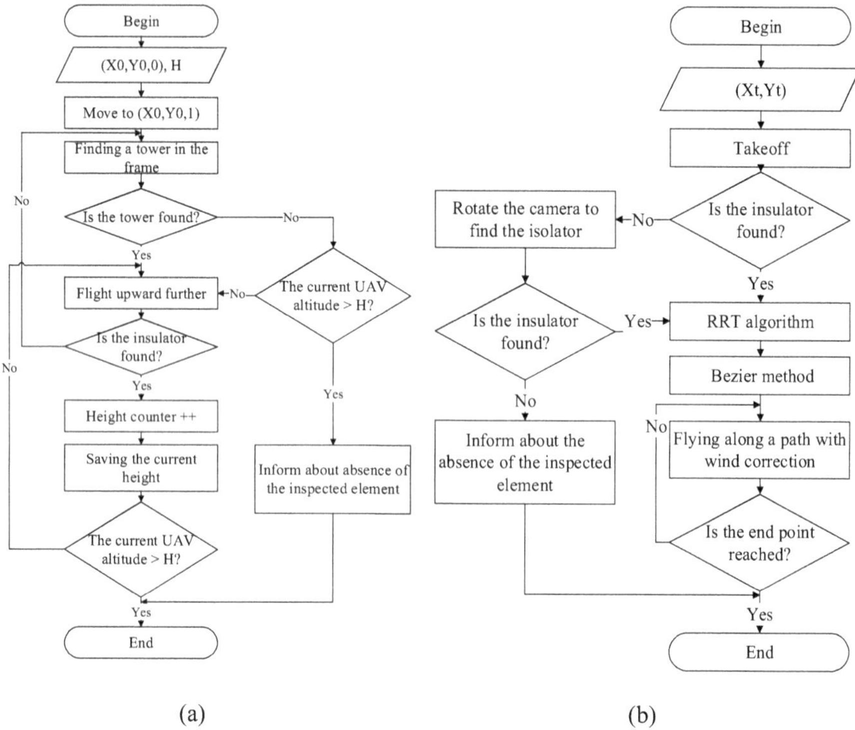

Fig. 1 Flowchart of the power line survey algorithm

operator receives an error message. The issue related to the camera rotation algorithm is not considered in this article. After detecting the isolator, the RRT algorithm is used to plan the trajectory of the UAV. Further, the planned path is smoothed by the Bezier method and sent to the UAV flight controller to move further. In the course of flight missions, photo capture and video recording are performed at certain control points or along the entire path. Flight tasks are associated with the movement of the UAV at given points, take-off, landing, etc. (such tasks are performed on the flight controller), and photo and video imaging are additional workflows that are performed simultaneously with the flight tasks but rely on different onboard equipment (onboard microcomputer). After receiving the trajectory to the flight controller, the UAV starts moving along the power line. At the same time, the position of the UAV in space is being adjusted, which ensures not only the stabilization of the vehicle against wind gusts, but also the retention of the objects under examination in the field of view. If the UAV reaches the end point, the algorithm ends.

In paper [2], it is proposed to detect the wires of power transmission lines for conducting an automated survey. The disadvantage of this solution is that during rotation of the transmission line the camera cannot detect the wires of the next section of the transmission line behind the corner tower. A section of a power transmission line is treated as a structure containing two power transmission towers and wires between them (Fig. 3). Accordingly, the solution in [2] is suitable only for one straight section of the PL. The advantage of the proposed solution is that the work of the developed survey algorithm is not limited to one straight section of the PL. With the help of the proposed method, it is possible to inspect several sections of power lines that follow each other and include angular towers (i.e., turns of the power line). (Fig. 5). A feature of the proposed method is the construction of a trajectory in two-dimensional space. On long routes of power transmission lines, towers of the same height are installed; therefore, we can assume that the UAV operates at the same height, i.e., in the plane where the insulators are located.

5 Simulation Results

To simulate the proposed method, the Gazebo environment was chosen. There is a section of a power transmission line on the stage. The map of the surveyed section of the power transmission line is a standard two-dimensional map with the indicated obstacles on it. To test the proposed method, a standard UAV model "iris" was used. The simulations were carried out on a computer with the following hardware: Intel Core i5-8250U processor (1.6 GHz), memory (RAM) 8 GB DDR4-2400 MHz, nVidia GeForce MX 150 2 GB GDDR5 video card, 256 GB M.2 SSD main hard disk (3300 MB/s—read from disk; 1200 MB/s write to disk).

Figure 2a shows a map of the surveyed section of the power transmission line. This section consists of two towers and overhead power lines above them. On the map,

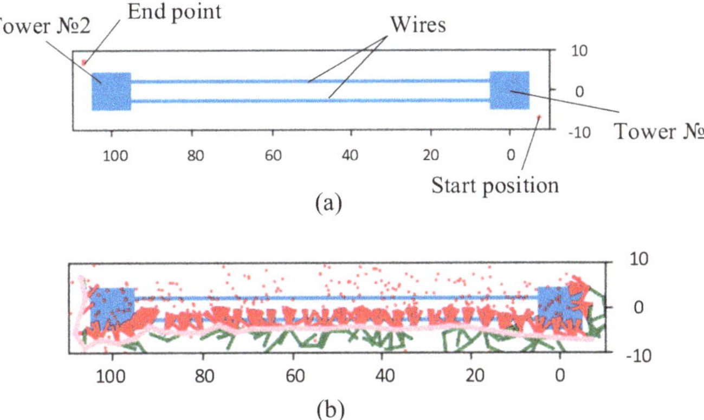

Fig. 2 Map of the surveyed section of the power transmission line: **a** a map with signed elements of the power transmission line section, **b** a map with the constructed trajectory for the UAVб obtained with the RRT algorithm. All red points are randomly generated; green lines are visited edges of a tree; red lines are edges of a tree that do not fit; pink line is a found path; blue geometric shapes are obstacles

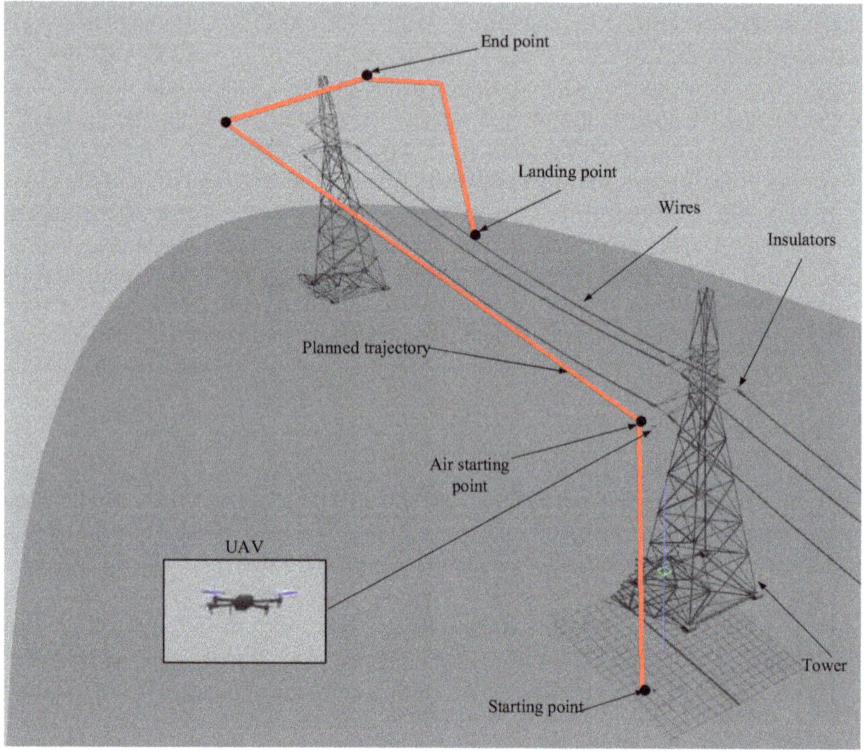

Fig. 3 Test bench in Gazebo modeling environment

the transmission line section includes two squares connected by straight lines, which corresponds to the towers of the transmission line section and wires. The distance between two neighboring towers is 100 m. The starting point has coordinates (– 7; – 7), and the end point is (7; 107).

Following the algorithm of UAV movement along the power line (Fig. 1b), the operator enters the coordinates of the end point, the height of the tower and the obstacle map (Fig. 2a). Once at the height of the insulators, the UAV holds its launch air position for a while. This time is equal to the time spent planning the trajectory. Using the RRT algorithm, the flight trajectory for the UAV is planned from the start point to the end point (Fig. 2b). The trajectory is planned in such a way that the UAV moves at a given distance along the power line. The green lines in Fig. 2b show the path generation tree. The red lines in Fig. 2b show the path leading to collisions with the power transmission line towers or wires. The pink line in Fig. 2b is the found planned trajectory. Then the resulting trajectory is smoothed using the Bezier method and sent to the UAV flight controller for use in the Gazebo simulation environment.

Figure 3 shows a simulated scene consisting of a surveyed section of a power transmission line and an iris UAV with a Logitech c920 camera, modeled in Gazebo. After receiving the planned trajectory, the UAV starts its movement from the starting point. After vertical take-off to the first operating altitude, the UAV moves along the planned trajectory. After completing the survey along a given trajectory, the UAV performs a vertical landing at the target point.

Figure 4 shows the experiments, simulated in Gazebo. The average speed of the UAV is set by the operator in advance and is 5.1 km/h (~ 1.4 m/s). The graphs show the flight paths of the UAV. The green line is the trajectory found using the RRT algorithm. The red line is the adjusted trajectory based on the readings of the UAV's environmental sensors and its flight controller. This trajectory is calculated onboard the UAV after receiving the computed trajectory, corrected for external conditions. The blue line shows the actual trajectory which the UAV follows.

Figure 4a shows the results of the experiment without adjustments for wind in the simulation. The deflections of the trajectories from the found ones are minimal, and this is due to the technical characteristics of the UAV in use. Figure 4b shows the flight

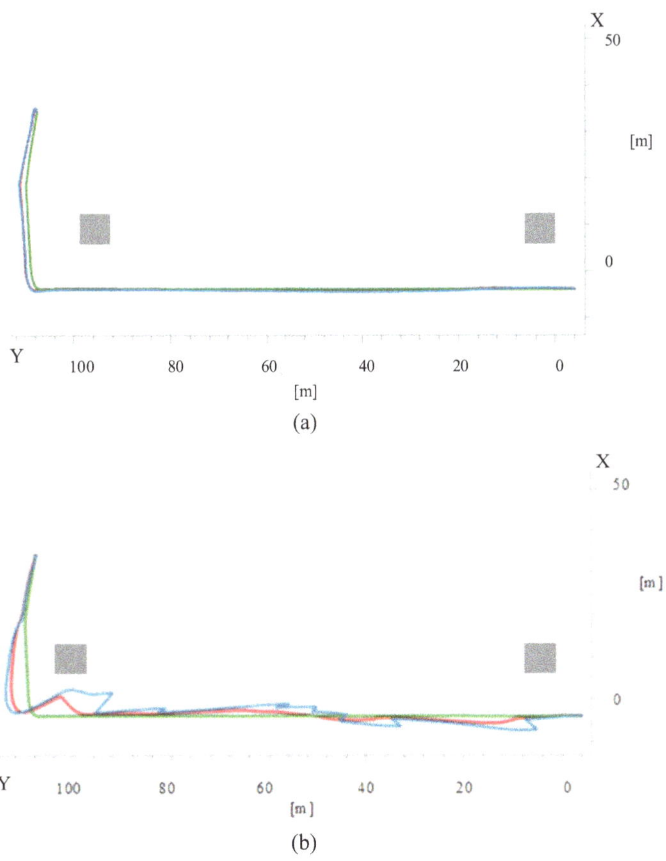

Fig. 4 Flight trajectory in Gazebo simulator in coordinates (x, y): the green line is the trajectory found using the RRT algorithm; the red line is the corrected trajectory based on the readings of the UAV's flight controller; the blue line shows the actual trajectory along which the UAV moves

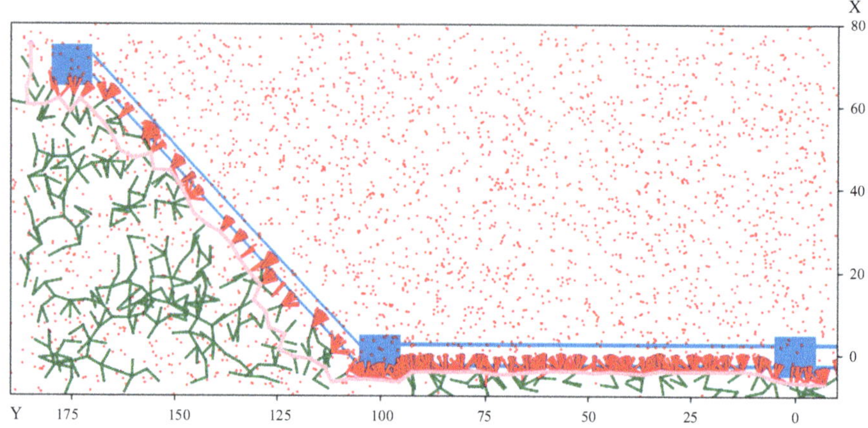

Fig. 5 Planned trajectory for two transmission lines with one corner tower: red points are all randomly generated points; green lines are explored edges of a tree; red lines are edges of a tree that do not fit; pink line is a path found; blue geometric shapes are obstacles

path when connected to a wind simulation. The wind is simulated from a random side at random times and its speed is 15 m/s. In the case shown in Fig. 4, the flight duration is 1 min 40 s and the path length is 214.2 m. In the case shown in Fig. 4b, the flight duration is 2 min 41 s and the path length is 224.4 m. The wind causes difficulties during the flight, but the flight controller provides for compensation of these disturbances and follow the planned trajectory. The maximum deviation of the actual trajectory in a strong wind of 15 m/s is 5 m, which is acceptable for this simulation, since this deviation did not lead to collisions with power transmission line elements. The results of the experiments show that the previously found flight trajectory of the UAV using the developed algorithm is suitable for examining the power transmission line section even in the windy conditions.

The time required for calculating and searching for the trajectory for the UAV directly depends on the number of nodes studied, the step of the algorithm, and the size of the operational area.

The average operating time of the RRT algorithm for inspection of one and two sections of the power transmission line (Fig. 5) is no more than 0.3 μs. Consequently, the algorithm for the movement of UAVs along the power lines can plan the trajectory of movement on large sections of power lines with angled towers. In further research, the problem of using the proposed survey method on real power transmission lines will be considered.

6 Conclusion

The advantages of using UAVs in power transmission line non-destructive testing consist in that the UAV utilization can ensure monitoring in many hard-to-reach places and under various circumstances (emergency situations, bad weather, lack of access); helps to obtain significantly more accurate data for further analysis. This solution is often cheaper than ground-based survey methods.

In this paper, all tasks of inspection of power lines are conditionally divided into two types: detecting changes in the geometric parameters of power lines and detecting damage of the integrity of power lines elements. In this paper, to solve problems of the second type, it is proposed to establish key survey points above the power line and on one of the sides of the power line parallel to the wire.

In the case of a flight from one of the sides of a power transmission line along a wire, it is necessary to plan the route in advance so that the UAV operates at a safe distance from the towers and wires. To develop a method for inspecting power lines using UAVs, an RRT algorithm was chosen, since it does not require much computing resources, is flexibly customizable for any task, and is easy to implement.

The proposed method for the survey of power transmission lines includes two algorithms. The first algorithm provides for finding all operational heights at which the feature points of the power transmission line are located. The second is responsible for the movement of the UAV along the power line.

The advantage of this method is that the survey is possible on long transmission lines with corner towers. The peculiarity of the method is that the path planning using the RRT algorithm along the power transmission line occurs in two-dimensional space at a given working height. The path search time in the survey method considered here does not exceed 0.3 μs for two sections of power lines with an angular tower and a total path length of 200 m.

To test the performance of the proposed algorithm for the movement of the UAV along the power transmission line, the Gazebo software environment was chosen. The simulation results showed that the average path planning time does not exceed 0.3 μs. Moreover, such an algorithm allows the UAV to fly around the towers at a safe distance, subject to a predetermined terrain map, and the presence of strong winds. This means that the proposed survey method can be applied in real-world conditions on long power transmission lines with corner towers.

References

1. Baikov, I., Golubev, P., Sizykh, Yu.: Application of remote sensing methods in the survey of overhead power transmission lines. Electr. Trans. Distrib. **1**, 54–57 (2016). (in Russian)
2. Shabanova, A., Tolstoi, I., Lebedev, I.: The mode of constructing safe trajectories of motion of the unmanned aerial vehicle while monitoring power lines considering the influence of their electromagnetic fields. Problemele Energeticii Regionale **44**(3), 17–30 (2019)
3. Yap, P.: Grid-based path-finding. In: Conference of the Canadian Society for Computational Studies of Intelligence, pp. 44–55. Springer, Berlin, Heidelberg (2002)

4. Yakovlev, K.S.: Graphs of a special structure in the problems of trajectory planning. In: Proceedings of the III international conference "System analysis and information technologies SAIT-2009". ISA RAS (2009) (in Russian)
5. Wooden, D.T.: Graph-based path planning for mobile robots. Dissertation Georgia Institute of Technology, p. 100 (2006)
6. Kavraki, L.E., Svestka, P., Latombe, J.C., Overmars, M.H.: Probabilistic roadmaps for path planning in high-dimensional configuration spaces. IEEE Trans. Robot. Autom. **12**(4), 566–580 (1996)
7. Yakovlev, K.S., Baskin, E.S.: Graph models in the problem of planning a trajectory on a plane. Artif. Intell. Decis. Making **1**, 5–12 (2013). (In Russ.)
8. Dijkstra, E.W.: A note on two problems in connexion with graphs. Numerische Math. **1**(1), 269–271 (1959)
9. Pearl, J.: Heuristics. AddisonWesley Publishing Company, Reading, Massachusetts (1984)
10. Hart, P.E., Nilsson, N.J., Raphael, B.: A formal basis for the heuristic determination of minimum cost paths. IEEE Trans. Syst. Sci. Cybern. **4**(2), 100–107 (1968)
11. Likhachev, M., Stentz, A.R.: Search. In: Proceedings of the Twenty-Third AAAI Conference on Artificial Intelligence, pp. 344–350 (2008)
12. Yakovlev, K.S.: HGA*: an efficient trajectory planning algorithm on a plane. Artif. Intell. Decis. Making **2**, 16–25 (2010). (in Russian)
13. Xiaolong, H., Jiang, B.: Deep-learning-based autonomous navigation approach for UAV transmission line inspection. In: 2018 Tenth International Conference on Advanced Computational Intelligence (ICACI), pp. 455–460 (2018)
14. Zhang, H., Wang, Y., Zheng, J., Yu, J.: Path planning of industrial robot based on improved RRT algorithm in complex environments. IEEE Access **6**, 53296–53306 (2018)
15. Svenstrup, M., Bak, T., Andersen, H.J.: Minimising computational complexity of the RRT algorithm a practical approach. In: 2011 IEEE International Conference on Robotics and Automation, pp. 5602–5607. IEEE (2011)
16. Noreen, I., Khan, A., Habib, Z.: A comparison of RRT, RRT* and RRT*-smart path planning algorithms. Int. J. Comput. Sci. Netw. Secur. (IJCSNS) **16**(10), 20 (2016)
17. LaValle, S.M.: Rapidly-exploring random trees: a new tool for path planning (1998)
18. LaValle, S.M., Kuffner, J.J., Jr.: Randomized kinodynamic planning. Int. J. Robot. Res. **20**(5), 378–400 (2001)
19. Schwarzer, F., Saha, M., Latombe, J.C.: Adaptive dynamic collision checking for single and multiple articulated robots in complex environments. IEEE Trans. Rob. **21**(3), 338–353 (2005)
20. Choi, J., Curry, R., Elkaim, G.: Path planning based on Bézier curve for autonomous ground vehicles. Adv. Electr. Electron. Eng.-IAENG Special Edn. World Cong. Eng. Comput. Sci. **2008**, 158–166 (2008)
21. Lin, C.C., Chuang, W.J., Liao, Y.D.: Path planning based on Bezier curve for robot swarms. In: 2012 Sixth International Conference on Genetic and Evolutionary Computing, pp. 253–256 (2012)

Approach to Image-Based Segmentation of Complex Surfaces Using Machine Learning Tools During Motion of Mobile Robots

Julia Rubtsova

Abstract In development of formation control systems for modular robotic devices, especially relevant is the problem of analysis and, particularly, segmentation of complex surfaces, down which the robotic device will move. An approach based on fine-tuning of a neural network model HRNet was developed to solve the problem of segmentation of complex surfaces. Model fine-tuning was performed based on a custom dataset, which included 15,000 labeled images. The training dataset included the scenes of the following types: scenes with stairways, scenes with even surfaces, scenes with isolated obstacles and with groups of obstacles. Approbation and functional quality assessment of the developed approach were performed based on the test dataset, which included 3000 images with different levels of scene illumination. According to the results of the testing, the developed approach shows decent quality of segmentation on images with even surfaces (IoU = 90.2%) and with stairways (IoU = 71.3%) as well maintains some resilience to the variations in the scene luminosity levels. After fine-tuning, the averaged performance metrics of this neural network model on images with luminosity levels of 100% and 70% increased by 9.1% and 7.3% at average and resulted in 65.5% and 52%, respectively.

1 Introduction

There is currently increased engagement in solving of problems, related to formations of modular robotic systems (RS). The formation control system for a modular robotic system is intended analyze the surface, along which the robotic device should go, to transform the robot into the configuration, most suitable for the given surface.

During development of formation control systems for a modular RS, there is a relevant problem of image-based segmentation of complex surfaces. Most existing approaches to solve this problem are intended for processing of object groups that do not include complex surfaces. The scope of this research is to develop an approach

J. Rubtsova (✉)
St. Petersburg Federal Research Center of the Russian Academy of Sciences (SPC RAS), St. Petersburg Institute for Informatics and Automation of the Russian Academy of Sciences, 39, 14th Line, 199178 St. Petersburg, Russia

A. Ronzhin and V. Shishlakov (eds.), *Electromechanics and Robotics*, Smart Innovation, Systems and Technologies 232, https://doi.org/10.1007/978-981-16-2814-6_17

to image-based segmentation of complex surfaces during robot motion along these surfaces. The most existing approaches to solve the presented problem concern with the segmentation of a group of objects, which does not include complex surfaces. This research is aimed to develop an approach to image-based segmentation of complex surfaces during motion of RS through an uncharted area. The proposed approach should fit for various practical settings, while retaining the required level segmentation accuracy, as well enable the image-based segmentation of complex surfaces in real-time mode.

2 Related Works

Currently, there exist various methods and approaches to the solution of image-based segmentation problem [1–3]. The most promising solutions in this domain are the approaches to the segmentation problem, which involve neural networks [4–9]. Let us consider the most recent neural network-based approaches to the segmentation problem. In [4], a neural network model Mask R-CNN is presented, which builds upon the Faster R-CNN architecture [5] and has been trained on the COCO dataset [6]. This neural network model solves the problem of instance-segmentation, i.e., detection of pixels, belonging to each instance, belonging to each class, separately. The types of instances that this network allows detection include various kinds of office furniture, other objects, and people.

In [7] a multi-scale deep neural network, HRNet is presented, which ensures solution of semantic segmentation problem. According to the model, presented in this paper, segmentation is performed in different scales, and the results are then averaged [7]. This model is trained on the Cityscapes dataset [8] and is used for detection of road surface marking, vehicles and people. Having applied this model to the datasets Cityscapes [8] and Mapillary Vistas [9], the authors obtained the value of Intersection over Union (IoU) metric, equal to 85.1%. Moreover, according to the data, featured in the research, this neural network model enables image segmentation in real-time mode.

Authors of [10] proposed a CPN neural network model, which was applied for semantic segmentation of images. This neural network model is based on a convolutional neural network architecture ResNet [11] and trained on the Cityscapes dataset [8]. This neural network model extracts low-level, medium-level and high-level features in a cross-cutting multi-layer manner. According to the model, proposed by the authors, image segmentation is achieved via increase in number of the stacked layers. Having applied this neural network model to the ImageNet dataset [12], authors obtained the value of the IoU metric, equal to 81.13%.

In [13], a neural network model EfficientPS is presented, intended for the panoptic segmentation of objects on images. The architecture of this neural network model includes the following components: common reference model, based on the EfficientNet architecture [14]; two-sided convolutional neural network FPN [15]; semantic head; convolutional instance head, based on the Mask R-CNN architecture

[4]; panoptic module, which aggregates outcomes of semantic and instance heads. Having applied this neural network model to the Cityscapes dataset [8], authors obtained the value of the IoU metric, equal to 84.21%.

Authors of [16] proposed a neural network model DKNAS, which enables automated lookup of network architecture for image segmentation. Authors propose the Densely Connected NAS (DCNAS) framework, which provides for deeper search of suitable architectures. The model directly searches for an optimal network structure, which would enable adequate representation of images from the target dataset. The DCNAS architecture delivers state-of-the-art results, solving the semantic segmentation problem. Having applied this neural network model to the Cityscapes dataset [8], authors obtained the value of the IoU metric, equal to 83.6%.

Hence, the most promising solution for image segmentation seems the neural network model HRNet [7]; hence, it is characterized by the highest value of the IoU metric (85.1%), compared to the others solutions considered. Besides, according to the data, presented in [7], such neural network model provides for image segmentation in real-time mode.

Further, we proceed to the development of an approach to image-based segmentation of complex surfaces using machine learning tools during motion of mobile robots.

3 Development of an Approach to Image-Based Segmentation of Complex Surfaces Using Machine Learning Tools During Motion of Mobile Robots

According to the results of the performed analysis of methods and approaches for image-based segmentation method for complex surfaces with the neural network model, HRNet [7] was chosen as reference. The principal features of this network are: the highest accuracy of image segmentation among the all solutions considered, as well the possibility for image segmentation in real-time mode. The HRNet network utilizes a coding–decoding structure with a complex separable convolution to capture more precise object outlines via gradual recovery of spatial data. This network contains a coder module that gradually reduces the features and captures higher-level semantic data, as well as a decoder module for further refinement of segmentation results, particularly, along the edges of the objects.

Neural network model HRNet is based on the convolutional neural network architecture DeepLab V3+ [17]. This architecture is intended for multi-scale processing of images to obtain segmentation maps. This network enables encoding of multi-scale contextual data, reasoning on the incoming functions with filters or with multi-speed join operations with many effective vision fields (ASPP). Besides common convolutions, this network utilizes dilated convolutions as well, which enable to respect a broader contextual information, without increasing the parameter count. In this network, such convolutions are utilized in the pooling module of the spatial pyramids. Further the bilinear interpolation is employed, which increases the resolution of the segmentation map. The neural network of such architecture is applicable for processing of three-channel color images from the RGB color space. Processing the image, the neural network returns the number of objects revealed, as well the set of arrays (classes and masks). The masks represent the target set of points, belonging to the considered object on image.

Neural network model HRNet [7] has been trained on a large-scale dataset Cityscapes [8], which is intended for solution of the image segmentation problem. The Cityscapes dataset contains a number of video sequences, including outdoor scenes from 50 different cities, captured in different seasons under different weather conditions. Every frame in a video sequence is a color three-channel image, where the pixels are labeled as belonging to 30 different object classes. Authors of [7] utilized the automatic labeling approach with the Cityscapes dataset to improve the quality of the training dataset. For automatic labeling, the authors of [7] chose the hard labeling strategy, according to which the luminosity value of each pixel is mapped to the specified threshold value. Hence, the authors of [7] trained the neural network model HRNet on the dataset, which contained 3500 thoroughly annotated images and 20,000 automatically annotated images.

Since the neural network model HRNet was initially trained on a dataset, consisting exclusively of outdoor scenes, it was decided within the present research to fine-tune this model on a dataset, containing indoor scenes with annotated segmentation masks of floor surfaces from office rooms. The process of fine-tuning of the neural network model HRNet was performed on an established and manually labeled dataset, containing 15,000 color three-channel images from the RGB color space. The dataset was established using the RealSense D435 camera [18]. This dataset includes the following indoor scenes: (1) with stairways; (2) with even surfaces; (3) with isolated obstacles; with groups of obstacles, including slopes, hollow spots and other surface irregularities, where the overturning of robotic device is possible. The obtained images were manually annotated with Computer Vision Annotation Tool (CVAT) [19]—a piece of software, developed by the Intel corporation. CVAT is intended for data labeling and enables data annotation for various machine learning-related purposes: object detection, image classification, as well as image segmentation.

Sample images from the prepared dataset as well as their segmentation masks obtained using the CVAT software [19] are presented in Fig. 1: scenes with a stairway (Fig. 1a); scenes with even (Fig. 1b); scenes with a single obstacle, where a piece of furniture located near the camera lens acts as a single obstacle (Fig. 1c); scenes with a set of obstacles, where a set of chairs acts as obstacles located near the camera lens (Fig. 1d).

Further, we proceed to the estimation of the proposed approach implementation, applied to the segmentation of complex surfaces on image during motion of modular robots.

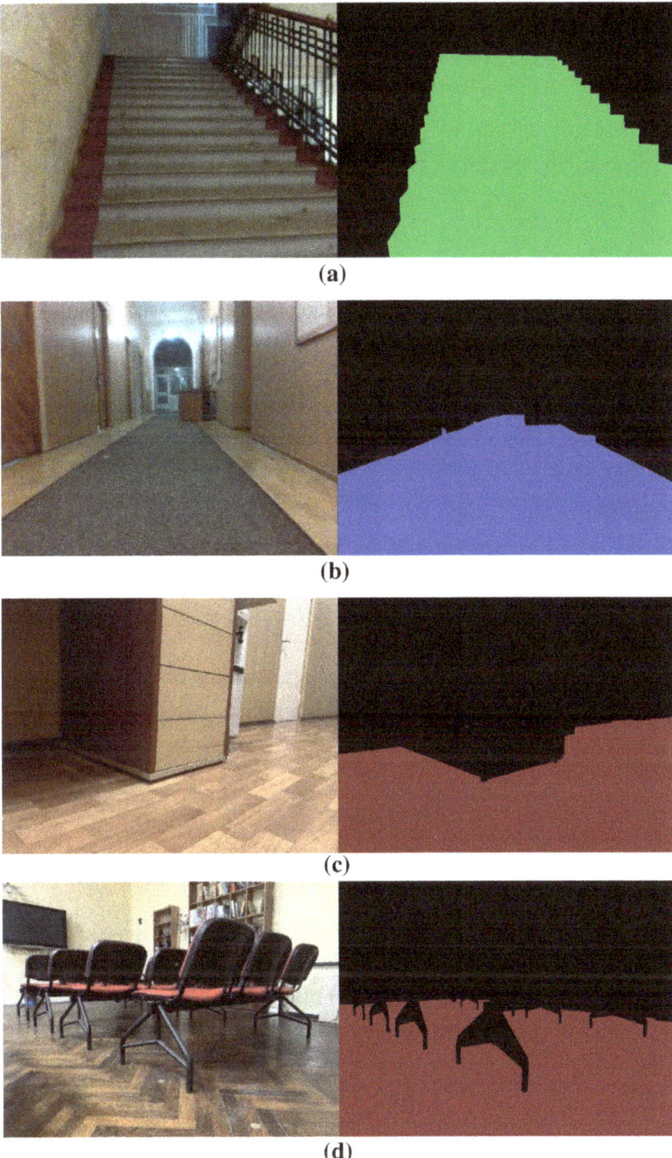

Fig. 1 Sample images from the dataset, established for the fine-tuning of neural network model HRNet, as well their segmentation masks: scenes with stairways; scenes with even surfaces; scenes with isolated obstacles; scenes with groups of obstacles

4 Testing of the Developed Approach to the Segmentation of Complex Surfaces

Approbation and assessment of performance quality of the neural network model HRNet and fine-tuned model of the neural network HRNet was performed on the test dataset, which included 3000 images. This dataset was established using a depth camera RealSense D435. Every image $\boldsymbol{I}_i$ from the test dataset is associated with a certain set of parameter values $\boldsymbol{P}_i$. These parameters are described here:

- Scene type T. Includes four different scene categories: scenes with stairways (1), scenes with even surfaces (2), scenes with an isolated obstacle in each scene (3) and scenes with groups of obstacles (4).
- Scene luminosity level L. Test dataset includes images, obtained under different lighting conditions: 50, 70 and 100% luminosity level, where 100% corresponds to the regulatory level of luminosity for office rooms [20].

Hence, the test dataset was divided into three subsets of images, which were different in the scene luminosity level: 50, 70, and 100%. Each of these subsets includes 1000 images, containing the following number of scenes: 200 scenes with stairways; 300 scenes with even surfaces; 300 scenes with an isolated obstacle each; 200 scenes with groups of obstacles.

To ensure direct assessment of performance quality of the developed approach, the metric Intersection over Union (IoU) was chosen, which is defined according to the following expression:

$$\frac{1}{k}\sum_{i}^{k}\frac{n_{ij}}{t_i \cup t_j} * 100\%,$$

where t_i is set of pixels in the target segment i, t_j is set of pixels assigned by the neural network model to the target segment; n_{ij} is set of pixels, correctly attributed by the model to the target segment; k is number of classes, being segmented. Therefore, the higher is the value of the IoU metric, the better is the segmentation quality.

In the following Fig. 2, we present the assessment of performance quality of the neural network model HRNet [7] and fine-tuned neural network model HRNet on the subset of the test dataset with scene luminosity level of 100%.

From the data, presented in Fig. 2, it can be concluded that the performance quality of the fine-tuned neural network HRNet is 15.2% higher for the image class, which includes scenes with stairways; 12.3% higher for the scenes, including the groups of obstacles and 7.9% higher for the image class, which includes scenes with even surfaces, compared to the solution, presented in [7]. For the class of scenes with an isolated obstacle on each, both solutions show an approximately equal value of the IoU metric. It should also be noted that the most successful results were shown by the fine-tuned neural network model HRNet for the images, which include scenes with even surfaces IoU = 90.2%, as well for the image class, containing the scenes with

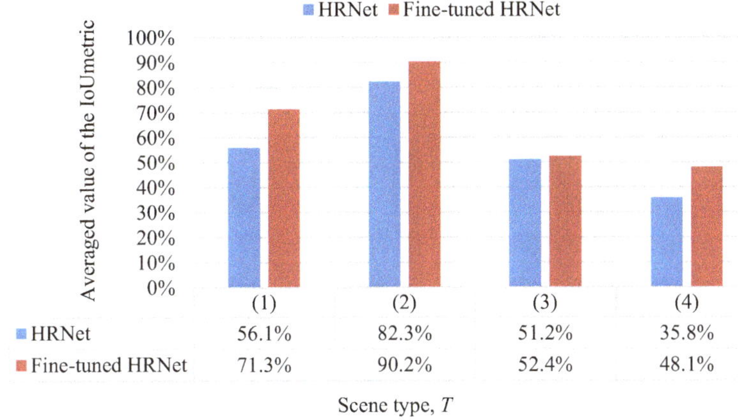

Fig. 2 Assessment of performance quality of the neural network model HRNet [7] and fine-tuned neural network model HRNet on the subset of the test dataset with scene luminosity level of 100%

stairways IoU = 71.3%. The poorest segmentation quality within this experiment in complex surface segmentation by the fine-tuned neural network HRNet was shown on images, which include scenes with groups of obstacles (48.1%). Presumably, such results can be explained by geometric features of the obstacles on the images. Therefore, the averaged performance quality of the initial model on this part of the test dataset was 56.4%, and averaged performance quality of the fine-tuned model was 65.5%.

Further, let us consider the influence of the scene luminosity level on the performance quality of the developed solution. In Fig. 3, the estimation of the performance quality of the neural network model HRNet [7] (Fig. 3a) and fine-tuned model HRNet (Fig. 3b) on the test dataset, containing images, which were composed under different luminosity levels is shown

From the charts, presented in Fig. 3, it can be concluded that irrespective of the type of the scene under consideration and its inherent luminosity, the best results are shown by the fine-tuned model of the neural network HRNet. The values averaged by the scene types of the IoU on images with luminosity levels of 100% and 70% for the neural network model HRNet [7] are equal to 56.4% and 44.7%, respectively. The averaged values by the scene types of the IoU on images with luminosity levels of 100% and 70% for the fine-tuned neural network model HRNet are equal to 65.5% and 52%, respectively. The fine-tuned neural network model HRNet performs significantly better in good lighting conditions, but this effect becomes less pronounced in darker environments. For the neural network model HRNet [7], the value of the IoU metric (averaged by scene type) for images with luminosity level of 50% is equal to 24.6%, whereas for the fine-tuned model of the neural network HRNet, it is equal to 26.8%.

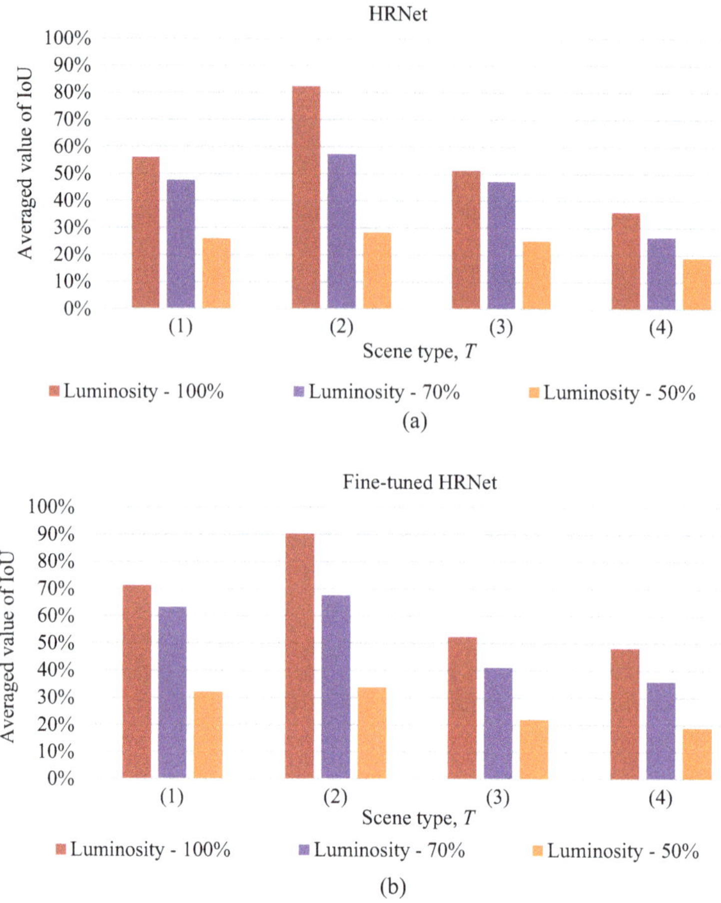

Fig. 3 Estimation of influence of the luminosity level L of a scene, as per the test dataset, which includes images, composed in different lighting conditions: **a** neural network model HRNet [7]; **b** fine-tuned model of neural network HRNet

Therefore, the developed approach to image-based segmentation of complex surfaces, involving fine-tuning of the neural network model HRNet, enabled significant increase of the segmentation quality of complex surfaces. The developed solution demonstrates decent segmentation accuracy for the images with 100 and 70% scene luminosity levels.

5 Conclusion

Upon the results, the proposed approach approbation to the segmentation of complex surfaces on the test dataset of 3000 images was concluded that the fine-tuning of the neural network model HRNet, performed within this study, enabled to increase sufficiently the quality of solutions of image-based segmentation of complex surfaces. The developed solution demonstrates decent quality of image segmentation for images, captured under good lighting conditions: averaged by the scene types, on images with luminosity levels of 100% and 70% for the fine-tuned neural network model HRNet were equal to 65.5% and 52%, respectively. It should be noted that the best segmentation quality was shown by the fine-tuned neural network model HRNet on images, containing even surfaces IoU = 90.2%, and stairways (IoU = 71.3%). Therefore, the developed approach is to a certain extent resilient to the variations of the scene luminosity level and enables successful solution of image-based segmentation of complex surfaces problem during path planning for the robot motion.

Within further research, it is anticipated to enhance the proposed approach in terms of utilization of computational resources for subsequent integration of it into the system that controls formations of a modular robotic device.

Acknowledgements This research is supported by the RFBR Project No. 20-08-01109_A.

References

1. Grady, L.: Random walks for image segmentation. IEEE Trans. Pattern Anal. Mach. Intell. **28**(11), 1768–1783 (2006)
2. Kharinov, M.V., Khanykov, I.G.: Optimization of piecewise constant approximation for segmented image. SPIIRAS Proc. **3**(40), 183–202 (2015). https://doi.org/10.15622/sp.40.12
3. Wu, Z., Leahy, R.: An optimal graph theoretic approach to data clustering: theory and its application to image segmentation. IEEE Trans. Pattern Anal. Mach. Intell. **15**(11), 1101–1113 (1993)
4. He, K., Gkioxari, G., Dollár, P., Girshick, R.: Mask r-cnn. In: Proceedings of the IEEE International Conference on Computer Vision, pp. 2961–2969 (2017)
5. Ren, S., He, K., Girshick, R., Sun, J.: Faster r-cnn: towards real-time object detection with region proposal networks. IEEE Trans. Pattern Anal. Mach. Intell. **39**(6), 1137–1149 (2016)
6. Lin, T.Y., Maire, M., Belongie, S., Hays, J., Perona, P., Ramanan, D., Dollár, P., Zitnick, C.L.: Microsoft coco: common objects in context. In: European Conference on Computer Vision, pp. 740–755. Springer, Cham (2014)
7. Tao, A., Sapra, K., Catanzaro, B.: Hierarchical Multi-Scale Attention for Semantic Segmentation. arXiv:2005.10821 (2020)
8. Cordts, M., Omran, M., Ramos, S., Rehfeld, T., Enzweiler, M., Benenson, R., Franke, U., Roth, S., Schiele, B.: The cityscapes dataset for semantic urban scene understanding. In: Proceedings of the IEEE Conference on Computer Vision And Pattern Recognition, pp. 3213–3223 (2016)
9. Neuhold, G., Ollmann, T., Rota Bulo, S., Kontschieder, P.: The mapillary vistas dataset for semantic understanding of street scenes. In: Proceedings of the IEEE International Conference on Computer Vision, pp. 4990–4999 (2017)
10. Yu, C., Wang, J., Gao, C., Yu, G., Shen, C., Sang, N.: Context prior for scene segmentation. In: Proceedings of the IEEE/CVF Conference on Computer Vision and Pattern Recognition, pp. 12416–12425 (2020)
11. Szegedy, C., Ioffe, S., Vanhoucke, V., Alemi, A.: Inception-v4, Inception-Resnet and the Impact of Residual Connections on Learning. arXiv:1602.07261 (2016)

12. Deng, J., Dong, W., Socher, R., Li, L.J., Li, K., Fei-Fei, L.: Imagenet: a large-scale hierarchical image database. In: 2009 IEEE Conference on Computer Vision and Pattern Recognition, pp. 248–255 (2009)
13. Mohan, R., Valada, A.: Efficientps: Efficient panoptic segmentation. arXiv:2004.02307 (2020)
14. Tan, M., Le, Q.V.: Efficientnet: Rethinking Model Scaling for Convolutional Neural Networks. arXiv:1905.11946 (2019)
15. Lin, T.Y., Dollár, P., Girshick, R., He, K., Hariharan, B., Belongie, S.: Feature pyramid networks for object detection. In: Proceedings of the IEEE Conference on Computer Vision and Pattern Recognition, pp. 2117–2125 (2017)
16. Zhang, X., Xu, H., Mo, H., Tan, J., Yang, C., Ren, W.: DCNAS: Densely Connected Neural Architecture Search for Semantic Image Segmentation. arXiv:2003.11883 (2020)
17. Chen, L.C.: Encoder-decoder with atrous separable convolution for semantic image segmentation. In: Proceedings of the European Conference on Computer Vision (ECCV), pp. 801–818 (2018)
18. Keselman, L., Iselin Woodfill, J., Grunnet-Jepsen, A., Bhowmik, A.: Intel realsense stereoscopic depth cameras. In: Proceedings of the IEEE Conference on Computer Vision and Pattern Recognition Workshops, pp. 1–10 (2017)
19. McLure, C.E.: Implementing subnational value added taxes on internal trade: the compensating VAT (CVAT). Int. Tax Public Financ. **7**(6), 723–740 (2000)
20. GOST R. 55710-2013 Illumination of workplaces inside buildings. In: Standards and Measurement Methods (2013) (in Russian)

Algorithm of Georeferencing and Optimization of 3D Terrain Models for Robot Path Planning

Egor Aksamentov and Valeriia Izhboldina

Abstract When developing navigation systems for ground-based robotic vehicles, particular attention is paid to path planning algorithms. The performance quality of these algorithms directly depends on the accuracy of the terrain maps. This paper proposes an algorithm for reconstructing a 3D terrain map model from images obtained using an unmanned aerial vehicle. The algorithm using reference points proposed here enables to associate the orthophotomap and 3D terrain model to GPS coordinates as per the WGS 84 standard. By optimizing the model tuning for the characteristics of robotic vehicles, it was possible to delete redundant information on the terrain maps. Testing of the algorithm showed that as a result of post-processing of the 3D model, the number of vertices decreased by an average of 9.45%. The presented 3D model is a graph, the edges of which form polygons. Reducing the number of vertices in the graph will reduce the size of the 3D model, and as a result, positively influence the running time of the path planning algorithms. The presented algorithm does not depend on the type of camera in use, the type of unmanned aerial vehicle in use, and the size of the considered area.

1 Introduction

For autonomous robotic navigation, it is necessary to solve the problems of localization, mapping, and path planning. These problems are inherently related to each other, and the quality of every one of them affects the implementation of the others. When developing algorithms for planning the path of movement of ground-based robots that specialize in operational missions in large open areas, problems arise related to the size of the terrain. The main problem common to all autonomous robotic vehicles (RV) is related to their limited operating time. To increase the performance efficiency, it is sensible to provide the robots with a navigational system, containing information about the path of the RV movement in advance [1]. To construct a path, the

E. Aksamentov (✉) · V. Izhboldina
St. Petersburg Institute for Informatics and Automation of the Russian Academy of Sciences, St. Petersburg Federal Research Center of the Russian Academy of Sciences (SPC RAS), 39, 14th Line, 199178 St. Petersburg, Russia

A. Ronzhin and V. Shishlakov (eds.), *Electromechanics and Robotics*, Smart Innovation, Systems and Technologies 232, https://doi.org/10.1007/978-981-16-2814-6_18

navigation system of the robot requires input data about the area and should ensure determination of its current location. One of the most common options for solving of the problem of simultaneous localization and mapping is the use of algorithms based on the SLAM method (simultaneous localization and mapping) [2–4]. To construct a map of the environment, the SLAM method uses data from various devices: 2D sensors [5], 3D scanners [6], monocular cameras [7], stereo cameras [8] installed on a RV. This method does not allow determining in advance the location of large areas, impassable for the robot, such as water bodies, hills, ravines. Since the autonomous operation of the RV depends on the volume of the battery, avoiding such obstacles reduces the useful work time. Mapping a 3D model of such a terrain area may require several duty cycles of the ground-based RV.

The simplest and cheapest way to obtain information about the area is to survey it using an unmanned aerial vehicle (UAV). A feature of using UAVs is the ability to obtain images from places where movement is obviously impossible. In addition, the mobility and relatively high speed of movement of the UAV allows you to explore large areas of the terrain in a fairly short time. Images obtained with the UAV can be stitched into an orthophotomap [9]. Orthophotomap is a 2D image of the terrain, and it is used to represent objects relative to each other.

The main idea of the algorithms for path planning using 2D maps assumes that the territory along which the RV will move is flat. If it is possible to abstract away such a characteristic as elevation difference, it can entail a revolution of the RV or significantly increase the time for the implementation of tasks. An alternative to orthophotomap can represent 3D models of terrain maps. They allow you to display all the features of the terrain, such as irregularities and elevation differences [10]. There are several methods for creating 3D maps of the terrain, the most popular of which are: lidar [11] and processing of aerial photography materials using photogrammetry algorithms [12].

To solve the problem of choosing a tool for creating 3D terrain models, the authors of the paper [13] empirically compared the quality of the output data of the methods lidar and processing of aerial photography materials using photogrammetry algorithm when surveying a gravel pit. At the output of the methods, a 3D terrain model is established, described by point clouds. The experiment showed that both methods ensure minimal displacement between two point clouds; however, point clouds obtained with lidar have a higher density and provide accuracy up to several millimeters. The disadvantages of using lidar include rather expensive equipment and a limited range of the device. When exploring large areas of terrain, the device needs to remain in constant motion, which is time-consuming as such. In this case, photogrammetric processing of images from UAVs is a more sensible option, since it provides a sufficient level of detail and accuracy for RV path planning, while the process of surveying the terrain is performed several times faster and cheaper than with lidars.

A heightmap [14] also called a 2.5D map usually is a grayscale image, where each pixel corresponds to the landscape height in the certain area. Authors of [15] showed that such map presentation approach is practical with RV operation, provided that the dangling obstacles, such as treetops, bridge spans, and similar structures are positioned higher than the robot top point. Numerous papers exist, considering as optimization of such mapping approach, as well its practical applicability in solving of specific problems [16, 17]. Though, virtually any methods, concerned with 2.5D mapping, suffer from a following downside: the terrain maps are interpreted in top-down dimension, depending on the height of RV position. Such perspective does not provide comprehensive representation of environment, what directly limits the applicability of the maps, composed this way.

The 3D model composed from the images from the UAV will be a digital representation of the area of interest with all the features of the terrain preserved. But the analysis of such models can be difficult, since with an increase in the size of the study area, the size of the digital model proportionally grows, and large computing resources may be required to process it. This model may contain some redundancy, for example, areas where the motion of RV is obviously impossible, which can be neglected. This paper proposes an algorithm for optimizing 3D models of terrain maps composed from images obtained from UAVs and automatically associated with a geodetic coordinate system for RV motion. Optimization consists in removing redundant information about impassable areas. By associating the model to geodetic coordinates, it is possible to localize all terrain features of the terrain and determine their actual size. Depending on the cross-country properties of the RV, the algorithm independently optimizes the model, leaving only passable places.

2 3D Terrain Map Composition

In this paper, the photogrammetry method was used to compose 3D terrain models from 2D images captured with UAVs. The main idea of photogrammetry is to search for feature points in several images of the same area (see Fig. 1), which were captured under different angles. Each such accordance is called a feature point.

The quality of the reconstructed terrain model directly depends on the UAV's flight path. The larger the area of overlap between neighboring images is, the more

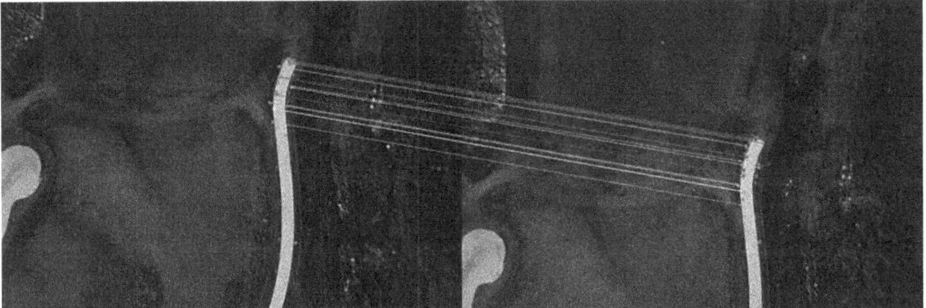

Fig. 1 Feature points of several images

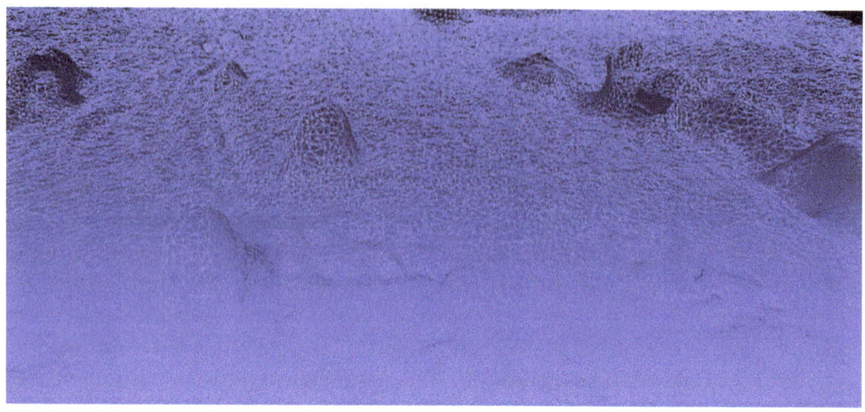

Fig. 2 Terrain relief obtained by photogrammetry

accurately objects on the scene will be restored. Before starting the terrain exploration, it is necessary to set the UAV flight path in such a way that the overlap between adjacent frames would be at least 40%. This percentage of overlap provides a sufficient number of reference objects on the scene for the most complete reconstruction of the terrain model [18]. The final model is a sparse point cloud. This point cloud is densified and adjacent points are treated as polygon vertices. As a result, the 3D model is a graph, the edges of which correspond to the sides of the polygons. An example of generating a 3D terrain model by the photogrammetry method is shown in Fig. 2.

At this step, the resulting map of the area is given for illustrative purposes only. On its basis, it is possible to identify objects with which a RV may collide, but it is impossible to localize these areas on the ground. To reveal such areas, it is necessary to represent the resulting polygonal model in a geodetic coordinate system.

3 Algorithm for Associate 3D Model to Geodetic Coordinates

When exploring the terrain, the onboard computer of the UAV records the global positioning system (GPS) coordinate of the place where the frame was taken. This GPS coordinate corresponds to the location of the central pixel in the image. When constructing a 3D terrain model, due to an insufficient degree of overlap of adjacent images, the location of this point may not be restored correctly. This situation can occur when the central pixel does not have duplicates in adjacent images.

For accurate association of the model with geodetic coordinates, it is necessary to correlate each pixel in each image with the corresponding GPS coordinates. The authors of the paper [19] presented an algorithm for fast association of images to geodetic coordinates. This algorithm was developed to quickly detect structures that have suffered from any cataclysm. To create the orthophotomap, the authors used Photoscan software and the GDAL [20] geospatial data processing library. Since this method was developed to be used in emergency situations, where every minute is critical to save some survivors, the resulting accuracy of localization of destroyed structures of several tens of meters was treated as a quite acceptable. However, for RV navigation, such accuracy is not acceptable.

To associate of the of the polygonal terrain model with geodetic coordinates, it is necessary to composition an orthophotomap. Orthophotomap is built by stitching neighboring images and is a 2D high-resolution map of the area. An example of a stitched orthophotomap of the area from 7 images is presented in Fig. 3.

Preparation of an orthophotomap of the terrain is an intermediate step in the developed algorithm and serves only to obtain the GPS coordinates of feature points, which were used to composition a 3D model. Further calculation of geospatial coordinates for each pixel is performed using the GDAL library. GDAL is a library for processing geospatial data formats. It contains production-ready tools for converting and associate data in various coordinate systems. As mentioned earlier, when capturing images from an UAV, each central pixel in the image is associated with a GPS coordinate. Since an orthophotomap is a set of such images, the resulting map contains many pixels with known GPS positioning. Based on this data, GPS coordinates are calculated for each pixel in the image.

The flowchart of the final algorithm is presented in Fig. 4. The images, obtained from the UAV, are used as input for this algorithm. On the basis of the images obtained, an orthophotomap and a 3D terrain model are constructed. With the help of GPS coordinates of each image, the orthophotomap is associated with the geodetic

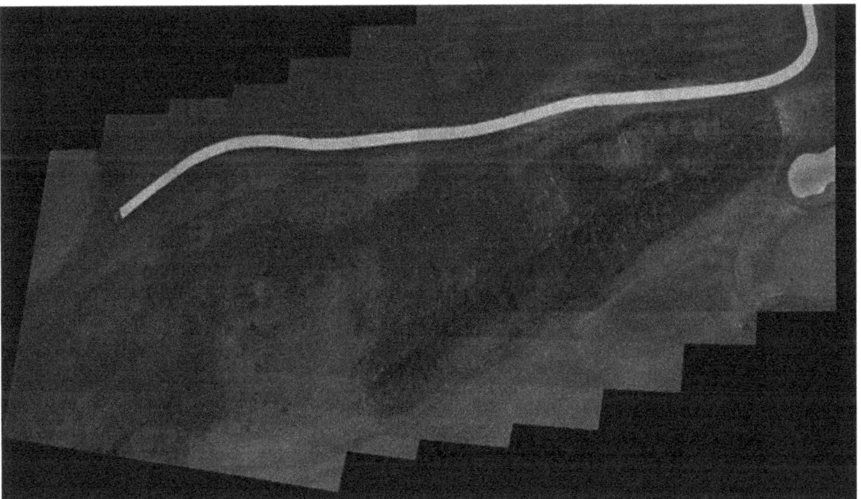

Fig. 3 Example of orthophotomap

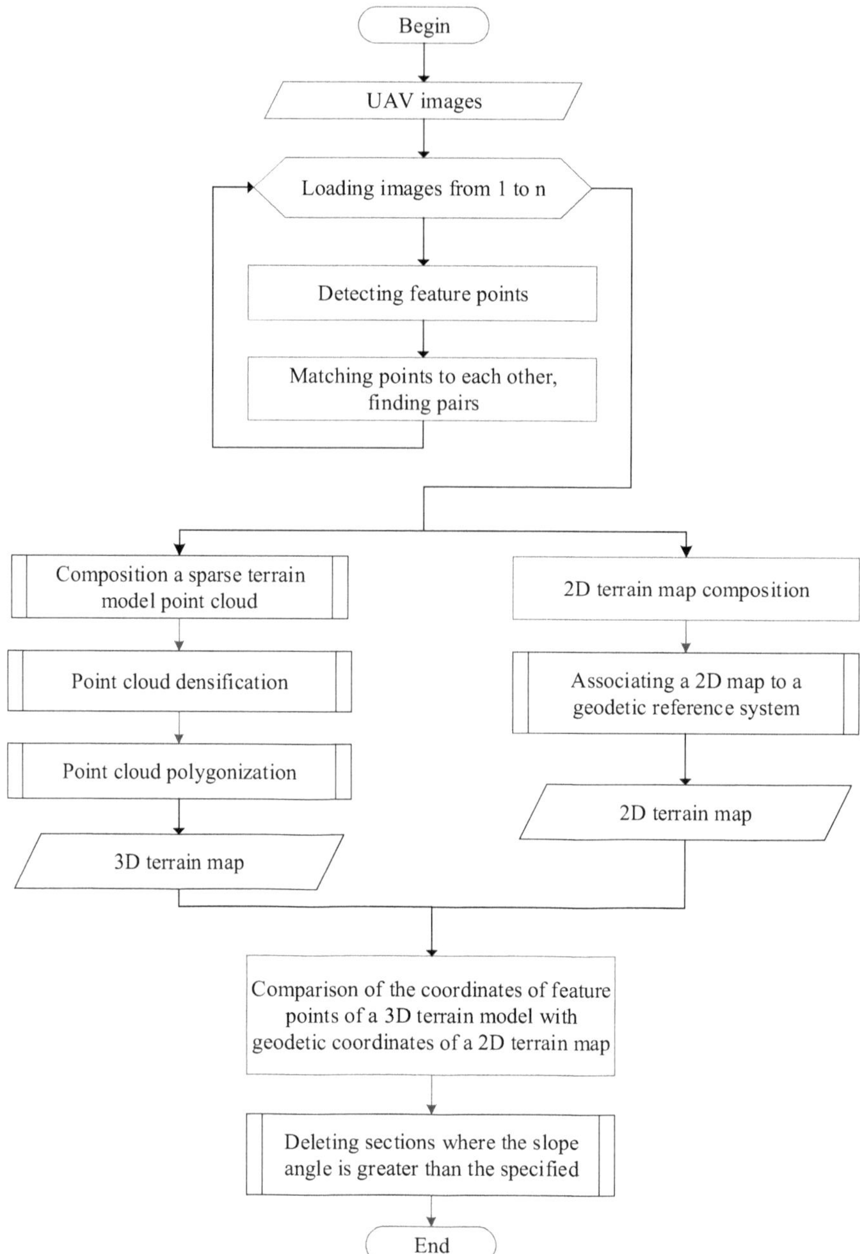

Fig. 4 Algorithm for geospatial association of a 3D model with geodetic coordinates

coordinate system. The geocentric global system WGS 84 was chosen as the coordinate system, whose error is about 5 cm and remains the same in the planetary scale [21]. Further, based on the points that were used to compose a 3D scene and their relevant coordinates, this point cloud is tied to coordinate systems using the PDAL [22] library. PDAL is analogous to the geospatial library GDAL, but for 3D data. As a result, each polygon is mapped to the corresponding geocoordinate.

Figure 2 shows the resulting 3D terrain map with some redundancy. It manifests itself in the form of groups of polygons located at a large angle relative to the ground level, on which the movement of RV is obviously impossible. Removing such redundancy at the stage of creating a 3D terrain model allows to reduce further calculations required to find the trajectories of the RV movement.

Since each polygon is represented by a triangle, the stitched image can be represented as a plane, and then the problem of determining the angle of inclination of the plane is limited to the problem of finding the angle between two normal vectors of these planes and is solved by the following equation:

$$\cos\alpha = \frac{A_1 * A_2 + B_1}{\sqrt{A_1^2 + B_1^2 + C_1^2}\sqrt{A_2^2 + B_2^2 + C_2^2}},$$

where *A*, *B*, and *C* are coefficients of the plane equation.

Since the critical angle of inclination of the surface, where the RV can move, depends on the design of the vehicle, the values of this angle are set separately for each specific case. An example of displaying an optimized 3D terrain map is presented in Fig. 5. The critical angle was set to 30°.

In Fig. 5, blue color indicates excluded areas of the terrain that will not be considered in path planning on this 3D model. As a result of the optimization of the model,

Fig. 5 Optimized terrain map

Table 1 Experimental results

Parameter	Data set number		
	1	2	3
Number of images (ea)	28	56	367
Flight altitude (m)	195	32	42
Map size (m)	1227 × 495	615 × 440	232 × 312
Number of vertices before processing (ea)	196,924	147,622	78,830
Number of vertices after processing (ea)	183,421	127,631	74,218
Model optimization result (%)	6.85	15.66	5.85

the terrain relief that pose a danger to RV is removed, and the final digital model of the surface can be used as a basis for further tasks in the field of RV movement.

4 Results

To test the developed algorithm, three datasets were used. The sets were obtained from three UAVs of different configurations. For the sake of experiment integrity, the flight location, altitude, and types of cameras were also different. For each experiment, the critical angle was set at 30 degrees, the area of overlap between adjacent images horizontally and vertically was about 40–60%, the flight altitude of the UAV was from 32 to 195 m above sea level. The first dataset came from a golf course, an open area with virtually no trees or buildings. This area, without any sharp relief, features with a slight height difference of about 3 m. The second dataset was collected from a small field with many single trees. The third set, like the second, contains images of areas with rare trees. But this area has a hilly relief. The result of testing the algorithm is presented in Table 1.

The results of the experiment showed that the algorithm reduced the number of model vertices by an amount from 5.85 to 15.66% of the total number of vertices. Testing on data, obtained under various conditions, indicates that this algorithm does not depend on the characteristics of the specific UAV and of the type of area considered, which indicates its versatility.

5 Conclusion

This paper presented the algorithm for the optimization and geospatial associating of 3D terrain models, which serves as reference for solving the problems of navigation and motion of RV in large open areas. The 3D model generated by the algorithm is a digital representation of the terrain under consideration, and all surface features are localized due to the automatic associating of the model with the geodetic coordinate system. By optimizing of the model, namely the delete of obviously impassable areas, it was possible to reduce the size of the model by an average of 9.45%. The compactness of the model allows to reduce the amount of input data required to build paths, and, as a consequence, reduce the time, potentially required for path planning.

References

1. Lavrenov, R.O., Magid, E.A., Matsuno, F., Svinin, M.M., Suthakorn, J.: Development and implementation of spline-based path planning algorithm in ros/gazebo environment. SPIIRAS Proc. **18**(1), 57–84 (2019)
2. Kim, C., Lee, S., Kim, H.J.: Convergence-enhanced dense RGB-D odometry with a rotational motion prior from a gyroscope. In: 2017 11th Asian Control Conference (ASCC), pp. 2528–2533. IEEE (2017)
3. Mur-Artal, R., Tardós, J.D.: Orb-slam2: an open-source slam system for monocular, stereo, and rgb-d cameras. IEEE Trans. Rob. **33**(5), 1255–1262 (2017)
4. Mur-Artal, R., Montiel, J.M.M., Tardos, J.D.: ORB-SLAM: a versatile and accurate monocular SLAM system. IEEE Trans. Rob. **31**(5), 1147–1163 (2015)
5. Li, R., et al.: A novel RGB-D SLAM algorithm based on points and plane-patches. In: 2016 IEEE International Conference on Automation Science and Engineering (CASE), pp. 1348–1353. IEEE (2016)
6. Zhou, Y., Kneip, L., Rodriguez, C., Li, H.: Divide and conquer: Efficient density-based tracking of 3D sensors in Manhattan worlds. In: Asian Conference on Computer Vision, pp. 3–19. Springer, Cham (2016)
7. Davison, A.J., Reid, I.D., Molton, N.D., Stasse, O.: MonoSLAM: real-time single camera SLAM. IEEE Trans. Pattern Anal. Mach. Intell. **29**(6), 1052–1067 (2007)
8. Elinas, P., Sim, R., Little, J.J: /spl sigma/SLAM: stereo vision SLAM using the Rao-Blackwellised particle filter and a novel mixture proposal distribution. In: Proceedings 2006 IEEE International Conference on Robotics and Automation (ICRA), pp. 1564–1570. IEEE (2006)
9. Tchernykh, V., Beck, M., Janschek, K.: Optical flow navigation for an outdoor UAV using a wide angle mono camera and DEM matching. IFAC Proc. Vol. **39**(16), 590–595 (2006)
10. Muchiri, N., Kimathi, S.: A review of applications and potential applications of UAV. In: Proceedings of Sustainable Research and Innovation Conference, pp. 280–283 (2016)
11. Vosselman, G.: Slope based filtering of laser altimetry data. Int. Arch. Photogramm. Remote Sens. **33**(B3/2), 935–942 (2000)
12. Nex, F., Remondino, F.: UAV for 3D mapping applications: a review. Appl. Geomat. **6**(1), 1–15 (2014)
13. Weber, A., Lerch, T.: Point clouds: laser scanning versus UAS photogrammetry. GIM Int. Worldwide Mag. Geomat. **32**(4), 35–38 (2018)
14. Hadsell, R., Bagnell, J.A., Hebert, M.: Accurate rough terrain estimation with space-carving kernels. Robot. Sci. Syst. **2009**, 62 (2009)
15. Kweon, I.S., Kanade, T.: Extracting topographic terrain features from elevation maps. CVGIP Image Understand. **59**(2), 171–182 (1994)
16. Gutmann, J.S., Fukuchi, M., Fujita, M.: A floor and obstacle height map for 3D navigation of a humanoid robot. In: Proceedings of the 2005 IEEE International Conference on Robotics and Automation, pp. 1066–1071. IEEE (2005)
17. Gu, J., Cao, Q.: Path planning for mobile robot in a 2.5-dimensional grid-based map. Ind. Robot. Int. J. (2011)
18. Agüera-Vega, F., Carvajal-Ramírez, F., Martínez-Carricondo, P.: Assessment of photogrammetric mapping accuracy based on variation ground control points number using unmanned aerial vehicle. Measurement **98**, 221–227 (2017)

19. Wang, S., Ding, L., Chen, Z., Dou, A.: A rapid UAV image georeference algorithm developed for emergency response. J. Sens. **2018** (2018)
20. GDAL. https://gdal.org/. Last accessed 2021/02/23
21. Slater J.A., Malys S.: WGS 84—past, present and future. In: Advances in Positioning and Reference Frames, pp. 1–7. Springer, Berlin, Heidelberg (1998)
22. PDAL. https://pdal.io/. Last accessed 2021/02/23

Algorithm for Edge Detection of Floodable Areas Based on Heightmap Data

Konstantin Zakharov and Anton Saveliev

Abstract Path planning for outdoor operation of ground-based robots, with respect to surface features of the terrain involves recognition of various obstacles. Besides the areas with complicated surface features (elevations, ravines), buildings, and other human-made objects, floodable areas can be potentially dangerous for operation of outdoor robots in irregular terrain. This paper presents the algorithm for floodable areas edge detection (FAED) through analysis of surface features on heightmap. The algorithm respects height values from any points of the map, as well precipitation data in the area of interest, depending on which the wetness of certain areas can increase or decrease. Experiments were performed to compare number of cells in areas, outlined manually and detected with FAED. The experimental results showed that the total number of cells, revealed with FAED, is 17% less at average, than in areas, outlined manually. Hence, the cell array in areas, whose edges were detected with FAED, is more accurate, than with manual edge detection, what enables more efficient heightmap utilization in path planning for ground-based robotic vehicles.

1 Introduction

The problems, related to autonomous navigation of robotic vehicles (RV) in uneven terrain, represent a great scientific interest today. The solutions of these problems are highly relevant in agricultural applications of these vehicles, exploration of remote, or difficult grounds or in landscape mapping. Navigational system of such robots should enable the operator to analyze the surface, where the vehicle moves, for static obstacles, such as elevations or other impassable areas. When the robot moves over irregular terrain, areas in which water accumulates are critical: ravines, pits, and other dangerous objects. Labeling these areas as obstacles in advance will allow the path planning algorithm to build a route, bypassing these sections.

K. Zakharov (✉) · A. Saveliev
St. Petersburg Institute for Informatics and Automation of the Russian Academy of Sciences, St. Petersburg Federal Research Center of the Russian Academy of Sciences (SPC RAS), 39, 14th Line, St. Petersburg 199178, Russia

A. Ronzhin and V. Shishlakov (eds.), *Electromechanics and Robotics*, Smart Innovation, Systems and Technologies 232, https://doi.org/10.1007/978-981-16-2814-6_19

To locate floodable areas, you need elevation data, which can be obtained from a heightmap. A heightmap is usually composed as a grayscale image in which the intensity of each pixel represents a specific elevation value. For example, in the Gazebo simulation environment, the heightmap is displayed as a set of cells, which together represent a 3D terrain model. Each cell is a square with an area of 1×1 m and a specific height value. To obtain a heightmap of any area of the Earth, special services can be employed, for example Tangram Heightmapper [1].

This paper presents the floodable areas edge detection (FAED) algorithm for detecting the edges of floodable areas by analyzing the topography of a heightmap. This algorithm detects special points on the heightmap by analyzing the height value of each cell and detects the edges of floodable areas using the Moore neighborhood tracing algorithm. The algorithm considers rainfall data, which affects the occupancy of certain areas with water.

2 Related Work

Since the heightmap is initially presented as an image, the methods of detecting the edges of a connected pattern in the image can be used to outline potentially floodable areas. In digital images, expressed in binary values, a pixel can have one of the following values: 1—belonging to the pattern, or 0—belonging to the background, i.e., no grayscale. To reveal objects in a digital pattern, it is necessary to find sets of such black pixels that are "connected" to each other. On a heightmap, the edges of areas can be detected, which are limited by special points. This approach is similar to detecting the edges of patterns in image. Further, we use the following definitions: white cell is a cell whose height value is within the normal range and which is not included in the area edge; black cell is the height value of which is above or below a certain threshold, and which is included in the edge of the area.

The objects in a given digital pattern are the connected components of that pattern. Two pixels can be considered adjacent, if they have only a common edge (4-connectivity), and pixels that have a common vertex or edge (8-connectivity). To detect the edge of the image where the pixels have 4-connectivity, there is a square tracing algorithm [2]. This algorithm detects the initial (black) pixel of the image and makes it current, after which it bypasses adjacent pixels according to the following rules: if the current pixel is black (1), then a transition is made to the pixel on the left, which is added to the set of pixels of the border B and becomes the new current pixel; if the current pixel is white (0), then a transition is made to the pixel to the right, which becomes the new current pixel.

One of the disadvantages of the algorithm is the choice of the stopping criterion. In the original description of the square tracing algorithm, the termination condition is to hit the starting pixel a second time. If the algorithm depends on such a criterion, then it is not be able to detect the edges of a large family of patterns [3]. An effective criterion is proposed by Jacob Eliosoff, the essence of which is to stop after hitting the initial pixel for the second time in the same way as we hit it initially (through the same edge).

Changing the stopping criterion generally improves the efficiency of the square tracing algorithm but does not allow to overcome other disadvantages that it reveals in processing of patterns with special types of connectivity.

To detect the edge of the image, where the pixels have 8-connectivity, there exists a tracing algorithm for the Moore neighborhood [4]. Moore neighborhood of pixel P is a set of 8 pixels that have a common vertex or edge with this pixel. This algorithm detects the initial (black) pixel of the image and makes it the current one, after which it bypasses adjacent pixels according to the following rules: if the current pixel C is black, then it is added to the set B and becomes the current black pixel P, after which the transition to the previous pixel is performed. Then the algorithm starts to traverse clockwise all pixels from the Moore neighborhood $M(P)$. If the current pixel C is white, then a transition is made to the next clockwise pixel from the Moore neighborhood $M(P)$. The algorithm exits when the starting pixel becomes the current peak for the second time.

In addition to detect edges connected by a pattern, there are also methods for detecting edge topographic lines [5–8]. In [9], a modification of the Moore neighborhood tracing algorithm is presented. The main essence of the algorithm is when the current pixel is black, its Moore neighborhood is checked clockwise until no more black pixels are encountered. Then the initial pixel is again treated as the current pixel, and then the Moore neighborhood of each black pixel is checked counterclockwise until there are no more black pixels.

There is also a method for detecting area edges that uses local geometric features of a topographic map in conjunction with its general presentation [10]. However, the methods for edge detection are poorly suited for outlining of closed areas, since their stopping criterion is redundant for this task.

There is also the Radial Sweep Algorithm to detect the shape of an 8-connected pattern [11]. Its essence is similar to the Moore neighborhood tracing algorithm. Every time a new edge pixel is detected, it becomes the current pixel P. Next, an imaginary line segment is drawn that connects P to the previous boundary pixel. Then the segment rotates clockwise about P until it hits a black pixel in the vicinity of Moore neighborhood of pixel P. The rotation of the bar is identical to checking each pixel in Moore neighborhood P. In [12], an algorithm for performing Delaunay triangulation is presented for a set of points in a 2D plane in combination with a Radial Sweep Algorithm. The results of the experiments showed that the presented algorithm works approximately two times faster than the algorithm for detecting the convex hull when performing Delaunay triangulation for a randomly generated set of points.

Another algorithm for detecting the edge of an 8-connected pattern was developed by Pavlidis [13]. Any black edge pixel can be selected as the initial pixel under the following condition: if you initially stand on it, the left adjacent pixel is not black. In other words, it is necessary to "enter" the initial pixel in such a direction that the left adjacent pixel is white ("left" is taken here relative to the direction in which the initial pixel is "entered"). The algorithm of Pavlidis operates on three adjacent pixels: a pixel in front (P_2) and two pixels on its sides (P_1, P_3).

The algorithm sequentially visits pixels P_1, P_2, P_3 from left to right. The first found black pixel is added to the set B and becomes the current boundary pixel, after which the adjacent peak-selves P_1, P_2, P_3 are also found for it. If there are no black pixels among P_1, P_2, P_3, then rotate by 90° clockwise to get a new set of pixels P_1, P_2, P_3.

The algorithm finishes in two cases: the algorithm allows three rotations (each time 90° clockwise), then ends execution and declares the pixel isolated, or when the current boundary pixel is the starting pixel, the algorithm terminates execution, "declaring" that was found the edge of the pattern.

The algorithm of Pavlidis is a little more complicated than algorithm for tracing of Moore neighborhood, in which there are no special cases requiring separate processing, but it will not be able to determine the edges of a large group of patterns that have a certain kind of connectivity. The algorithm works well on 4-connected patterns. Problems with its performance arise when tracing some 8-connected patterns that are not 4-connected.

For a more accurate determination of the edges of floodable areas, in this paper, it is proposed to consider the data on the amount of precipitation in the terrain under consideration. In [14–16], a system for flood monitoring and forecasting is described. The system is based on a suite of hydrological and hydrodynamic models, integrated Earth remote sensing data, and is implemented on the basis of a service-oriented architecture. The test results show that the use of this system fully implements the necessary functionality of operational flood forecasting systems, and the fulfillment of the basic requirements for such systems indicates the possibility of widespread use of such systems in government sector and in emergency services.

As we know, algorithms for edge detection of floodable areas on a height map have not yet been developed. However, there exists an algorithm for constructing floodable areas of the area based on satellite images and analysis of feature points of terrain [17]. This algorithm uses digital topographic maps that are not accommodated for path planning activity of a ground-based robot.

To solve the problem of edge detection in floodable areas, the Moore neighborhood algorithm was chosen. The use of criterion of Jacob significantly improves the tracing performance of the Moore neighborhood, making it the best algorithm for determining the edge of any pattern, regardless of its connectivity features. Also, the tracing algorithm for Moore neighborhood is comparable in efficiency with the Radial Sweep Algorithm.

3 Detection of Floodable Areas

The initial data for the problem of detecting the edges of floodable areas are a heightmap, represented by many cells, the amount of precipitation in a certain area,

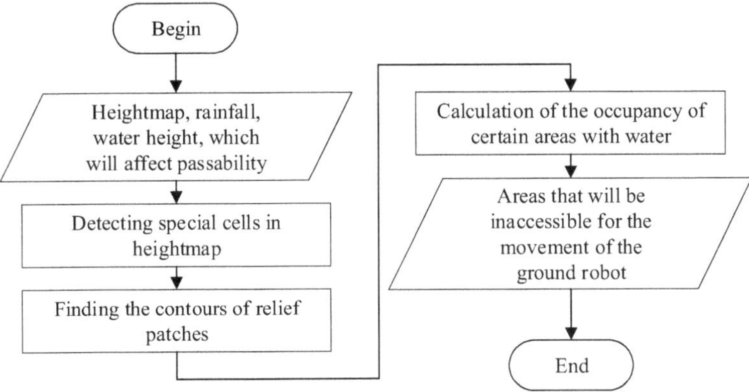

Fig. 1 Detection of floodable areas

the water height level, which will affect the terrain crossing capacity of the ground-based robot. Edge detection (see Fig. 1) begins with discovery of special cells whose height value is above or below a certain threshold.

The edges of the floodable areas are then drawn on a heightmap. For this, a starting black cell is detected, from which the construction of the edge begins using the tracing algorithm of the Moore neighborhood. The search for edges of areas ends when there are no unchecked black cells left on the heightmap. The output data of the algorithm for detecting the edges of floodable areas are a set of sets of cells included in the edges of special areas, and information on the filling of cells with water.

3.1 Algorithm for Detection of Specific Cells on Heightmap

Before starting to detect the edges of special areas to need to detect special cells on the heightmap, the height value of which is outside the norm.

Let $h_{\min}$, $h_{\max}$ be the lower and upper thresholds of the normal value of the cell height *h*(*p*), respectively. The value of these parameters is determined in advance; however, the value of $h_{\min}$ can still be influenced by the amount of precipitation that falls in a given area. Equation for calculating the value of the parameter $h_{\min}$:

$$h_{\min} = h^{b}_{\min} + \text{rf} * k_{\text{rf}},$$

where $h^{b}_{\min}$ is initial lowest threshold of the normal cell height value *h*(*p*), rf is rainfall per day (mm), k_{rf} is coefficient of influence of rainfall data.

To detect special cells on the heightmap, the height value of each cell is sequentially analyzed from bottom to top from right to left. If the value of the cell height is greater than $h_{\max}$, then it is added to the set B_h. If the value of the cell height is less than h_{min}, then it is added to the set B_l. Cells from the sets B_l and B_h will be used further when searching for the edges of the terrain regions. Figure 2 shows a flowchart of the algorithm for detecting special cells on a heightmap.

The algorithm sequentially iterates over all the cells of the heightmap and analyzes their height value. Initially, all cells are marked as white. If $h(p(i, j)) > h_{max}$ or $h(p(i, j)) < h_{min}$, then the cell is marked as black and added to the corresponding set.

This algorithm is necessary to find the heightmap cells, the height value of which is outside the norm. The found sets of cells will then be used as input to detect the edges of floodable areas that are impassable for a ground-based robot.

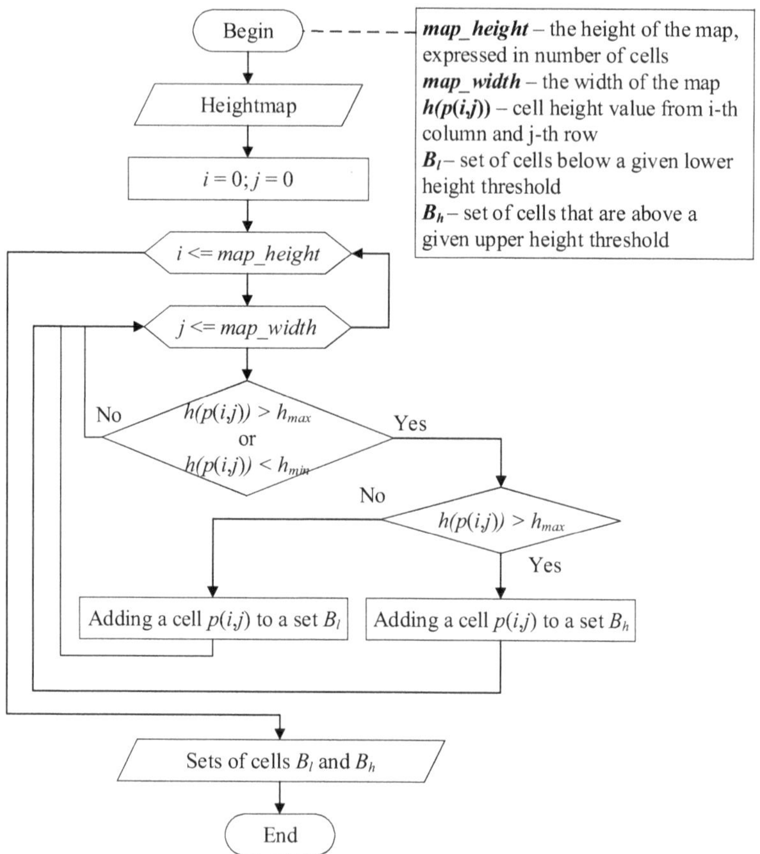

Fig. 2 Algorithm for detection of specific cells on heightmap

3.2 Algorithm for Edge Detection of Areas in Terrain

The Moore neighborhood tracing algorithm is used to detect the edges of the regions. Input data are: heightmap H, with black and white cells marked on it. Output data are: the set of sets B_{all} (B_1, B_2,…, B_k) of boundary cells, i.e., circuit. Figure 3 shows a flowchart of the algorithm for detecting the edges of terrain areas.

The conceptual idea of the algorithm is as follows: every time we hit the black cell *P*, we go back, that is, to the white cell in which we were situated earlier. Then we go around cell *P* clockwise, visiting each cell in its Moore neighborhood, until we get into a black cell. To detect the starting cell, the following sequence is used: the heightmap cells are iterated sequentially from bottom to top from left to right until a black cell is detected, which is assigned as the starting one. If the initial cell is in the set B_l, then only cells from the set B_l will be tracked. Accordingly, if the initial cell is from the set B_h, then only cells from the set B_h will be tracked.

This is done in order to avoid the problem of detecting incorrect edges of the area. In the algorithm for tracing the Moore neighborhood, stopping criterion of Jacob is used, the essence of which is to stop after hitting the initial pixel for the second time in the same way as we hit it initially (through the same edge). The use of criterion of Jacob significantly improves the tracing performance of the Moore neighborhood and allows detecting the edges of the largest number of pattern groups.

After the completion of one iteration of the algorithm, aa edge of one closed area will be created, within which the value of the cell height is higher or lower than the specified threshold. Further, the cells from the set B_i, which make up the edge of the *i*-th region, are marked as checked, and a new iteration of the algorithm starts from detecting a new initial cell. The algorithm finishes its work when there are no unviewed black cells left on the heightmap.

3.3 Calculation of Flooding Patterns for Specific Areas

When the robot moves over irregular terrain with floodable areas, it is important to consider the water level at each segment of the path. If the water level in the heightmap cell exceeds the threshold, deemed appropriate according to the technical characteristics of the robot, this can result in robot damage. The filling of certain areas with water is calculated as the sum of the volumes of all cells included in this area. The volume of the cell is calculated by the following equation:

$$V_{i,j} = s * (h_{\min} - p(i, j).z),$$

where *i*, *j* are the column and row of the cell, *s* is the area of the cell (1 × 1 m). The information about the water level in the cell can later be used when planning the path of the ground-based robot on the heightmap. The water level can be used as one of

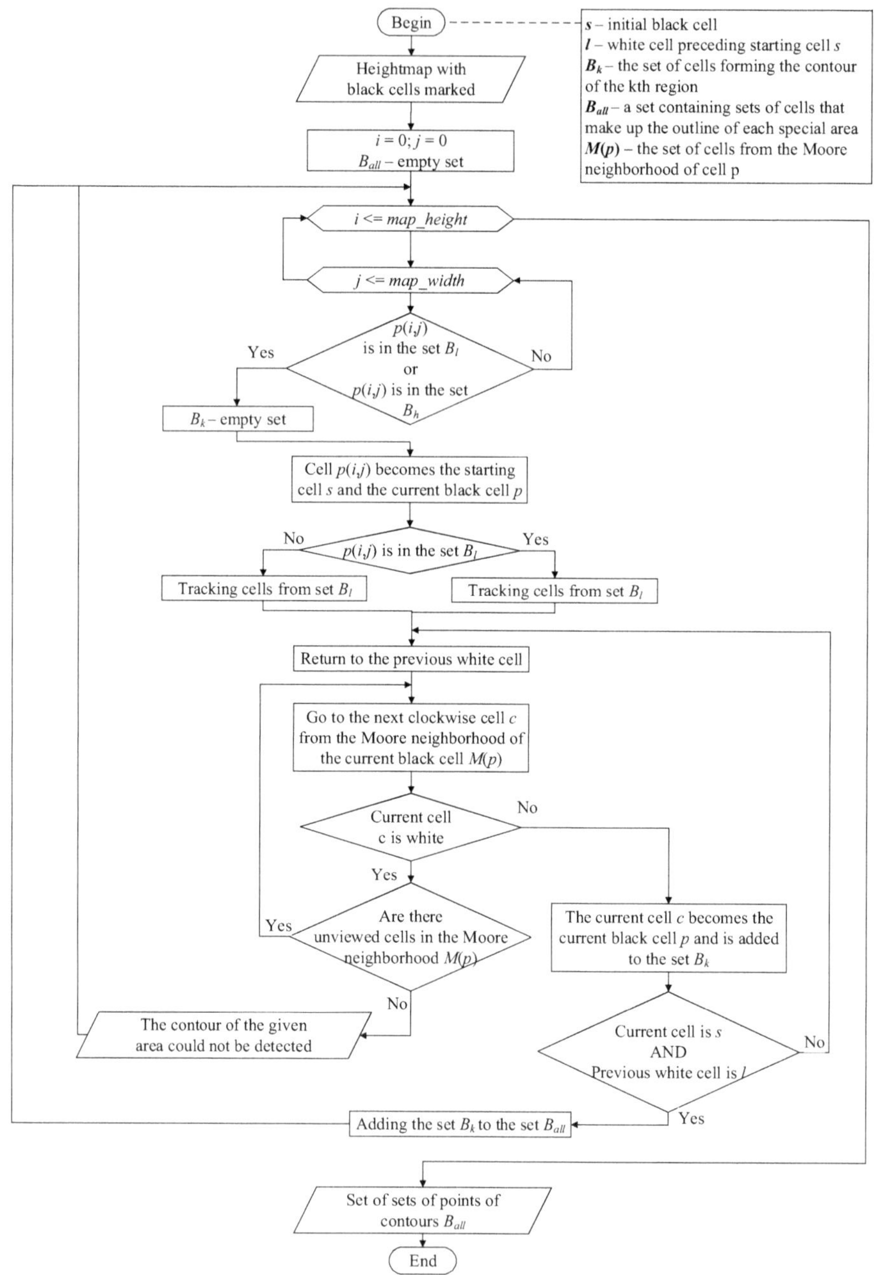

Fig. 3 Flowchart of the algorithm for detecting the edges of terrain sections

the parameters on which the cost of switching between cells depends. Cells where the water level exceeds a certain threshold will be marked on the map as obstacles.

The total amount of water in a certain area depends on the amount of precipitation, because with the increase of this parameter, the edge of each area, the value of the cell height in which is lower than $h_{\min}$, will increase. With an increase in the value of $h_{\min}$, the volume of water in each cell of the flooded area will also increase.

To detect all the cells included in a certain area, the Moore neighborhood of each cell from the edge is analyzed for the presence of unvisited black cells, which are added to the general set of edge cells. Figure 4 shows an algorithm for detecting the cells included in a certain area.

This algorithm detects all cells within a certain area bounded by a path. Preprocessing of the heightmap when detecting the edges of special areas excludes the addition of erroneous cells to the set of all cells of the area, because all circuits are isolated from each other. Detecting the cells included in a certain area will help to accurately calculate the water occupancy of certain areas.

4 Results

To test the effectiveness of the developed FAED approach, experiments were performed to compare the number of cells in the edges detected manually and using FAED. For the experiments, 5 heightmaps of 127 $\times$ 127 pixels were generated. In Fig. 5, in the form of grayscale images, heightmaps are presented, on which the edges of special areas are outlined with red lines (above are the edges of FAED, below are edges, detected manually).

Figure 6 shows a comparison of the total number of cells in the edges detected in different ways.

Figure 6 shows that the areas detected using FAED, on average, contain 17% fewer cells than the areas detected manually. This is because the FAED algorithm analyzes each cell in the heightmap to accurately outline each area. This makes the path cell array more accurate than the approach with manual path outlining. Detecting of floodable areas can help the path planning algorithm find a more efficient and safer route for the ground-based robot to move over irregular terrain depending on rainfall and other natural factors.

5 Conclusion

In this paper, an algorithm FAED was presented to detect edges of floodable areas on the heightmap. The approach is based on heightmap analysis and detection of such cells on this map, whose height value of is outside of a certain range. After analyzing the values of the heights of the cells, the tracing algorithm of the Moore neighborhood is applied to detect the edges of each floodable area. Experiments were

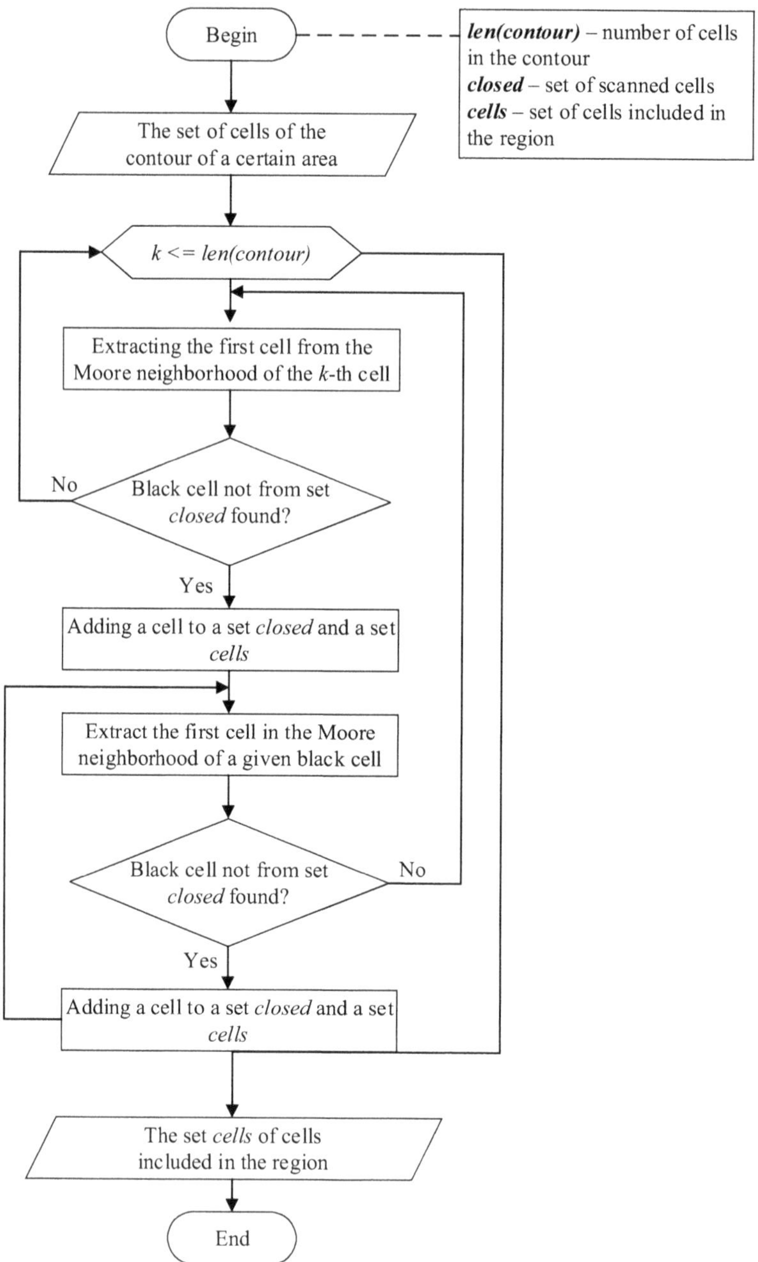

Fig. 4 Algorithm for detection of cells, which belong to the specific area

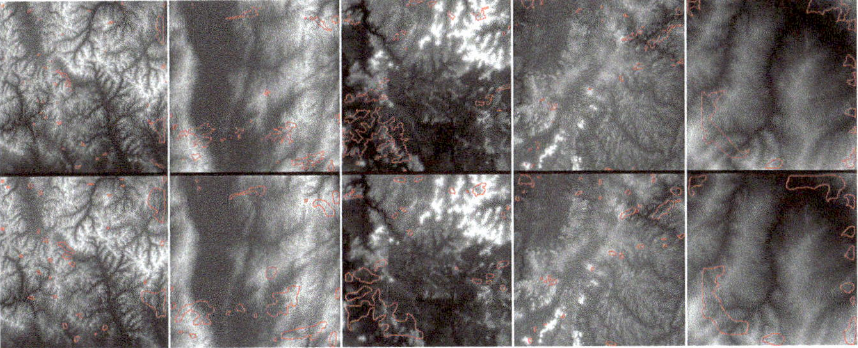

Fig. 5 Matching of edges, detected in different ways

Fig. 6 Comparison of total number of cells in edges, detected in different ways

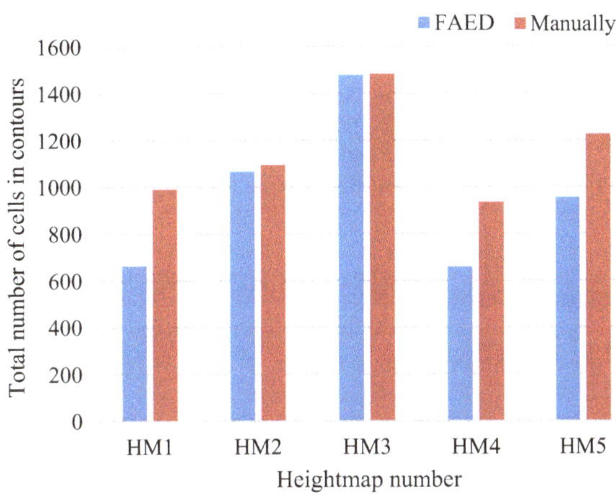

carried out to compare the total number of cells in the edges, detected manually and using an algorithm.

The results of the experiment showed that the total number of cells in the edges detected by the algorithm is, on average, 17% less than that of the edges detected manually. This is due to the higher accuracy of the FAED algorithm relative to manual marking when detecting special points and edges of areas. This will help reduce the amount of memory for storing path cells, which will be a big advantage when zooming in on the map. In the future, this approach is planned to be combined with the algorithm for path planning of a ground-based robot in a 3D environment [18] to build path considering the influence of the environment on the parameters of the terrain.

References

1. Tangrams. https://tangrams.github.io/heightmapper/. Last accessed 2021/01/11
2. Wakayama, T.: A core-line tracing algorithm based on maximal square moving. IEEE Trans. Pattern Anal. Mach. Intell. **1**, 68–74 (1982)
3. Reddy, P.R., Amarnadh, V., Bhaskar, M.: Evaluation of stopping criterion in contour tracing algorithms. Int. J. Comput. Sci. Inf. Technol. **3**(3), 3888–3894 (2012)
4. Moore, F.R., Langdon, G.G.: A generalized firing squad problem. Inf. Control **12**(3), 212–220 (1968)
5. Seo, J., et al.: Fast contour-tracing algorithm based on a pixel-following method for image sensors. Sensors **16**(3), 353 (2016)
6. Carlson, D.A., et al.: Diagnosis of esophageal motility disorders: esophageal pressure topography versus conventional line tracing. Am. J. Gastroenterol. **110**(7), 967 (2015)
7. Kim, K., et al.: Comparison of divertor heat flux splitting by 3D fields with field line tracing simulation in KSTAR. Phys. Plasmas **24**(5), 052506 (2017)
8. Richiusa, M.L., et al.: Bare and limiter DEMO single module segment concept first Wall misalignment study by 3D field line tracing. Fusion Eng. Des. **160**, 111839 (2020)
9. Pradhan, R., et al.: Contour line tracing algorithm for digital topographic maps. Int. J. Image Process. (IJIP) **4**(2), 156–163 (2010)
10. Salvatore, S., Guitton, P.: Contour line recognition from scanned topographic maps. J. WSCG **12**(1–3), 419–426 (2004)
11. Mirante, A., Weingarten, N.: The radial sweep algorithm for constructing triangulated irregular networks. IEEE Ann. Hist. Comput. **2**(3), 11–21 (1982)
12. Sinclair, D.: S-hull: A Fast Radial Sweep-Hull Routine for Delaunay Triangulation. arXiv: 1604.01428 (2016)
13. Pavlidis, T.: Algorithms for Graphics and Image Processing. Springer Science & Business Media (1982)
14. Krylenko, I., et al.: Modeling ice-jam floods in the frameworks of an intelligent system for river monitoring. Water Resour. **47**, 387–398 (2020)
15. Zelentsov, V.A., Potryasaev, S.A., Pimanov, I.Y., Ponomarenko, M.R.: Integrated use of GIS, remote sensing data and a set of models for operational flood forecasting. Int. Arch. Photogramm. Remote Sens. Spatial Inf. Sci. **XLII-3/W8**, 477–483 (2019)
16. Zelentsov, V.A., et al.: A model-oriented system for operational forecasting of river floods. Her. Russ. Acad. Sci. **89**(4), 405–417 (2019)
17. Efremova, O., Kunakov, J., Pavlov, S., Sultanov, A.: Development of a processing method of digital maps and satellite images for solving problems of emergencies. Int. Soc. Opt. Photon. **10774**, 1077419 (2018)
18. Zakharov, K., Saveliev, A., Sivchenko, O.: Energy-efficient path planning algorithm on three-dimensional large-scale terrain maps for mobile robots. In: International Conference on Interactive Collaborative Robotics, pp. 319–330. Springer, Cham (2020)

External RGB-D Camera Based Mobile Robot Localization in Gazebo Environment with Real-Time Filtering and Smoothing Techniques

Kirill Kononov, Roman Lavrenov, Lilia Gavrilova, and Tatyana Tsoy

Abstract In previous work, we successfully studied the possibility of a mobile robot localization using an external RGB-D camera. We conducted virtual experiments in a Gazebo simulator using ROS with a Turtlebot3 Waffle Pi as a mobile robot. find_object_2d package was used to localize the Turtlebot3 and send its computed position to ROS. Thus, we extended our research and ratchet that result up by implementing the algorithm for filtering and smoothing the computed mobile robot position. Our task was to develop a thin library that is a wrapper over the find_object_2d package. The filtering algorithm and smoothing can compute the supposed robot position or predict it in cases when the robot disappears from the camera's field of view. We conducted virtual experiments over again and draw a comparison between previous results (without filtering and smoothing algorithm) and current results (using filtering and smoothing algorithm with slight improvements).

1 Introduction

Correct robot localization is important for any mobile robot tasks, including path planning, mapping [1], SLAM, performing critical tasks, operating in special environments, etc. [2–4]. In most cases, an inaccurate localization becomes a serious problem that might cause a wrong robot behavior [5]. An incorrect mobile robot localization is an considerable problem that arises due to accumulation of odometry errors, harsh environment, sliding and slippage of wheels on an underlying support surface, noisy or unstable GPS signal, and other different problems [6]. This way, a robot transmits less and less relevant data about its location over time.

Integration of filtering and smoothing methods into localization algorithms could significantly improve a localization accuracy [7]. In [8] Kalman filter [9] was used

K. Kononov · R. Lavrenov (✉) · L. Gavrilova · T. Tsoy
Laboratory of Intelligent Robotics Systems (LIRS), Intelligent Robotics Department, Institute of Information Technology and Intelligent Systems, Kazan Federal University, Kazan, Russia
e-mail: lavrenov@it.kfu.ru
URL: https://www.kpfu.ru/eng/itis/research/laboratory-of-intelligent-robotic-systems.com

A. Ronzhin and V. Shishlakov (eds.), *Electromechanics and Robotics*, Smart Innovation, Systems and Technologies 232, https://doi.org/10.1007/978-981-16-2814-6_20

for filtering and smoothing of a mobile robot localization based on odometric and sonar sensors. In [10] a Particle filter [11] was employed for smoothing robot self-localization, which was based on a strength of a WLAN signal.

In our previous work, we had studied a possibility of tracking a mobile robot position within the Gazebo simulator [12, 13] using an external RGB-D camera and find_object_2d package [14]. find_object_2d package [14] is a simple application that allows to detect a particular object within an image from a pre-created dictionary of objects using different types of OpenCV [15] detectors and descriptors. We employed ORB (Oriented FAST [16] and rotated BRIEF [17]) algorithm [18] since it provides a good combination of the detector and the descriptor for a task, which requires to localize the mobile robot rapidly and accurately. The resulting trajectory of the robot computed positions was not smooth and formed a chaotic zigzag line, the robot frequently failed to calculate its position (due to find_object_2d calculations' failures in about 20% of cases) and an average localization inaccuracy was 0.14 m while traveling within a 6 × 6 m room. A new algorithm presented in this paper performs filtering and smoothing of a moving robot localization data, successfully solves the trajectory smoothness issue, failures of a position calculation, and improves the average localization inaccuracy in approximately 2.25 times.

2 Related Work

The usage of filtering and smoothing algorithms for localization were described in [8] and [10]. In [8] Kalman filter [9] was used for filtering and smoothing of the mobile robot localization based on odometric and sonar sensors. In [10] the realization of Particle filter [11] was used for smoothing robot localization which is based on the strength of WLAN signal. In our work, we introduce an algorithm-helper that assist the robot to localize itself indoor using an external RGB-D camera. Thus, the developed method of filtering and smoothing of the mobile robot localization will increase its accuracy and ensure a smooth trajectory of its movement.

3 Proposed Approach

In our research we used the Gazebo simulator [13] with ROS [19] to simulate mobile robot movements in an arbitrary empty room. Without loss of generality was selected a room of size 6 ×6 m and was connected a RGB-D camera to the ceilings at a geometrical center of the room. This allows the camera to capture an entire floor surface and localize the robot. To track the robot position, find_object_2d package provided its coordinates through tf [20] and publish them in ROS [19]. Inaccuracy of localization was calculated as a translation between robot base_link coordinates and the obtained from find_object_2d coordinates. One of the issues that arose during virtual experiments was a piece-wise trajectory of the computed robot coordinates,

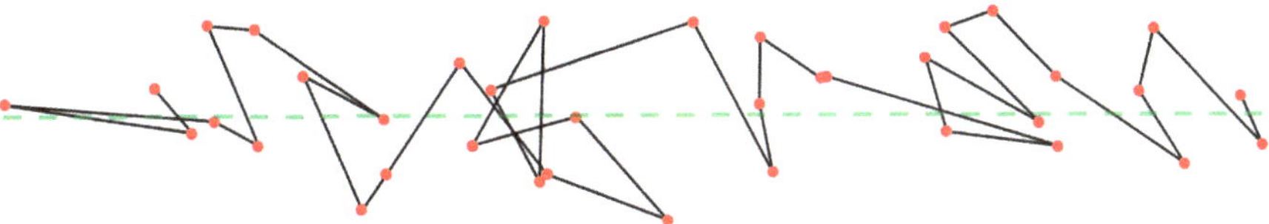

Fig. 1 An example of the robot coordinates trajectory computed with find_object_2d package [14]. Black dots are the coordinates, red line is the trajectory constructed from these points, and green dotted line is the real trajectory of the robot

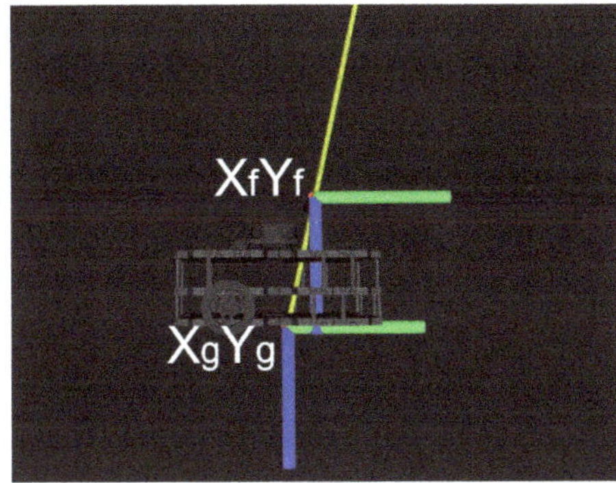

Fig. 2 XfYf frame origin corresponds to (x_f, y_f) point, $x_g y_g$ frame origin corresponds to (x_g, y_g) point

which was obviously incorrect (Fig. 1). A natural approach to this problem was to apply a filter that could smooth a trajectory and thus increase the accuracy of the localization.

Package find_object_2d computes robot position (x_f, y_f) as a closest to the camera point of the robot. In [21] obtained (x_f, y_f) was further projected orthogonally onto the XY plane of a support surface (floor) and next a distance from that projected point to base_link of the mobile robot was calculated. First improvement in the current approach is a straight line from the camera frame origin to (x_f, y_f). The line was further extended until it reached XY plane. The intersection point was labeled as (x_g, y_g) (Fig. 2).

Another improvement is switching the coordinate frame that we accepted as the correct robot position from base_link frame origin (in [21]) to base_scan frame origin. We use the base_scan as the closest coordinate frame to the robot center (Fig. 3). It was done due to the previously [21] calculated by us localization inaccuracy as the translation between computed robot position frame and base_link frame. The last one is placed between robot center and its front border. Thus, we use base_scan frame to calculate the inaccuracy of the localization, since it locates quite close to the robot center.

Before comparing the difference between localization accuracy before and after improvements, we want to explain the filtering algorithm we used in our previous work [21]. We introduced the α-value that means the distance between real robot position and robot position computed with find_object_2d [14]. If in a certain iteration

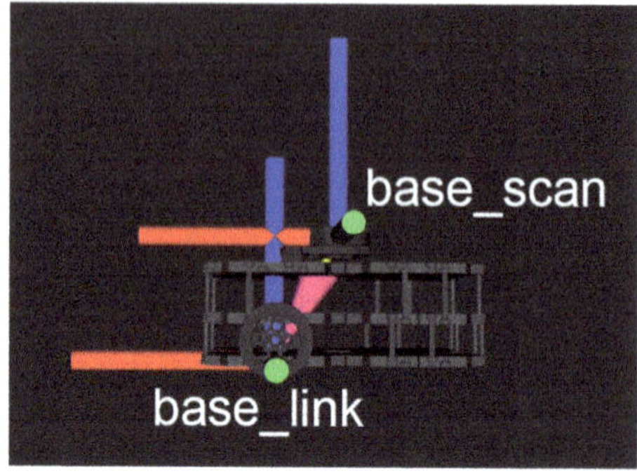

Fig. 3 Difference between positions of base_link and base_scan frames. base_scan frame is quite close to the robot center instead of base_link frame

α was more than 0.25 m, we would not accept this computed robot position. We undertook a small research and found out that such value of α is the most optimal and effective since it provides about 81–82% of acceptable robot position computations and the average value of inaccuracy was 0.13 m. Increasing the α value would also increase the percentage of acceptable robot position computations and as well it would increase the average inaccuracy of robot localization. Since we are interested in increasing the percentage of acceptable computed positions and decreasing the average computation inaccuracy, in our previous work we chose the α value equal to 0.25 m as the most effective [21]. However, those two new improvements make it possible to change the α value to get better results that are presented in Table 1.

These two improvements had a positive effect on the overall localization accuracy. Finally, our main improvement was the development of the algorithm which filters and smooths post-data from the find_object_2d package [14] and sends the processed predicted position of the mobile robot back to ROS [19]. We used tf package [20] to compare a predicted position with a real position, analyze received results, and to compare them to the previous research results.

Table 1 Comparison of the different configuration of the α value

Configuration	Acceptable computations (%)	AVG inaccuracy (m)
$\alpha = 0.25$ m (Before improvements)	82	0.14
$\alpha = 0.25$ m (After improvements)	97	0.09
$\alpha = 0.14$ m (After improvements)	82	0.075

The implemented improvements make it possible to obtain better results with the same value of α: acceptable computations increased from 82 to 97%, and the average inaccuracy decreased from 0.14 to 0.09 m. Reducing α to 0.14 m gives an average inaccuracy of 0.075 m, while saving percentage of acceptable computations at 82% as it was before the improvements

4 Algorithm Explanation

Developed filtering and smoothing algorithm uses data from find_object_2d package [14] which sends its data about a detected robot to ROS [19] through tf [20]. Then, our algorithm processes that data to compute the supposed position of the detected robot (Fig. 4).

Parameters are used in our algorithm:

- **window_size:** size of the array which contains the history of previous coordinates. The larger this value, the smoother the trajectory becomes, if so the algorithm reacts more slowly to sudden changes in speed and direction of the robot movement. Further denotes as "Window size".
- **alpha:** value for filtering input coordinates. If the distance between current and previous coordinates is more than this value, the current coordinate is not accepted and replaced by the predicted position which is calculated based on calibration. Further denotes as "α".

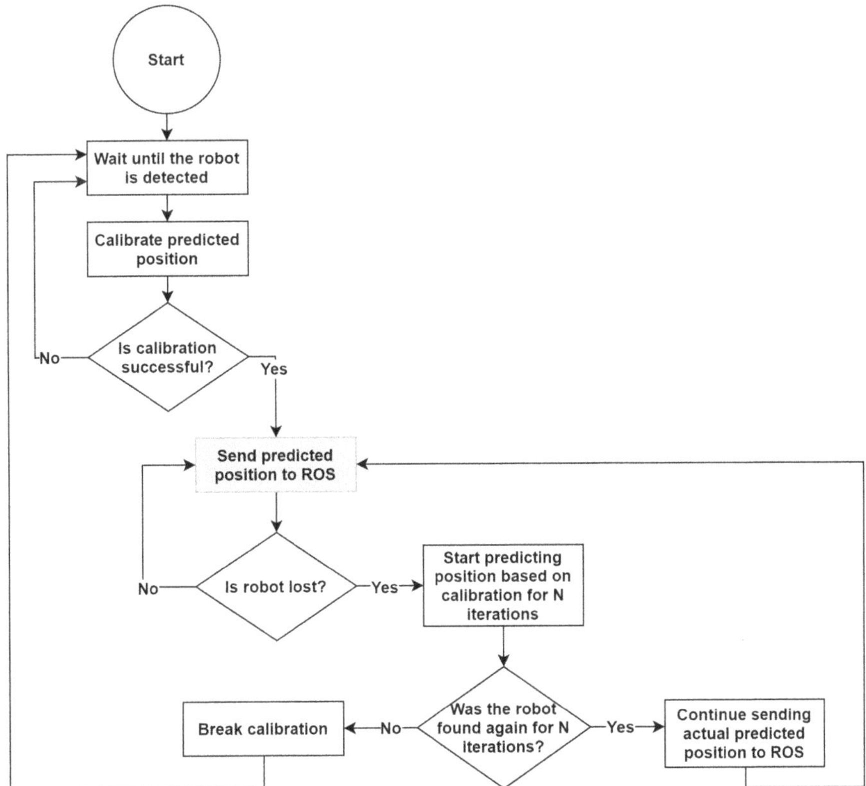

Fig. 4 Flow diagram which explains our filter and smoothing algorithm

- **calibration_threshold:** delay before the start of the calibration. This delay is necessary because the find_object_2d package [14] is very unstable to localize a robot that enters the camera‘s field of view when its body is not fully visible. Further denotes as “Calibration threshold”.
- **frame_loss_threshold:** delay before resetting of the calibration. Used while the algorithm is predicting the movement of a robot that cannot be localized by the find_object_2d [14] package after a successful calibration decrements at each iteration if the robot was not detected. When this parameter reaches 0, the calibration is reset and the robot position prediction stops. Further denotes as “Frame loss threshold”.

The developed algorithm contains the following parts: (a) calibration of predicted position, (b) filtering and smoothing the trajectory of robot movement based on calibration, (c) predicting the position of the robot that has disappeared from the camera‘s field of view.

Calibration of the predicted position is the process which tries to capture robot position for a certain count of iterations and then analyze received localization data. Analyzing of that data means the calculation of the average offset between computed coordinates (See Eq. 1):

$$\text{Average Offset} = \left(\sum_{i=2}^{\text{WindowSize}} (\text{Buffer}_i - \text{Buffer}_{i-1}) \right) / \text{WindowSize}, \quad (1)$$

where WindowSize specifies the size of the buffer. Buffer is the coordinate container of previously computed coordinates, i is the index of the coordinate in the buffer.

After the successful calibration, we receive average offset between successive coordinates and initialize a new instance of the “Predicted position” which contains (X; Y) coordinate of the last robot position received from find_object_2d package [14]. At this moment, algorithm has an array-buffer of last X (X is equal to the Window size of the Buffer) computed coordinates of the robot. During the next iteration, we receive new computed robot coordinates from find_object_2d package [14] and delete the oldest element from the buffer. Further, we calculate average offset between all remaining elements in the buffer including the last computed one, add this offset to the predicted position coordinate and push this coordinate to the buffer (See equation to compute predicted position):

$$\begin{aligned}\text{Average Offset} = &\left(\sum_{i=2}^{\text{WindowSize}-1} (\text{Buffer}_i - \text{Buffer}_{i-1}) \right. \\ &\left. + (\text{ComputedPos} - \text{Buffer}_{\text{WindowSize}}) \right) / \text{WindowSize},\end{aligned} \quad (2)$$

$$\text{PredictedPos} = \text{PredictedPos} + \text{AverageOffset}, \quad (3)$$

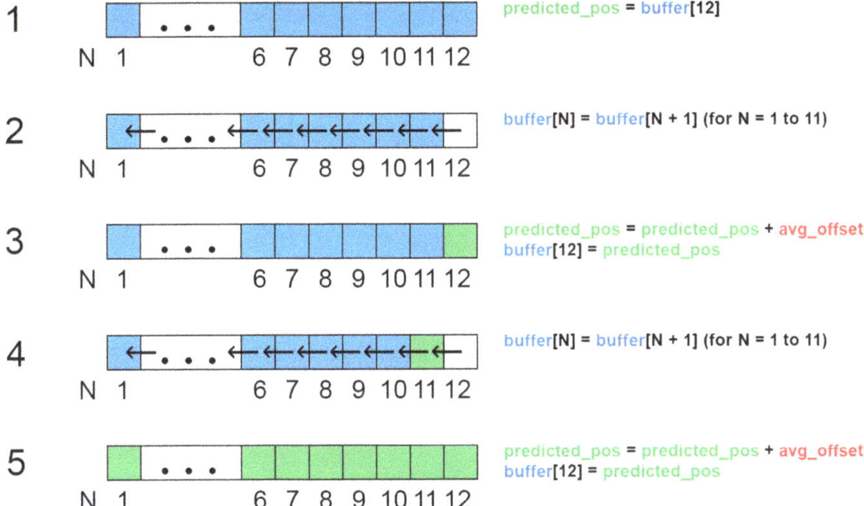

Fig. 5 Part of algorithm's work after successful calibration (Window size = 12). Blue cells are coordinates that are computed with find_object_2d package [14] and green cells are predicted (smoothed) coordinates. Step 1: set predicted position (predicted_pos) as the last coordinate from the buffer. Step 2: shift coordinates left by 1 (at this moment algorithm received a new computed coordinate). Step 3: calculate average offset (avg_offset) between coordinates in buffer (11 elements) including last computed coordinate (1 element). Add calculated avg_offset to predicted_pos and push updated predicted_pos in the end of the buffer. Step 4: repeat actions from Step 2. Step 5: after repeating Steps 3 and 4 for X iterations (X is equal to the Window size parameter) the buffer contains only smoothed coordinates

where WindowSize specifies the size of the buffer, Buffer is the coordinate container of previously computed coordinates, i is the index of the coordinate in the buffer, ComputedPos is the last computed position of the robot that is received from find_object_2d package [14].

These steps are repeated at every iteration and the algorithm produces a smooth trajectory of robot movement (Fig. 5).

Calibration could be failed in certain cases and our algorithm tries to predict the possible position of the robot:

- **Accumulated computation errors If find_object_2d** [14] computes the wrong robot position for a certain number of iterations in a row, the algorithm breaks, and the predicted position stops at a certain point. The distance between computed robot position and predicted position is greater than the α value, so our algorithm can't accept new computed coordinates due to this fact, and after a certain number of iterations, which is set by the value of the frame loss threshold parameter, calibration is canceled. Further algorithm waits for detection of the robot and begins a new calibration.
- **Robot loss from the camera's field of view** If the robot drives into the places where it cannot be detected by a camera, the algorithm tries to predict its possible

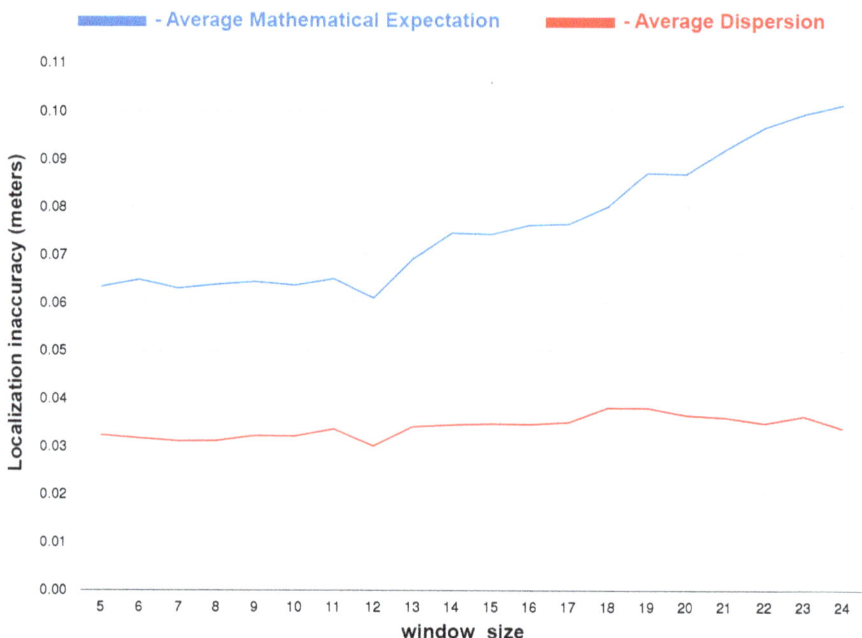

Fig. 6 Results of finding the most optimal Window size parameter for our algorithm. X-axis means the value of this parameter, Y-axis means inaccuracy of the robot localization in meters. Window size = 12 provides the most accurate filtering and smoothing results

movement trajectory for a certain number of iterations, which is set by the value of the frame loss threshold parameter. If the robot is not detected again, the calibration is canceled too and the algorithm waits for detection of the robot and begins a new calibration.

We conducted the experiments to define the most optimal and effective Window size and α parameters. Started with the first one (α was set at 0.25 m as the most optimal value in our previous work) we conducted 10 experiments for each value from 5 to 24, and found that 12 is the most efficient value for this parameter (Fig. 6), since it provides minimal average mathematical expectation and minimal average dispersion. So, we have found the most optimal value for the Window size parameter and further proceeded to determine the most optimal value for the α parameter. We repeated our experiments with Window size = 12, but changing α parameter from 0.05 to 0.5 with a step of 0.025. The experiments result show that all values that are more than 0.2 m give almost the same results (Fig. 7), but we decided to use α equal to 0.3 m.

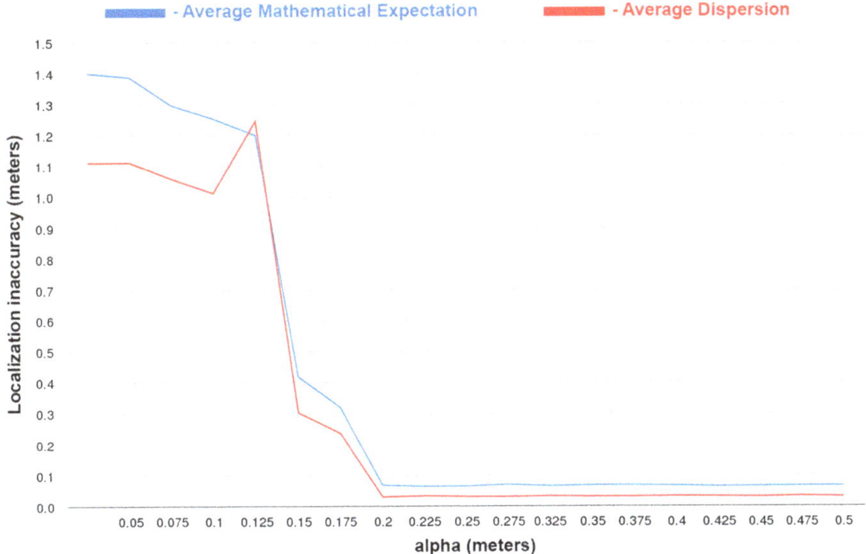

Fig. 7 Results of finding the most optimal α parameter for Window size parameter set at 12. X-axis means the value of the α parameter, Y-axis means inaccuracy of the robot localization in meters. With α greater than 0.2 m we get similar results

5 Experimental Results

We conducted the identical experiments as it was in our previous research [21] and compared old and new results for the same parameters. Our experiments contained linear and curvilinear routes:

- **Linear movement** 8 direct routes covering all areas of the room. Each route was tested 10 times to obtain average results.
- **Curvilinear movement** large circle (radius = 2 m) movement, small circle (radius = 1.25 m) movement, and 3 different chaotic routes. Each route was tested 10 times to obtain average results.

The analysis of the results (presented in Table 2) showed that using our algorithm for filtering and smoothing of the mobile robot localization data gives more accurate and stable average values.

The first method ("previously" in Table 1) supposes localizing of the mobile robot without using the developed filtering and smoothing algorithm. This method is based on a simple filter implemented on accepting computed coordinates only with an α less than 0.25 m (more details are available in our previous work [21]).

The second method ("currently" in Table 1) supposes the use of the new filtering and smoothing algorithm presented in this paper. The algorithm developed by us is based on the calibration of the predicted robot position. During the calibration, the algorithm calculates the average coordinate offset for certain iterations and then

Table 2 Comparison of the previous and current results obtained by experiments

	Mathematical expectation (m)	Dispersion (m)	Minimum (m)
Linear movement (previously)	0.136	0.045	0.028
Linear movement (currently)	0.044	0.015	0.010
Curvilinear movement (previously)	0.125	0.055	0.006
Curvilinear movement (currently)	0.072	0.029	0.030
Average value (previously)	0.130	0.049	0.017
Average value (currently)	0.058	0.022	0.020

All values calculated as the average of a specific type of experiment

uses this data to compute the smoothed predicted position of the robot. The average coordinate offset is updated at each new iteration, so the algorithm can react to sudden changes in the robot movements. Furthermore, the new algorithm can predict the approximate robot position in cases where the find_object_2d package [14] localizes the robot with significant errors or in other cases where the robot disappears from the camera's field of view (for example, there is some obstacle between the robot and the camera).

According to expectation, the linear movement gives the most accurate localization result, while curvilinear movement localization also becomes more accurate, but less accurate than linear movement localization. We can conclude that our algorithm reduces the localization inaccuracy by more than a half. However, the average minimum is still not close to zero, so we can confirm that our algorithm has a guaranteed error which is about 0.02 m.

6 Conclusions

It follows that the conducted experiments were successful and developed algorithm for filtering and smoothing the computed robot localization works properly. This method makes it possible to localize a mobile robot using the find_object_2d package [14] more accurately, avoid critical computational errors, successfully smooth the trajectory of the mobile robot movement, and predict the possible robot position when it drives into a place where the camera cannot detect it. The developed algorithm improves the average localization inaccuracy in approximately 2.25 times. Our future task is to develop a new method of mobile robot localization using an external RGB or RGB-D camera which will allow to detect and track several robots properly in difficult tasks, for example during the localization of the identical robots or/and different robots using only one camera [22].

Acknowledgements This work was supported by the Russian Foundation for Basic Research (RFBR), project ID 19-58-70002.

References

1. Iakovlev, R., Saveliev, A.: Approach to implementation of local navigation of mobile robotic systems in agriculture with the aid of radio modules. Telfor J **12**(2), 92–97 (2020)
2. Giesbrecht, J.: Global path planning for unmanned ground vehicles. Technical Report Defence Research And Development, Suffield (Alberta) (2004)
3. Lavrenov, R., Magid, E.: Towards heterogeneous robot team path planning: acquisition of multiple routes with a modified spline-based algorithm. In: MATEC Web of Conferences. vol. 113, p. 02015. EDP Sciences (2017)
4. Magid, E., Tsubouchi, T.: Static balance for rescue robot navigation: discretizing rotational motion within random step environment. In: International Conference on Simulation, Modeling, and Programming for Autonomous Robots. pp. 423–435. Springer (2010)
5. Bai, Y., Wang, Y., Svinin, M., Magid, E., Sun, R.: Function approximation technique based immersion and invariance control for unknown nonlinear systems. IEEE Control Syst Lett **4**(4), 934–939 (2020)
6. Panov, A.I., Yakovlev, K.S., Suvorov, R.: Grid path planning with deep reinforcement learning: preliminary results. Procedia Comput Sci **123**, 347–353 (2018)
7. Ronzhin, A., Rigoll, G., Meshcheryakov, R.: Interactive Collaborative Robotics. Springer (2016)
8. Jetto, L., Longhi, S., Venturini, G.: Development and experimental validation of an adaptive extended Kalman filter for the localization of mobile robots. IEEE Trans Robotics and Automat **15**(2), 219–229 (1999)
9. Welch, G., Bishop, G., et al.: An introduction to the kalman filter (1995)
10. Nurminen, H., Ristimäki, A., Ali-Löytty, S., Piché, R.: Particle filter and smoother for indoor localization. In: International Conference on Indoor Positioning and Indoor Navigation. pp. 1–10. IEEE (2013)
11. Nummiaro, K., Koller-Meier, E., Van Gool, L.: An adaptive color-based particle filter. Image Vision Comput **21**(1), 99–110 (2003)
12. Abbyasov, B., Lavrenov, R., Zakiev, A., Yakovlev, K., Svinin, M., Magid, E.: Automatic tool for gazebo world construction: from a grayscale image to a 3D solid model. In: 2020 IEEE International Conference on Robotics and Automation (ICRA). pp. 7226–7232. IEEE (2020)
13. Koenig, N., Howard, A.: Design and use paradigms for gazebo, an open-source multi-robot simulator. In: 2004 IEEE/RSJ International Conference on Intelligent Robots and Systems (IROS)(IEEE Cat. No. 04CH37566). vol. 3, pp. 2149–2154. IEEE (2004)
14. Labbé, M.: Find-Object. http://introlab.github.io/find-object (2011), Accessed 14 May 2020
15. Noble, F.K.: Comparison of opencv's feature detectors and feature matchers. In: 2016 23rd International Conference on Mechatronics and Machine Vision in Practice (M2VIP). pp. 1–6. IEEE (2016)
16. Viswanathan, D.G.: Features from accelerated segment test (fast). In: Proceedings of the 10th workshop on Image Analysis for Multimedia Interactive Services, London, UK. pp. 6–8 (2009)
17. Calonder, M., Lepetit, V., Strecha, C., Fua, P.: Brief: Binary robust independent elementary features. In: European Conference on Computer Vision, pp. 778–792. Springer (2010)
18. Rublee, E., Rabaud, V., Konolige, K., Bradski, G.: Orb: An efficient alternative to sift or surf. In: 2011 International Conference on Computer Vision, pp. 2564–2571. IEEE (2011)

19. Quigley, M., Conley, K., Gerkey, B., Faust, J., Foote, T., Leibs, J., Wheeler, R., Ng, A.Y., et al.: Ros: an open-source robot operating system. In: ICRA Workshop on Open Source Software. vol. 3.2, p. 5. Kobe, Japan (2009)
20. Foote, T.: tf: the transform library. In: 2013 IEEE Conference on Technologies for Practical Robot Applications (TePRA), pp. 1–6. IEEE (2013)
21. Kononov, K., Larvenov, R., Tsoy, T., Martínez-García, E.A., Magid, E.: Virtual experiments on mobile robot localization with external smart RGB-D camera using ROS. In: IEEE Eurasia Conference on IOT, Communication and Engineering (2021)
22. Safin, R., Garipova, E., Lavrenov, R., Li, H., Svinin, M., Magid, E.: Hardware and software video encoding comparison. In: 2020 59th Annual Conference of the Society of Instrument and Control Engineers of Japan (SICE). pp. 924–929. IEEE (2020)

Investigation of the Possibility of Vector-Command Control Based on Forearm EMG

Natalia Budko, Mikhail Medvedev, Artem Budko, and Raisa Budko

Abstract The work is devoted to solving the problems of increasing the depth and increasing the long-term stability of communication channels in human–machine interfaces, built on the electrical activity databases of the forearm muscles. For this, a method for analyzing electromyogram (EMG) signals is proposed, which combines vector and command control. The mathematical model for vector analysis of EMG is built in spherical coordinates based on the real spatial arrangement of the electrodes on the forearm, taking into account the possibility of a random phase shift during operation. Vector analysis of EMG is used to solve the calibration task of EMG sensors channels by the spatial arrangement of the electrodes and calculating the resultant vector of muscular forces. These forces are used as an additional information channel to set the movement direction of the control object operating point. Command control is based on gesture recognition by means of a pretrained artificial neural network (ANN) of the multilayer perceptron type. Processing results of actually recorded EMG signals by the proposed method are presented. There are given research results of correlation between processed signal fragments duration and the process of extracting the hand rotational movement information. It is proposed to use the signal duration of 250 ms. An algorithm of reassigning and calibrating the EMG channels amplification is proposed, which makes it possible to use further a once trained ANN for recognition and classification of gestures, while the position changing of electrodes between operation sessions is allowed. The work results can be used for calibration algorithms development, gesture recognition, and control of technical objects based on electromyographic human–machine interfaces.

1 Introduction

The task of controlling mechatronic and robotic systems is especially relevant for the development of human–machine interfaces for people with disabilities [1–7]. The number of disabled people in November 2020 is more than 11 million people, which

N. Budko (✉) · M. Medvedev · A. Budko · R. Budko
Research and Development Institute of Robotics and Control System (RDIRCS), Southern Federal University, 2, Shevchenko str., Taganrog 347928, Russia

A. Ronzhin and V. Shishlakov (eds.), *Electromechanics and Robotics*, Smart Innovation, Systems and Technologies 232, https://doi.org/10.1007/978-981-16-2814-6_21

corresponds to 8% of the total population of Russia. More than 200 thousand are in need of hand or leg prosthetics [8]. One of the possible examples is the control of an assistant robot with a manipulator fixed on a wheelchair or other device as an actuator.

One of the main problems of bionic control is the non-stationarity of EMG signals and the change in the position of the electrodes relative to the muscles during operation [9, 10]. The classification accuracy varies significantly over time, since the data recorded on the same day have characteristics that differ from the data recorded on another day, due to the influence of real conditions, such as displacement of sensors, changes in the state of the skin–electrode contact, and changing patterns of muscle activity in the process of developing a movement stereotype. Now one of the key problems is not laboratory, short-term conditions, but daily use [11, 12]. Daily displacement of the electrodes during operation can lead to differences in signal properties, making them unrecognizable to a pretrained classifier. Thus, the problem arises of assessing the spatial position of the electrodes according to the EMG data for reassigning the channels, before submitting the information to the input of the gesture classifier. In this paper, it is proposed to perform a calibration procedure based on the direction of the maximum vector of muscle efforts during a circular motion of the hand, since the position of such a vector is due to the anatomical features of the structure of the forearm and is invariant to the position of the electrodes.

Another problem of the classical approach to the analysis of EMG, based on the classification and recognition of gestures, is the limitation on the number of degrees of freedom for a one-time movement of the operation point [13–16]. For example, a gesture classifier-based operating point position control system may be based on four gestures, brush up, down, left, and right. Then, each hand gesture will correspond to a movement along the coordinate axes, so the command set will consist of four directions. For clarity, consider an example based on the task of moving the operating point (position of the manipulator or cursor) from point A to point B, Fig. 1.

In a rectangular coordinate system, the shortest path will be the trajectory shown on Fig. 1a. However, such a multistage control strategy is not inherent in humans, and with a high probability, with a limited choice of the direction of movement of the operating point, he will solve the task of moving along a non-optimal trajectory, an example of which shown on Fig. 1b, c. This leads to an increase in the time to reach the result and negatively affects its accuracy. It is more natural for a person to move to the target point along a vector with an arbitrary direction, rather than along

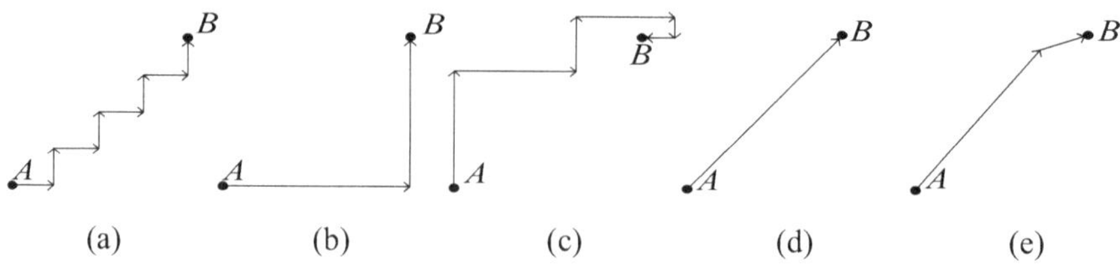

Fig. 1 **a–e** Examples of solving the task of operating point moving

a sequence of movements directed along the axes of a rectangular system. Examples of such trajectories are shown in Fig. 1d, e. At the same time, there is an additional opportunity to reduce the time to reach the target by changing the speed of movement of the operation point in proportion to the muscle effort.

Thus, the task of developing a control system based on EMG is actual, which allows to set an arbitrary direction of the displacement vector of the operation body, since it improves the usability, reduces the time for performing the final action, and increases the positioning accuracy. Notable research in this regard is the two-dimensional EMG guided pointer invented by Rosenberg, which is known as the biofeedback pointer [17, 18]. The main problem of the method, noted by the authors of the study, is that the user's movements may not be adequately synchronized with the cursor [19]. It should also be noted the need for precise positioning of the electrodes in this method.

In a study by Tavakoli et al., the classification of hand gestures by two channels of the EMG signal is described. The support vector method (SVM) is proposed for classification [20]. The advantage of this method is a high speed of calibration and classification accuracy, as well as operation stability in a multidimensional feature space. The SVM algorithm shows relevant results when there is a small sample for training. The accuracy of the classifier reaches from 95 to 100% for five output classes.

This paper presents the results of analyzing the possibility of vector control of the operating point position in an arbitrary two-dimensional space of spherical or Cartesian coordinates. Based on the spatial position of the electrodes, the resulting vector of muscle efforts was calculated. Some aspects of using this information to calibrate the sensor system are also given.

2 Methodology

To record EMG of the forearm muscles, the MYO Thalmic Labs armband is used, Fig. 2a. The MYO armband has eight sections. When putting on, the armband sensors are evenly spaced around the circumference of the forearm. Consider the arrangement of the sensors on the forearm, Fig. 2b.

Figure 2c shows that precise positioning of the armband over the same muscles can be difficult due to the complex structure of the forearm. It is advisable to focus on the muscle groups involved in the direction of the arm in different directions. For example, moving the wrist up and holding this gesture activates the muscles on the outside of the forearm. This is true for other areas as well.

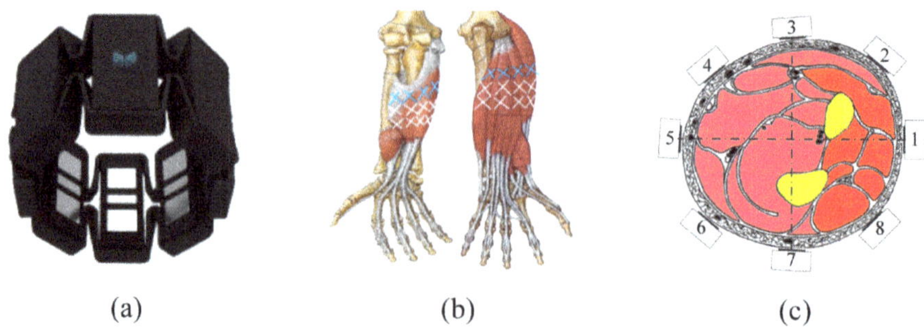

Fig. 2 MYO armband appearance and the location of the sensors on the forearm: **a** the appearance of the armband; **b** a location of the muscles and sensors, and **c** an example of the mutual position of the sensors and muscles on the section of the forearm

2.1 *Preliminary Data Evaluation*

Signal records are obtained to analyze the initial data. The direction of movement selected at an angle of 90° to each other. The following directions of the hand correspond to such positions: from oneself ("Up"), toward oneself ("Down"), left, and right. Figure 3 shows the EMG signals for the studied gestures in the time domain.

Figure 4 shows the images of the studied gestures in pie charts.

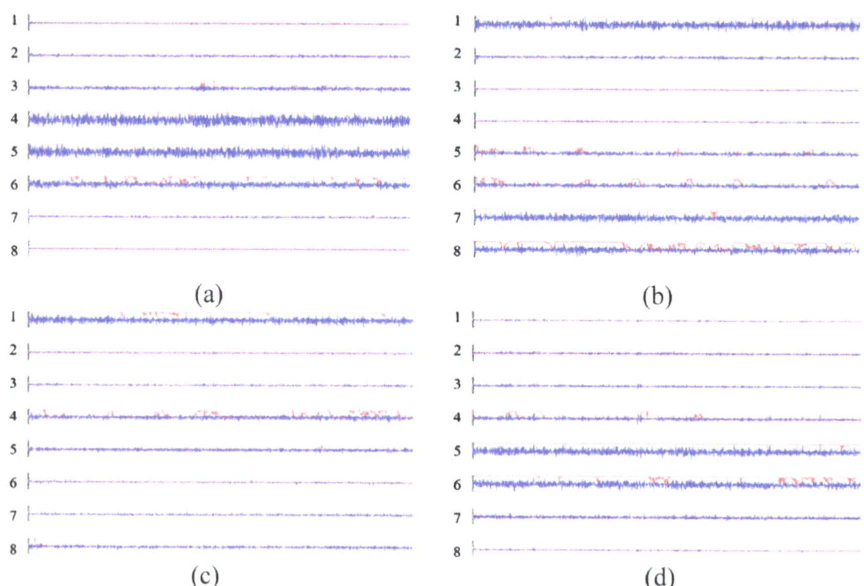

Fig. 3 EMG signals for various gestures: **a** from oneself ("Up"), **b** toward oneself ("Down"), **c** brush to the left, **d** brush to the right

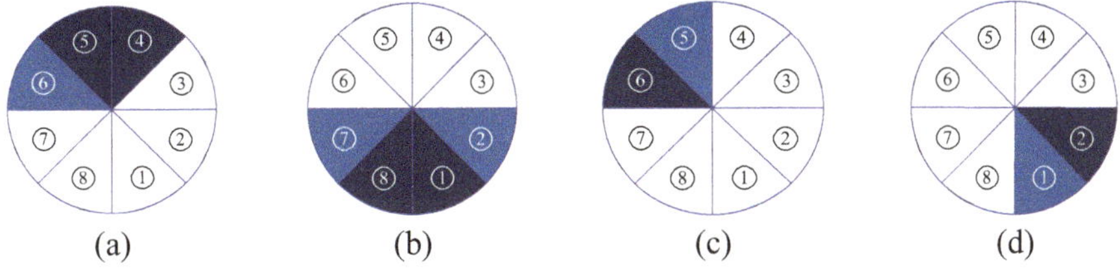

Fig. 4 Images of the studied gestures in pie charts: **a** from oneself ("Up"), **b** toward oneself ("Down"), **c** brush to the left, and **d** brush to the right

The color saturation of the sector in Fig. 4 is determined by the averaged signal amplitude of the corresponding channel. Analysis of the data shows that, in general, the directions of the resultant vectors coincide with the direction of the operator's effort, but they have an asymmetric picture.

2.2 *Calculation of the Resultant Vector of Muscle Efforts*

The MYO armband has eight EMG sensors spaced evenly around the circumference of the forearm. The output data of the armband can be represented as a vector diagram of eight coplanar vectors originating at one point and located at an angle of $360/8 = 45°$ to each other, as shown on Fig. 5a. Figure 5b shows an example of an instantaneous vector diagram with gesture from oneself ("Up"). An example of the result of performing calculations in accordance with expressions (1)–(3) for actually recorded signals is shown in Fig. 5b.

A polar coordinate system is selected. The origin of the φ coordinate is chosen arbitrarily by binding to any channel, and later it is redefined during the calibration

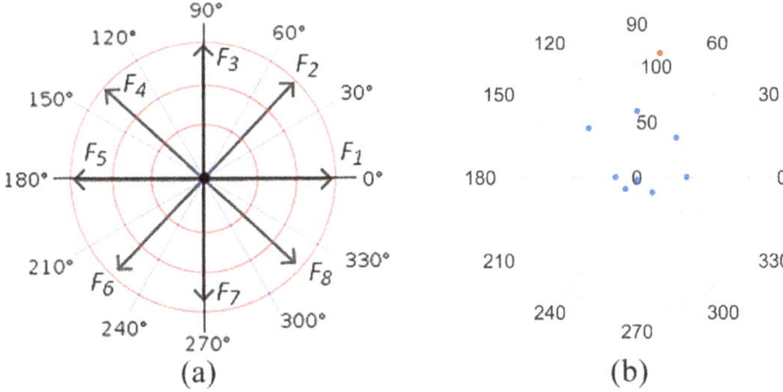

Fig. 5 Representation of output signals in vector diagram form: **a** vector construction scheme and **b** an example of calculating the resultant vector for actually recorded signals: the resultant vector is marked with a red marker

procedure based on the direction of the maximum vector of muscle forces for circular motion. Vector of the signal on the polar plane of coordinates is determined by its angular coordinate φ_i, which is a constant for each channel and amplitude of the signal F_i.

The process of calculating the resultant vector repeats for all channels. Let us draw an analogy with the system of forces acting on a rigid body, used in theoretical mechanics. Then, for a converging system of forces, shown in Fig. 5, the resultant vector will be sum of vectors force acting on the body at the point of application corresponding to the origin of the radial coordinates (1):

$$\vec{R} = \vec{F}_1 + \vec{F}_2 + \vec{F}_3 + \vec{F}_4 + \vec{F}_5 + \vec{F}_6 + \vec{F}_7 + \vec{F}_8. \quad (1)$$

To calculate R, it is necessary to add all eight vectors F_1–F_8 sequentially. The resultant addition vector of two coplanar vectors is calculated by trigonometrically using the cosine theorem (2):

$$\vec{R}_1 = \sqrt{\vec{F}_1 + \vec{F}_2 + 2\vec{F}_1\vec{F}_2\cos(180 - \alpha)}, \quad (2)$$

where $\alpha =$ the angle between the initial vectors.

The angle between the resultant vector and one of the original vectors calculate by use the sine theorem (3):

$$\varphi_{R_1} = \arcsin\left[\vec{F}_2 \sin\left(\frac{180^\circ - \alpha}{\vec{R}_1}\right)\right], \quad (3)$$

where $\alpha =$ the angle between the initial vectors.

For Eq. (3), at each step of the vectors, the condition of inadmissibility of dividing 0 is checked.

2.3 Rotational Motion Analysis

To assess the possibility of controlling the position point on the basis the resultant vector of muscle efforts, an EMG analysis was performed during rotational movement of the hand. During the experiment, a signal was recorded when the right handmade five rotational movements in a clockwise direction with a period of about 2 s per one full revolution (0.5 Hz). The recording duration was 13 s, which at a sampling rate of 200 Hz is 2600 reports. Since the muscles of the forearm are unevenly developed, the pattern of the sample without calibration is asymmetric, as shown in Fig. 6a. To improve user convenience, at this stage, it is necessary to calibrate the gains of each channel in order to obtain the most symmetrical picture. The vector coordinates after the procedure of calibration gain by channel are shown in Fig. 6b.

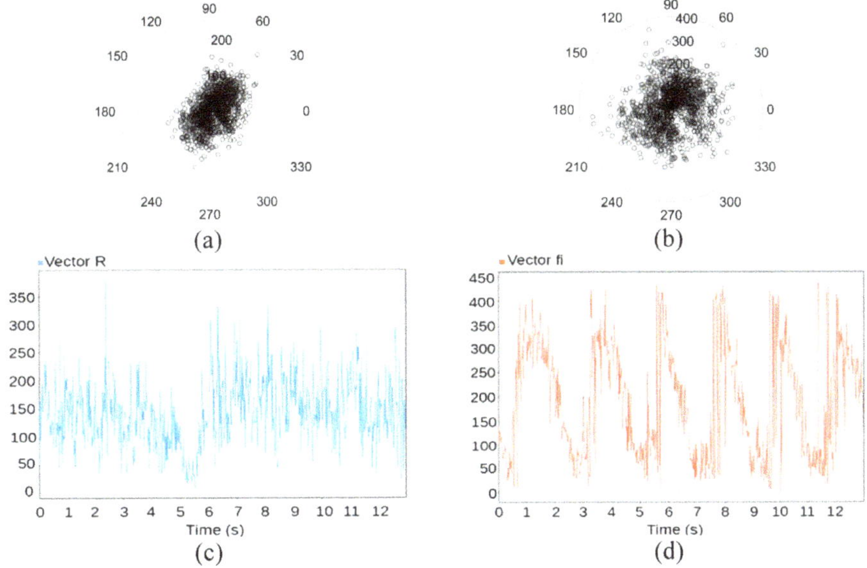

Fig. 6 Calculated vectors for dataset: **a** without calibration of gain by channels, **b** with calibration of gain by channels, **c** graph of radius R, and **d** graph of angle φ versus time

In this study, the gain coefficients are calculated from the averaged values of the amplitudes of the entire vector in proportion to the averaged value of an arbitrarily selected channel, taken as the reference one. The task of adjusting the signal amplification factors across the channels to ensure maximum symmetry of the vector diagram is one of the main tasks for obtaining a high-quality control process.

To isolate the useful part of the signal, it is necessary to carry out preliminary filtering and processing. After recording the experimental sample, it is necessary to split the signal into fragments (frames) of finite duration and highlight informative features. Preprocessing for calculating the resultant vector of muscle efforts carried out the basis of averaging over the modulus of all signal values within one frame. The range of frame durations from 10 ms to 1.5 s was investigated, which corresponds to averaging over 2–300 reports at a sampling rate of 200 Hz. The step of changing the frame duration was 25 ms or 5 reports. Figure 7 shows examples of the calculated coordinates of the resultant vector for various frame durations.

Analysis showed that using a frame duration of more 400 ms, information about the nature of the action (brush rotation) is lost. In this case, rotation of the brush with a higher frequency than in this experiment can additionally limit the maximum frame length for averaging. Using shorter length signal frames allows to raise of operating speed of the system, but leads to noise in the output coordinates. For development of analysis system of EMG, compromise between the speed of the system and the smoothness of the change in coordinates depend on the targets and characteristics of the control object. In this paper, optimal frame duration is 250 ms. The graphs of the

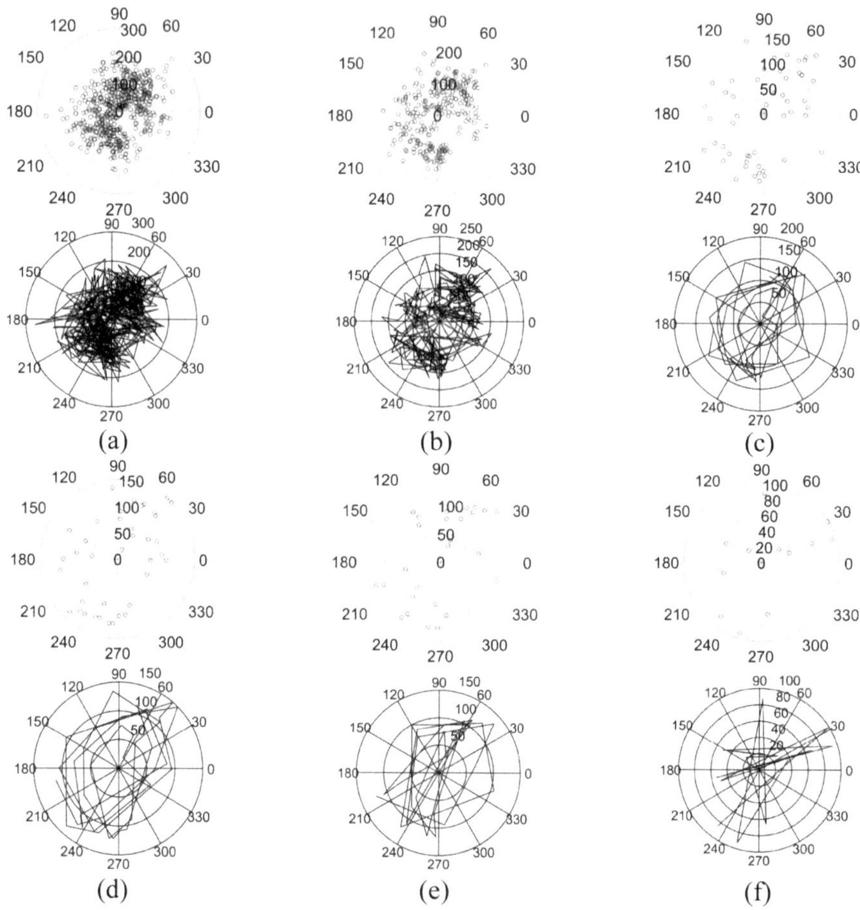

Fig. 7 Calculated coordinates of the resultant vector for five rotational movements in different frame durations: **a** 25 ms, **b** 50 ms, **c** 250 ms, **d** 350 ms, **e** 1 s, and **f** 1, 5 s

movement of the operation point in the polar coordinate system and the coordinates themselves in the time domain are shown in Fig. 8.

In Fig. 8b, for clarity of the trajectory of the point movement, the coordinate of the radius of the vector cumulatively is calculated from the beginning to the end of the movement. If necessary, the transition to the Cartesian coordinate system is carried out according to Formula (4):

$$X_1 = F_1 \sin(\varphi_{12}), \;\; Y_1 = F_1 \cos(\varphi_{12}). \tag{4}$$

To improve the quality of control, it is advisable to approximate and smooth the output coordinates of the resulting vector. Figure 9 shows the trajectory of the

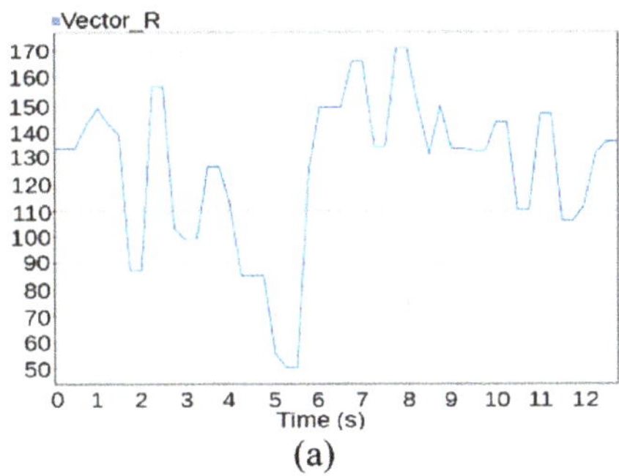

(a)

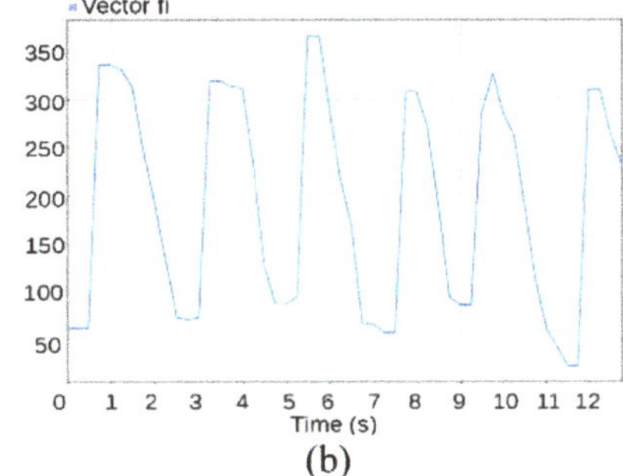

(b)

Fig. 8 Graphs of the operation point movement in the polar coordinate system and the coordinate values in the time domain: **a** graphs of the vector R and **b** the angle φ depending on time in polar coordinate system

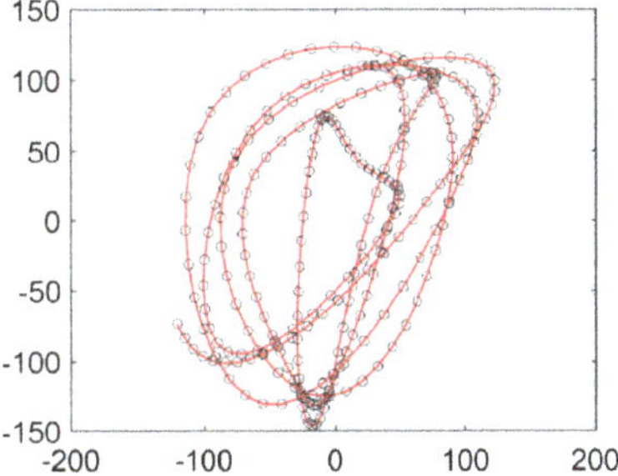

Fig. 9 Smoothed trajectory of the operation point in the Cartesian coordinate system

operating point in the Cartesian coordinate system, where the coordinates of vector are smoothed by splines.

Analysis of the trajectory of the operation point in Fig. 9 shows that, despite the entered gains, and the smoothing procedure, the graph remains asymmetric relative to the beginning of the report. This indicates the anatomical features of the human hand, in which in a certain direction there may be more large and strong muscles than in another. Further improvement of the result is possible by the use of ANN algorithms with reinforcement to refine the gain factors, as well as the use of algorithms for processing vectors of input and output values, etc. Information about the anatomical features obtained from such analysis can be used for determine the spatial position of the sensors during use.

3 Channel Calibration Algorithm Based on Vector Analysis of Forearm EMG Image

A calibration algorithm for everyday use based on the obtained data is shown in Fig. 10.

Figure 10a shows the algorithm for the initial configuration of the interface. EMG signals read and write. The average resultant vector for three rotations is calculated and stored in memory. Then, the coordinate system is redefined in the direction of the resultant vector of muscle effort and the channels linked. Next, the ANN-based gesture classifier is trained and tested, and the ANN coefficients are stored in memory. In daily use, Fig. 10b, at the beginning of the work, for calibration purposes, the average resultant vector for the three rotations is calculated and compared with the reference obtained during the initial setup. If the data does not match, the channels are calibrated first, which includes redefining the channel numbers and adjusting the gains. The procedure allows you to bring the data on the EMG channels to

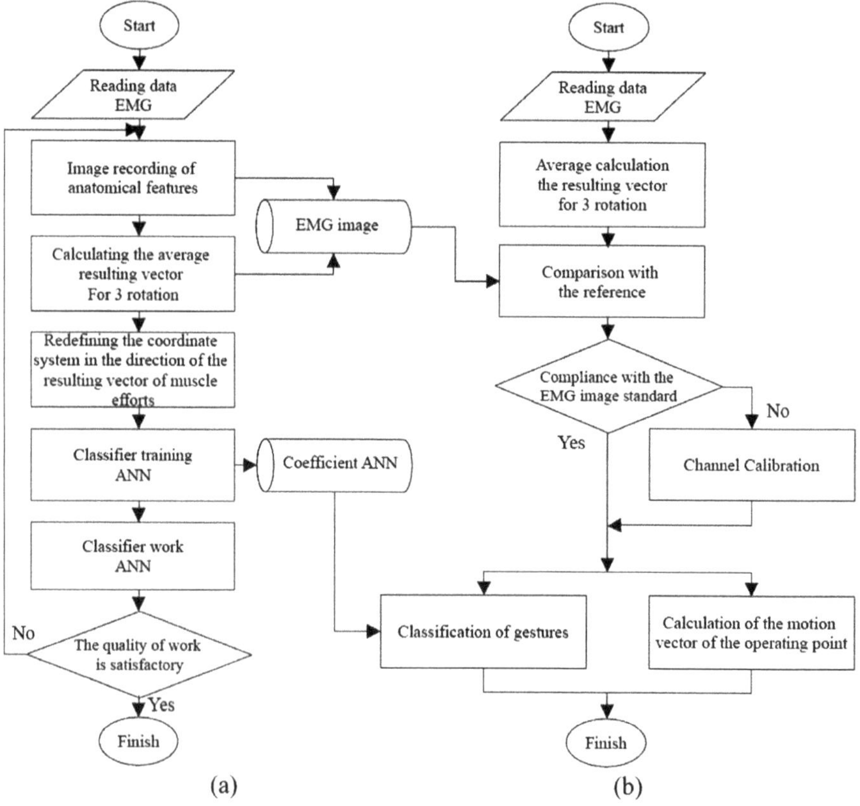

Fig. 10 **a** Algorithm during initial setup, **b** daily use

the reference configuration during training and use the previously obtained ANN weights. The calibration algorithm is shown in more detail in Fig. 11.

To calibrate the channels, the resulting vector and phase difference are calculated based on vector analysis of the forearm EMG dataset. Then, the channels redefined and the gain calibrated taking into account the reference values. The received data are saved in tables of reassignment and gain factors. Next, the coordinate system is redefined according to the muscle force vector and the calibration controlled. If compliance with the EMG image standard observed, then the calibration program ends, if not, the signal is reregistered and the calibration procedure repeated.

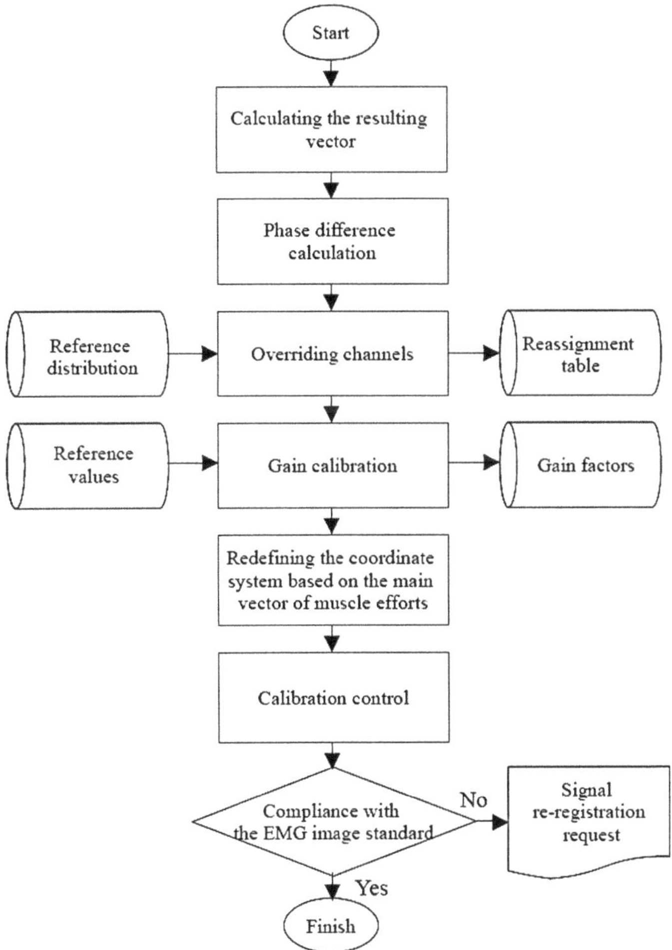

Fig. 11 Calibration algorithm

4 Conclusion

The paper proposes a method based on vector EMG analysis and allows extracting useful information about the resultant vector direction of muscular effort and anatomical features of the forearm. The research results for the case of rotational motion are presented. Proposed method of EMG processing based on a vector diagram and averaging over frames allows to obtain data on the spatial position of sensors based on the asymmetry of the structure of the forearm. The possibility to use the obtained data for calibration of the signal pickup system and control the position of the operating point is shown. Proposed algorithm for the operation of the human–machine interface for daily use is based on the calibration of channels by vector analysis. Algorithm is resistant to the electrodes position changing.

References

1. Oskoei, M.A., Hu, H.: Myoelectric control systems—a survey. Biomed. Signal Process. Control **2**(4), 275–294 (2007). https://doi.org/10.1016/j.bspc.2007.07.009
2. Budko, N.A., Budko, R.Yu., Budko, A.Yu.: Application of artificial neural networks in human-machine interfaces. Model. Optim. Inf. Technol. **7**(1), 328–340 (2019)
3. Budko, R.Yu., Chernov, N.N., Budko, N.A., Budko, A.Yu.: Recognition of the electromyogram of the forearm and the choice of gestures to control the prosthesis. Model. Optim. Inf. Technol. **7**(1), 54–66 (2019)
4. Budko, R.Yu., Chernov, N.N., Budko, N.A.: Research of methods of EMG classification in the problem of control of hand gestures. Proc. High. Educ. Inst. Instrum. Mak. **62**(12), 1098–1104 (2019). https://doi.org/10.17586/0021-3454-2019-62-12-1098-1104
5. Budko, R.Yu., Chernov, N.N., Budko, N.A.: Method of control of devices for replacement of lost functions based on myosignal and its verification in real time. Electron. Sci. J. Bull. Youth Sci. Russ. **6** (2019)
6. Gorokhova, N.M., Golovin, M.A., Chezhin, M.S.: Management methods for upper limb prostheses. Sci. Tech. Bull. Inf. Technol. Mech. Opt. **19**(2), 314–325 (2019). https://doi.org/10.17586/2226-1494-2019-19-2-314-325
7. Karpov, V.E., et al.: Architecture of a wheelchair control system for disabled people: towards multifunctional robotic solution with neurobiological interfaces. Mod. Technol. Med. **11**(1), 90–100 (2011)
8. Federal State Information System: Federal register of the disabled. https://sfri.ru/analitika/chislennost. Accessed 2020/12/01
9. Person, R.S.: Electromyography in Human Research. Nauka, Moscow (1969)
10. Pylatiuk, C., et al.: Comparison of surface EMG monitoring electrodes for long-term use in rehabilitation device control. In: IEEE 11th International Conference on Rehabilitation Robotics, vols. 1 and 2, pp. 348–352 (2009)
11. Rajulu, S.L.: Decomposition of electrical signals for biomechanical interpretation. PhD thesis, The Ohio State University (1990)
12. Vigreux, B., Cnockart, J.C., Pertuzon, E.: Factors influenced quantified surface EMG. J. Appl. Physiol. **41**(4), 119–129 (1979)
13. Ogiri, Y.: Development of an upper limb neuroprosthesis to voluntarily control elbow and hand. In: Proceedings 26th IEEE International Symposium on Robot and Human Interactive Communication (RO-MAN), Lisbon, Portugal, pp. 298–303 (2017). https://doi.org/10.1109/ROMAN.2017.8172317

14. Yusevich, Yu.S.: Essays on Clinical Electromyography. Medicine, Moscow (1972)
15. Cowan, H., Brumlick, J.: Guide to electromyography and electrodiagnostics. Medicine, Moscow (1975)
16. Gekht, B.N.: Theoretical and Clinical Electromyography. Nauka, Moscow (1990)
17. Ivanovsky, Yu.V.: Principles of using the biofeedback method in the system of medical rehabilitation. Biol. Feedback **3**, 2–9 (2000)
18. Kunelskaya, N.L., Rezakova, N.V., Gudkova, A.A., Hecht, A.B.: Biofeedback method in clinical practice. S.S. Korsakov J. Neurol. Psychiatry **114**(8), 46–50 (2014)
19. Rosenberg, R.: The biofeedback pointer: EMG control of a two-dimensional pointer, wearable computers. In: Digest of Papers. Second International Symposium, pp. 162–163 (1998)
20. Tavakoli, M., et al.: Robust hand gesture recognition with a double channel surface EMG wearable armband and SVM classifier. Biomed. Signal Process. Control **46**, 121–130 (2018). https://doi.org/10.1016/j.bspc.2018.07.010

Approach to Image-Based Recognition of User Face in Setting of Partial Face Occlusion by Personal Protective Equipment

Maksim Letenkov, Roman Iakovlev, and Alexey Karpov

Abstract Within this research, aimed to solve the problem of face recognition in setting of partial face occlusion by personal protective equipment an approach was proposed, which assumed training of a neural network model FaceNet on a specialized dataset Masked VGGFace2. This paper presents a strategy of training dataset augmentation based on the convolutional neural network model MTCNN and MaskTheFace software. Using this suite, based on the VGGFace2 dataset, an extended dataset Masked VGGFace2 was established, including images of human faces, partially occluded by personal protective equipment. Within this research, a series of experiments was performed, aimed to qualitative assessment of predictions, obtained with this solution, in comparison to the FaceNet model, which was trained on the original VGGFace2 dataset. The performed experiments revealed, that the proposed solution demonstrates a significantly higher recognition quality of partially occluded faces (AP = 0.9488)—recognition accuracy increase as per average precision (AP) metric was over 24%, compared to the original model. On test subset from the original dataset VGGFace2, the value of the AP metric was 0.9635, what outperformed the respective metric of the original model by 2.8% and justified a high genericity of the proposed solution in terms of face recognition problem.

1 Introduction

Generally, the image-based recognition of user face is among the relevant actual problems in the computer vision domain. Currently, many approaches exist for solution of problems, related to face detection and recognition in video sequences or in isolated images [1–7]. However, the most of such solutions lack genericity and require specific imaging conditions, as well are prone to high error rate in the settings with partial occlusion of the object being imaged [1–3, 8]. The scope of this research is to develop an approach to user face recognition, when the face is partially occluded by

M. Letenkov (✉) · R. Iakovlev · A. Karpov
St. Petersburg Federal Research Center of the Russian Academy of Sciences (SPC RAS), St. Petersburg Institute for Informatics and Automation of the Russian Academy of Sciences, 39, 14th Line, 199178 St. Petersburg, Russia

A. Ronzhin and V. Shishlakov (eds.), *Electromechanics and Robotics*, Smart Innovation, Systems and Technologies 232, https://doi.org/10.1007/978-981-16-2814-6_22

the personal protective equipment. The proposed approach is intended to be utilized in mobile robotic devices and is applicable for solution of a common face recognition problem, as well as of recognition of faces, partially occluded by personal protective equipment, thereby meeting requirements regarding system performance and resource consumption limit.

2 Methods of Face-Based Human Identification

Among the existing approaches to image-based user identification, the most relevant ones seem the identification methods, based on deep learning models [1, 2], hence such models ensure decent quality of prediction and can be applied in design of distributed systems of biometric identification. They are also suitable for real-time assessment of the environment, even being deployed on relatively low-powered hardware platforms, e.g., on devices from the NVIDIA Jetson line-up, what is especially handy in designing of a biometric identification system, where the hardware layer is completely or partially implemented as mobile robotic devices or with other non-static devices for video fixation [9, 10].

Considering the existing machine learning models, being used for facial biometric identification of users, the particular attention should be paid to such models as AlexNet [11], VGG-Face [2, 12], SqueezeNet [13], and FaceNet [1, 14, 15].

Currently, the convolutional neural network AlexNet is one of the most prominent models in the computer vision domain [16]. The distinctive feature of this model is fast downsampling of latent representations, implemented with stride-based convolutions and max-pooling layers. The principal downside of this model is a relatively low functional accuracy compared to more recent solutions [16].

The deep learning model VGG-Face is based on the architecture of a 16-layer convolutional neural network VGG-16, proposed in [2], which has a deeper convolutional structure, than AlexNet. The initial VGG-16 model showed a 92.7% recognition accuracy, being assessed on the ImageNet dataset [17], consisting of over 14 million of images. These images are divided into 1000 different classes, what supports the high generalization performance of this model. VGG16 is an improved version of AlexNet: inside of this model the large-scale filters are substituted by several consecutive filters of 3×3 dimension. The VGG-Face model was trained on a dataset, containing 2.6 million face images [2], divided into 2622 classes. It is worth noting, that, unlike VGG-16, the set of output fully connected layers of VGG-Face model is limited by 4096 neurons in each layer, except the last one, which is defined by 2622 neurons. The proposed architecture enabled to achieve greater recognition accuracy, compared to the AlexNet model [18].

The architecture of the SqueezeNet model was initially proposed in 2016 in [13]. One of the research objectives, pursued by the authors, was to develop a relatively compact model, having fewer training parameters, compared to the AlexNet model, what would enable to reduce the computational complexity of the resulting model. The model, proposed by the authors, being assessed on the ImageNet dataset, showed resulting accuracy, comparable with the prediction accuracy of the neural network model AlexNet.

The deep learning model FaceNet [1, 14, 15] is based on the GoogleNet architecture [19], and the corresponding model is a prize-winning solution from the ImageNet 2014 contest. Within this contest, it outperformed more common convolutional neural networks. Compared to the models AlexNet and VGG-Face, FaceNet shows greater prediction accuracy, as well has lower computational complexity. One of the most important specific features of the FaceNet model is its applicability on low-powered devices with hardware acceleration, thereby a relatively high frequency of prediction generation is retained, what enables design of biometric identification systems, applicable in real-time mode [9].

Based on the specific features of deep learning models, presented above, in terms of development of user face recognition approach, with respect to possible partial occlusion of the face and utilization on mobile robotic devices, the most promising solution seems to be the deep learning model FaceNet.

To implement the approach to user face recognition, the most relevant datasets have also to be defined, suitable for training of the chosen neural network model and subsequent testing of the obtained solution. Currently exist several widely known datasets, intended for training and validation of different face recognition models. The following ones are in public domain: VGGFace2 [20], Labeled Faces in the Wild (LFW) [21], and set MS-Celeb-1M [22].

The VGGFace2 dataset [20] contains over 3.3 million images, divided into 9131 classes, each represented by a series of face images of an individual human, taken in different non-deterministic conditions, including: different lighting levels of images in the scene, different occlusion levels in the area of interest, and variability of spatial positions of faces on image. Additional input to the variability of this dataset is caused by the different age of persons, whose faces are contained in the dataset. Considering decent level of representativeness of VGGFace2, as well the sheer number of the presented classes, as well images, defining them, this dataset is suitable as for training purposes, as well for fine-tuning and for validation of already trained deep learning models.

The LFW dataset [21] contains over 13,000 images, divided into 5749 classes, each of them associated with an individual person. However, this dataset shows extreme imbalance among different classes, what impairs its applicability for model training purposes. The LFW dataset is utilized mainly for assessment of already trained neural network models.

One of the most extensive public labeled face datasets is the MS-Celeb-1M [23]. This dataset contains over 100 thousand classes, represented by over 10 million images. Having such large volume and high level of representativeness, this dataset suffers mainly from noisiness of data, contained in it. Many images, presented in this dataset, actually do not relate to their nominal classes or cannot be attributed to any of the classes in this dataset. Therefore, it can be concluded, that this dataset is not suitable for validation of trained models, but it can be useful to obtain a reference pretrained model, later to be fine-tuned on alternative, less noisy datasets.

Therefore, in terms of this research, upon analysis of publicly available datasets, intended for training and validation of neural network models for face recognition, it was decided to use VGGFace2 as a reference dataset [20]. Further, we consider the proposed approach to user face recognition in the setting of partial occlusion of the face by personal protective equipment.

3 Description of the Developed Approach

Upon results of the analysis of the related methods and approaches, for the recognition of users the setting of partial occlusion of the face by personal protective equipment, the deep learning neural network model FaceNet [4] was chosen as reference. The highlights of this model are high recognition accuracy, as well applicability in real-time mode. High-level architecture of the neural network model FaceNet is presented in Fig. 1.

Within this high-level architecture diagram, the following units can be distinguished: batch input layer, deep learning unit, including the reference deep architecture, as well the L2-regularization (L2). In this unit, the GoogleNet deep learning model is utilized as reference [19], specifically its implementation, based on the Inception-Resenet-v1 [22] architecture. It is important to note, that the interaction with the deep learning unit within the FaceNet model is performed in black-box manner.

To reduce spatial dimensionality, the convolution operations with filters 1×1 are employed in GoogleNet. Utilization of these operations allows to reduce the number of channels, while retaining the remaining features (height and width) of input data. Hence, the RGB image with dimensions $256 \times 256 \times 3$ can be transformed into the $256 \times 256 \times 1$ image using the inception module of GoogleNet architecture.

Within the neural network model FaceNet the loss function is defined as an elemen-twise sum of values of the triplet loss function and of the L2-regularization function. In the course of model training, the function triplet loss aims, on the one part, to minimize distances between every image being evaluated (anchor) and the images, treated as positive relative to it, and, on the other part, to maximize distances between every image being evaluated and the images, treated as negative relative to it. This process is implemented for every class, included into the training subsetб whereas the L2-regularization function is used to compensate for model overfitting.

Fig. 1 High-level architecture of the neural network model FaceNet [1]

Fig. 2 Results of face detection on images from the VGGFace2 dataset with subsequent alignment using MTCNN [4]

Upon results of the analysis, performed for training of neural network model FaceNet within this research, the VGGFace2 dataset was chosen. However, the VGGFace2 dataset does not contain any human faces, partially occluded by personal medical protective equipment. Therefore, to obtain the enough representative dataset without much noisy data, but applicable to solve the present problem, it was decided to augment the original VGGFace2 dataset with artificially generated half-synthetic images, containing human faces, partially occluded by personal protective equipment. Composition of such images was performed based on the source data from the VGGFace2 dataset, according to the following strategy.

At first, according to the approaches to training data preparation, outlined in [14, 15], detection of face parts and feature points was performed with subsequent image alignment using affine transforms. In this manner, an intermediary dataset was obtained, consisting of aligned RGB images of extracted user faces with dimensions of 160 $\times$ 160 pixels (Fig. 2). This stage of data preparation was completed using multi-task CNN (MTCNN) [4].

At the next stage of data preparation, direct generation of semi-synthetic images of human faces was performed with artificially overlayed personal protective equipment on them. The personal protective equipment was overlayed onto aligned face images, which were obtained on the previous stage. To compose the relevant image set, the MaskTheFace was utilized [24]. MaskTheFace is a package of computer vision scripts, which implement the routines of overlaying of artificial personal protective equipment on face images. At its core, this package employs the dlib framework [3], which enables overlaying of protective equipment, with respect of face rotation and tilt, making associations with the feature points on face. MaskTheFace provides some templates of personal protective equipment, mainly as medical face masks, as well enables adjustment of template color, as well the filling texture of it. The results of completion of this stage during preparation of training dataset are given in Fig. 3.

Utilization of the strategy, outlined above, enabled to compose an extended VGGFace2 dataset: for every image from the set a new version was generated, with protective equipment, overlaid on face, retaining the original aligned image. It is worth noting, that during augmentation data loss was revealed, consisting in skipping of certain images by the protective equipment overlaying tool. It was caused by occurrence of statistical errors due to incorrect detection of face area or facial

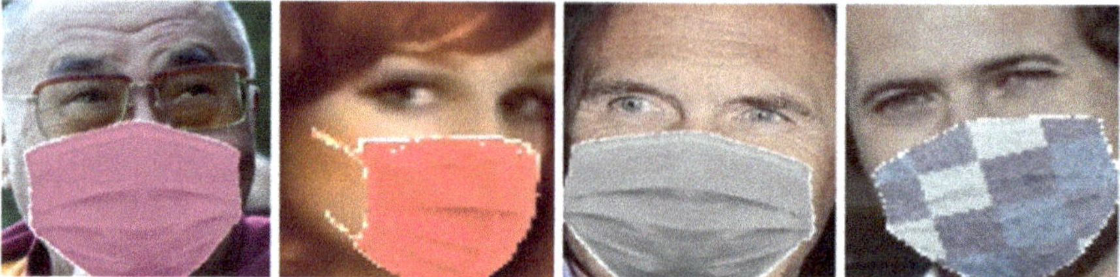

Fig. 3 Results of generation of semi-synthetic data samples based on the aligned dataset VGGFace2 using the MaskTheFace package [24]

feature points. Images, which were omitted in the process of augmentation, were not included into the extended dataset VGGFace2.

In the following, we assess the implementation of the proposed approach to image-based user face recognition in the settings of partial occlusion by personal protective equipment.

4 Experiments

Within this research, training, and validation of the chosen neural network model, FaceNet was performed on the extended dataset VGGFace2 (further Masked VGGFace2), which included 5,830,200 images. For validation and performance quality assessment of the neural network model, trained on the Masked VGGFace2 dataset, training set and deferred validation set were established. The size of the training set, established in such manner, amounted to 4,951,566 images. The training of the neural network model was performed using a cross-validation. To facilitate this, the set was divided into five equally sized image blocks. Fragmentation was performed in such manner, that all image subsets (blocks) would contain the images of the same users, without multiple inclusions of the same data units into different subsets of images. Cross-validation enabled to determine the number of model training epochs, which would enable to minimize the loss function value and no model overfitting would occur. According to several experiments in training of FaceNet model, the best results are usually achieved during the 8th training epoch. Figure 4 shows the charts of loss function and values of the accuracy metric, obtained during training of FaceNet model instances on training subsets from the Masked VGGFace2 dataset with cross-validation employed.

It is evident, that the instances of the neural network, trained on different subsets, show similar values of the loss function and of chosen quality metric, what confirms the decent genericity of the model and lack of overfitting-related issues. The values of the accuracy metric are given in Table 1. These values were obtained during assessment of models, which were trained using cross-validation approach.

Based on the values of quality metrics, presented above, and values, obtained during validation of models, it can be concluded, that the extended dataset Masked

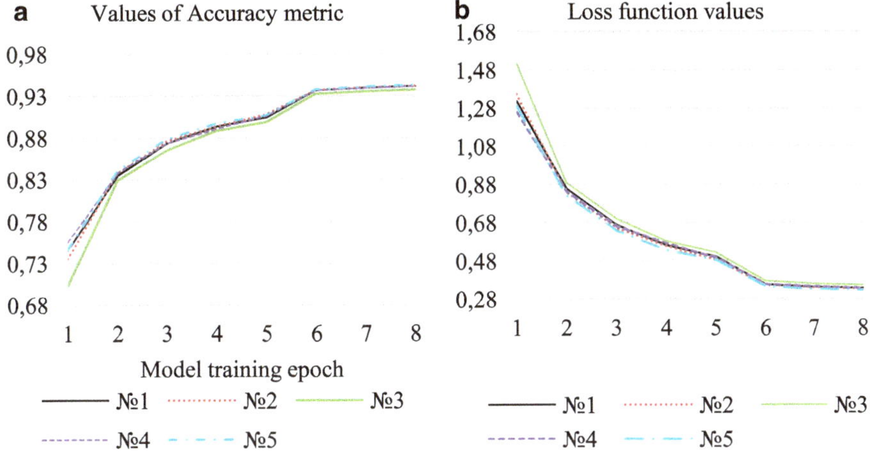

Fig. 4 Results of training of FaceNet model instances on subsets from the Masked VGGFace2 dataset with cross-validation employed: **a** values of accuracy metric and **b** loss function values depending on the number of training epochs

Table 1 Values of accuracy metric obtained during assessment of models, which were trained using cross-validation

Image subset	Accuracy	Loss	Precision
No. 1	0.9413	0.3358	0.8903
No. 2	0.9418	0.3341	0.8912
No. 3	0.9372	0.3544	0.8830
No. 4	0.9412	0.3378	0.8901
No. 5	0.9427	0.3265	0.8927

VGGFace2 was composed correctly. Because the predictions, obtained with the model, trained on the data subset No. 5, feature the highest prediction accuracy, this one implementation of the FaceNet model was employed in further experiments.

To assess the prediction accuracy of the obtained model, a series of experiments was performed, which were intended to assess: the original FaceNet model, trained on VGGFace2—(1); proposed model, trained on Masked VGGFace2—(2). The validation of both models was performed as on a deferred test subset from the Masked VGGFace2 dataset, as on a subset from the original VGGFace2 dataset. As target metrics the following ones were used: average accuracy over classes (AAc), average precision over classes (AP), and average recall over classes (AR) [25]. The obtained results are given in Table 2.

Based on the presented results of the model assessment, it can be concluded, that the model, trained on the Masked VGGFace2 dataset, shows the optimal results as on the test subset, which includes faces, partially occluded by personal protective equipment, as well on the test subset without such occlusions: quality metrics AAc, AP and AR reached the values 0.999989, 0.9488, 0.9457 in the former case and

Table 2 Results of prediction accuracy estimation for models FaceNet (1) and (2) on test subsets from the original and extended dataset

Deferred test subset (model)	AAc	AP	AR
VGGFace2 (1)	0.999987	0.9354	0.9306
VGGFace2 (2)	0.999992	0.9635	0.9606
Masked VGGFace2 (1)	0.999930	0.7649	0.6556
Masked VGGFace2 (2)	0.999989	0.9488	0.9457

0.999992, 0.9635, 0.9606 in the latter case, respectively. Here, it is important to note, that the model, trained on the original dataset, showed inacceptable results in the experiment on a test subset from VGGFace2, because values of quality-related metrics AP and AR proved to be 25% lower at average. It is noteworthy, that the model, obtained in this paper, demonstrates better recognition quality on the test subset from the original dataset VGGFace2 as well, because values of metrics AP and AR were 0.9635 and 0.9606, respectively, what is by 3% more at average, than in the model, trained on the original dataset. The obtained results allow to conclude, that the solution, proposed in this paper, has high degree of genericity in terms of the face recognition problem and can be successfully employed in settings with partial covering or occlusion of the object being imaged.

Therefore, within this paper, through training of neural network model FaceNet on the enhanced dataset Masked VGGFace2, composed using the MTCNN model and MaskTheFace software, an approach was developed and successfully approbated for image-based user face recognition in settings of partial occlusion of face by personal protective equipment.

5 Conclusion

The solution, developed within this research, showed significantly better face recognition performance as in settings without partial occlusion (AP = 0.9635), as well in settings with partial occlusion of faces by personal protective equipment (AP = 0.9488). Compared to the original model, the recognition accuracy increase by the AP metric was over 24% in the former case and about 3% in the latter case. The obtained results confirm higher genericity degree of the proposed solution and lead to conclude, that it is sensible to use the FaceNet model, trained on the extended Masked VGGFace2 dataset, to solve the problem of user face recognition, particularly in settings with partial occlusion of faces with personal protective equipment. It is also important to note, that, because of the relatively low computational complexity of the employed model, the proposed solution could be applicable in wide range of robotic devices.

Within further research, it is considered to perform approbation of the proposed solution on different robotic platforms [26, 27] to assess accuracy and operation speed of this solution on low-powered computational devices.

Acknowledgements This research is supported by the RFBR (project no. 20-04-60529).

References

1. Schroff, F., Kalenichenko, D., Philbin, J.: FaceNet: a unified embedding for face recognition and clustering. In: Proceedings of the IEEE Conference on Computer Vision and Pattern Recognition, pp. 815–823 (2015)
2. Parkhi, O.M., Vedaldi, A., Zisserman, A.: Deep face recognition (2015)
3. Dlib C++ Library: http://dlib.net/. Accessed 2021/01/12
4. Zhang, K., Zhang, Z., Li, Z., Qiao, Y.: Joint face detection and alignment using multitask cascaded convolutional networks. IEEE Signal Process. Lett. **23**(10), 1499–1503 (2016)
5. Deng, J., Guo, J., Zhou, Y., Yu, J., Kotsia, I., Zafeiriou, S.: Retinaface: single-stage dense face localisation in the wild. arXiv preprint arXiv:1905.00641 (2019)
6. Zhang, F., Fan, X., Ai, G., Song, J., Qin, Y., Wu, J.: Accurate face detection for high performance. arXiv preprint arXiv:1905.01585 (2019)
7. Li, J., Wang, Y., Wang, C., Tai, Y., Qian, J., Yang, J., Wang, Ch., Li, J., Huang, F.: DSFD: dual shot face detector. In: Proceedings of the IEEE Conference on Computer Vision and Pattern Recognition, pp. 5060–5069 (2019)
8. He, Y., Xu, D., Wu, L., Jian, M., Xiang, S., Pan, C.: LFFD: a light and fast face detector for edge devices. arXiv preprint arXiv:1904.10633 (2019)
9. Letenkov, M., Levonevskiy, D.: Fast face features extraction based on deep neural networks for mobile robotic platforms. In: International Conference on Interactive Collaborative Robotics, pp. 200–211 (2020)
10. Jose, E., Greeshma, M., Haridas, M.T., Supriya, M.H.: Face recognition based surveillance system using FaceNet and MTCNN on Jetson TX2. In: 2019 5th International Conference on Advanced Computing & Communication Systems (ICACCS), pp. 608–613 (2019)
11. Krizhevsky, A., Sutskever, I., Hinton, G.E.: Imagenet classification with deep convolutional neural networks. Commun. ACM **60**(6), 84–90 (2017)
12. Keras-vggface: https://github.com/rcmalli/keras-vggface. Accessed 2021/01/11
13. Iandola, F.N., Han, S., Moskewicz, M.W., Ashraf, K., Dally, W.J., Keutzer, K.: SqueezeNet: AlexNet-level accuracy with 50x fewer parameters and <0.5 MB model size. arXiv preprint arXiv:1602.07360 (2016)
14. Facenet: https://github.com/davidsandberg/facenet. Accessed 2021/01/10
15. Facenet-pytorch: https://github.com/timesler/facenet-pytorch. Accessed 2021/01/11
16. Grm, K., Štruc, V., Artiges, A., Caron, M., Ekenel, H.K.: Strengths and weaknesses of deep learning models for face recognition against image degradations. IET Biom. **7**(1), 81–89 (2017)
17. Deng, J., Dong, W., Socher, R., Li, L.J., Li, K., Fei-Fei, L.: ImageNet: a large-scale hierarchical image database. In: 2009 IEEE Conference on Computer Vision and Pattern Recognition, pp. 248–255 (2009)
18. Sarkar, E., Korshunov, P., Colbois, L., Marcel, S.: Vulnerability analysis of face morphing attacks from landmarks and generative adversarial networks. arXiv preprint arXiv:2012.05344 (2020)
19. Szegedy, C., Liu, W., Jia, Y., Sermanet, P., Reed, S., Anguelov, D., Erhan, D., Vanhoucke, V., Rabinovich, A.: Going deeper with convolutions. In: Proceedings of the IEEE Conference on Computer Vision and Pattern Recognition, pp. 1–9 (2015)
20. Cao, Q., Shen, L., Xie, W., Parkhi, O.M., Zisserman, A.: VGGFace2: a dataset for recognising faces across pose and age. In: 2018 13th IEEE International Conference on Automatic Face & Gesture Recognition (FG 2018), pp. 67–74 (2018)
21. Huang, G.B., Mattar, M., Berg, T., Learned-Miller, E.: Labeled faces in the wild: a database for studying face recognition in unconstrained environments. In: Workshop on Faces in 'Real-Life' Images: Detection, Alignment, and Recognition (2008)

22. Szegedy, C., Ioffe, S., Vanhoucke, V., Alemi, A.: Inception-v4, inception-ResNet and the impact of residual connections on learning. Proc. AAAI Conf. Artif. Intell. **31**(1) (2017)
23. Guo, Y., Zhang, L., Hu, Y., He, X., Gao, J.: MS-Celeb-1M: a dataset and benchmark for large-scale face recognition. In: European Conference on Computer Vision, pp. 87–102 (2016)
24. MaskTheFace: https://github.com/aqeelanwar/MaskTheFace. Accessed 2021/01/11
25. Padilla, R., Netto, S.L., da Silva, E.A.B.: A survey on performance metrics for object-detection algorithms. In: 2020 International Conference on Systems, Signals and Image Processing (IWSSIP), pp. 237–242 (2020)
26. Ryumin, D., Kagirov, I., Axyonov, A., Pavlyuk, N., Saveliev, A., Kipyatkova, I., Zelezny, M., Mporas, I., Karpov, A.: A multimodal user interface for an assistive robotic shopping cart. Electronics **9**(12), 2093 (2020)
27. Pavliuk, N., Kharkov, I., Zimuldinov, E., Saprychev, V.: Development of multipurpose mobile platform with a modular structure. In: Proceedings of 14th International Conference on Electromechanics and Robotics "Zavalishin's Readings", pp. 137–147 (2020)

Developing of a Software–Hardware Complex for Automatic Audio–Visual Speech Recognition in Human–Robot Interfaces

Denis Ivanko, Dmitry Ryumin, and Alexey Karpov

Abstract In recent years, audio speech has become more and more popular and often used in modern human–robot interfaces. Such natural form of communication is highly appreciated by users. There is no doubt that in the nearest future, alongside with the technology development, we will encounter the development of such "native" human–robot interfaces. In this paper, we propose the architecture and develop the software–hardware complex designed for automatic speech recognition with a dictionary of small and medium size and to be used in robots. A distinctive feature of the developed software–hardware complex is the presence of an audio–visual speech synchronization module, which allows both (1) to detect a speech signal in audio data and (2) to take into account the natural asynchrony between acoustic and visual speech. Based on this, it is possible (3) to synchronize the speech sections of audio and video streams in time. Another distinctive feature is the presence of a modality combining module, which allows (1) to combine informative data from audio and video signals and (2) to adjust the weights of each modality depending on the SNR level, which allows achieving optimal recognition accuracy even in acoustically noisy conditions.

1 Introduction

Speech recognition is one of the most interesting and complex tasks of artificial intelligence. Here, the achievements of very different fields are involved: from computer linguistics to digital signal processing and machine learning.

Nowadays, for our devices, sounding speech represents a digital signal. And if we look at the recording of this signal, then we will not see either words or pronounced phonemes (phoneme is the minimal meaningful unit of the language, i.e., a sound

D. Ivanko (✉) · D. Ryumin · A. Karpov
St. Petersburg Federal Research Center of the Russian Academy of Sciences (SPC RAS), St. Petersburg Institute for Informatics and Automation of the Russian Academy of Sciences, 39, 14th Line, 199178 St. Petersburg, Russia

A. Karpov
e-mail: karpov@iias.spb.su

A. Ronzhin and V. Shishlakov (eds.), *Electromechanics and Robotics*, Smart Innovation, Systems and Technologies 232, https://doi.org/10.1007/978-981-16-2814-6_23

whose replacement may change the meaning of a word or phrase)—different "speech events" smoothly flow into each other without forming clear boundaries. The same phrase, uttered by different people or in different settings, will vary at the signal level. At the same time, people somehow recognize each other's speech: therefore, there are invariants according to which the signal can be restored. The search for such invariants is the task of acoustic modeling.

At the same time, the acoustic model is just one of the components of the speech recognition system. What should we do if the recognition dictionary consists of a large number of words (hundreds of thousands or even millions), because many of them will be very similar in pronunciation or even coincide. However, in the presence of context, the role of acoustics falls: slurred, noisy, or ambiguous words can be restored "by meaning." Language models are used to account for the context. The task of the language model is to determine the probability of a sequence of words. The most common approach to language modeling is n-gramm-based statistical models, which are sequences of n words that assumed that the probability of a word depends only on $n - 1$ words preceding it.

Thus, significant successes have been achieved in the field of automatic speech recognition over the past decades—there are many commercial applications that make investments in this area profitable. Automatic speech recognition systems are widely used in smart virtual assistants (e.g., Siri from Apple, Alexa from Amazon, Cortana from Microsoft, Alisa from Yandex, etc.), including voice control, voice search, voice text input, and other applications. Speech recognition technologies are widely used in various areas of business, for example, in telephony by creating voice self-service systems and automating the processing of incoming and outgoing calls, which allow receiving reference information and consultations, ordering goods and services, conducting surveys, questionnaires, informing, and any other scenarios. Automatic speech recognition systems are embedded in the voice interfaces of smart home systems, voice interfaces of electronic robots, household appliances, and even car navigation systems. With the help of automatic speech recognition technologies, social services to organize contactless interaction are created for people with disabilities.

Nevertheless, despite significant successes, the main goal of research, which was originally implied in ASR field—the free communication of man and the "machine"—has not yet been achieved. The development of the field has revealed new difficulties that challenge researchers at the present stage.

In quiet and cozy office conditions, automatic recognition systems have already reached a certain level and are used in a number of applications; however, the quality and reliability of computer-based speech analysis in real-world applications remain insufficient. One of the most difficult tasks in the field of ASR is the recognition of continuous speech. The complexity of the speech recognition problem is mainly related to the variability of its main parameters, which are influenced by many factors. First of all, this is a random component of the process of speech formation, which leads to a variety of descriptions of the same word spoken by the same speaker. More significant variability is associated with individual differences in the speech apparatus of different speakers due to the influence of the speaker's gender, age, accents, and the emotional and physical state of the speaker. In addition, the acoustic aspect, i.e., microphone change, its location relative to the mouth, surrounding acoustic environment also has a strong influence on recognition accuracy. It should be noted that the existing systems and models of automatic speech recognition are still significantly inferior to human speech abilities, especially in real conditions of functioning, which indicates their insufficient adequacy and makes the use of speech technologies in industry and everyday life ineffective. In many operating conditions (in particular, with a low sound signal quality, the presence of external noise or extraneous conversations), automatic recognition systems cannot provide the required quality of operation even when using various filtering and noise reduction methods.

To date, there is no generally accepted approach for developing audio–visual speech recognition systems. There are no representative databases for training models that have all the necessary parameters, such as a sufficient number of speakers, phoneme-viseme time labeling, an adequate dictionary size, etc. (there are practically no databases for languages other than English). There are no studies on the impact of video recording speed on speech recognition accuracy. There is little research on the effect of acoustically noisy conditions on the performance of audio–visual speech recognition systems. Very few works are devoted to the study of inflectional languages (such as Russian). However, there is a huge difference between the recognition of analytical languages (e.g., English) and inflectional languages, due to the presence of a much larger number of word forms and grammatical rules in the latter.

The complex of the above facts allows us to confidently state a significant gap in scientific research with regard to automatic recognition of audio–visual speech. To fill it, we decided to develop a novel software–hardware complex for automatic audio–visual Russian speech recognition to be used in modern human–robot interfaces.

2 Backgrounds

After the discovery of the McGurk effect in 1976 [1], it became obvious that humans themselves pay attention to the lip's movement of the interlocutor in order to better understand the meaning of the statement during conversation.

In general, researchers divide the solution of the audio–visual speech recognition problem into two parts: first, in extracting the most informative features from each modality, and second, in the most successful way of fusion both modalities [2]. A typical structure of bimodal speech recognition is shown in Fig. 1.

Thus, the first important step in the design of AV recognition systems is to create the best possible unimodal recognizers. And the second step is to determine how to integrate knowledge from both modalities (in our case, audio and video) in order to preserve the useful information of each modality, and at the same time get rid of the shortcomings of both.

Representing modalities (audio and visual) in an appropriate and efficient feature space is an important step before their fusion. To ensure this, the constructed information features should contain the valuable characteristic of the speech signal.

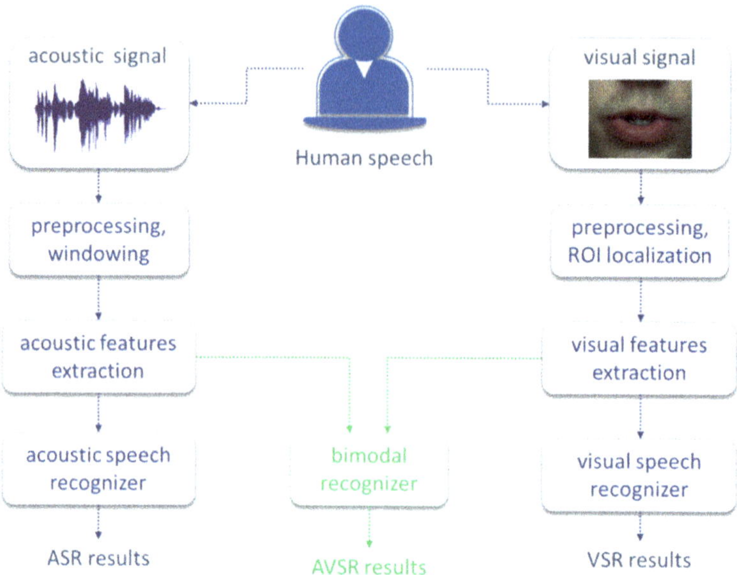

Fig. 1 Basic structure of audio–visual speech recognizer

One of the most commonly used features in modern speech recognition systems is Mel-frequency cepstral coefficients (MFCC) [3].

Visual information plays a key role in automatic speech recognition when audio is corrupted by background noise, or even inaccessible [4]. Speech recognition using visual information is called lip-reading. The initial idea of visual speech recognition comes from humans' experience: we are able to recognize spoken words from the observation of a speaker's face without or with limited access to the sound part of the voice [5]. During the last two decades, there have been significant advances in the research of audio-based ASR. Initially, researchers expected that automatic visual speech recognition would be easily accomplished based on the progress achieved in development of audio-based ASR. However, early attempts did not yield good results. Despite the poor performance, visual features still helped boost the ASR performance on some low-quality audio data through audio–visual (AV) speech information fusion [6]. Initial visual speech analysis also provided a lot of interesting information about human ability to lip-read [7]. For example, it was discovered that visual features provide very weak speech information in a large vocabulary continuous speech recognition (LVCSR) task. The basic unit that describes how speech conveys linguistic information is the phoneme. Similarly, the basic visually distinguishable unit, used in the audio–visual speech processing and corresponding literature [8, 9], is the viseme. Phonemes represent the manner of articulation, and visemes represent the place of articulation [10]. Among researchers, there is no universal consensus on the exact grouping of phonemes into visemes, although some groups are clearly defined [11].

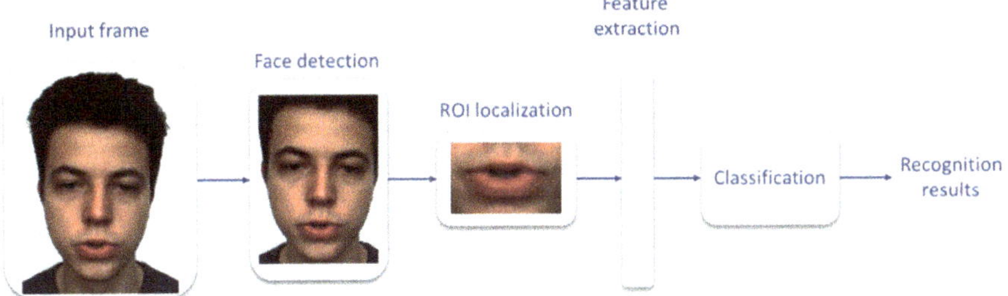

Fig. 2 Automated lip-reading pipeline example

According to modern approaches to automated visual speech recognition, the task can be divided into three sequential stages: region-of-interest (ROI) detection, visual features extraction, and visual speech recognition. The visualization of this pipeline is presented in Fig. 2.

Detecting a region-of-interest that contains the speech-related mouth motion is the first and very important step in the construction of a reliable visual speech recognition system. This is explained by the fact that the quality of ROI significantly affects the resulting speech recognition accuracy. To extract ROIs, many researchers relied on the active appearance models (AAM) [12, 13], Haar-like feature-based boosted classification framework [14, 15], skin color thresholding [16], etc.

In comparison with well-known MFCC features for acoustic speech, researchers were not able to find the best combination of informative features universally accepted for representing visual speech (despite significant amount of studies in the recent years). To date, there are several basic types of visual features existing in scientific literature. The most widespread of them are: pixel-based features [17]—raw pixel data used directly or after some image processing; geometry-based features [18]—geometric information of the talking mouth is extracted as features; motion-based features [19]—features designed to describe the motion; model-based features [20]—a model of the mouth motion is built, and model parameters are used as visual features, or a combination of mentioned above features [21, 22].

There are several state-of-the-art methods for model training. Initially, the most widespread methods were based on the use of hidden Markov models (HMM) for visual speech recognition and their coupled or multistream versions for audio–visual speech recognition [23]. However, at present, the approaches based on the use of neural networks of different architectures have become increasingly popular [24]. Some methods based on SVM can also be found in the literature [25].

To sum up, the problem of automatic audio–visual speech recognition lies, firstly, in extracting the most informative features from each modality and, secondly, in the most successful way of fusion both modalities. We also noted the fact, that despite significant successes recently achieved in the field of acoustic speech recognition, the task of automated lip-reading still remains underdeveloped. Therefore, in our work, we will address two of these mentioned problems, especially with the regard of developing proper software–hardware complex for audio–visual speech recognition. In the next section, we will consider it in more detail.

3 Data

The speech datasets are mandatory for training any modern speech recognition system based on probabilistic models and machine learning techniques. For the training of audio-only-based speech recognition systems, there are already a number of large commercial or free corpora available. They have all the desired parameters such as adequate data size, realistic variability, standard experimental settings, and evaluation measures. However, when it comes to audio–visual or visual-only databases, the situation is rapidly changing for the worse.

The vast majority of existing datasets are designed for the English language, and no more than 1–2 datasets for other languages, including German, French, Czech, Japanese, Chinese, Spanish, Polish, and Russian. Thus, databases of other languages are almost non-existent, which indicates a significant technology gap that needs to be filled.

Russian is one of the most widely spoken languages in the world, with up to 300 million native speakers (more than 140 million in Russia). However, robust and reliable automatic Russian speech recognition systems, practically, do not exist. The development of Russian speech technologies is heavily influenced by the nature of the language, such as absence of strict grammatical constructions in sentences, huge amount of word formation rules, large number of exceptions, and the variability of Russian speech in the presence of dialects and accents [26].

The speech recognition task is characterized by many parameters, such as: speech modality (acoustic/visual), the size of recognition dictionary (small/medium/large), the ambient noise level (SNR value), and type of speech input (letter-wise/keywords/continuous speech). Word boundaries in a continuous speech can be determined only in the process of decoding, by selecting the optimal sequence of words that best matches the input speech stream according to acoustic and language models.

Speech recognition accuracy is strongly connected with the size of recognition dictionary. There are several possible classifications of the size of a recognized dictionary exist in scientific literature. A small dictionary is a dictionary containing single words up to several dozens. Such a dictionary is suitable for recognizing sequences of numbers (phone numbers, numerical codes, etc.), voice commands for moving technical objects (car, airplane, etc.), control systems for various equipment (e.g., medical) and robots' remote control, etc. A recognizable medium-sized dictionary contains hundreds of words. Such a dictionary is sufficient for most dialog systems. A large dictionary contains thousands of words, and such recognition systems can be used in automated assistive systems or speech dictation systems in a limited subject area. A dictionary of over one hundred thousand words in size is considered super-large and allows to create shorthand systems for almost any synthetic language (such as all Slavic languages, including Russian).

Unlike analytic natural languages, such as English or French, in which grammatical meanings are unambiguously expressed by the word itself, the Russian language belongs to the family of inflected languages. Inflective languages are distinguished by a rich morphology and a developed system of word formation. Such languages can synthesize fairly long word forms from several components (morphemes or syllables) according to grammatical rules. Moreover, in fluent speech, the end of a word is not pronounced as clearly as the initial part, which leads to acoustic uncertainty and, on average, lower recognition accuracy compared to analytical languages.

Based on the researched features of the Russian language, to create a recognition system for audio–visual speech, initially we need a small- or medium-sized dictionary, constructed in a way to cover all the phoneme–viseme variety of the language. With this configuration each speaker has to pronounce at least 100–150 phrases of continuous Russian speech.

Thus, having determined the basic requirements to the number of speakers, dictionary size, quality of the video, and recording parameters, we decided to use HAVRUS [27] corpus collected in 2016–17 in SPIIRAS since it is the only one that satisfies the given parameters.

4 Software–Hardware Complex for Audio–Visual Speech Recognition

The developed software–hardware complex for audio–visual speech recognition runs on x64-compatible computers running operating systems (OS) of the Windows family (64 bit), developed in the C++ programming language in the Microsoft Visual Studio development environment and performs the following functions:

- parametric representation of audio and video signals for automatic audio–visual speech recognition;
- automatic recognition of continuous Russian speech with a dictionary of small and medium size (up to a thousand words of the subject area), obtained both from the files of the speech database, and directly from the microphone and high-speed video camera.

The architecture of the developed software–hardware complex is shown in Fig. 3.

A distinctive feature of the developed software–hardware complex is the presence of an audio–visual speech synchronization module, which allows both (1) to detect a speech signal in audio data and (2) to take into account the natural asynchrony between acoustic and visual speech, on the basis of which it is possible (3) to synchronize the speech sections of audio and video streams in time.

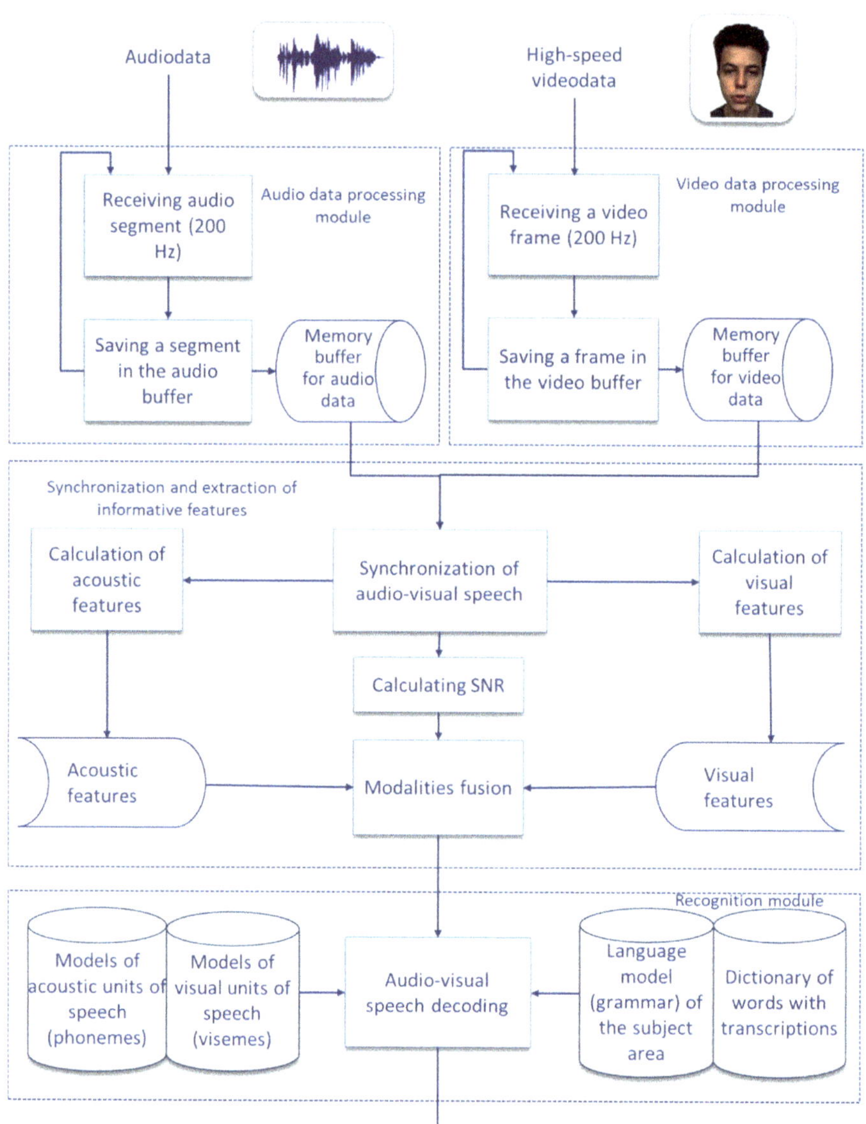

Fig. 3 Architecture of the developed software–hardware complex for automatic Russian audio–visual speech recognition

Another distinctive feature is the presence of a modality combining module, which allows both (1) to combine informative data from audio and video signals, and (2) to adjust the weights of each modality depending on the signal-to-noise (SNR) level, which allows achieving optimal recognition accuracy even in acoustically noisy conditions.

The developed software–hardware complex for audio–visual speech recognition performs the following tasks:

- calculating of informative features from time-synchronized audio and video signals;
- audio–visual Russian speech recognition;
- displaying the result of audio–visual speech recognition in text form in a dialog box;
- calculating the accuracy of audio–visual speech recognition;
- calculating the speed of audio–visual speech recognition.

The input data is the audio signal from the microphone and the video signal from the video camera. The audio signal must have the following parameters: sampling frequency 16–44 kHz, 16 bits per digital sample, and mono format. The video signal must have the following parameters: resolution 640 × 480 pixels at 30–200 frames per second and color—24 bits per pixel. The output data is the recognition results in the text form.

5 Evaluation Experiments

Table 1 shows the results of the speech recognition system under various noisy conditions. Two types of acoustic noises (babble noise and white noise) with different signal-to-noise ratios (SNR from 0 to 40 dB) were added to the test audio data.

In the audio-only recognition systems, the WRR starts to degrade at signal-to-noise level lower than 15–20 dB to almost zero at SNR of 0 dB, because at the SNR level of <5 dB it is literally impossible to distinguish a useful acoustic signal from a noise. The visual-only-based speech recognition system is not affected by any acoustic noises, so the WRR remains constant. Thus, it can be concluded that

Table 1 WRR versus SNR comparison for different speech recognition models under two types of noises: babble noise and white noise

SNR, dB	0	5	10	15	20	25	30	40
WRR, audio + «white noise»	3.3	29.8	51.5	62.77	71.01	75.6	78.94	81.82
WRR, audio + «babble noise»	1.3	25.8	47.5	64.25	75.12	78.63	80.57	81.82
WRR, CHMM + «white noise»	26.8	37.4	53.4	64.26	71.95	76.06	79.2	81.82
WRR, CHMM + «babble noise»	26.1	35.4	48	64.37	75.56	78.95	80.99	81.82
WRR, video	25.9	25.9	25.9	25.9	25.9	25.9	25.9	25.9

Table 2 Modalities weights, depending on the SNR level

SNR, dB	0	5	10	15	20	25	30	40
Weights (audio: video)	0:2.0	0.5:1.5	1.0:1.0	1.1:0.9	1.2:0.8	1.4:0.6	1.5:0.5	1.5:0.5

the best recognition results can be achieved combining both audio and visual information streams. In severe acoustically noisy conditions, the weight of the audio modality should be minimized and the recognition system should rely only on the video modality (SNR < 10 dB). However, if SNR > 10 dB, the video modality can no longer provide a significant increase in accuracy and its weight must be reduced.

In the range from 0 to 10–15 dB SNR, the recognition accuracy of audio–visual systems already exceeds the recognition accuracy of audio-only systems. For 15 dB, this difference results in 2–3% WRR, and at 0 dB, SNR level this difference even reaches 25.9% in our experiments.

The optimal weights depend on the data and the recording conditions. Table 2 shows the weights used during the experiments. Given to correct use of these features, the audio–visual system becomes more robust and it is possible to obtain the best possible WRR in any acoustic conditions by modifying the weights of the modalities. To improve performance in real applications, this method can also be combined with noise filtration algorithms.

6 Conclusions

In this paper, we proposed the architecture and develop the software–hardware complex designed for automatic audio–visual speech recognition and to be used in robots. A distinctive feature of the developed software–hardware complex is the presence of an audio–visual speech synchronization module, which allows both (1) to detect a speech signal in audio data and (2) to take into account the natural asynchrony between acoustic and visual speech. Based on this, it is possible (3) to synchronize the speech sections of audio and video streams in time. Another distinctive feature is the presence of a modality combining module, which allows (1) to combine informative data from audio and video signals and (2) to adjust the weights of each modality depending on the SNR level, which allows achieving optimal recognition accuracy even in acoustically noisy conditions.

Based on the conducted studies, it can be concluded that the use of the developed software–hardware complex makes it possible to improve the speech recognition accuracy on continuous Russian speech, especially under noisy conditions when the accuracy of the audio-only speech recognition system is degraded by acoustic noise.

Acknowledgements This research is supported by the Russian Foundation for Basic Research (project No. 19-29-09081 мк).

References

1. McGurk, H., MacDonald, J.: Hearing lips and seeing voices. Nature **264**, 746–748 (1976)
2. Ivanko, D., Ryumin, D., Axyonov, A., Zelezny, M.: Designing advanced geometric features for automatic Russian visual speech recognition. In: Proceedings of 20th International Conference on Speech and Computer, SPECOM 2018, pp. 245–255. Springer, Cham (2018)
3. Davis, S., Mermelstein, P.: Comparison of parametric representations for monosyllabic word recognition in continuously spoken sentences. IEEE Trans. Acoust. Speech Signal Process. **28**(4), 357–366 (1980)
4. Potamianos, G., Neti, C., Matthews, I.: Audio-visual automatic speech recognition: an overview. In: Bailly, G., Vatikiotis-Bateson, E., Perrier, P. (eds.) Issues in Visual and Audio-Visual Speech Processing. MIT Press, Cambridge (2004)
5. Hlavac, M.: Automated lipreading with LipsID features. PhD thesis, Pilsen (2019)
6. Zhou, Z., Zhao, G., Hong, X., Pietikainen, M.: A review of recent advances in visual speech decoding. Image Vis. Comput. **32**(9), 590–605 (2014)
7. Ivanko, D., Ryumin, D., Kipyatkova, I., Axyonov, A., Karpov, A.: Lip-reading using pixel-based and geometry-based features for multimodal human-robot interfaces. In: 14th International Conference on Electromechanics and Robotics "Zavalishin's Readings", ERZR 2019, pp. 197–207. Springer, Cham (2019)
8. Ivanko, D., Ryumin, D., Karpov, A.: An experimental analysis of different approaches to audio-visual speech recognition and lip-reading. In: 15th International Conference on Electromechanics and Robotics "Zavalishin's Readings", ERZR 2020, pp. 1–13. Springer, Cham (2020)
9. Casanovas, A.L., Vandergheynst, P.: Audio-Visual Object Extraction Using Graph Cuts. Ecole Polytechnique Federale de Lausanne (EPFL), Lausanne, Switzerland (2012)
10. Nock, H.J., Iyengar, G., Neti, C.: Speaker localisation using audio-visual synchrony: an empirical study. In: International Conference on Image and Video Retrieval, pp. 488–499. Springer, Berlin, Heidelberg (2003)
11. Katsaggelos, K., Bahaadini, S., Molina, R.: Audiovisual fusion: challenges and new approaches. Proc. IEEE **103**(9), 1635–1653 (2015)
12. Newman, J., Cox, S.: Language identification using visual features. IEEE Audio Speech Lang. Process. **20**(7), 1936–1947 (2012)
13. Lan, Y., Theobald, B., Harvey, R.: View independent computer lip-reading. In: Proceedings of International Conference Multimedia Expo (ICME), pp. 432–437. IEEE (2012)
14. Zhou, Z., Hong, X., Zhao, G., Pietikainen, M.: A compact representation of visual speech data using latent variables. IEEE Trans. Pattern Anal. Mach. Intell. **36**(1), 181–187 (2014)
15. Zhao, G., Barnard, M., Pietikäinen, M.: Lipreading with local spatiotemporal descriptors. IEEE Trans. Multimedia **11**(7), 1254–1265 (2009)
16. Estellers, V., Gurban, M., Thiran, J.: On dynamic stream weighting for audio-visual speech recognition. IEEE Trans. Audio Speech Lang. Process. **20**(4), 1145–1157 (2012)
17. Hong, S., Yao, H., Wan, Y., Chen, R.: A PCA based visual DCT feature extraction method for lip-reading. In: Proceedings of Intelligent Informatics and Hiding Multimedia and Signal Processing, pp. 321–326 (2006)
18. Cetingul, H., Yemez, Y., Erzin, E., Tekalp, A.: Discriminative analysis of lip motion features for speaker identification and speech reading. IEEE Trans. Image Process. **15**(10), 2879–2891 (2006)
19. Yoshinaga, T., Tamura, S., Iwano, K., Furui, S.: Audio-visual speech recognition using lip movement extracted from side-face images. In: Proceedings of International Conference Auditory-Visual Speech Processing (AVSP), pp. 117–120 (2003)
20. Lan, Y., Harvey, R., Theobald, B., Ong, E., Bowden, R.: Comparing visual features for lipreading. In: Proceedings of International Conference on Auditory-Visual Speech Processing (AVSP), pp. 102–106 (2009)

21. Rahmani, M.H., Alamsganj, F.: Lip-reading via a DNN-HMM hybrid system using combination of the image-based and model-based features. In: 3D International Conference on Pattern Recognition and Image Analysis, pp. 195–199. IEEE (2017)
22. Radha, N., Shahina, A., Khan, A.: An improved visual speech recognition of isolated words using combined pixel and geometric features. Indian J. Sci. Technol. **9**(44), 83–93 (2016)
23. Ivanko, D., Karpov, A., Fedotov, D., Kipyatkova, I., Ryumin, D., Ivanko, D., Minker, W., Zelezny, M.: Multimodal speech recognition: increasing accuracy using high-speed video data. J. Multimodal User Interfaces **12**(4), 319–328 (2018)
24. Chung, J.S., Zisserman, A.: Lip reading in the wild. In: Proceedings of ACCV, pp. 171–182 (2016)
25. He, J., Zhang, H.: Lipreading recognition based on SVM and DTAK. In: 4th International Conference on Bioinformatics and Biomedical Engineering, pp. 321–324 (2010)
26. Karpov, A., Kipyatkova, I., Zelezny, M.: A framework for recording audio-visual speech corpora with a microphone and a high-speed camera. In: International Conference on Speech and Computer (SPECOM 2014), pp 50–57. Springer, Cham (2014)
27. Verkhodanova, V., Ronzhin, A., Kipyatkova, I., Ivanko, D., Karpov, A., Železný, M.: HAVRUS corpus: high-speed recordings of audio-visual Russian speech. In: Speech and Computer (SPECOM 2016), vol. 9811, pp. 338–345. Springer, Cham (2016)

Pilot Studies on Avrora Unior Car-Like Robot Control Using Gestures

Nikita Nikiforov, Tatyana Tsoy, Ramil Safin, Yang Bai, Mikhail Svinin, and Evgeni Magid

Abstract Gesture recognition is not only an important communication channel in human-human interaction but it also allows a human to communicate with other intelligent devices. This paper presents a concept for controlling the car-like robot Avrora Unior locomotion using gestures. We created a list of 18 control commands that contains basic and compound commands. A group of 17 volunteers used this list to create individual control gestures independently. A small part of the obtained dataset of gestures was used with the Teachable machine service in order to preliminary evaluate a possibility of constructing a full-scale model and to train it appropriately.

N. Nikiforov (✉) · T. Tsoy · R. Safin · E. Magid
Laboratory of Intelligent Robotics Systems (LIRS), Institute of Information Technology and Intelligent Systems, Kazan Federal University, Kazan, Russia
URL: https://www.kpfu.ru/eng/itis/research/laboratory-of-intelligent-robotic-systems.com

T. Tsoy
e-mail: tt@it.kfu.ru

R. Safin
e-mail: safin.ramil@it.kfu.ru

E. Magid
e-mail: magid@it.kfu.ru

Y. Bai · M. Svinin
Information Science and Engineering Department, College of Information Science and Engineering, Ritsumeikan University, 1 -1 -1 Noji -higashi, Kusatsu, Shiga 525-8577, Japan
e-mail: yangbai@fc.ritsumei.ac.jp
URL:
http://www.en.ritsumei.ac.jp/academics/college-of-information-science-and-engineering/.com

M. Svinin
e-mail: svinin@fc.ritsumei.ac.jp

A. Ronzhin and V. Shishlakov (eds.), *Electromechanics and Robotics*, Smart Innovation, Systems and Technologies 232, https://doi.org/10.1007/978-981-16-2814-6_24

The obtained model demonstrated acceptable recognition rate. We also attempted to apply SURF and FLANN techniques for matching with the direct matching approach and the skeleton-based approach, but the matching results were not satisfactory.

1 Introduction

Nowadays, computer vision is used in various types of activities, including such complicated tasks as automatic object detection, search, and recognition. Recognition systems might concentrate on such objects as car plates [1], component labels [2], animals [3], and many others. In security field, human face recognition is used to ensure security of ATMs [4] and personal gadgets [5]. In the medical field, computer vision helps to detect leukocytes [6], cancer [7] and other diseases and disorders. In search and rescue operations, when it is physically dangerous for a person to operate within a dangerous zone, robots that replace people in victim search actively apply computer vision techniques that allow autonomous functions of a vehicle [8] and teleoperation remote control [9]. The list of objects of interest constantly replenishes as new problems arise that are difficult for human precise and fast detection and recognition capabilities.

Gesture recognition is often used as an important communication channel not only in human-human and human-animal interaction, but in human-robot interaction [10] and in a human communication with other intelligent devices and objects, for example, in operating household devices [11] or in interactive game-based learning [12]. In [13] authors demonstrated a hand gesture interface system for appliances' control in smart home environments [14]. In [15], a depth camera extracted a hand depth silhouette and the obtained images were recognized using a trained random forest[16]. In [17], a gesture recognition system generated an appropriate command, which allowed a selection, a mouse control [18], an exit and other additional functionalities.

In this paper, we focus on human-robot interaction using gestures [19]. Our long-term project's goal is to attempt enabling the car-like robot Avrora Unior (Fig. 1) control with user gestures, while these gestures are not completely predefined in advance [20]. To develop a gesture control concept for the Avrora Unior robot, we carefully studied and tested locomotion capabilities of the robot and created an exhaustive list of basic commands that are required to control the robot. A group of students were asked to provide their own unlimited gestures, which in their opinion would correspond to each command from the list. We attempted to use SURF [21] and FLANN [22] approaches to allow gesture recognition without constructing a skeleton of a user. This paper presents preliminary results of the pilot study.

The rest of the paper is organized as follows. Section 2 describes gesture control concept for the Avrora Unior robot. Section 3 explains process of collecting dataset and its characteristics. In Sect. 4 we present pilot studies. Finally, we conclude in Sect. 5.

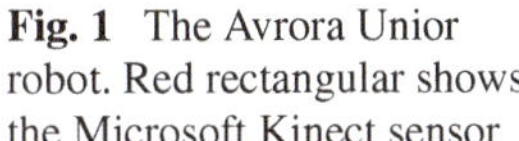

Fig. 1 The Avrora Unior robot. Red rectangular shows the Microsoft Kinect sensor

2 Gesture Control Concept

Use of gestures to interact with robots and smart devices has been explored in many works. Gestures allow controlling industrial robots via gesture-based user-friendly interface [23], Leap Motion technology [24] or Microsoft Kinect Controller [25]. Even a low-cost USB camera could successfully recognize and track user's hand movement and allow controlling simple activities [26, 27]. Phyo et.al. demonstrated an interaction with a humanoid NAO robot assistant with static hand gestures [28]. Gao et.al. presented a gesture-based smart wheelchair control for aged and disabled that was successfully validated within indoor environment [29]. Recently, Zhang et.al. demonstrated a gesture-based control of a real unmanned vehicle with Kinect-V2 sensor that employs upper body pose recognition for 13 joints and a dynamic time warping [30].

Our car-like robot Avrora Unior (Fig. 1) is equipped with the Microsoft Kinect sensor [31] that primarily targets for environment monitoring, mapping and obstacle avoidance. An important characteristic of this controller is the range at which it guarantees correct values. Considering the depth sensor, the maximum distance between an object and the sensor is limited to 3.5 m [32]. This limitation should be taken into account when control gestures are selected and when a dataset for machine learning of possible gestures is constructed [33].

The Avrora Unior robot could be controlled in teleoperational mode using a special one-hand held motion controller or locomote autonomously [34]. The primary purpose of introducing additional gesture control was to enable hands-free convenient testing process of new motion and interaction algorithms, which would allow to correct the robot movement and to avoid accidents in a much more robust and fast way. A user should approach the robot at a distance of 3–4 m and, while staying in front of the robot and entirely within the Microsoft Kinect sensor field of view, show a particular control gesture, which triggers a corresponding command execution. Currently, we selected control gestures to be static in order to reduce a complexity of their recognition.

Since for full-size autonomous vehicles safety is critical, typically all control gestures are predefined in advance and then the robot control system is taught to recognize these gestures using an exhaustive set of examples [35]. However, such approach requires an operator to carefully study the gestures and to be always concentrated in order to use a proper one. In our case would like to allow an unprepared user to control the robot intuitively, which implies an unconstrained control that could guess the user intentions and operate accordingly.

At the first step of the project, we created a broad list of remote control commands. The selection was based on several factors: a convenience of using a command, an importance of the command, and a possibility of using it remotely. A basic set contained just forward and backward motion, left and right turns. Next, several more complicated commands were added to the set, e.g., turning 180 degrees, approaching a static person and automatic parallel parking [36]. The set of the control commands is presented in Table 1.

3 Dataset

For dataset collection it was necessary to record a 30–40 s video with a static gesture for each command keeping a distance from a camera to a user within a range of 3–4 m, which matches the Kinect sensor capabilities [37]. A group of 17 people, students and employees of Laboratory of Intelligent Robotic Systems,[1] were asked to propose a gesture for each of the 18 control commands independently of the others. None of the participants knew what gestures the other participants had selected. This was done in order to analyze and select the most appropriate gesture for a particular command based on the received variety of gestures and their statistical distribution. Frames were extracted from each video using Matlab software. Table 1 presents a number of frames that were extracted from the collected set of 306 (17 people, 18 gestures each) short video sequences for each gestures.

Unfortunately, the similarity between gestures from different users was rather small. Only one command, "Full stop", had a small variety of patterns and a sigh similarity of gestures within a pattern in many cases (see Fig. 2).

4 Pilot Studies

4.1 Optimal Gestures Selection

A selection of optimal gestures for each command was based on 3 criteria: a size (of a bounding box for entire user's body, while demonstrating a command), an ease

[1] https://kpfu.ru/eng/itis/research/laboratory-of-intelligent-robotic-systems.

Table 1 The control commands list and a size of datasets (frames) for each command

No.	Command	Comments	Frames
1	Moving forward	At a predefined constant speed	30,755
2	Backward movement	At a predefined constant speed	29,011
3	Turn the wheels to the right	The wheels are turned, while the vehicle is static, the turning angle gradually increases	30,450
4	Turn the wheels to the left	The wheels are turned, while the vehicle is static, the turning angle gradually increases	28,990
5	Increase speed	Speed gradually increases by a predefined value	29,048
6	Decrease speed	Speed gradually decreases by a predefined value	28,829
7	Full stop	Emergency braking	30,233
8	A mode of ignoring a user is on	Ignore all commands, except for a full stop and a command to disable this mode	30,615
9	A mode of ignoring a user is off	Disable Ignore Mode	29,427
10	Move forward while turning the wheels to the right	With a predefined constant speed and a predefined angle of rotation of the wheels	28,781
11	Move forward while turning the wheels to the left	With a predefined constant speed and a predefined angle of rotation of the wheels	30,021
12	Move backward while turning the wheels to the right	With a predefined constant speed and a predefined angle of rotation of the wheels	26,581
13	Move backward while turning the wheels to the left	With a predefined constant speed and a predefined angle of rotation of the wheels	26,939
14	Automatic 180° turn		30,941
15	Automatic 90° turn to the right		29,691
16	Automatic 90° turn to the left		29,892
17	Drive to a user		31,205
18	Automatic parallel parking	The only complicated command within the set, which implies a closest parking spot search [38] and further parking [36]	31,847

Fig. 2 Proposed gestures for "Full stop" command by two users

of use, and an absence of overlaps of upper limbs' links of a user's skeleton. It is important that a bounding box for a gesture does not exceed a predefined threshold, which depends on a distance from a user to the robot (in our pilot study selection we targeted for distance of 3–4 m) and the user's height; otherwise, the gesture might be out of field of view of the Kinect sensor, which is located at a height of 0.35 m from the ground [39]. For example, Fig. 3 presents unsuitable gestures that would be outside of field of view.

Some gestures might be difficult to perform due to physical limitations of an average human skeleton and joints' flexibility. This means it is highly likely that such gestures would be rarely used by a typical user. Figure 3 also presents an example of a gesture that might be difficult to repeat as this posture is rather uncomfortable for a typical human.

In order to analyze, a posture of a user and to extract a control signal from the posture a basic skeleton extraction approach was selected. To draw a basic skeleton, the OpenPose system [40] was used. It was decided to avoid using gestures that may not be recognized correctly due to overlapping upper limbs' links of a basic skeleton (see Fig. 4).

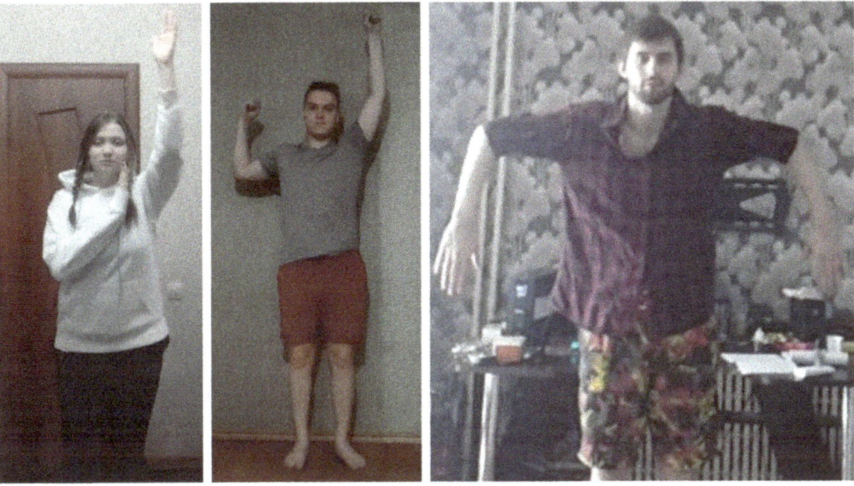

Fig. 3 Examples of unsuitable gestures due to their bounding box height (left and center) and uncomfortable posture (right)

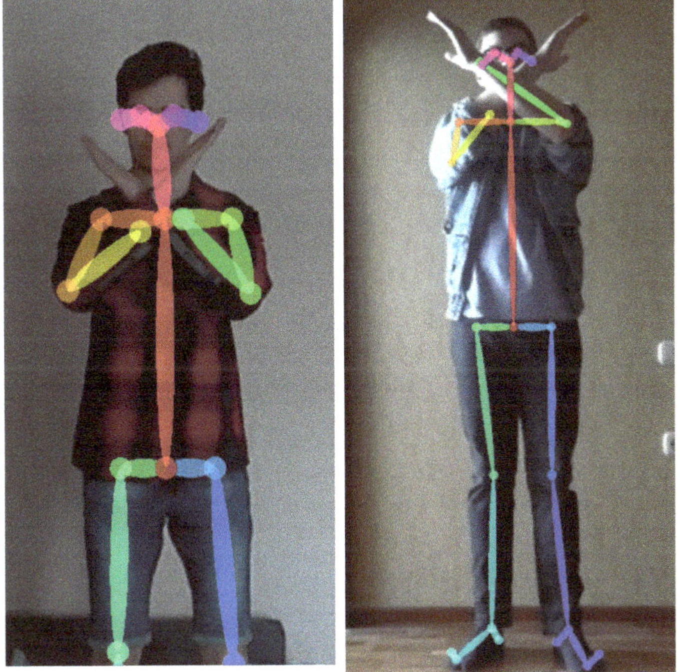

Fig. 4 Examples of unsuitable gestures due to overlapping of basic skeleton links

4.2 Training on Datasets

To verify if the constructed dataset of images would further allow us to create a full-scale model for machine learning, we used the Teachable machine service [41], which draws skeletons for dataset images and allows constructing and training a new model according to the selected training parameters. At this initial stage, for each command we used only 50 images from 5 participants to train a very basic model, which was taught to distinguish all 18 classes (all possible control commands of Table 1). Yet, even for a such small dataset the trained model demonstrated acceptable results (Fig. 5) that would obviously improve with the training set growth.

In order to check whether it is possible to detect correctly displayed gestures without constructing a skeleton and whether the skeleton use could improve the recognition precision we used Speeded-Up Robust Features (SURF [21]) and Fast Library for Approximate Nearest Neighbors (FLANN [22]) techniques. SURF was used to search for key features of the images (frames from the video sequences). Next, in order to obtain a quick and efficient matching, the comparison was performed using key feature comparator FLANN. The experiments obviously demonstrated that images with a similar posture of skeletons have more matches than in other cases. Figure 6 demonstrates an example of key feature points matching with two approaches: direct matching (the upper row in the figure; only 67 matches were successful) and using a basic skeleton (the central row in the figure; 103 matches were successful). Yet, for similar postures a skeleton-based approach also provided

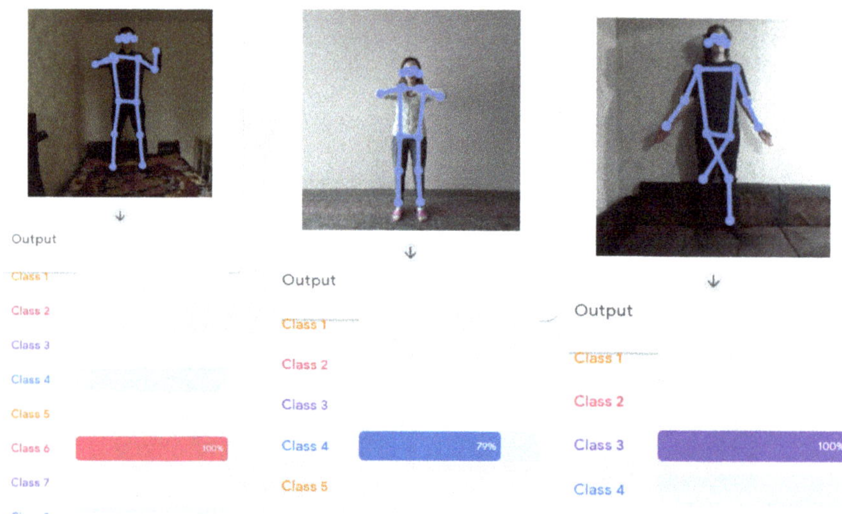

Fig. 5 An example of model training results using the Teachable machine service: 100% success for Class 6, 79% for Class 4, and 100% for Class 3. *Note* a mistake in the skeleton extraction for the image on the right

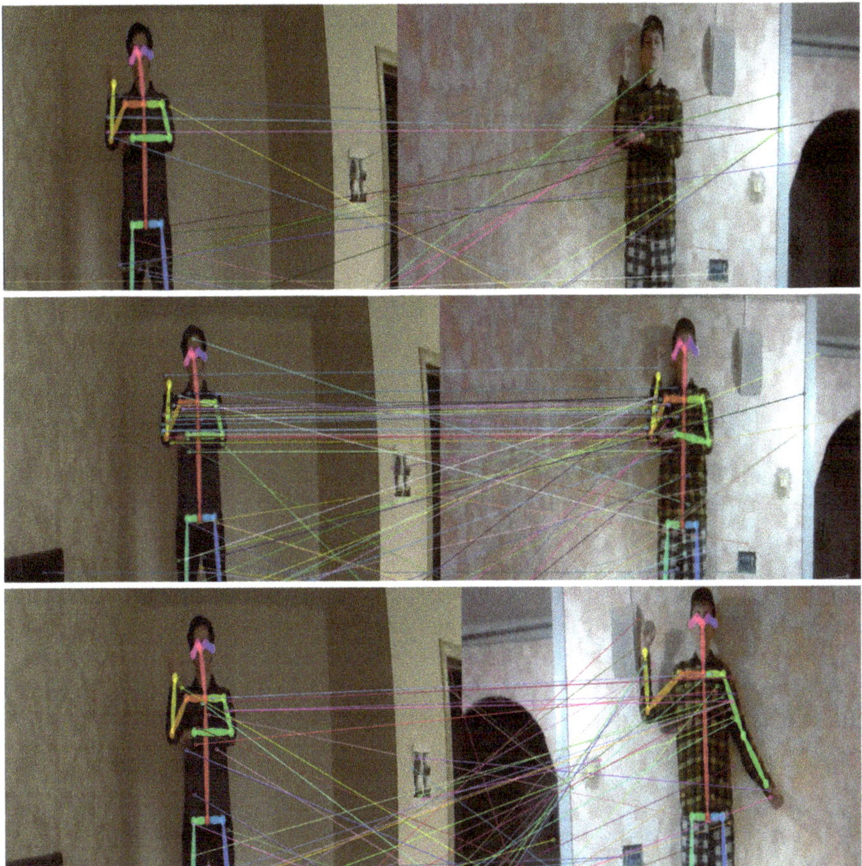

Fig. 6 A direct matching of postures without skeleton use (the upper row, 67 matches), a skeleton-based matching (the central row, 103 matches) and a skeleton-based matching of slightly different gesture (the bottom row, 91 matches)

a good level of matching (the bottom row in the figure; 91 matches were successful); however, in this example the two postures are objectively similar and thus a high level of matching features was appropriate.

Matching of some other images demonstrated that using a skeleton might not always cause a quantitative improvement relatively to the direct matching. For example, Fig. 7 demonstrates a particular case where the direct matching of postures without a skeleton quantitatively (technically) outperformed the skeleton-based matching (106 matches vs. 101 matches respectfully). Yet, a close look at the suggested corresponding matches demonstrates that in both cases a vast majority of the matches was wrong while the situation with proper matches was slightly better for the skeleton-based approach. The same issues with wrong matches applies to Fig. 6, but again the skeleton-based approach (Fig. 6, central row) demonstrated a significantly better

Fig. 7 A direct matching of postures without skeleton use (the upper row, 106 matches) and a skeleton-based matching (the bottom row, 101 matches)

amount of properly matched key features that the direct approach. The experimental results demonstrated that using SURF and FLANN techniques for matching, both with the direct approach and the skeleton-based approach, failed to provide acceptable level of matching for two similar human postures. Therefore, constructing a new classifier and teaching it appropriately becomes inevitable.

5 Conclusions

In this paper, we presented a concept for controlling the car-like robot Avrora Unior locomotion using human gestures. The list of 18 control commands contained basic and compound commands. A group of 17 volunteers used the commands' list to create individual control gestures independently. A small part of the obtained dataset of gestures (less than 0.2%) was used with the Teachable machine service in order to preliminary evaluate the possibility of constructing a full-scale model and to train it appropriately. The obtained model demonstrated acceptable recognition rate. We also attempted to apply SURF and FLANN techniques for matching with the direct matching approach and the skeleton-based approach, but they demonstrated an insufficient matching quality. Finally, we concluded that the collected dataset will allow constructing a good model that could be taught to successfully distinguish locomotion control gestures. The gestures from the developed datasets were not studied in any statistical aspects yet, which is a part of our ongoing work that will allow using the most statistically significant gestures and demonstrate reliability and validity of the proposed gesture selection criteria.

Acknowledgements This work was supported by the Russian Foundation for Basic Research (RFBR), project ID 19-58-70002. Forth and fifth authors acknowledge the support of the Japan Science and Technology Agency, the JST Strategic International Collaborative Research Program, Project No. 18065977.

References

1. Qadri, M.T., Asif, M.: Automatic number plate recognition system for vehicle identification using optical character recognition. In: 2009 International Conference on Education Technology and Computer, pp. 335–338. IEEE (2009)
2. He, L., Chao, Y., Suzuki, K., Wu, K.: Fast connected-component labeling. Pattern Recognit **42**(9), 1977–1987 (2009)
3. Nguyen, H., Maclagan, S.J., Nguyen, T.D., Nguyen, T., Flemons, P., Andrews, K., Ritchie, E.G., Phung, D.: Animal recognition and identification with deep convolutional neural networks for automated wildlife monitoring. In: 2017 IEEE international conference on data science and advanced Analytics (DSAA), pp. 40–49. IEEE (2017)
4. Ray, S., Das, S., Sen, A.: An intelligent vision system for monitoring security and surveillance of atm. In: 2015 Annual IEEE India Conference (INDICON), pp. 1–5. IEEE (2015)
5. Sutoyo, R., Harefa, J., Chowanda, A.: Unlock screen application design using face expression on android smartphone. In: MATEC Web of Conferences. vol. 54, p. 05001, EDP Sciences (2016)
6. Cuevas, E., Díaz, M., Manzanares, M., Zaldivar, D., Perez-Cisneros, M.: An improved computer vision method for white blood cells detection. Computational and Mathematical Methods in Medicine (2013)
7. Lee, H., Chen, Y.P.P.: Image based computer aided diagnosis system for cancer detection. Expert Syst Appl **42**(12), 5356–5365 (2015)
8. Al-Kaff, A., Moreno, F.M., de la Escalera, A., Armingol, J.M.: Intelligent vehicle for search, rescue and transportation purposes. In: 2017 IEEE International Symposium on Safety, Security and Rescue Robotics (SSRR), pp. 110–115. IEEE (2017)
9. Perez-Grau, F., Ragel, R., Caballero, F., Viguria, A., Ollero, A.: Semi-autonomous teleoperation of uavs in search and rescue scenarios. In: 2017 International Conference on Unmanned Aircraft Systems (ICUAS), pp. 1066–1074. IEEE (2017)
10. Shirwalkar, S., Singh, A., Sharma, K., Singh, N.: Telemanipulation of an industrial robotic arm using gesture recognition with kinect. In: 2013 International Conference on Control, Automation, Robotics and Embedded Systems (CARE), pp. 1–6. IEEE (2013)
11. Rashid, M., Han, X.: Gesture control of zigbee connected smart home internet of things. In: 2016 5th International Conference on Informatics, Electronics and Vision (ICIEV), pp. 667–670. IEEE (2016)
12. Hsiao, H.S., Chen, J.C.: Using a gesture interactive game-based learning approach to improve preschool children's learning performance and motor skills. Comput Educat **95**, 151–162 (2016)
13. Rahman, A.M., Hossain, M.A., Parra, J., El Saddik, A.: Motion-path based gesture interaction with smart home services. In: Proceedings of the 17th ACM international conference on Multimedia, pp. 761–764 (2009)
14. Hussain, S., Schaffner, S., Moseychuck, D.: Applications of wireless sensor networks and rfid in a smart home environment. In: 2009 Seventh Annual Communication Networks and Services Research Conference, pp. 153–157. IEEE (2009)
15. Muñoz-Salinas, R., Medina-Carnicer, R., Madrid-Cuevas, F.J., Carmona-Poyato, A.: Depth silhouettes for gesture recognition. Pattern Recognit Lett **29**(3), 319–329 (2008)
16. Pal, M.: Random forest classifier for remote sensing classification. Int J Remote Sensing **26**(1), 217–222 (2005)

17. Rautaray, S.S.: Real time hand gesture recognition system for dynamic applications. Int J UbiComp (IJU) **3**(1) (2012)
18. Vivek Veeriah, J., Swaminathan, P.: Robust hand gesture recognition algorithm for simple mouse control. Int J Comput Commun Eng **2**(2), 219–221 (2013)
19. Galin, R., Meshcheryakov, R.: Review on human–robot interaction during collaboration in a shared workspace. In: International Conference on Interactive Collaborative Robotics, pp. 63–74. Springer (2019)
20. Malov, D., Edemskii, A., Saveliev, A.: Architecture of proactive localization service for cyber-physical system's users. In: International Conference on Interactive Collaborative Robotics, pp. 10–18. Springer (2019)
21. Bay, H., Ess, A., Tuytelaars, T., Van Gool, L.: Speeded-up robust features (surf). Comput Vision Image Understanding **110**(3), 346–359 (2008)
22. Goel, A., Saxena, S.C., Bhanot, S.: Modified functional link artificial neural network. Int J Electri Comput Eng **1**(1), 22–30 (2006)
23. Tang, G., Webb, P.: The design and evaluation of an ergonomic contactless gesture control system for industrial robots. J Robotics (2018)
24. Chen, S., Ma, H., Yang, C., Fu, M.: Hand gesture based robot control system using leap motion. In: International Conference on Intelligent Robotics and Applications, pp. 581–591. Springer (2015)
25. Mikadlicki, K., Pajor, M.: Real-time gesture control of a CNC machine tool with the use Microsoft Kinect sensor. Int J Sci Eng Res **6**(9), 538–543 (2015)
26. Grif, H.S., Farcas, C.C.: Mouse cursor control system based on hand gesture. Procedia Technol **22**, 657–661 (2016)
27. Song, S., Yan, D., Xie, Y.: Design of control system based on hand gesture recognition. In: 2018 IEEE 15th International Conference on Networking, Sensing and Control (ICNSC), pp. 1–4. IEEE (2018)
28. Phyo, A.S., Fukuda, H., Lam, A., Kobayashi, Y., Kuno, Y.: A human-robot interaction system based on calling hand gestures. In: International Conference on Intelligent Computing, pp. 43–52. Springer (2019)
29. Gao, X., Shi, L., Wang, Q.: The design of robotic wheelchair control system based on hand gesture control for the disabled. In: 2017 International Conference on Robotics and Automation Sciences (ICRAS), pp. 30–34. IEEE (2017)
30. Zhang, B., Yang, M., Yuan, W., Wang, C., Wang, B.: A novel system for guiding unmanned vehicles based on human gesture recognition. In: 2020 IEEE International Conference on Real-time Computing and Robotics (RCAR), pp. 345–350. IEEE (2020)
31. Zhang, Z.: Microsoft Kinect sensor and its effect. IEEE Multimedia **19**(2), 4–10 (2012)
32. Han, J., Shao, L., Xu, D., Shotton, J.: Enhanced computer vision with Microsoft Kinect sensor: a review. IEEE Transa Cybernet **43**(5), 1318–1334 (2013)
33. Safin, R., Lavrenov, R., Tsoy, T., Svinin, M., Magid, E.: Real-time video server implementation for a mobile robot. In: 2018 11th International Conference on Developments in eSystems Engineering (DeSE), pp. 180–185. IEEE (2018)
34. Magid, E., Lavrenov, R., Khasianov, A.: Modified spline-based path planning for autonomous ground vehicle. In: ICINCO (2), pp. 132–141 (2017)
35. Lavrenov, R., Zakiev, A.: Tool for 3d gazebo map construction from arbitrary images and laser scans. In: 2017 10th International Conference on Developments in eSystems Engineering (DeSE), pp. 256–261. IEEE (2017)
36. Imameev, D., Shabalina, K., Sagitov, A., Su, K.L., Magid, E.: Modelling Autonomous Parallel Parking Procedure for Car-Like Robot Avrora Unior in Gazebo Simulator, pp. 428–431 (2020)
37. Safin, R., Garipova, E., Lavrenov, R., Li, H., Svinin, M., Magid, E.: Hardware and software video encoding comparison. In: 2020 59th Annual Conference of the Society of Instrument and Control Engineers of Japan (SICE), pp. 924–929. IEEE (2020)
38. Imameev, D., Zakiev, A., Tsoy, T., Bai, Y., Svinin, M., Magid, E.: Lidar-based parking spot search algorithm. In: Thirteenth International Conference on Machine Vision. vol. 11605, p. 1160502. International Society for Optics and Photonics (2021)

39. Shabalina, K., Sagitov, A., Su, K.L., Hsia, K.H., Magid, E.: Avrora unior car-like robot in gazebo environment. In: International Conference on Artificial Life and Robotics, pp. 116–119 (2019)
40. Cao, Z., Hidalgo, G., Simon, T., Wei, S.E., Sheikh, Y.: Openpose: realtime multi-person 2d pose estimation using part affinity fields. IEEE Trans pattern Anal Mach Intell **43**(1), 172–186 (2019)
41. Carney, M., Webster, B., Alvarado, I., Phillips, K., Howell, N., Griffith, J., Jongejan, J., Pitaru, A., Chen, A.: Teachable machine: Approachable web-based tool for exploring machine learning classification. In: Extended Abstracts of the 2020 CHI Conference on Human Factors in Computing Systems, pp. 1–8 (2020)

Electromechanics and Electric Power Engineering

Diagnosis of Moisture Content in Oil-Paper Bushings Using Statistical Indicators Based on Frequency Domain Spectroscopy

Hossein Taghizade Ansari, Abolfazl Vahedi, Pavel Alexandrovich Khlyupin, and Nami Mahmoudi

Abstract Condenser bushings are considered the Achilles' heel of power transformers because most of the explosions of transformers relate to their bushings' failure. Typical bushings monitoring methods measure the tan delta, power factor, and bushings' capacitance in line frequency. In contrast, the dielectric frequency response (DFR) or the frequency domain spectroscopy (FDS) measures these parameters over a wide frequency range. Some studies have revealed that the FDS curves have distinguished data that help detect bushing failures in their onset. However, there has not been a standard FDS interpretation method. Human experts' opinions play a considerable role in the interpretation process. In this paper, a finite element method model of a 96 kV oil-impregnated paper bushing is built with different OIP moisture content. The FDS test is simulated under these moisture levels. Seven two-array and six one-array statistical indices are introduced to interpret the FDS results. The results have shown that all six one-array statistical indices, apart from average deviation (AVEDEV) and standard deviation (SD), can be perfect indicators of moisture levels of the OIP. Moreover, they can be used to assess the moisture content of the OIP bushings, and human errors are eliminated using this method.

1 Introduction

Capacitive-grading or condenser bushings are essential parts of power transformers, which are both the most expensive and most essential power systems apparatus [1]. More than 10% of failures of all transformers relates to bushing failures. Moreover,

H. Taghizade Ansari · A. Vahedi (✉)
Electrical Engineering Department, Iran University of Science and Technology, Narmak, Tehran, Iran
e-mail: avahedi@iust.ac.ir

P. A. Khlyupin
Ufa State Petroleum Technical University (USPTU), 1 Cosmonavtov st., Ufa 450062, Russian Federation

N. Mahmoudi
PENCOPOWER, Tehran, Iran

A. Ronzhin and V. Shishlakov (eds.), *Electromechanics and Robotics*, Smart Innovation, Systems and Technologies 232, https://doi.org/10.1007/978-981-16-2814-6_25

a bushing failure can cause an explosion and blast that put the other apparatus and buildings at risk. Typical methods measure the dissipation factor, power factor, and capacitance of bushings at line frequency. There are boundaries for these parameters in different standards that help the maintenance of bushings [2, 3]. There are several reports in [4, 5] about failures and explosions of oil-impregnated paper (OIP) bushings without any indication of abnormal status in their last typical test results. For instance, it has been reported in [4] that during regular operation of 250 MVA power transformer, the 400 kV OIP bushing in the L2 phase was exploded and caused the transformer fire. It was stated that the exploded porcelain shield of the OIP bushing caused damage to the tap changer conservator and caused the oil leakage, which caused a quick spread of transformer fire. However, it has been reported that the analysis of the results of dielectric losses factor and capacitance measurements of the damaged bushing from the last years of operation did not reveal anything suspicious. This report and the other related reports reveal that the information extracted from the typical tests is not sufficient, and the FDS test can be a solution to these situations.

Dielectric response measurement is a powerful method for evaluating dielectric properties, which has been developed since the 1900s. This method is categorized into time-domain measurements and frequency domain measurements. The latter is the focused measurement in this paper and called frequency domain spectroscopy (FDS). The dissipation factor and the insulation system's capacitance within a wide range of frequency, usually from 0.1 MHz to 1 kHz, are measured in the FDS method [6].

Typically, this method is used for the insulation system monitoring of transformers [7–11]. However, some works discussed the FDS method's feasibility on condenser bushings and reached convincible results in recent years. In [12, 13], the authors built a model of OIP bushing using the finite element method. Using the FEM model and complex permittivity, the complex capacitance of the OIP bushing was calculated. Moreover, the dissipation factor (tan delta) and the OIP bushing capacitance over a wide range of frequencies were extracted from the complex capacitance. They calculated the dissipation factor and the bushing's capacitance for different moisture levels of OIP, and therefore, the different curves for different levels of OIP moisture were obtained. Analyzing these curves, they noticed that the dissipation factor's sensitivity and the capacitance of the bushing in line frequency to the OIP moisture level is low that causes moisture of the OIP which is undetected when only the typical tests of the bushing are done.

Furthermore, the authors noticed that the moisture of the OIP has a significant impact on the dissipation factor and the capacitance in the lower frequencies (<0.1 Hz). The authors in [14] built several OIP bushings in which the moisture level of OIP is different. They tested these bushings with the FDS method at different temperatures. The results were transferred to the reference temperature by their proposed method. Afterward, an estimation of moisture content in the OIP based mathematical equation named Havriliak–Negami equation was proposed and validated. The results indicate that the moisture content in OIP bushings can be estimated by their proposed method. Several kinds of literature like [15–21] dispute the FDS method's preference over the typical tests. However, none of those works establish a comprehensive instruction for FDS result interpretation. In this paper, several FEM models of OIP bushing under different moisture levels of the OIP are built, and the FDS test is simulated. Afterward, the dissipation factor and the capacitance of these bushings over a wide range of frequencies are obtained. Afterward, thirteen statistical indices are introduced, and they are applied to the FDS curves to interpret the results. The results of this paper establish a new interpretation method of FDS curves for the condenser bushing monitoring.

2 FEM Model of OIP Bushing

A 96 kV OIP bushing consisting of 17 layers of OIP is modeled using the Flux software. The kraft paper and the oil used for this bushing are Munksjö Thermo 70 and Nynas Nitro 10X, respectively. Because of the OIP bushing geometry's existing symmetry, only half the bushing is drawn in the software. Figure 1 shows the regularly

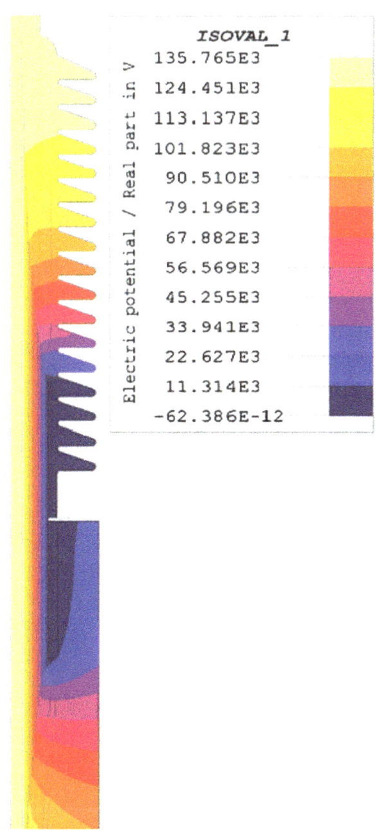

Fig. 1 FEM model of 96 kV OIP bushing

graded electric potential distribution of this OIP bushing, which is a salient feature of condenser bushings.

The kraft paper and the oil were manufactured by the Ahlstrom Munksjö and the Nynas, respectively. Moreover, the complex permittivity values for the OIP in different frequencies were taken from [22, 23]. Afterward, the FDS test is simulated, and the tan-delta and the capacitance curves for different moisture levels of OIP are obtained. The tan-delta curves and the capacitance curves for different moisture content are shown in Figs. 2 and 3, respectively. It can be understood from Figs. 2 and 3 that the sensitivity of the dissipation factor and the capacitance to the moisture content of the OIP in lower frequencies is higher than the line frequency. This fact highlights the importance of the FDS test for condenser bushings monitoring. Moreover, it is essential to note that all the simulation is done at a reference temperature, which is 20 °C. The obtained results in this section are along with [13–18]. As mentioned earlier, there are other researches in which the impact of the OIP moisture

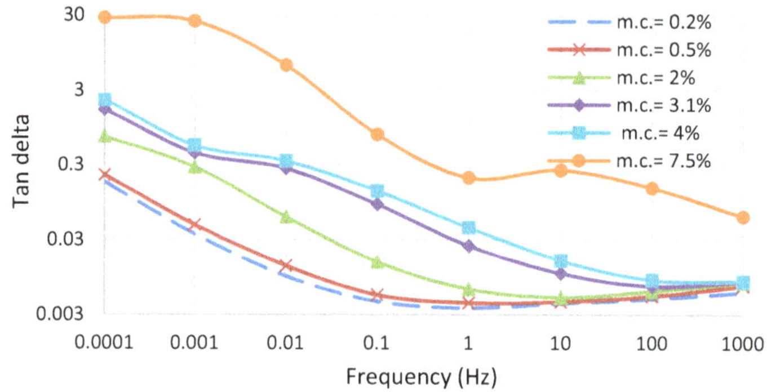

Fig. 2 Tan-delta curves for different moisture content

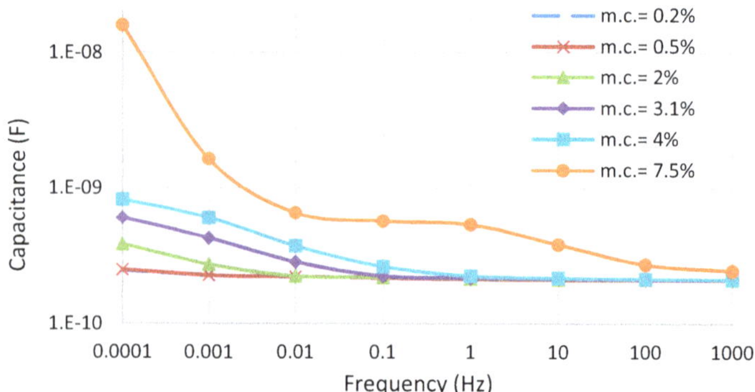

Fig. 3 Capacitance curves for different moisture content

content on FDS curves is investigated, or the moisture content of the OIP is estimated using a mathematical equation. In the next section, several statistical indices are introduced, and they are applied to the FDS curves.

3 Statistical Indices

Several studies like [24–28] have been conducted to interpret the frequency response analysis (FRA) results of transformers using statistical and numerical indices. Under normal and healthy conditions, the FRA characteristic, called a transformer's fingerprint, is needed to use this method. The numerical and statistical indices calculate the differences between the FRA fingerprint and the other FRA characteristics in the form of one number. In other words, the values of numerical and statistical indices are the features of these characteristics which been extracted. As far as the authors know, this method has not been used to interpret the FDS results of condenser bushings. Table 1 shows the names of statistical indices using in this paper and their abbreviation.

Obtained results from the FDS test of condenser bushings can be categorized into dissipation factor and capacitance curves. Each of these curves contains essential information that should be considered. Hence, two sets of reference curves, one for tan-delta and the other for capacitance curves, are considered for this study till the interpretation has an accurate and worthy result. The equations of seven two-array (need two signal) statistical indices and six one-array (need one signal) statistical indices used in this paper are summarized below. In the following equations, X and Y are the vector of a new FDS curve and the FDS reference curve for each set, respectively, $X(i)$ and $Y(i)$ are the ith elements of these vectors, $\bar{X}$ is the mean of the $X(i)$, and n is the number of samples in a vector. It is worth saying that the value of the correlation coefficient is one, and the other two-array indices' values are zero for normal status:

Table 1 Abbreviation definition of the statistical indices

Abbr.	Definition	Abbr.	Definition
CC	Correlation coefficient	AVE	Average
ED	Euclidean distance	AVEDEV	Average deviation
MAX	Maximum of difference	VAR	Variance
SSE	Sum squared error	SEM	Standard error of mean
SSMMRE	Sum squared max–min ratio error	SD	Standard deviation
RMSE	Root mean square error	HAR	Harmonic mean
DABS	Absolute difference		

$$\mathrm{CC} = \frac{\sum_{i=1}^{n} X_i Y_i}{\sqrt{\sum_{i=1}^{n}[X_i]^2 \sum_{i=1}^{n}[Y_i]^2}},$$

$$\mathrm{ED} = \sqrt{\sum_{i=1}^{n}(Y_i - X_i)^2},$$

$$\mathrm{MAX} = \max(Y_i - X_i),$$

$$\mathrm{SSE} = \frac{1}{n}\sum_{i=1}^{n}(Y_i - X_i)^2,$$

$$\mathrm{SSMMRE} = \frac{1}{n}\sum_{i=1}^{n}\left(\frac{\max(Y_i, X_i)}{\min(Y_i, X_i)} - 1\right)^2,$$

$$\mathrm{RMSE} = \sqrt{\frac{1}{n}\sum_{i=1}^{n}\left(\frac{|Y_i| - |X_i|}{\frac{1}{n}\sum_{i=1}^{n}|X_i|}\right)^2},$$

$$\mathrm{DABS} = \frac{1}{n}\sum_{i=1}^{n}|Y_i - X_i|,$$

$$\mathrm{AVE} = \frac{1}{N}\sum_{i=1}^{N} X_i,$$

$$\mathrm{AVEDEV} = \frac{1}{N}\sum_{i=1}^{N}\left|X_i - \bar{X}\right|,$$

$$\mathrm{CVAR} = \sigma = \frac{1}{N}\sum_{i=1}^{N}\left(X_i - \bar{X}\right)^2,$$

$$\mathrm{SEM} = \frac{\sigma}{\sqrt{N}},$$

$$\mathrm{SD} = \sqrt{\frac{1}{N-1}\sum_{i=1}^{N}\left(X_i - \bar{X}\right)^2},$$

$$\mathrm{HAR} = \frac{N}{\sum_{i=1}^{N}\frac{1}{X_i}}.$$

4 Results and Discussion

The results of applying two-array statistical indices on the tan-delta and the capacitance curves are described in Tables 2 and 3, respectively. All two-array statistical indices show a regular upward trend and good sensitivity toward the OIP moisture level increases, apart from the CC. The CC shows a linear correlation between two sets of data, and its value is between 1 and −1. According to the results, the CC changes as the moisture level of OIP changes. However, CC changing is not well ordered entirely, and it cannot be a perfect indicator of the moisture level of the OIP.

It is evident from Figs. 2 and 3 that the sensitivity of the OIP bushing dissipation factor to the moisture content of OIP is higher than the bushing's capacitance. The results of applying one-array statistical indices on the tan-delta and the capacitance curves are described in Tables 4 and 5, respectively. Values of all the six one-array indices show an upward trend toward the OIP moisture level increasing, except the HAR, which shows a downward trend.

It can be understood from Tables 4 and 5 that the changes in values of the one-array indices related to the tan-delta curves are more significant than the capacitance curves. This fact yields that the dissipation factor is more sensitive to the OIP moisture

Table 2 Results of two-array statistical indices for tan-delta curves

Indices	Different moisture content compared to m.c. = 0.2%					
	0.2%	0.5%	2%	3.1%	4%	7.5%
CC	1	0.99	0.93	0.78	0.75	0.82
MAX	0	0.1	0.54	1.4	2.1	28.1
ED	0	0.71	5.5	18.9	27.4	529
SSE	0	2e−5	0.01	0.01	0.03	9.4
DABS	0	2e−3	0.01	0.06	0.09	1.2
SSMMRE	0	0.13	6.5	175	391	5.6e4
RMSE	0	0.71	5.5	18.8	27.3	528

Table 3 Results of two-array statistical indices for capacitance curves

Indices	Different moisture content compared to m.c. = 0.2%					
	0.2%	0.5%	2%	3.1%	4%	7.5%
CC	1	0.99	0.8	0.87	0.91	0.58
MAX	0	16.5	137	357	576	1.5e4
ED	0	186	1.1e3	5.3e3	1.1e4	8e4
SSE	0	1.2	38	928	4.2e3	2.1e5
DABS	0	0.68	2.1	14.1	33.4	252
SSMMRE	0	2e−5	7e−4	0.02	0.08	4
RMSE	0	5e−3	0.03	0.14	0.3	2.2

Table 4 Results of one-array statistical indices for tan-delta curves

Indices	Different moisture content					
	0.2%	0.5%	2%	3.1%	4%	7.5%
AVE	6e−3	8e−3	0.02	0.06	0.09	1.2
AVEDEV	2e−3	3e−3	0.02	0.07	0.1	1.5
VAR	3e−5	8e−5	1e−3	0.01	0.02	7.9
SEM	3e−5	5e−5	2e−4	6e−4	8e−4	0.02
SD	0.04	0.05	0.13	0.26	0.32	1.2
HAR ($\times 10^3$)	5.3	4.8	4.3	4.1	3.9	3.5

Table 5 Results of one-array statistical indices for capacitance curves

Indices	Different moisture content					
	0.2%	0.5%	2%	3.1%	4%	7.5%
AVE	215	216	217	229	248	466
AVEDEV	3.6	4	4.8	20	42	151
VAR	17	23	73	931	3.6e3	1.5e5
SEM	0.02	0.03	0.05	0.18	0.35	2.3
SD	1.9	2	2.2	4.4	6.4	12.3
HAR ($\times 10^6$)	7.35	7.3	7.24	7.2	7.18	7.13

level than the capacitance. Hence, the interpretation of the results using the tan-delta curves is efficient. Moreover, two-array indices' values depend on the reference curve, which might be inaccessible because the FDS test is optional for condenser bushing. Thus, the interpretation of the FDS results using one-array indices is reasonable.

Figures 4, 5, 6, 7, 8, and 9 show the relationship of all six one-array indices with different moisture contents of the OIP, apart from the 3.1% moisture content.

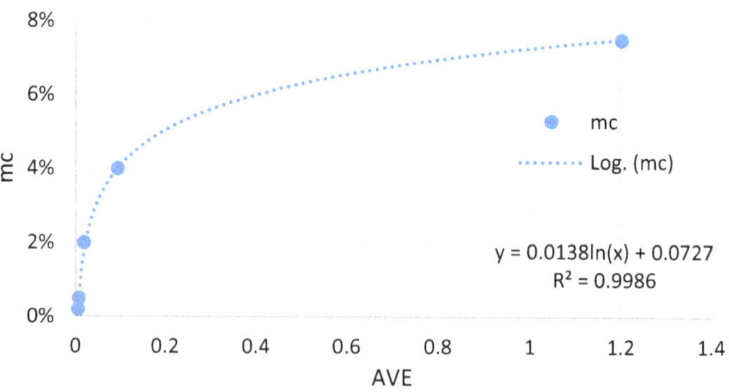

Fig. 4 Relationship between AVE and moisture content of the OIP

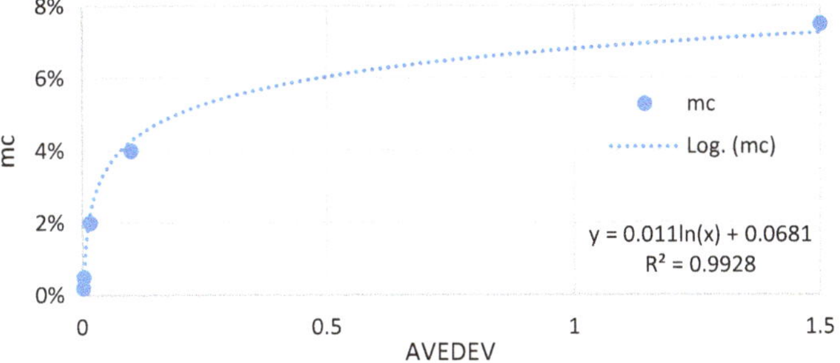

Fig. 5 Relationship between AVEDEV and moisture content of the OIP

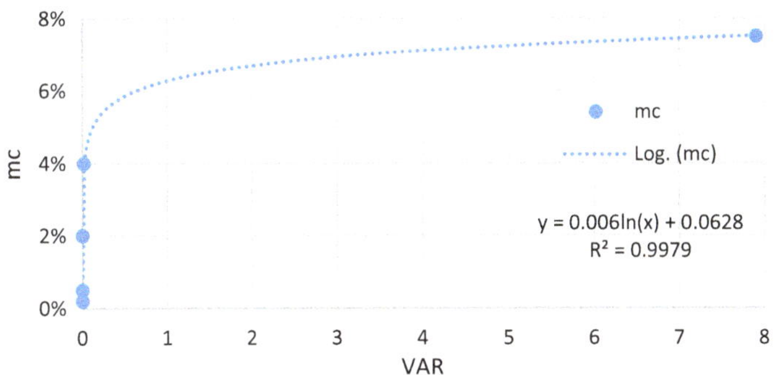

Fig. 6 Relationship between VAR and moisture content of the OIP

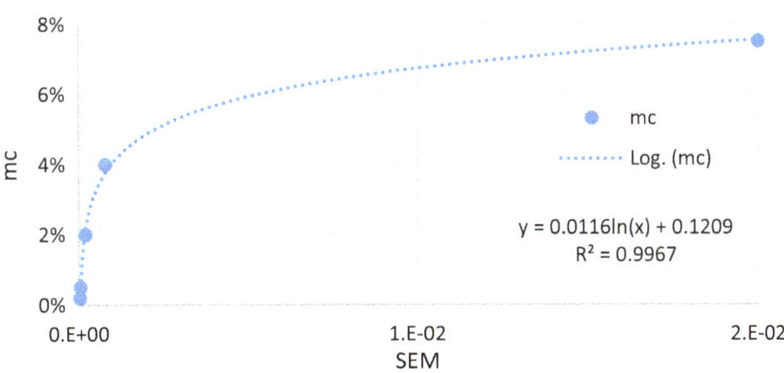

Fig. 7 Relationship between SEM and moisture content of the OIP

All six one-array indices vary with moisture content and follow a logarithmic law. Each one-array index has an exclusive relationship with the OIP moisture content, written in these figures. *R*-squared is an indicator of how well data fit the regression model, which its ideal value is one. From this point of view, all six one-array indices can be used to assess the moisture content of the OIP bushing.

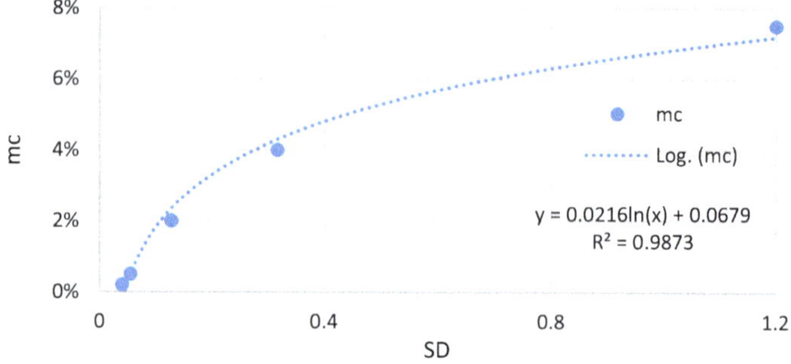

Fig. 8 Relationship between SD and moisture content of the OIP

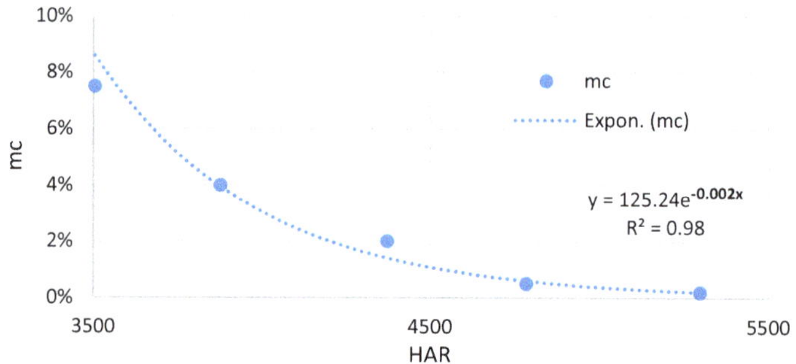

Fig. 9 Relationship between HAR and moisture content of the OIP

Table 6 shows the estimated values of the real 3.1% moisture content based on fitting equations of Figs. 4, 5, 6, 7, 8, and 9 and the values of six one-array indices related to 3.1% moisture content. According to Table 6, the AVEDEV and the SD

Table 6 Estimated moisture content values

Fitting equation	Estimated m.c. (%)	Error (%)
$mc = 0.0138\ln(AVE) + 0.0727$	3.4	9.7
$mc = 0.011\ln(AVEDEV) + 0.0681$	3.8	22
$mc = 0.006\ln(VAR) + 0.0628$	3.4	9.7
$mc = 0.0116\ln(SEM) + 0.1209$	3.4	9.7
$mc = 0.0216\ln(SD) + 0.0679$	3.9	26
$mc = 125.24e^{-0.002HAR}$	3.4	9.7

show significant errors of moisture content estimation. From this perspective, the AVE, VAR, SEM, and HAR can be a perfect indicator of the OIP moisture content.

5 Conclusion

Using the FEM model of a 96 kV OIP bushing, the tan-delta and capacitance curves were simulated over a wide frequency band for different moisture levels of OIP. Moreover, two sets of statistical indices, including seven two-array and six one-array indices, were introduced to interpret the FDS results. It was evident from the results that all six one-array indices can be perfect indicators of the OIP moisture level, apart from the AVEDEV and the SD. These indices' values show an upward trend toward the OIP moisture level increases in a logarithmic way, apart from the HAR, which shows a downward trend. Moreover, these four indices' estimation errors for the real 3.1% moisture content are below 10%, which substantiates the efficiency of the presented method. This paper emphasizes the interpretation of the FDS results of the OIP bushings with different moisture levels using the statistical indices. It establishes a novel and quantitative method to interpret the FDS results. Furthermore, this paper's results improve condenser bushing maintenance and eliminate human errors, which can be a forward step to standardize the FDS results of condenser bushings interpretation.

References

1. Taghizade Ansari, H., Vahedi, A.: Effect of the asymmetrical axial displacement of transformer windings on FRA characteristics. J. Crit. Rev. **7**(2), 469–476 (2020)
2. IEEE Standard C57.19.01: Performance Characteristics and Dimensions for Outdoor Apparatus Bushings (2000)
3. International Electrotechnical Commission, IEC 60137: Insulated Bushings for Alternating Voltages Above 1000 V. IEC, Geneva (2008)
4. Jan, K.: Operating damages of bushings in power transformers. Trans. Electr. Eng. **1**(3), 89–93 (2012)
5. Bahr, P., Christensen, J., Brusetti, R.: On-line diagnostic case study involving a general electric type U bushing. In: 74th Annual Double Client Conference, pp. 1–11 (2007)
6. Dhlamini, S.M.: Bushing diagnosis using artificial intelligence and dissolved gas analysis. PhD, University of the Witwatersrand (2007)
7. Jadav, R.B., Ekanayake, C., Saha, T.K.: Understanding the impact of moisture and aging of transformer insulation on frequency domain spectroscopy. IEEE Trans. Dielectr. Electr. Insul. **21**(1), 369–379 (2014)
8. Zhou, L., Wang, D., Guo, L., Wang, L., Jiang, J., Liao, W.: FDS analysis for multilayer insulation paper with different aging status in traction transformer of high-speed railway. IEEE Trans. Dielectr. Electr. Insul. **24**(5), 3236–3244 (2017)
9. Supramaniam, G.K., Hussien, Z.F., Aizam, M.: Application of frequency domain spectroscopy (FDS) in assessing dryness and aging state of transformer insulation systems. In: 2008 IEEE 2nd International Power and Energy Conference, pp. 55–62 (2008)

10. Liu, J., Fan, X., Zhang, Y., Zheng, H., Zhang, C.: Condition prediction for oil-immersed cellulose insulation in field transformer using fitting fingerprint database. IEEE Trans. Dielectr. Electr. Insul. **27**(1), 279–287 (2020)
11. Gao, J., Yang, L., Wang, Y., Liu, X., Lv, Y., Zheng, H.: Condition diagnosis of transformer oil-paper insulation using dielectric response fingerprint characteristics. IEEE Trans. Dielectr. Electr. Insul. **23**(2), 1207–1218 (2016)
12. Smith, D.J., McMeekin, S.G., Stewart, B.G., Wallace, P.A.: A variable frequency model of a transformer bushing with localized moisture content. In: Annual Report Conference on Electrical Insulation and Dielectric Phenomena, pp. 451–454 (2012)
13. Smith, D.J., McMeekin, S.G., Stewart, B.G., Wallace, P.A.: A dielectric frequency response model to evaluate the moisture content within an oil-impregnated paper condenser bushing. IET Sci. Meas. Technol. **7**(4), 223–231 (2013)
14. Wang, D., Zhou, L., Liao, W., Wang, A., Xu, X., Guo, L.: Moisture estimation for oil-immersed bushing based on FDS method: at a reference temperature. IET Gener. Transm. Distrib. **12**(11), 2762–2769 (2018)
15. Bouaicha, A., Fofana, I., Farzaneh, M., et al.: Dielectric spectroscopy techniques as quality control tool: a feasibility study. IEEE Electr. Insul. Mag. **25**(1), 6–14 (2009)
16. Buchacz, J., Cichoń, A., Skubis, J.: Detection of conductive layers short circuit in HV condenser bushings using frequency domain spectroscopy. IEEE Trans. Dielectr. Electr. Insul. **24**(1), 552–558 (2017)
17. Zhang, D., Zhao, H., Yun, H., Liu, X., Han, Y., Mu, H., Zhang, G.J.: Study on FDS characteristics of oil-immersed paper insulation bushing under non-uniform moisture content. IET Sci. Meas. Technol. **12**(5), 691–697 (2018)
18. Yang, K., Dong, M., Hu, Y., Xie, J., Xu, G., Ren, M.: Dielectric response analysis of oil-immersed paper insulated bushing under moisture. In: Proceedings of the 2nd International Conference on Information Technologies and Electrical Engineering, pp. 1–6 (2019)
19. Zhiming, H., Mingli, F., Ran, Z., et al.: Research on the wetting mechanism of oil-paper insulated bushings and the application of FDS method. In: 2019 IEEE Sustainable Power and Energy Conference (iSPEC), pp. 1903–1908 (2019)
20. Smith, D.J., McMeekin, S.G., Stewart, B.G., Wallace, P.A.: The modeling of electric field, capacitance and dissipation factor of a high voltage bushing over varying frequency. In: 2012 47th International Universities Power Engineering Conference (UPEC), pp. 1–6 (2012)
21. Yang, F., Du, L.: A circuital model-based analysis of moisture content in oil-impregnated-paper insulation using frequency domain spectroscopy. IEEE Access **8**, 47092–47102 (2020)
22. Linhjell, D., Lundgaard, L., Gafvert, U.: Dielectric response of mineral oil impregnated cellulose and the impact of aging. IEEE Trans. Dielectr. Electr. Insul. **1**(14), 156–169 (2007)
23. Linhjell, D., Hestad, O.L., Gafvert, U., Lundgaard, L.E.: Dielectric response of oil-impregnated cellulose from 0.1 mHz to 3 MHz. In: IEEE International Conference on Dielectric Liquids, pp. 277–280 (2005)
24. Behjat, V., Mahvi, M.: Statistical approach for interpretation of power transformers frequency response analysis results. IET Sci. Meas. Technol. **9**(3), 367–375 (2015)
25. Behjat, V., Mahvi, M., Rahimpour, E.: New statistical approach to interpret power transformer frequency response analysis: non-parametric statistical methods. IET Sci. Meas. Technol. **10**(4), 364–369 (2016)
26. Paleri, A., Preetha, P., Sunitha, K.: Frequency response analysis (FRA) in power transformers: an approach to locate inter-disk SC fault. In: IEEE PES Asia-Pacific Power and Energy Engineering Conference (APPEEC), pp. 1–5 (2017)
27. Wesley, N.K., Bhandari, S., Subramaniam, A., Bagheri, M., Panda, S.K.: Evaluation of statistical interpretation methods for frequency response analysis based winding fault detection of transformers. In: 2016 IEEE International Conference on Sustainable Energy Technologies (ICSET), pp. 36–41 (2016)
28. Samimi, M.H., Tenbohlen, S.: FRA interpretation using numerical indices: state-of-the-art. Int. J. Electr. Power Energy Syst. **89**, 115–125 (2017)

Common-Mode Voltage Elimination of Three-Phase Multilevel Voltage Source Inverter by Means of Quarter-Wave-Symmetric Space Vector PWM Approach

Nikolay Lopatkin

Abstract The new common-mode-voltage-eliminating (CMVE) space vector PWM algorithm for three-phase multilevel voltage source inverter (MLVSI) of any arbitrary topology with any arbitrary number of generated equal voltage levels is proposed. This technique employs all the approaches and attributes of the recently offered quarter-wave-symmetric space vector PWM algorithm but, unlike the latter, operates with quantities in the coordinate frame of the relative values of the MLVSI phase load voltages, referred to the three-phase load common star point (phase-to-neutral load voltages). The applied MLVSI voltages space vectors are the selected space vectors which produce the zero value of the three-phase common-mode voltage, understood as the voltage across the neutral point of the load and the inverter system ground point. Similar to the baseline algorithm, intended for shaping the high-quality MLVSI output voltages and currents, the new CMVE algorithm generates enough high-quality output power, thanks to keeping the quarter-wave symmetry of the star (phase-to-neutral) and delta (phase-to-phase) voltages. The dependences of the THD and low-order weighted THDs values on the phase voltage amplitude modulation index are obtained by PSIM-simulation for the three low, multiple of 6, values of frequency modulation index.

1 Introduction

A common-mode voltage (CMV) in both single-phase and three-phase AC power systems is the source of many undesirable phenomena, such as leakage currents through parasitic elements, shaft voltage and bearing currents in AC motors that reduce the efficiency and reliability of the power supplying and service lives of the system components. In some instances, there may be improper operations of automation due to a CMV, and, in consequence, decline in the quality of the generated voltage and a risk of electric shock to service personnel. The CMV problem is especially

N. Lopatkin (✉)
Shukshin Altai State University for Humanities and Pedagogy, Korolenko str., 53, 659333 Biysk, Altai Region, Russia

A. Ronzhin and V. Shishlakov (eds.), *Electromechanics and Robotics*, Smart Innovation, Systems and Technologies 232, https://doi.org/10.1007/978-981-16-2814-6_26

actual in inverter-based systems [1, 2], including transformerless photovoltaic AC power supplies [3].

The CMV of a three-phase AC system is generally defined as the voltage across the neutral point of the load *n* and the system ground point *G*. Such the system ground node could be the neutral point *N* of the three-phase star-connected AC source, as it is shown in Fig. 1a. So, u_{AN} ($u_{AN} = u_{AG}$) and u_{An} are the phase A voltages, which are referred to the three-phase source common star point *N* (source voltage), and to the three-phase load common star point *n* (load voltage), u_{nG} is the considered common-mode voltage. Here, $\dot{z}_A$, $\dot{z}_B$, $\dot{z}_C$ and $\dot{z}_N$ are the load impedances of the phases A, B, C and the neutral wire impedance, respectively. If the neutral wire is absent, then the circuit becomes the three-wire one.

In this case, the CMV concept is coinciding with the neutral bias voltage concept and with the zero-sequence voltage concept in the sense of an absolute value of the latter. With regard to the subject of inverter-based power supply systems, such the kind of system grounding and the CMV interpretation appears in many multilevel voltage source inverters' (MLVSI) circuits, having a separated per phase supplying. Probably, the most often applied MLVSI circuit of them is cascaded H-bridge (CHB)

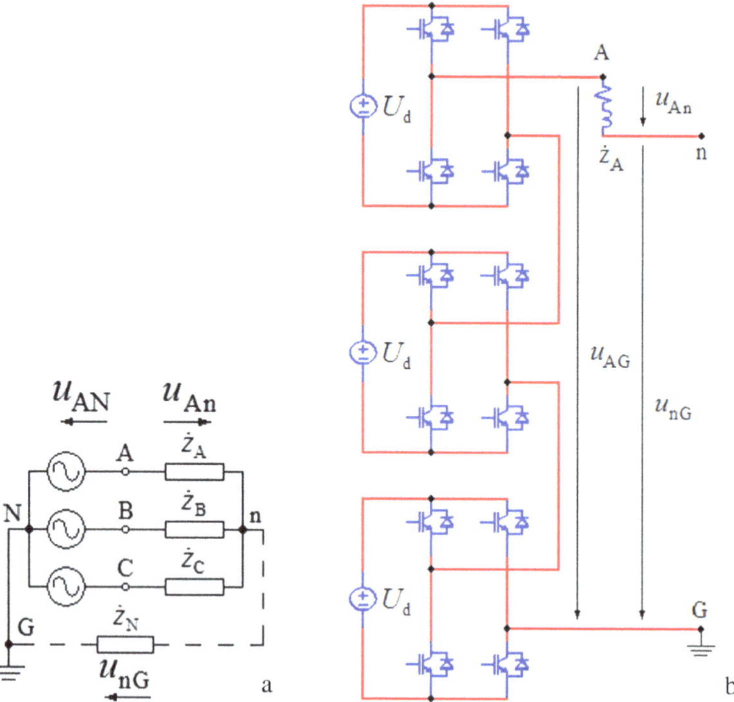

Fig. 1 Common-mode voltage: in simplified three-phase four-wire or three-wire star-connected circuit (**a**); in three-phase seven-level cascaded H-bridge inverter (**b**)

topology. The phase A simplified equivalent circuit of the three-phase CHB MLVSI power circuit is shown in Fig. 1b.

The other MLVSI circuits, using the same power source DC voltage levels for all the three phases, have the central point of the sectioned DC voltage source (DC bus mid-point) as the system ground point *G*. The most known and used types of such MLVSIs are the diode clamped or neutral point clamped (NPC) and flying capacitors (FC) topologies, see the simplified equivalent circuits in Fig. 2a, b, respectively.

Thus, the CMV u_{nG} reduction and elimination are the problem, common to all of the three-phase inverter topologies, and attracting many researchers. Like to solutions for single-phase inverters, all the CMV-mitigating measures in three-phase inverter-based systems can be roughly divided into circuitry, control and mixed solutions [4]. Among the applied hardware means there are special-designed external passive and active common-mode filters and special inverter topologies [5].

But typical common-mode noise filters are bulky, expensive and reducing a system efficiency. The special topologies are mostly oriented to low-level inverters, since

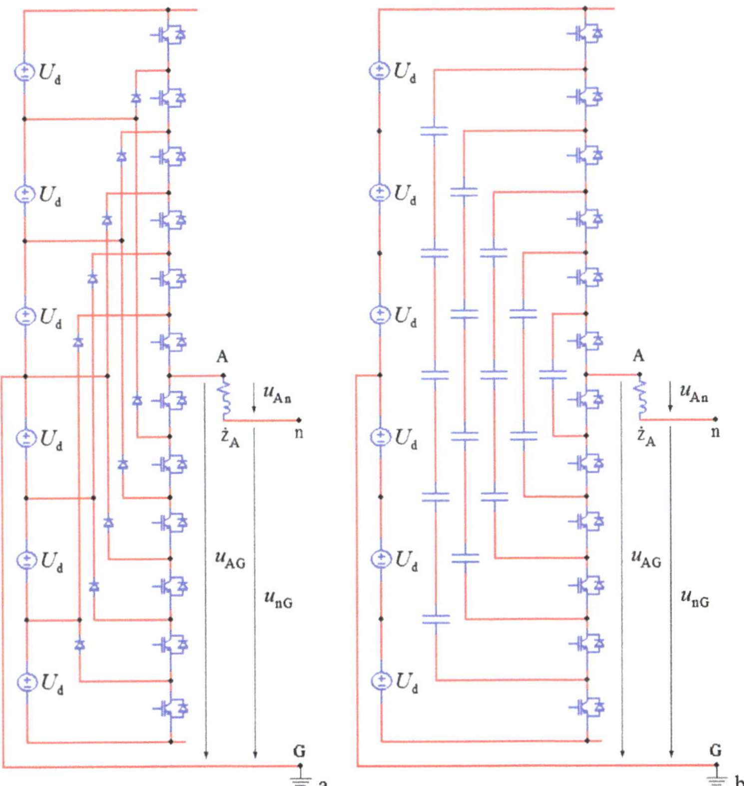

Fig. 2 Common-mode voltage: in three-phase seven-level neutral point clamped inverter (**a**); in three-phase seven-level flying capacitors inverter (**b**)

MLVSI circuits a priori have a reduced amplitude and voltage transition step of the CMV.

The control solutions are generally based on advanced pulse-width modulation (PWM) techniques [6–9], including modified sinusoidal PWM technologies. In case of MLVSI applying, space vector PWM (SVPWM) techniques provide a rich choice of switching states and generated voltage space vectors (SV), therefore the CMV can be completely eliminated due to selecting and using particular SVs.

The purpose of this paper is to offer a new, general and simple common-mode-voltage-eliminating SVPWM algorithm for three-phase MLVSI of any arbitrary topology with any arbitrary number of generated equal voltage levels. It can be treated as some modification or customization of the before proposed author's quarter-wave-symmetric SVPWM (QWS-SVPWM) technique [10, 11].

2 New Common-Mode-Voltage-Eliminating Space Vector PWM Technique

Despite the proposed common-mode-voltage-eliminating (CMVE) space vector PWM technique is really new, its underlying principle is known at least since the early 2000's (see, for example, [12]), and that is that the applied MLVSI voltages space vectors are the zero-common-mode-voltage space vectors (ZCMVSV). The ZCMVSV are the selected space vectors which produce the zero value of the three-phase common-mode voltage, understood as the voltage across the neutral point of the load n and the inverter system ground point G. One of the recent most advanced and carefully scrutinized techniques, which are using the ZCMVSV, is presented in [13].

For all the three-phase MLVSI circuits, taking into account of the different location of their nodes G (see the previous section), the executed three-phase CMV is calculated as follows:

$$u^*_{\mathrm{nGexe}}(t) = \frac{u^*_{\mathrm{AGexe}}(t) + u^*_{\mathrm{BGexe}}(t) + u^*_{\mathrm{CGexe}}(t)}{3}, \quad (1)$$

where «exe» in the subscripts designate the executed signals and power circuit voltages, whose waveforms are the results of the technique applying; asterisks in the superscripts designate the relative (normalized) values, $u^*_{\mathrm{X}} = u_{\mathrm{X}}/U_{\mathrm{d}}$, $u^*_{\mathrm{XY}} = u_{\mathrm{XY}}/U_{\mathrm{d}}$; U_{d} is the direct voltage of the unit (base) level.

Through the example of the seven-level inverter, the MLVSI whole voltage space vectors and its zero-common-mode-voltage voltage space vectors in the oblique-angled delta voltages' coordinates $\left(u^*_{\mathrm{AB}}, u^*_{\mathrm{BC}}\right)$ are shown in Fig. 3a, b, respectively. The above-mentioned paper [13] operates with triangles of the three nearest ZCMVSVs (see Fig. 3b), but the resulting algorithm, offered there, seems to be of quite high computation cost compared with the QWS-SVPWM approach.

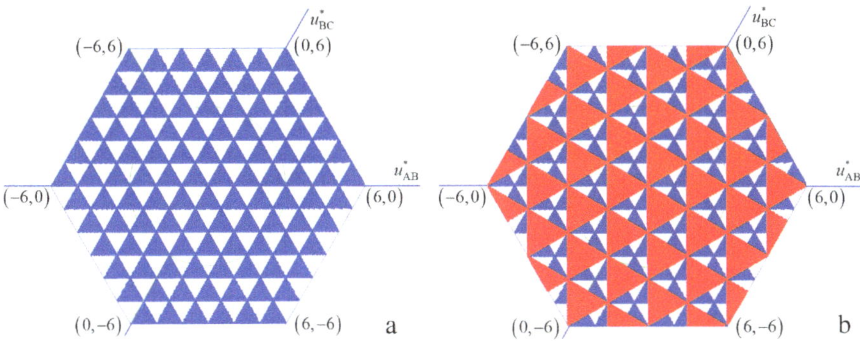

Fig. 3 Space vector diagram of three-phase seven-level inverter in coordinate frame of relative values of MLVSI delta voltages: triangles of the three nearest voltage space vectors (**a**); triangles of the three nearest zero-common-mode-voltage space vectors (**b**)

The essential simplification is achieved in the presented here technique. The zero-value condition for the executed CMV voltage, $u^*_{\mathrm{nGexe}}(t) = 0$, means that the voltage referred to the system ground point *G* should be equal to the respective phase load voltage, i.e. to the voltage relative to the three-phase load common star point *n*, for each phase *X*. Thus, in view of the fact that the executed phase-to-neutral point-*n* voltages $u^*_{\mathrm{Xnexe}}(t)$, *X* {A, B, C}, are assumed to be symmetric, i.e., the three-phase load is supposed to be balanced, the following conditions should be valid:

$$u^*_{\mathrm{XGexe}}(t) = u^*_{\mathrm{Xnexe}}(t),\;\; u^*_{\mathrm{AGexe}}(t) + u^*_{\mathrm{BGexe}}(t) + u^*_{\mathrm{CGexe}}(t) = 0. \tag{2}$$

If the initial values of $u^*_{\mathrm{XGexe}}(t)$ voltages have been chosen arbitrarily, the ZCMVSVs in Fig. 3b correspond to CMV voltage $u^*_{\mathrm{nGexe}}(t)$ values, equal to integer numbers (the sums $u^*_{\mathrm{AGexe}}(t) + u^*_{\mathrm{BGexe}}(t) + u^*_{\mathrm{CGexe}}(t)$ are multiples of 3), as we can conclude from Eq. (1). The second equation of (2) helps disambiguate the choice of voltages $u^*_{\mathrm{XGexe}}(t)$ values.

With the assumption of balanced load, it is advisable to consider the ZCMVSVs in coordinate frame of relative values of MLVSI phase-to-neutral-point-*n* load voltages $\left(u^*_{\mathrm{Cn}}, u^*_{\mathrm{An}}\right)$, as it is shown in Fig. 4a. Here, the initial space vector diagram of Fig. 3b is rotated counterclockwise by 90°. Also, the scale of the diagram is increased $2\big/\sqrt{3}$ times so that the side of the new elementary triangle corresponds to 1 of the relative phase voltage. The u^*_{Cn} and u^*_{An} values are projected to the new oblique-angled 60° axes in a similar way as it had done in Fig. 3a for the u^*_{AB} and u^*_{BC} values. The CMVE-SVPWM algorithm employs all the attributes and steps of the baseline QWS-SVPWM algorithm [10], just replacing the delta voltages with the relevant star voltages, namely u^*_{AB} with u^*_{Cn}, u^*_{BC} with u^*_{An} and u^*_{CA} with u^*_{Bn}.

In particular, the coordinates and duty cycles of the modulating triangles' vectors are the integer and fractional parts of the samples u^*_{Xnrefs} of the reference star voltages and their inverse values, see Fig. 4b ("refs" in the subscripts are omitted in the figure).

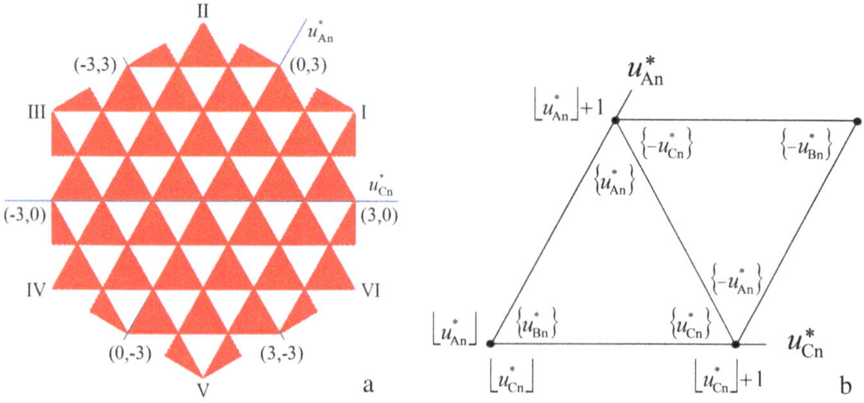

Fig. 4 Zero-common-mode-voltage space vectors of three-phase seven-level inverter: space vector diagram in coordinate frame of relative values of MLVSI phase-to-load-neutral-point voltages (**a**); coordinates and duty cycles of vectors in modulating triangles (**b**)

Here, $\lfloor x \rfloor$ means the rounding down x to the closest integer number, taking into account the sign (the «floor» function), so $\lfloor x \rfloor$ is the integer part of x; $\{x\}$ is the fractional part of x. Again, the fractional parts of the above star voltages and of their inverse values are the barycentric coordinates on the upward-pointing and on the downward-pointing elementary triangles, respectively, for the sampled positions of the reference voltage space vector (RVSV) inside of these triangles.

The equations, summarized in Table 1 for the above star voltages' samples inside the base rhombus, are valid including through the following equation:

$$\left\{u^*_{\text{Xnrefs}}\right\} + \left\{-u^*_{\text{Xnrefs}}\right\} = \begin{cases} 0 \text{ if } u^*_{\text{Xnrefs}} \text{ is integer value,} \\ 1 \text{ if } u^*_{\text{Xnrefs}} \text{ is non-integer value.} \end{cases} \tag{3}$$

Table 1 Values of sums of fractional values and of integer values for star normalized voltages' samples and their inverse values

Expression	Modulating triangle	
	Δ Upward-pointing, including perimeter	∇ Downward-pointing, excluding perimeter
$\{u^*_{\text{Cnrefs}}\}+\{u^*_{\text{Anrefs}}\}+\{u^*_{\text{Bnrefs}}\}$	1	2
$\lfloor u^*_{\text{Cnrefs}}\rfloor+\lfloor u^*_{\text{Anrefs}}\rfloor+\lfloor u^*_{\text{Bnrefs}}\rfloor$	−1	−2
$\{-u^*_{\text{Cnrefs}}\}+\{-u^*_{\text{Anrefs}}\}+\{-u^*_{\text{Bnrefs}}\}$	2	1
$\lfloor -u^*_{\text{Cnrefs}}\rfloor+\lfloor -u^*_{\text{Anrefs}}\rfloor+\lfloor -u^*_{\text{Bnrefs}}\rfloor$	−2	−1

The equations of Table 1 and Eq. (3) make it possible to compose the triangle type identifiers and to deal with the two coordinates $\left(u^*_{\mathrm{Cn}}, u^*_{\mathrm{An}}\right)$ instead of dealing with the three ones $\left(u^*_{\mathrm{Cn}}, u^*_{\mathrm{An}}, u^*_{\mathrm{Bn}}\right)$ and the three inverse values of the latters.

In rare occasions, when all the relative values of the star voltages have the just integer values, the presented SVPVM techniques designate the obtained space vector as the cornerstone rhombus vector $\left(\lfloor u^*_{\mathrm{Cnrefs}}\rfloor, \lfloor u^*_{\mathrm{Anrefs}}\rfloor\right)$ in Fig. 4b. This vector is easily defined as belonging to the upward-pointing triangle by means of a conventional triangle type identifier.

The following identifiers of the upward-pointing and downward-pointing triangles have been applied:

$$F_{\Delta\mathrm{s}} = \frac{\mathrm{sgn}\left(1 - \left\{u^*_{\mathrm{Cnrefs}}\right\} - \left\{u^*_{\mathrm{Anrefs}}\right\}\right) + 1}{2}, \quad F_{\nabla\mathrm{s}} = 1 - F_{\Delta\mathrm{s}}. \tag{4}$$

Here, the resulting 1 confirms that the tested triangle belongs to the matching type, and $\mathrm{sgn}(x)$ is the standard three-valued signum function.

The sector number for the SVPVM clock cycle interval k can be calculated as follows:

$$S_{\mathrm{k}} = \left\lfloor 6(k-1)/m_{\mathrm{f}}\right\rfloor + 1, \tag{5}$$

where $k = 1, 2, \ldots, m_{\mathrm{f}}$; m_{f} is the frequency modulation index,

$$m_{\mathrm{f}} = f_{\mathrm{c}}/f = T/T_{\mathrm{c}}. \tag{6}$$

Here, f_{c}, T_{c} and f, T are the clock cycle frequency and period and the modulating (MLVSI output voltage) frequency and period, respectively. Traditionally, to provide all the possible kinds of symmetry, the values of m_{f} have been chosen multiple of 6.

Like the baseline QWS-SVPWM algorithm, the new CMVE-SVPWM algorithm relies on the three-segment switching sequence scheme, producing the quarter-wave-symmetric output voltages' waveforms. The moving of the executed voltage space vector (EVSV) is non-returnable on each clock cycle interval and follows the same vector sequences as in [10], but the coordinates $\left(u^*_{\mathrm{Cn}}, u^*_{\mathrm{An}}\right)$ are itself new and correspond to Fig. 4a. The set (in ascending order of the sector number) of the sequences of the being executed vectors inside the triangles for the first ring of modulating triangles is {(1, 0), (0, 0), (0, 1); (0, 1), (0, 0), (−1, 1); (−1, 1), (0, 0), (−1, 0); (−1, 0), (0, 0), (0, −1); (0, −1), (0, 0), (1, −1); (1, −1), (0, 0), (1, 0)}. With the appearance of the downward pointed triangle in the second ring of modulating triangles, the moving of the EVSV endpoint has performed along the zigzag line. So, the set of the vector sequences in sector I of the second ring is {(2, 0), (1, 0), (1, 1); (1, 0), (1, 1), (0, 1); (1, 1), (0, 1), (0, 2)}. The same order is kept at every transition to the subsequent sector. Therefore, the set of the vector sequences in sector II of the second ring is {(0, 2), (0, 1), (−1, 2); (0, 1), (−1, 2), (−1, 1); (−1, 2), (−1, 1), (−2, 2)} and so on. Every subsequent ring adds two triangles (one triangle of each type) per sector, keeping both the EVSV trajectory of the moving inside the triangle of each type and the type of the initial (and the final) triangle in each sector. Thus, the increase in the numbers of the rings and inverter levels entails no changes in the pattern of EVSV moving.

The durations of the being executed space vectors on the k clock cycle interval are determined through processing the preliminary sampled (for the mid-point of the interval) values of the relative reference star voltages u^*_{Xnrefsk}, namely

$$u^*_{\mathrm{Cnrefsk}} = m_{\mathrm{aY}} \sin\big((2\pi k - \pi)/m_{\mathrm{f}} + 2\pi/3\big),\ \ u^*_{\mathrm{Anrefsk}} = m_{\mathrm{aY}} \sin\big((2\pi k - \pi)/m_{\mathrm{f}}\big), \tag{7}$$

where m_{aY} is the star voltage (phase voltage) amplitude modulation index,

$$m_{\mathrm{aY}} = U/U_{\mathrm{d}} = U^*, \tag{8}$$

U and U^* are the value and relative value of the RVSV magnitude.

In accordance with the main principle of the two voltages (the coordinates of the EVSV) shaping in the base QWS-SVPWM algorithm, each of the directly executed on the clock cycle interval k output phase-to-ground voltages are being formed as the two-component one:

$$u^*_{\mathrm{XGexek}}(t_{\mathrm{kc}}) = u^*_{\mathrm{Xnexek}}(t_{\mathrm{kc}}) = \lfloor u^*_{\mathrm{Xnrefsk}} \rfloor + f_{\mathrm{Xnexek}}(t_{\mathrm{kc}}), \tag{9}$$

where t_{kc} is the current time from the start of the clock cycle with number k, $t_{\mathrm{kc}} = t - (k - 1) \cdot T_{\mathrm{c}}$; t is the total current time;

$f_{\mathrm{Xnexek}}(t_{\mathrm{kc}})$ is the pulse function, possessing the only values 0 and 1.

The pulses of the value 1 of function $f_{\mathrm{Xnexek}}(t_{\mathrm{kc}})$ have been found for the both of triangles' types and all the sectors by the nature of changes in the variable $u^*_{\mathrm{Xnexek}}(t_{\mathrm{kc}})$ during the transition from one executed space vector to the other in the vectors' switching sequence. The durations of $f_{\mathrm{Xnexek}}(t_{\mathrm{kc}})$ pulses and the instants of time of their rising and falling edges are defined via the fractional values of the sampled relative reference star voltages u^*_{Xnrefsk}, taking into account the clock cycle interval k. Namely, in addition to S_{k}, the values of $F_{\Delta \mathrm{sk}}$ and $F_{\nabla \mathrm{s}}$ are depending on k and understood as related to u^*_{Cnrefsk} and u^*_{Anrefsk} when calculating them by (4). The pulse functions $f_{\mathrm{Cnexek}}(t_{\mathrm{kc}})$ and $f_{\mathrm{Anexek}}(t_{\mathrm{kc}})$ are clearly defined in Table 2 for sectors I and II.

Due to the taking place symmetry kinds, including the three-phase symmetry, the $f_{\mathrm{Anexek}}(t_{\mathrm{kc}})$ description for RVSV argument's value range 0°…90° is sufficient for both $u^*_{\mathrm{AGexe}}(t)$ and $u^*_{\mathrm{CGexe}}(t)$ implementing, for instance, with use of lookup tables.

Also, the only function $f_{\mathrm{Anexek}}(t_{\mathrm{kc}})$ would be enough if we decided to generate the only $u^*_{\mathrm{AGexe}}(t)$ voltage function and obtain the rest two voltage functions through the shifts by $\pm 2\pi/3$.

In view of the simplicity of the operations in (7)–(9), as well as of the $f_{\mathrm{Cnexek}}(t_{\mathrm{kc}})$ and $f_{\mathrm{Anexek}}(t_{\mathrm{kc}})$ values finding, the consideration of the full 360° range could be even more effective. One can obtain the whole data on $f_{\mathrm{Xnexek}}(t_{\mathrm{kc}})$ in [10], replacing the

Table 2 Pulse functions of executed phase-to-ground voltages, defined via fractional parts of two sampled reference star voltages

RVSV's sector and argument's value range	Modulating triangle's type and EVSV's trajectory	Pulse functions of executed phase voltages on clock cycle k	
		$f_{\text{Cnexek}}(t_{kc})$	$f_{\text{Anexek}}(t_{kc})$
I 0°…60°	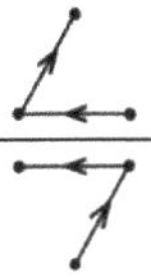		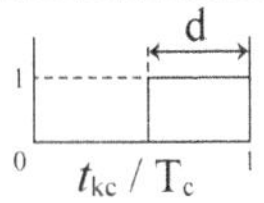 t_{kc}/T_c $d=\{u^*_{\text{Anrefsk}}\}$
II 60°…120°	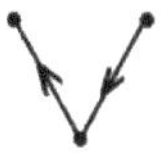	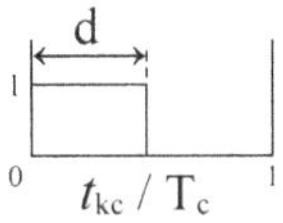 t_{kc}/T_c $d=\{u^*_{\text{Cnrefsk}}\}$	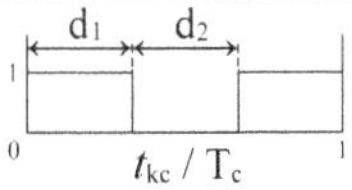 t_{kc}/T_c $d_1=\{u^*_{\text{Cnrefsk}}\}+\{u^*_{\text{Anrefsk}}\}-1$ $d_2=1-\{u^*_{\text{Anrefsk}}\}$
	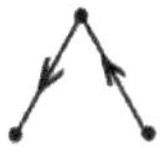		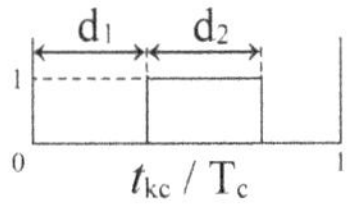 t_{kc}/T_c $d_1=\{u^*_{\text{Cnrefsk}}\}$ $d_2=\{u^*_{\text{Anrefsk}}\}$

variables, related to the delta voltages u^*_{AB} and u^*_{BC}, with $\{u^*_{\text{Cnrefs}}\}$ and $\{u^*_{\text{Anrefs}}\}$, respectively.

After determining the values of $u^*_{\text{CGexek}}(t_{\text{kc}})$ and $u^*_{\text{AGexek}}(t_{\text{kc}})$ according to (9), the value of the last voltage function on the clock cycle k can be calculated, taking into account (2): $u^*_{\text{BGexek}}(t_{\text{kc}}) = -\left(u^*_{\text{CGexek}}(t_{\text{kc}}) + u^*_{\text{AGexek}}(t_{\text{kc}})\right)$. At last, the obtained executed voltage functions $u^*_{\text{AGexe}}(t)$, $u^*_{\text{BGexe}}(t)$ and $u^*_{\text{CGexe}}(t)$ are being utilized as the reference functions to be reproduced at the MLVSI power circuit output.

As can be seen in the space vector diagram of the zero-common-mode-voltage space vectors for the seven-level inverter (Fig. 4a), the maximum relative magnitude of the RVSV and EVSV vectors $U^*_{\max}$ (the maximum phase voltage amplitude modulation index m_{aYmax}), which requires applying no unprogrammed vectors and employing no overmodulation strategy, is equal to 3. For the N-level three-phase MLVSI under the new CMVE-SVPWM control algorithm, the above value, indicating the limit of the linear range, can be calculated as follows:

$$U^*_{\max} = m_{\text{aYmax}} = \frac{1}{2}(N-1). \tag{10}$$

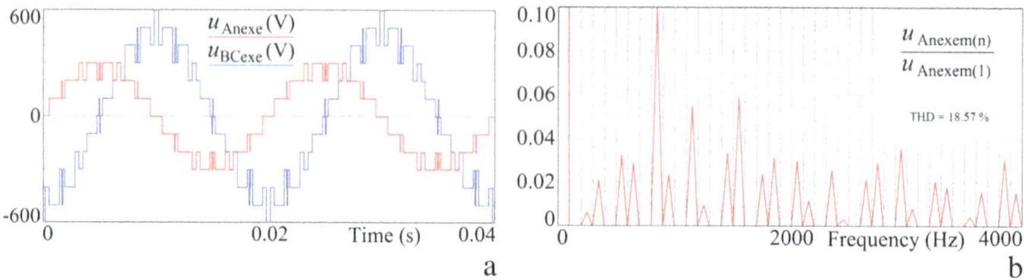

Fig. 5 PSIM-simulated results for particular control parameters: waveforms of output phase-to-neutral voltage and delta voltage (**a**); normalized phase-to-neutral voltage spectrum (**b**)

3 PSIM Simulations' Results

The PSIM-simulated voltages waveforms and spectrum of the CHB MLVSI idealized power circuit of Fig. 1b are shown in Fig. 5 for particular control parameters of the new CMVE-SVPWM, $m_f = 24$ and $m_{aY} = 2.8$. As can be seen, the presented here CMVE-SVPWM keeps the same waveforms' symmetry and harmonics content as the baseline QWS-SVPWM algorithm. Namely, for m_f values that are multiples of 6, the only odd harmonics present whose orders' numbers are as follows:

$$n = 6 \cdot h \pm 1, \tag{11}$$

where h belongs to the natural numbers.

The dependences of the output phase-to-neutral voltage's weighted THD factors of the zero to third orders (THD and the integrated voltage harmonics factors [14]) on the phase voltage amplitude modulation index are presented in Fig. 6. In view of the concept of a low-frequency space vector PWM for reducing the dynamic power losses in semiconductor switches, here, the simulation results for the new CMVE-SVPWM are shown for the three lowest, multiple of 6, values of frequency modulation index only, namely 12, 18 and 24. The QWS-SVPWM curves are included for comparison. As can be seen, the price of the common-mode-voltage eliminating is a large increase in the integrated voltage harmonics factors (IHF) values in the m_{aY} range of 0...1. In addition, the linear range of the phase voltage amplitude modulation index is decreased $2/\sqrt{3}$ times.

Nevertheless, the voltage quality is quite acceptable in the m_{aY} range of 1...3, and it can be improved in the values of IHF of specified order by m_f value matching [10].

4 Conclusion

The new common-mode-voltage-eliminating space vector PWM algorithm for three-phase multilevel voltage source inverter of any arbitrary topology with any arbitrary

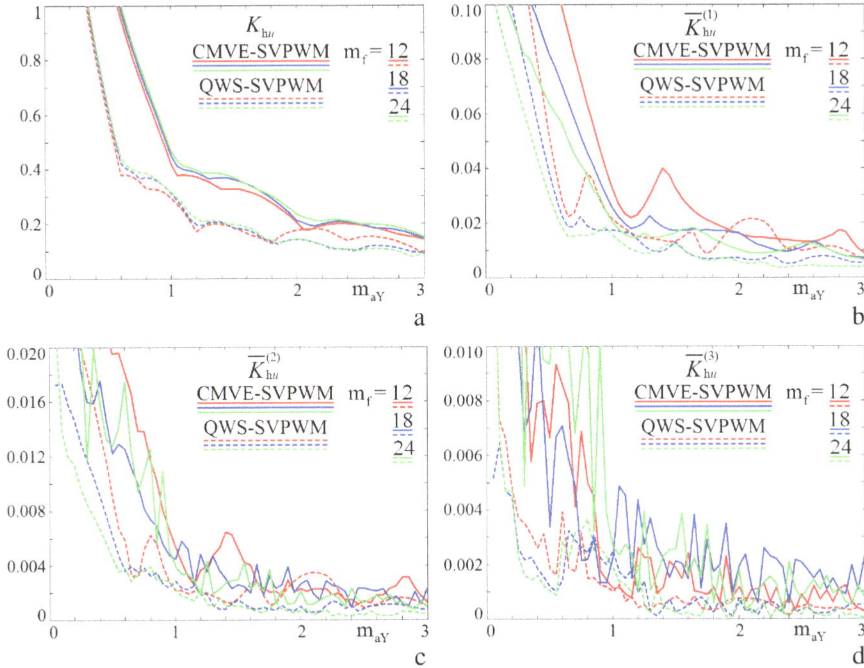

Fig. 6 Integrated voltage harmonics factors (IHF) versus phase voltage amplitude modulation index: zero-order IHF (THD) (**a**); first-order IHF (**b**); second-order IHF (**c**); third-order IHF (**d**)

number of generated equal voltage levels is proposed. The applied MLVSI voltages space vectors are the selected space vectors which produce the zero value of the three-phase common-mode voltage. The offered technique employs all the approaches and attributes of the recently offered quarter-wave-symmetric space vector PWM algorithm, but, unlike the latter, operates with quantities in the coordinate frame of the relative values of the MLVSI phase load voltages, referred to the three-phase load common star point. The dependences of the THD and low-order weighted THDs values on the phase voltage amplitude modulation index are obtained by PSIM-simulation for the three lowest, multiple of 6, values of frequency modulation index.

The unavoidable degradation of the quality of the generated output power can be reduced for some specific mode by the matching of the frequency modulation index value for the value minimization of the integrated voltage harmonics factor with the order appropriate to the particular kind of the generalized load. The proposed technique is promising for multilevel inverters with enough high number of supplying equal DC voltage sources, for example, for inverters integrated with photovoltaic systems.

References

1. Mutze, A.: Bearing Currents in Inverter-Fed AC-Motors (Berichte aus der Elektrotechnik). Shaker Verlag GmbH, Germany (2004)
2. Mütze, A.: Thousands of hits: on inverter-induced bearing currents, related work, and the literature. Elektrotech. Inf. Tech. **128**(11–12), 382–388 (2011)
3. Grishanov, E.V., Brovanov, S.V.: Theoretical aspects of the common-mode leakage current suppression in a photovoltaic power generation system based on multi-level H-bridge type converters. In: XIV International Scientific-Technical Conference on Actual Problems of Electronics Instrument Engineering (APEIE) Proceedings, pp. 32–37. IEEE, Novosibirsk (2018)
4. Wu, B., Narimani, M.: High-Power Converters and AC Drives, 2nd edn. Wiley-IEEE Press, Wiley, Hoboken (2017)
5. Han, D., Lee, W., Li, S., Sarlioglu, B.: Comparative performance evaluation of common mode voltage reduction three-phase inverter topologies. In: Applied Power Electronics Conference and Exposition (APEC) Proceedings, pp. 2625–2629. IEEE, San Antonio (2018)
6. Loh, P.C., Holmes, D.G., Fukuta, Y., Lipo, T.A.: Reduced common mode carrier-based modulation strategies for cascaded multilevel inverters. In: Conference Record of the 2002 IEEE Industry Applications Conference, 37th IAS Annual Meeting (Cat. No. 02CH37344), vol. 3, pp. 2002–2009 (2002)
7. Szymanski, J., Zurek-Mortka, M., Sadhu, P.K., Goswami, A.: Mitigation methods of ground leakage current caused by common-mode in voltage frequency drives. In: Sikander, A., et al. (eds.) Proceedings of 2nd International Conference on Energy Systems, Drives and Automations (ESDA 2019). Lecture Notes in Electrical Engineering, vol. 664, pp. 1-–11. Springer, Singapore (2020)
8. Jiang, D., Shen, Z., Li, Q., Chen, J., Liu, Z.: Advanced Pulse-Width-Modulation: With Freedoms to Optimize Power Electronics Converters. Springer, Singapore (2021)
9. Chen, H., Zhao, H.: Review on pulse-width modulation strategies for common-mode voltage reduction in three-phase voltage-source inverters. IET Power Electron. **9**(14), 2611–2620 (2016)
10. Lopatkin, N.: Quarter-wave symmetric space vector PWM with low values of frequency modulation index in control of three-phase multilevel voltage source inverter. In: Ronzhin, A., Shishlakov, V. (eds.) Proceedings of 15th International Conference on Electromechanics and Robotics “Zavalishin’s Readings”. Smart Innovation, Systems and Technologies, vol. 187, pp. 445-–457. Springer, Singapore (2021)
11. Lopatkin, N.: PSIM model of quarter-wave symmetric space vector PWM modulator for three-phase multilevel voltage source inverter. In: 2020 Ural Symposium on Biomedical Engineering, Radioelectronics and Information Technology (USBEREIT) Proceedings, pp. 0309–0312. IEEE, Yekaterinburg (2020)
12. Rodriguez, J., Pontt, J., Correa, P., Cortes, P., Silva, C.: A new modulation method to reduce common-mode voltages in multilevel inverters. IEEE Trans. Ind. Electron. **51**(4), 834–839 (2004)
13. Nguyen, T.K.T., Nguyen, N.V., Prasad, N.R.: Eliminated common-mode voltage pulsewidth modulation to reduce output current ripple for multilevel inverters. IEEE Trans. Power Electron. **31**(8), 5952–5966 (2016)
14. Zinoviev, G.S.: Fundamentals of Power Electronics. Textbook for Undergraduate Students, 5th edn. Jurajt, Moscow, p. 667 (2015)

Numerical Simulation of Speed-Torque Characteristics of Tape Winding Electromotor

Antonina Dolgih and Vladimir Martemyanov

Abstract The issues of speed-torque characteristics determination of the torque motor with the tape winding stator are considered. The complexity of analytical determination of such characteristics is associated with the distributed character of current flowing along the winding. It leads to use numerical research methods, in particular the finite element method software COMSOL Multiphysics. Attention is paid to the change in the motor-developed torque in the presence of the electromotive force (EMF) created in the winding during the rotor rotation. An accounting method of the rotation, EMF action on the torque value is proposed. It is shown that the torque depends mainly on the distributed current lines character when the rotor velocity changes. The obtained speed-torque characteristics are similar to the independent excitation DC motors. The obtained results make it possible to evaluate the characteristics of the motor model and are the basis for the tape winding torque motor calculating method development.

1 Introduction

Torque motors are widely used in the industry due to their ability to work in technical systems without the mechanical gearboxes use [1, 2]. The relevance of the torque motors use arises in a number of special devices, for instance, in spacecraft technology, where it is required to create an increased torque in the limited size of the device. Among the executive elements of such devices is the motor with the novel type of the winding which is proposed by the authors [3].

It is accepted that torque motor is an electric machine that creates a torque in a braked operation mode [4]. However, in recent times, electric motors are also referred to as torque ones, which are optimized for the specific developed torque. In any case, the torque motor operation is characterized by a large heat release, since most of the electrical power consumed from the source is released in the motor windings in the heat form. In this regard, torque motors have design features that allow providing the

A. Dolgih (✉) · V. Martemyanov
Tomsk Polytechnic University, 634050 Tomsk, Russia
e-mail: ivanovatonya@tpu.ru

A. Ronzhin and V. Shishlakov (eds.), *Electromechanics and Robotics*, Smart Innovation, Systems and Technologies 232, https://doi.org/10.1007/978-981-16-2814-6_27

intensive heat removal from the windings, protecting them from destruction [5, 6]. Among the design solutions that provide reliable and intensive, heat removal is the tape winding of the torque motor stator. The design features, operation principles, advantages and expected characteristics of such motor are detailed in [7, 8].

In the mentioned references, the braked operation mode of the motor is mainly considered. However, for practical application, it is necessary to obtain characteristics, evaluating the rotational operation mode of the motor. One of these is speed-torque characteristic, representing the dependence of the developed torque over the rotor velocity. The proposed work is devoted to the speed-torque characteristic determination of the motor with tape winding located on the stator.

2 Problem Statement

For a DC motor of independent excitation (in particular of a magnetoelectric type) with an armature winding of a traditional design, the torque on the shaft decreases as the rotor velocity increases due to the appearance and growth of the back-EMF induced in the armature winding and directed opposite to the applied source voltage. This interpretation of the reasons for the winding current decreases, and, consequently, the developed torque decreases and quite simply explains the process taking place in the winding conductors, through which the concentrated current flows.

The process of interaction between the main magnetic flux of the machine and the distributed current flowing through the tape winding is much more complicated. The tape winding can be considered as a set of separate conducting plates connected in series into the electric circuit. The plates are connected to each other in the contact zones obtained by making side cuts in the tape, alternating on both sides of the tape and made at a distance equal to the pole division of the rotor magnetic system.

Figure 1 shows the tape winding fragment—a plate with the current flowing through it. Current is distributed over the plate body and practically concentrated in the contact zones.

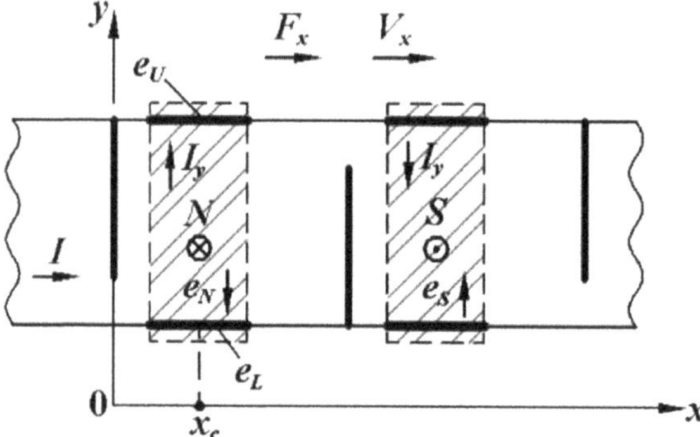

Fig. 1 Tape winding fragment

The components of the distributed current I_y directed perpendicular to the magnet pole possible displacement in the x-axis direction interact with the magnetic field and cause the appearance of the force F_x and the pole movement velocity V_x. Assuming that the magnetic fields created by the poles are uniform, we believe that in each of the elementary conductors set, located parallel to each other and directed perpendicular to the direction of the poles movement, an EMF will be induced. This EMF is proportional to the magnetic flux density, the tape width and the relative speed of the poles movement. The mentioned EMF is created by the electric potential difference of the upper e_{U} and lower e_{L} edges of the plate part, located in the zone of pole magnetic flux action:

$$e = B \cdot b \cdot V = B \cdot b \cdot R_{\mathrm{av}} \cdot \omega, \tag{1}$$

where B is the magnetic flux density; b is the plate (tape) width; R_{av} is the average winding radius; and ω is the angular velocity of motor rotor.

The EMF e is the opposite of the voltage applied to the plate; however, its effect is amphibological. In the case of the traditional winding, the applied voltage decreases algebraically due to the emerging back-EMF, while both the voltage and the back-EMF act along the entire length of the conductor located in the magnetic field. The resulting voltage difference creates less current and eventually less developed torque. In the case of the plate winding element, representing the conductor equivalent, the process of interaction of the applied voltage and the induced EMF on the current nature occurs in the plate section, which is penetrated by the magnetic flux. In this case, the developed torque decrease occurs mainly due to the distortion of the distributed current lines shape and decrease in its components I_y. Figure 2a shows the distributed current lines created by the voltage applied to the plate contacts, and the same, but in the presence of the induced EMF of rotation (Fig. 2b).

The task can be solved when firstly, the static (braked) operation mode is considered, and then, the influence of the EMF induced by the rotor rotation is taken into account.

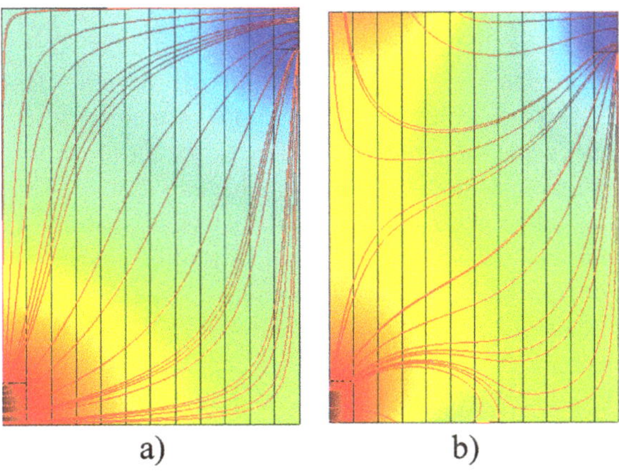

Fig. 2 Distributed current lines

3 Static Operation Mode

The results in Fig. 2 were obtained using the finite element simulation software COMSOL Multiphysics [9, 10]. Earlier in [7, 8], the approach to the torque determination of the tape winding motor operating in the braked mode was presented. First of all, the force F_1 acting from a single plate on the permanent magnetic field source is determined. In the COMSOL Graphics window, the plate geometry is built, then the boundary conditions and the plate material are set, and after that the plate area where the magnetic field source acts is indicated. To the magnet movement simulation relatively to the plate, its plane is divided into separate subregions, which are combined in a certain way in the process of calculating.

After solving the simulation task in the postprocessing mode of COMSOL Multiphysics, the current I_1 flowing through the plate is determined using the integral of the current density at the contact boundary. The double integral $D(x, y)$ of the components j_y of the distributed current density over the plate surface limited by the magnetic flux action is also determined. Also, the plate active resistance is defined by a given value of the voltage U_A applied to the plate contacts and the found value of the current I_1. The force created by this current and acting on the magnet will be detected as:

$$F_1 = B \cdot \Delta \cdot D(x, y), \tag{2}$$

where B is the uniform magnetic flux density индукция; Δ is the plate thickness; and $D(x, y)$ is the double integral of the distributed current density components.

The value of the torque developed by the motor, obtained earlier in [7, 8], can be represented as:

$$T = p \cdot N \cdot \Delta^2[N + 2R_{av}/\Delta] \cdot B \cdot D(x, y), \tag{3}$$

where p is the number of the rotor magnet pole pairs; N is the number of tape winding turns; and R_{av} is the winding average radius.

In the above expression (3), the voltage U_A applied to the plate is the electrical parameter that determines the motor operation. This voltage is set when defining the double integral $D(x, y)$ value by giving the contacts the corresponding potentials.

The task solution is carried out by numerical methods, with the setting of the design and magnetic parameters of the torque motor. The scheme used in flywheel motors and motor wheels of vehicles [7] was chosen as a constructive basis. The magnetic system has $p = 5$ pole pairs and consists of the outer and inner rotor, in the gap between which there is a tape winding connected to the motor stator. In the poles gap, the magnetic flux density with the induction $B = 0.2$ T is created. The magnetic circuit calculation was carried out using COMSOL Multiphysics. The tape winding contains $N = 50$ turns, and its average radius is $R_{av} = 0.06$ m. The geometrical dimensions of the plate provide the minimum active resistance [2]. In this case, the plate length a is determined by the pole division magnitude of the magnetic system

$a = \tau = 36$ mm. The tape width $b = 50$ mm, the plate is the element of tape. The ratio $a/b = 36/50 = 0.72$ is within the permissible limits providing the plate active resistance minimum. The contact width is $b_c = 5$ mm. The tape thickness is given by $\Delta = 10^{-4}$ m. This size represents a conditional, expected value, since the methods of winding manufacturing based on additive technologies [11] are under development.

The voltage U_A applied to the plate contacts is determined from the following conditions. It has been experimentally established that with a short-term current flow through the plate in the contact zone, the current density can reach a value of $j_p = 100$ A/mm^2. With the above geometric parameters, the contact cross-sectional area is:

$$S_c = b_c \cdot \Delta = 5 \cdot 0.1 = 0.5\,\text{mm}^2.$$

Therefore, the maximum current can be passed through the plate:

$$J_1 = j_p \cdot S_c = 100 \cdot 0.5 = 50\,\text{A}.$$

Then, after setting the boundary conditions for the contacts of the studied plate model in the form of some voltage Δu (e.g., on the left contact 0.001 V, on the right is ground mode), the solution of the problem is started, and the integral of the current density at the contact boundary is determined. Multiplying it by the plate thickness, we get the current i flowing through the contact. After that, the voltage U_A is determined:

$$U_A = \Delta u \cdot I_1 / i.$$

Substituting the numerical values into the above expression, we have:

$$U_A = 0.001 \cdot 50 / 1.6619 = 0.03\,\text{V}.$$

As a result, with $2p = 10$ and the winding turns number $N = 50$, the power supply voltage is determined:

$$U = 2p \cdot N \cdot U_A = 10 \cdot 50 \cdot 0.03 = 15\,\text{V}.$$

As shown above (3), the torque depends on the motor design parameters, which are constant, the flux density of the uniform magnetic field, which is considered stable in the plate overlap zone, and the double integral $D(x, y)$ magnitude. From all the factors included in the torque expression, the only double integral value is variable:

$$T = p \cdot N \cdot \Delta^2 \cdot \left[N + \frac{2R_{av}}{\Delta} \right] \cdot B \cdot D(x, y)$$

Table 1 Research results

x_c (mm)	D (A)	T (N m)	ΣD (A)	T_Σ (N m)
0	0	0	9475.0	5.922
3	3616.1	2.26	13,320.0	8.325
6	7442.8	4.6517	17,863.7	11.165
9	11,701.9	7.3137	23,403.7	14.627
12	10,420.9	6.513	17,863.5	11.165
15	9703.9	6.0649	13,320.0	8.325
18	9475.0	5.9219	9475.0	5.922
21	9703.9	6.0649	13,320.0	8.325
24	10,420.9	7.3136	17,863.7	11.165
27	11,701.8	7.3136	23,403.7	14.627
30	7442.6	4.6516	17,863.5	11.165
33	3616.1	2.26	13,320.0	8.325
36	0	0	9475.0	5.922

$$= 5 \cdot 50 \cdot 10^{-8} \cdot \left[50 + \frac{2 \cdot 6 \cdot 10^{-2}}{10^{-4}}\right] \cdot 0.2 \cdot D(x, y)$$
$$= 6.25 \cdot 10^{-4} \cdot D(x, y).$$

The torque determination was carried out at the voltage applied to the plate contacts $U_A = 0.03$ B.

Simulating the magnet center displacement along the plate at such voltage, the double integral values $D(x, y)$ were obtained. These are shown in the first column of Table 1. For convenient analysis, this table also contains the results obtained in further studies, which will be described below.

Figure 3 shows a graph of the double integral $D(x, y)$ changing based on the data in Table 1. The dependence of the torque over the pole center displacement will have the same form.

A motor with this characteristic can only operate in a limited range of rotor rotation angles. To expand this range, it is necessary to use a block of two winding sections, displaced around the axis by half the pole division of the rotor magnetic system [12].

The numerical simulation results of a two windings block action are also given in Table 1, where ΣD is the total value of the two windings current density integrals, and T_Σ is the resulting torque of the brushless DC electric motor. The form of the resulting torque dependence is shown in Fig. 4.

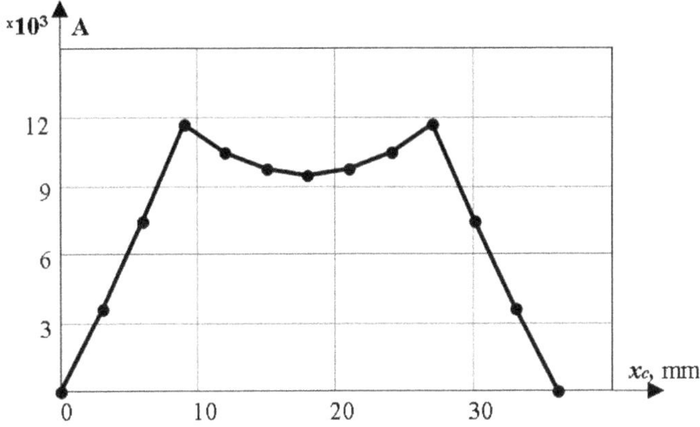

Fig. 3 Dependence of the double integral $D(x, y)$ over the pole center position

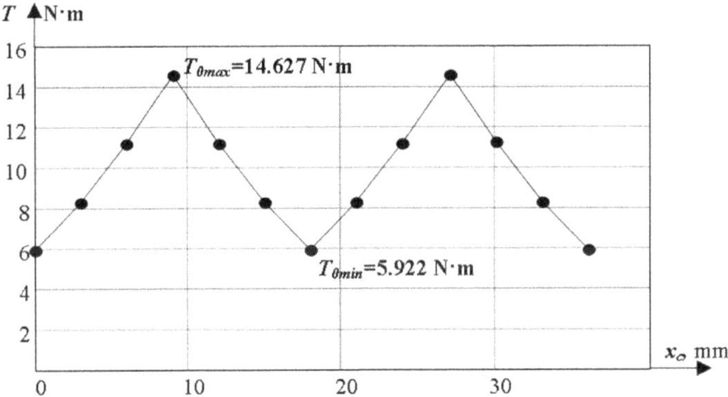

Fig. 4 Character of the torque changing within the pole division

4 Speed-Torque Characteristics

All the above characteristics refer to the braked operation mode, when the rotor velocity $\omega = 0$. If the rotor rotates, the potentials appear on the upper and lower plate edges (see Fig. 1), and their difference is determined by expression (1). Substituting the numerical values, we get:

$$\begin{aligned} e &= B \cdot b \cdot V = B \cdot b \cdot R_{\text{av}} \cdot \omega \\ &= 0.2 \cdot 0.05 \cdot 0.06 \cdot \omega = 0.6 \cdot 10^{-3} \cdot \omega. \end{aligned}$$

In further studies, it is necessary to set the boundary conditions corresponding to the EMF value of rotation when we determine the double integral value $D(x, y)$. In this case, it is unacceptable to set the boundary conditions using the electric potential mode in COMSOL, since the resulting value $D(x, y)$ is determined exclusively by the specified potential difference at the upper and lower plate edges and is completely independent of the supply voltage applied to the contacts, which contradicts the motor operation principle. The way out is to use the inward current flow mode. The required potential difference can be obtained indirectly by setting the density of the «internal» current $+j$ at the upper plate edge and $-j$ at the lower edge. Then, having received the integral value over the length of the area by the boundary integration mode and dividing it by the area length, we determine the average value of the potential in this area.

In general, the action of the induced EMF of rotation on the developed torque value can be described as follows. In the absence of supply voltage ($U_A = 0$), two or three values of the "internal" current density $\pm j$ are set in the selected area, and the average values of the potential differences that create this current are determined. The dependence $e(j)$ is determined, and usually, it is linear: $e = k_e \cdot j$. The potential difference e is calculated by expression (1) from a given value of ω, and then, from the dependence $e(j)$, the required values of the current density $\pm j_{\text{req}}$ are determined. It will be the boundary conditions of the upper and lower plate edges:

$$J_{\text{req}} = e/k_{\text{e}} = \left[0.6 \cdot 10^{-3}/k_e\right] \cdot \omega = k_\omega \cdot \omega.$$

Then, the value of the voltage U_A applied to the plate is set, and the value $D(x, y)$ is determined. These actions are carried out at all necessary positions of the pole center x_c over the plate. The results of coefficients k_e and k_ω determining are given in Table 2.

Let us refer to Fig. 4. The maximum developed torque will be at $x_c = 9$ mm, the minimum—at $x_c = 18$ mm. For these two pole positions, we determine the values of the integral $D(x, y)$, and taking into account that $T = 6.25 \times 10^{-4} \cdot D(x, y)$, we obtain the values for the speed-torque characteristics of the motor. The supply voltages applied to the winding are set as 5.0, 10.0, 15.0 V. The calculation results are shown in Tables 3 and 4.

The graphs of the motor speed-torque characteristics are built based on the calculated data obtained and illustrated in Fig. 5. The shaded areas of the graphs show the corridors in which the speed-torque characteristics are situated at the different pole positions relative to the plate. Note that the form of the obtained speed-torque characteristics is similar to the ones of independent excitation DC motors. These

Table 2 Determination of coefficients k_e and k_ω

x_c (mm)	$j = 10^6$ (A/m^2)	$j = 10 \times 10^6$ (A/m^2)	$j = 25 \times 10^6$ (A/m^2)	k_e	k_ω
9	4.384×10^{-4}	4.384×10^{-3}	0.01096	4.384×10^{-10}	1.3686×10^6
12	4.2305×10^{-4}	4.2331×10^{-3}	0.01058	4.231×10^{-10}	1.4181×10^6
15	4.149×10^{-4}	4.149×10^{-3}	0.01037	4.149×10^{-10}	1.4461×10^6
18	4.122×10^{-4}	4.125×10^{-3}	0.01031	4.12×10^{-10}	1.4563×10^6

Table 3 Torque values at $x_c = 9$ mm

ω (1/c)	0	2.5	5.0	10.0	15.0	20.0
$U = 5.0$ B	2.4378	1.427	0.4162	−1.6054	–	–
$U = 10.0$ B	4.8758	3.8649	2.8541	0.8325	−1.189	–
$U = 15.0$ B	7.3137	6.3029	5.292	3.2704	1.2488	−0.7728

Table 4 Torque values at $x_c = 18$ mm

ω (1/c)	0	2.5	5.0	7.5	10.0	12.5	15.0
$U = 5.0$ B	1.9739	0.9625	−0.0489	−1.0603	–	–	–
$U = 10.0$ B	3.9479	2.9365	1.925	0.9136	−0.0978		
$U = 15.0$ B	5.9219	4.9104	3.899	2.8876	1.8761	0.8647	−0.1467

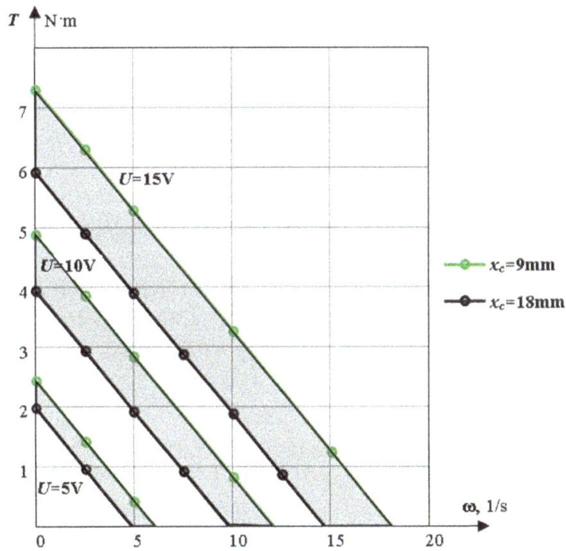

Fig. 5 Group of speed-torque characteristics

characteristics have high linearity and rigidity of about 0.4043 N m s over the entire supply voltage range.

Of particular, interest is the dependence of the power source current consumed by the winding at different rotor velocity ω. The results of numerical studies are shown in Table 5.

Here, we consider the case of the winding power supply with a voltage $U = 15.0$ V and at two positions of the pole center $x_c = 9$ mm and $x_c = 18$ mm. Unlike motors with a traditional winding, the current consumption does not decrease as the rotor velocity increases, but its growth is observed with a rate of (0.72–0.88) A s. This phenomenon is most likely explained by the fact that the induced EMF of rotation does not reduce the supply voltage applied to the plate, but distort the distributed

Table 5 Winding current consumption, A

x (mm)	ω (1/c)							
	0	2.5	5.0	7.5	10.0	12.5	15.0	20.0
9	49.8	52.03	54.2		58.54		62.9	67.2
18	49.8	51.6	53.5	55.3	57.1	58.9	60.7	

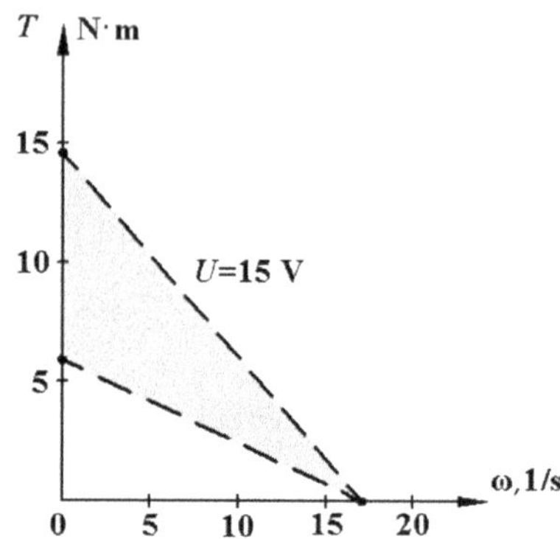

Fig. 6 Speed-torque characteristics of the motor with two windings

current lines, forcing the current to flow along the path of the least active resistance. The nature of the currents flow in this case is quite complicated and requires an additional study.

Proposed approach allows determining the type of the motor speed-torque characteristics with a block of two windings deployed by half the pole division. The initial values for the starting mode are given in Table 1. The ideal no-load speed is determined by the condition of equality of the positive torque at $x_c = 9$ mm of the first winding and negative torque at $x_c = 18$ mm of the second winding.

A general view of the motor speed-torque characteristics at the supply voltage of $U = 15$ V is shown in Fig. 6.

5 Conclusion

As a result, recommendations for the speed-torque characteristics determination of the tape winding torque motor with were obtained. The characteristics of the static motor operation mode are obtained by numerical methods. A method of the action of the induced EMF of rotation accounting at the motor-developed torque is proposed. The obtained speed-torque characteristics are highly linear. The winding current consumption increases to (20–30)% with the increase in the rotor velocity. The calculated data obtained allow us to evaluate the possible parameters of the torque model being created. The research results will form the basis of the developed methodology for calculating the torque motor with the tape winding stator.

Acknowledgements This study was supported by the grant from Russian Science Foundation, project no. 20-79-00055.

References

1. Dubrovskiy, G., Mikerov, A., Dzhankhotov, V., Pyrhonen, J.: General comparison of direct and geared drives for control applications. In: 2014 16th European Conference on Power Electronics and Applications, pp. 1–6 (2014)
2. Hughes, A., Drury, B.: Electric Motors and Drives. Fundamentals, Types and Applications, 4th edn. Elsevier (2013)
3. Martemyanov, V.M., Dolgih (Ivanova), A.G.: Momentnnyj dvigatel. In: Torque Motor. Patent RF, No. 2441310 (2010) (in Russ.)
4. Gieras, J.F.: Electrical Machines: Fundamentals of Electromechanical Energy Conversion, 3rd edn. CRC Press, Taylor and Francis Group (2017)
5. Faulhaber: Technical Information, 8th edn. Dr. Fritz Faulhaber GmbH & Co. KG, p. 71 (2016)
6. Torque Motors: ETEL Motion Technology, p. 15 (2017)
7. Dolgih, A.G., Martemyanov, V.M., Borikov, V.N.: Active element influence on the motor's torque. MATEC Web Conf. **113**, Article number 01013 (2017)
8. Dolgih, A.G.: Jelektromehanicheskij preobrazovatel's lentochnoj obmotkoj jakorja. Avtoref. diss. kand. tekhn. Nauk, Electromechanical Converter with Tape Winding Armature. Extended Abstract of Cand. Eng. Sci. Diss. (2017) (in Russ.)
9. Krasnikov, G.E., Nagornov, O.V., Starostin, N.V.: Modelirovanie fizicheskih processov s ispol'zovaniem paketa COMSOL Multiphysics: Uchebnoe posobie. In: Modeling Physics Using COMSOL Multiphysics: A Tutorial (2012) (in Russ.)
10. COMSOL AC/DC Module User's Guide. COMSOL 5.4 (2018)
11. Martemyanov, V.M., Dolgih, A.G.: Sposob izgotovlenija spiral'nogo aktivnogo jelementa statora momentnogo dvigatelja. In: Method of Manufacturing a Spiral Stator Active Element of the Torque Motor, Patent RF, No. 2713217 (2020) (in Russ.)
12. Martemyanov, V.M., Dolgih, A.G.: Ventil'nyj dvigatel'. In: Thyratron Motor, Patent RF, No. 2454776 (2012) (in Russ.)

Dynamic Properties of Technological Drive Operating in Acceleration Mode

Dmitry Ershov, Irina Lukyanenko, and Evgeny Zlotnikov

Abstract The study centers on dynamic transients resulting from jump-like changes in controller of machinery electrical drive. Discussion requires description of dynamic processes in various operation modes—acceleration, deceleration, reversing, and transition between different angular velocities given DC and AC engines momentums conditioned on angular velocity values. Conditions in which technological drives operate necessitate relevant transient modes aiming either to provide for maximization of operation speeds, or minimization of losses, or limitation of dynamic loads generated in components of kinematic chains which link electric drive and working element, etc. Control of transient processes in the electric drive guarantees maximum operation speeds and equipment performance given specific limitations. On the basis of the dynamic drive model with two degrees of freedom, the different stages of the run-up mode were analyzed. The characteristics of the static component of torque on the shaft in steady mode and the dynamic component caused by the transient process are determined. The ways of reduction of the dynamic component significant for the drive operation are offered.

1 Introduction

Electric drive is a principal component in manufacturing equipment which determines the equipment functional and operational parameters. Transient modes—which include start, acceleration, reversing, and deceleration of technological equipment drive—are the modes of equipment operation characterized by maximum loads and thus they necessitate coordination between the engine electro-mechanical properties and the equipment dynamic parameters given specific to certain mechanisms resistance forces and moment of inertia. The issues and problems related to modeling

D. Ershov (✉) · E. Zlotnikov
Department of Mechanical Engineering, St. Petersburg Mining University, 21 str. Line, 2, 199106 St. Petersburg, Russia

D. Ershov · I. Lukyanenko
Department of Advanced Mathematics and Mechanics, St. Petersburg State University of Aerospace Instrumentation (SUAI), Bolshaya Morskaya str., 67, 190000 St. Petersburg, Russia

A. Ronzhin and V. Shishlakov (eds.), *Electromechanics and Robotics*, Smart Innovation, Systems and Technologies 232, https://doi.org/10.1007/978-981-16-2814-6_28

various dynamic processes within electro-mechanical systems have been receiving major attention recently [1–4].

The acceleration mode of the electric drive is described as a jump-like change in control, i.e., change in the current supplied to the engine armature, given constant moment of load M_{L}^{0}. Limitation, i.e., reduction, of the engine starting momentum value, can be achieved by adding an extra resistance element and an induction coil to the engine circuit—such design allows higher values of inductance in the circuit. This circuit design, however, requires examination and adjustment for both mechanical and electro-magnetic parameters in the electric drive transient modes. The study of dynamic processes in acceleration mode of the electric drive operation can use the below dynamic parameters of engine [5]:

$$T_{\mathrm{e}}\dot{M}_{\mathrm{E}} + M_{\mathrm{E}} = \beta_{\mathrm{E}}\omega_{0} - \beta_{\mathrm{E}}\omega_{\mathrm{E}}, \tag{1}$$

where T_{e} stands for electro-magnetic time constant conditioned on inductance and resistance moment of engine armature circuit; M_{E} stands for engine momentum; β_{E} stands for static stiffness module; ω_{0} is angular velocity of engine idling mode; ω_{E} is angular velocity of engine running under load.

Computational model of technological equipment drive mechanical system is shown in Fig. 1 [5].

Figure 1 shows the diagram of a drive to consist of an engine rotor with the moment of inertia J_{1}, a spur gear of two gears 2 and 2* with moments of inertia J_{2} and J_{2*}, and an actuator or working element (e.g., main spindle) 3 with moment of inertia J_{3}. Moments of inertia of the elements shown in the diagram are calculated relative to the proper rotation axes. Engine momentum M_{E} is applied to the engine rotor and technological resistance forces moment M_{L} is applied to the working element.

Findings presented in [6–8] allow for discussion of specific parameters produced by drive operation in acceleration mode and examined with the electric dynamic model shown in Fig. 2.

First disk 1 with moments of inertia $J_{1} = J_{\mathrm{E}}$ refers to the engine rotor (Fig. 1) with engine momentum M_{E} applied to it. Second disk 2 with moments of inertia calculated as:

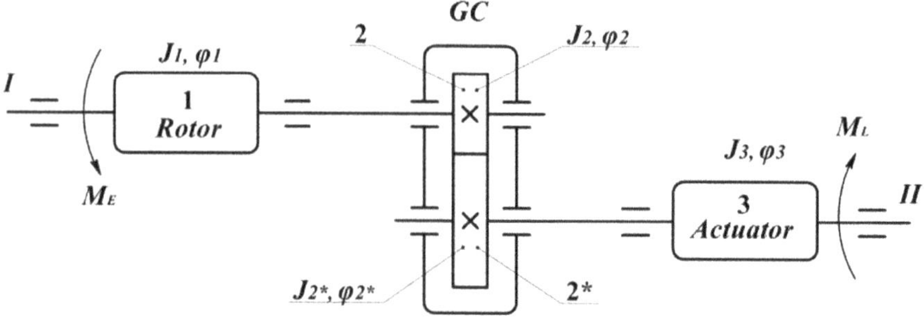

Fig. 1 Computational model of technological equipment drive

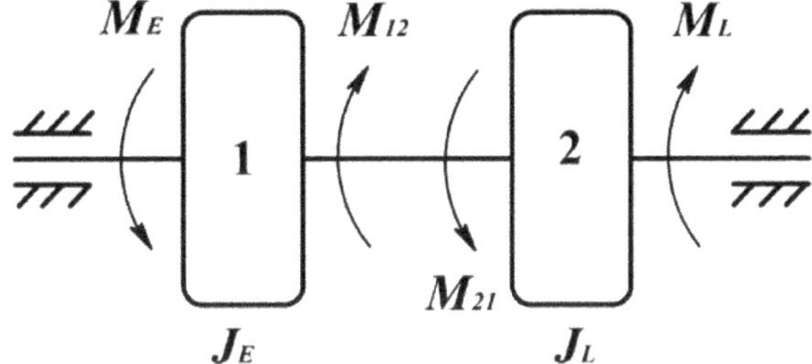

Fig. 2 Dynamic model of technological equipment drive

$$J_{\mathrm{L}} = J_2 + J_{2*}i^{-2} + J_3 i^{-2}, \tag{2}$$

allows for the inertia-related properties of the rest of mechanical system (drive) which, within the following discussion, is referred to as 'load' and in which i indicates the gear ratio of the drive spur gear. The reduced load moment $M_{\mathrm{L}}(t)$ is applied to this disk. The dynamic model presented allows for the torsion moment of the shaft connecting engine rotor 1 and engine load 2 to be introduced in the discussion. Further, M_{12} will stand for the torsion moment applied to disk 1 (engine rotor) and M_{21} will stand for the torsion moment applied to disk 2 (engine load). In the model presented, engine rotor with moments of inertia J_{E} is connected with a rigid shaft to the system working element with moments of inertia J_{L}. It can be stated that for any given time the equality $M_{12} = -M_{21}$ holds. The model described (Fig. 2) allows to study changes in the value of the shaft torsion moment while analyzing the effects of moments of inertia J_{E} and J_{L} relation on dynamic properties of technological equipment drive in various modes of operation.

Differential equations which describe electro-mechanical dynamic processes in the drive are as given below

$$\begin{cases} J_{\mathrm{E}}\dot{\omega}_{\mathrm{E}} = M_{\mathrm{E}} + M_{12} \\ J_{\mathrm{L}}\dot{\omega}_{\mathrm{L}} = M_{21} - M_{\mathrm{L}}^{0} \\ T_{\mathrm{e}}\dot{M}_{\mathrm{E}} + M_{\mathrm{E}} = \beta_{\mathrm{E}}\omega_0 - \beta_{\mathrm{E}}\omega_{\mathrm{E}} \end{cases}. \tag{3}$$

For the model studied, acceleration mode of the electric drive is divided into two stages. The first stage—following the start of the engine—is characterized by electro-magnetic processes which can be described by the third equation in the system (3). At this stage, engine momentum M_{E} increases in value in the range of zero to $M_{\mathrm{E}} = M_{\mathrm{L}}^{0}$. The second stage, at which $M_{\mathrm{E}} > M_{\mathrm{L}}^{0}$, is the stage of mechanical system acceleration.

2 Mathematical Description

2.1 Effect of Angular Velocity

The transient process in terms of engine angular velocity values $\omega_{\mathrm{E}}(t)$ can be presented in the following manner. When the engine starts, $\omega_{\mathrm{E}} = 0$, and given the third equation in the system (3) it can be described as:

$$T_{\mathrm{e}}\dot{M}_{\mathrm{E}} + M_{\mathrm{E}} = \beta_{\mathrm{E}}\omega_0 - M_{\mathrm{sc}},$$

where M_{sc} stands for electrical short moment.

General solution to this equation is done as:

$$M_{\mathrm{E}}(t) = M_{\mathrm{E}}^{(1)}(t) + M_{\mathrm{E}}^{(2)}(t).$$

Solution to the homogeneous equation is then:

$$M_{\mathrm{E}}^{(1)}(t) = Be^{\lambda t},$$

where B stands for amplitude value $M_{\mathrm{E}}^{(1)}(t)$.

Characteristic equation is $T_{\mathrm{e}}\,\lambda + 1 = 0$, which produces $\lambda = -\frac{1}{T_{\mathrm{e}}}$.

Particular solution can be given as $M_{\mathrm{E}}^{(2)}(t) = M_{\mathrm{sc}}$. In this case, $M_{\mathrm{E}}(t) = Be^{\lambda t} + M_{\mathrm{sc}}$.

Initial conditions produce the equality of $B = M_{\mathrm{sc}}$. Solution for the engine momentum at the first stage of engine operation can then be presented as:

$$M_{\mathrm{E}}(t) = M_{\mathrm{sc}}\left(1 - e^{-\frac{t}{T_{\mathrm{e}}}}\right). \tag{4}$$

Engine momentum value increases exponentially. The moment of time characterized by $t = \tau$, when the condition of $M_{\mathrm{E}} = M_{\mathrm{L}}^{0}$ holds, is calculated as:

$$\tau = -T_{\mathrm{e}}\ln\left[\frac{M_{\mathrm{sc}} - M_{\mathrm{L}}^{0}}{M_{\mathrm{sc}}}\right].$$

The second stage of acceleration mode is discussed given $t > \tau$, which requires the system of equations (3) to be reduced to one equation for variable ω_{E}, the resulting equation is as given below

$$(J_{\mathrm{E}} + J_{\mathrm{L}})\dot{\omega}_{\mathrm{E}} = M_{\mathrm{E}} + M_{\mathrm{L}}^{0}.$$

Solution to the last equation for M_{E} produces:

$$M_{\mathrm{E}} = (J_{\mathrm{E}} + J_{\mathrm{L}})\dot{\omega}_{\mathrm{E}} + M_{\mathrm{L}}^{0},$$

$$\dot{M}_{\mathrm{E}} = (J_{\mathrm{E}} + J_{\mathrm{L}})\ddot{\omega}_{\mathrm{E}}.$$

Substitution into the third equation of the system (3) of expression for calculation of M_{E} and $\dot{M}_{\mathrm{E}}$, values produces, given a few simple transformations, the following expression:

$$T_{\mathrm{e}} T_{\mathrm{m}} \ddot{\omega}_{\mathrm{E}} + T_{\mathrm{m}} \dot{\omega}_{\mathrm{E}} + \omega_{\mathrm{E}} = \omega_{\mathrm{E}}^{0}, \tag{5}$$

where T_{m} stands for mechanical constant of engine drive $T_{\mathrm{m}} = (J_{\mathrm{E}} + J_{\mathrm{L}})\beta_{\mathrm{E}}^{-1}$.

The general solution to Eq. (5) can be presented as a sum below:

$$\omega_{\mathrm{E}}(t) = \omega_{\mathrm{E}}^{(1)}(t) + \omega_{\mathrm{E}}^{(2)}(t).$$

Particular solution to the last equation $\omega_{\mathrm{E}}^{(2)}(t)$ is determined by expression of $\omega_{\mathrm{E}}^{(2)}(t) = \omega_{\mathrm{E}}^{0}$. General solution to homogeneous equation will take the following form

$$\omega_{\mathrm{E}}^{(1)}(t) = C\mathrm{e}^{\lambda t}.$$

Characteristic equation for expression (5) is $T_{\mathrm{e}} T_{\mathrm{m}} \lambda^2 + T_{\mathrm{m}} \lambda + 1 = 0$.

The roots in characteristic equation are calculated as:

$$\lambda_{1,2} = \frac{-T_{\mathrm{m}} \mp \sqrt{T_{\mathrm{m}}^2 - 4T_{\mathrm{e}} T_{\mathrm{m}}}}{2T_{\mathrm{e}} T_{\mathrm{m}}} = -\frac{1}{2T_{\mathrm{e}}} \mp \frac{1}{\sqrt{T_{\mathrm{e}} T_{\mathrm{m}}}} \sqrt{\frac{T_{\mathrm{m}}}{4T_{\mathrm{e}}} - 1}.$$

The above expression allows a conclusion that the character of roots, and hence, the parameters of drive motion in acceleration mode are determined by relation of time constants T_{e} and T_{m}. Three variations are possible in this case.

1. If $T_{\mathrm{m}} > 4T_{\mathrm{e}}$, the roots are two negative real roots:

$$\lambda_1 = -\frac{1}{2T_{\mathrm{e}}} - \frac{1}{\sqrt{T_{\mathrm{e}} T_{\mathrm{m}}}} \sqrt{\frac{T_{\mathrm{m}}}{4T_{\mathrm{e}}} - 1} \langle 0; \;\; \lambda_2 = -\frac{1}{2T_{\mathrm{e}}} + \frac{1}{\sqrt{T_{\mathrm{e}} T_{\mathrm{m}}}} \sqrt{\frac{T_{\mathrm{m}}}{4T_{\mathrm{e}}} - 1} \langle 0.$$

   General solution will be determined by sum of exponents:

$$\omega_{\mathrm{E}}^{(1)}(t) = C_1 \mathrm{e}^{\lambda_1 (t-\tau)} + C_2 \mathrm{e}^{\lambda_2 (t-\tau)}.$$

   Transient process in terms of angular velocity ω_{E} in acceleration mode of the engine drive can be described with the following formula

$$\omega_{\mathrm{E}}(t) = C_1 \mathrm{e}^{\lambda_1 (t-\tau)} + C_2 \mathrm{e}^{\lambda_2 (t-\tau)} + \omega_{\mathrm{E}}^0.$$

Such transient process is aperiodic and damped, since $\lambda_1 < 0$, $\lambda_2 < 0$. Angular velocity value in the set-steady mode equals $\omega_{\mathrm{E}} = \omega_{\mathrm{E}}^0$.

2. If $T_{\mathrm{m}} = 4T_{\mathrm{e}} = T_{\mathrm{m}}^*$, where T_{m}^* stands for critical value of mechanical time constant, two roots are equal:

$$\lambda_{1,2} = -\frac{1}{2T_{\mathrm{e}}}.$$

General solution to equation in this case will be:

$$\omega_{\mathrm{E}}^{(1)}(t) = (C_1 + tC_2)\mathrm{e}^{-\frac{1}{2T_{\mathrm{e}}}}.$$

Such transient process is also aperiodic and damped:

$$\omega_{\mathrm{E}}(t) = (C_1 + tC_2)\mathrm{e}^{-\frac{1}{2T_{\mathrm{e}}}} + \omega_{\mathrm{E}}^0.$$

3. If $T_{\mathrm{m}} < T_{\mathrm{m}}^* = 4T_{\mathrm{e}}$, the characteristic equation produces two complex-conjugate roots:

$$\lambda_{1,2} = -\frac{1}{2T_{\mathrm{e}}} \mp j\frac{1}{\sqrt{T_{\mathrm{e}}T_{\mathrm{m}}}}\sqrt{1 - \frac{T_{\mathrm{m}}}{T_{\mathrm{e}}^*}} = -\frac{1}{2T_{\mathrm{e}}} \mp j\Omega_{\mathrm{c}},$$

where $\Omega_{\mathrm{c}} = \frac{1}{\sqrt{T_{\mathrm{e}}T_{\mathrm{m}}}}\sqrt{1 - \frac{T_{\mathrm{m}}}{T_{\mathrm{e}}^*}}$ stands for oscillation frequency of the drive itself $j = \sqrt{-1}$.
In this case, general solution to homogeneous equation can be done as:

$$\omega_{\mathrm{E}}^{(1)}(t) = A\mathrm{e}^{-\frac{t-\tau}{2T_{\mathrm{e}}}} \sin[\Omega_{\mathrm{c}}(t-\tau) + \psi],$$

where A and ψ stand for constants of integration.
General solution to expression (5) can in this case be given as:

$$\omega_{\mathrm{E}}(t) = A\mathrm{e}^{-\frac{t-\tau}{2T_{\mathrm{e}}}} \sin[\Omega_{\mathrm{c}}(t-\tau) + \psi] + \omega_{\mathrm{E}}^0. \tag{6}$$

Function (6) describes transient process in terms of angular velocity values in acceleration mode of electric drive. This transient process is characterized as possessing oscillatory properties. Oscillation of the augend in (6) damps with time, and in set-steady mode of electric drive angular velocity value equals $\omega_{\mathrm{E}} = \omega_{\mathrm{E}}^0$.

Oscillatory transient process is more effective and preferable compared with aperiodic transient process as the latter is characterized by alternating-sign dynamic loads in the drive mechanisms and the phenomenon of overregulation which is caused by angular velocity and engine momentum being exceeded by values of ω_{E}^0 and M_{E}^0.

Oscillatory dynamic process can be described in the following manner. Starting or initial conditions can be set as: at a time point $t = \tau$ angular velocity ω_E and angular acceleration $\dot{\omega}_E$ both equal zero, since condition of $M_L = M_L^0$ holds. To determine the constants of integration, calculation of the derivative is required:

$$\dot{\omega}_E(t) = -\frac{1}{2T_e} A e^{-\frac{t-\tau}{2T_e}} \sin[\Omega_c(t-\tau)+\psi] + A\Omega_c e^{-\frac{t-\tau}{2T_e}} \cos[\Omega_c(t-\tau)+\psi]. \quad (7)$$

Initial conditions calculated with (6) and (7) produce expressions used in calculation of constants for integrating A and ψ:

$$0 = A \sin\psi + \omega_E^0;\ 0 = -\frac{1}{2T_e} A \sin\psi + A\Omega_c \cos\psi.$$

Solution to this equations system produces:

$$A = \frac{\omega_E^0}{\sqrt{1-\frac{T_m}{4T_e}}},\quad \tilde{\psi} = \pi + \mathrm{arctg}\sqrt{4\frac{T_e}{T_m} - 1} = \pi + \psi.$$

Particular solution to a differential Eq. (5) can be presented as:

$$\omega_E(t) = \omega_E^0 \left\{ 1 - \frac{e^{-\frac{t-\tau}{2T_e}}}{\sqrt{1-\frac{T_m}{4T_e}}} \sin[\Omega_c(t-\tau)+\psi] \right\}. \quad (8)$$

When mechanical time constant T_m of the electric drive increases the drive oscillation frequency Ω_c decreases. In this case, the drive can be regarded a damped oscillating link, and time constant T_m performs the function of dissipative factor characterizing oscillation damping process.

In case $T_m > T_m^*$, as it was demonstrated in the above discussion, the characteristic equation produces two negative real roots. The drive in this case is a link with aperiodic oscillation. Thus, the value T_m^* of time constant T_m is a critical factor. The ratio of $\frac{T_m}{T_m^*}$ determines the character of transient process in the drive.

2.2 *Effect of Engine Momentum*

The transient process in terms of engine momentum $M_E(t)$ can be presented in the following manner. The procedure requires transformation and reduction of system (3) to one equation in respect to variable $M_L(t)$. If joining left-hand and right-hand sides of the first and the second equations, then differentiating the third equation by time factor and further omitting $\dot{\omega}_E$ the resulting equation will be as:

$$T_e T_m \ddot{M}_E + T_m \dot{M}_E + M_E = M_L^0. \tag{9}$$

Equation (9) is similar to Eq. (5), thus the general integral for complex-conjugate roots, similarly to (6) can be calculated by the following formula:

$$M_E(t) = A_1 e^{-\frac{t-\tau}{2T_e}} \sin[\Omega_c(t-\tau) + \psi_1] + M_L^0, \tag{10}$$

where A_1, ψ_1 stand for constants of integration.

At the time point $t = \tau$, the engine momentum value, as it can be concluded from the analysis of the first stage ($0 \le t \le \tau$) of the drive acceleration mode, equals $M_E(t) = M_L^0$. Using (4) and given $t = \tau$, the derivative is calculated as:

$$\dot{M}_E(\tau) = \frac{M_{sc} - M_L^0}{T_e}.$$

Calculation of constants of integration A_1 and ψ_1 requires to determine the derivative function (10):

$$\dot{M}_E(t) = -\frac{1}{2T_e} A_1 e^{-\frac{t-\tau}{2T_e}} \sin[\Omega_c(t-\tau) + \psi_1] + A_1 \Omega_c e^{-\frac{t-\tau}{2T_e}} \cos[\Omega_c(t-\tau) + \psi_1].$$

Using the initial conditions for the second stage of drive acceleration mode and calculated functions $M_E(t)$ and $\dot{M}_E(t)$, the resulting equation is as:

$$M_L^0 = A_1 \sin\psi_1 + M_L^0,$$

$$\frac{M_{sc} - M_L^0}{T_e} = -\frac{1}{2T_e} A_1 \sin\psi_1 + A_1 \Omega_c \cos\psi_1.$$

Solution to this system of equations can be presented as:

$$A_1 = \frac{2(M_{sc} - M_L^0)}{\sqrt{\frac{4T_e}{T_m} - 1}}, \quad \psi_1 = 0.$$

Solutions substituted into expression (10) produce the particular solution to equation:

$$M_E(t) = \frac{2(M_{sc} - M_L^0)}{\sqrt{\frac{4T_e}{T_m} - 1}} e^{-\frac{t-\tau}{2T_e}} \sin[\Omega_c(t-\tau)] + M_L^0. \tag{11}$$

Functions (8) and (11) determine the drive transitive process conditioned on angular velocity $\omega_{\mathrm{E}}(t)$ and engine momentum $M_{\mathrm{E}}(t)$ when drive operates in acceleration mode. Graphs for these functions are shown in Fig. 3a, b. The axis of ordinates demonstrates dimensionless values $\frac{\omega_{\mathrm{E}}(t)}{\omega_{\mathrm{E}}^{0}}$ and $\frac{M_{\mathrm{E}}(t)}{M_{\mathrm{L}}^{0}}$.

Analysis of the functions obtained allows the following conclusion: at the first stage ($0 \leq t \leq \tau$) of acceleration mode, the mechanical system is motionless (area $0a$ shown on the graph in Fig. 3a), while the engine momentum value, according to (4), increases in the range of zero to value of M_{L}^{0} (area $0a_1$ shown on the graph in Fig. 3b).

Given $t > \tau$ the value of momentum $M_{\mathrm{E}}(t)$ satisfies the condition of $M_{\mathrm{E}}(t) > M_{\mathrm{L}}^{0}$ and mechanical system sets in motion (area ab on the graph shown in Fig. 3a). In point b of the graph, angular velocity $\omega_{\mathrm{E}}(t)$ reaches steady-state value of $\omega_{\mathrm{E}}(t) = \omega_{\mathrm{E}}^{0}$. However, since engine momentum value still satisfies the condition of $M_{\mathrm{E}}(t) > M_{\mathrm{L}}^{0}$ (point b_1 on the graph in Fig. 3b), area bc (Fig. 3a) is characterized by further increase in the values of angular velocity. In area b_1c_1, the engine momentum value

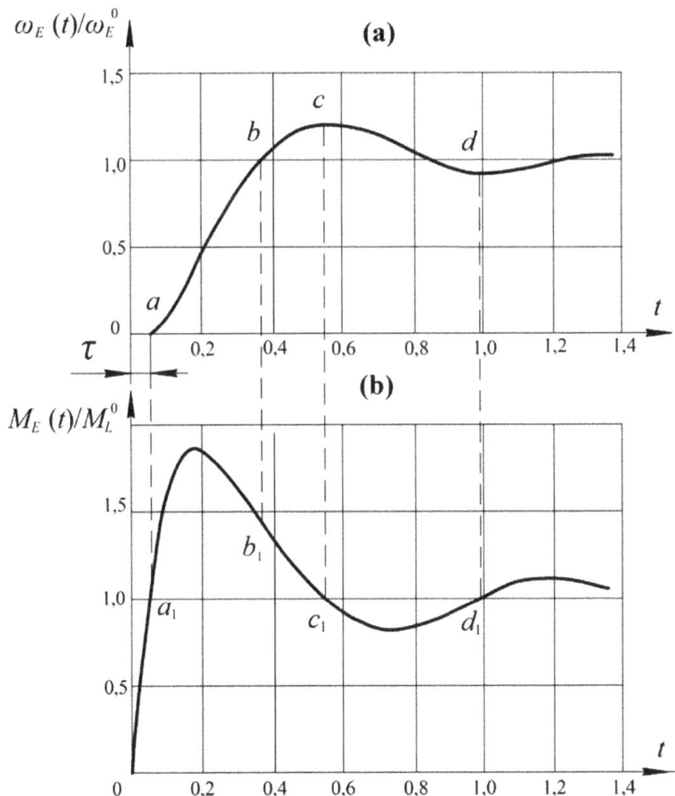

Fig. 3 Graph for transient process in drive acceleration mode **a** by angular velocity; **b** by engine momentum

decreases, and in point c_1, the equality of $M_E(t) = M_L^0$ is reached. Acceleration $\dot{\omega}_E$ becomes zero value and velocity ω_E reaches its maximum value (point c on the graph). In area c_1d_1 of the momentum graph (Fig. 3b), the condition of $M_E(t) < M_L^0$ holds, thus producing negative acceleration $\dot{\omega}_E$. In effect, area cd is characterized by decreasing values of velocity ω_E. In point d_1, the condition of equality $M_E(t) = M_L^0$ is reached again. Acceleration $\dot{\omega}_E$ becomes zero value, and angular velocity reaches its minimum value. Transient process in terms of angular velocity ω_E and engine momentum M_E is characterized with damped oscillations occurring in conditions close to ω_E^0 and M_L^0, which determine the electric drive set-steady motion.

3 Conclusion

Expressions (8) and (11) allow a conclusion that the higher the value of electromagnetic time constant T_e of the drive, the less intensive is the process of oscillations damping. Thus, adding armature circuit with an extra inductive coil, which serves to reduce the momentum peak values, results in increasing the drive transient process time, i.e., reduced performance.

The maximum increase in values of angular velocity or engine momentum by which they can exceed their set values is referred to as overregulation. This value can be used as an indirect parameter of the system dynamics.

Functions (8) and (11) produce in Fig. 4 a phase trajectory of $\omega_E = f(M_E)$, which is characteristic of the drive condition in acceleration mode, the latter determined by M_E, ω_E coordinates.

The mode of set-steady motion is determined with point A with coordinates M_L^0 and ω_E^0. This point is set as a point of intersecting between static parameters of the engine and the working element $M_E = M_L^0$.

At the first stage of acceleration mode ($0 \le t \le \tau$) given the mechanical system is completely motionless, the representative point S moves along the x-axis from the origin of coordinates to a point in which $M_E = M_L^0$. At the second stage of acceleration mode ($t > \tau$), the representative point moves along the curve asymptotically approaching point A. Phase trajectory provides graphic representation of the electric

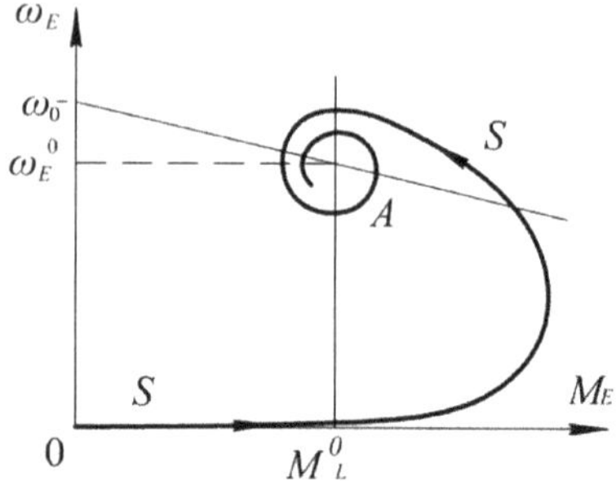

Fig. 4 Phase trajectory of angular velocity-to-engine momentum values relation in acceleration mode

drive dynamic properties. The parameter thus obtained cannot be used as a factor setting strict relation of angular velocity ω_E to momentum M_E as such strictness is characteristic of the drive static condition.

It should be also noted that the model discussed makes no allowance for the effects of internal losses caused by friction and possible production faults in the elements and components of the machinery drive mechanisms [9–12].

Application of rational and effective technologies allows increased accuracy and higher quality of the drive components, and such condition provides reduced vibration parameters during the operation of technological equipment and machinery.

References

1. Povetkin, V.: Design of dynamic model drive on ball mill given damping properties of its elements. Min. Inf. Anal. Bull. **5**, 184–192 (2018)
2. Kalinin, D.: Modeling non-linear oscillations in cylindrical spur gears for drives in aviation. Vestn. Samara Univ. Nat. Sci. Ser. **3**, 193–201 (2015)
3. Tarabarin, V.: Modeling dynamics of control drive electro-mechanical part with wave gear reducer. In: 2012 Studies in Machinery, Drive Systems and Machine Parts, Herald of the Bauman Moscow State Technical University, vol. 5, pp. 117–127 (2015)
4. Grudinin, V.: Dynamic analysis of angular oscillation damper. Proc. Irkutsk State Tech. Univ. **4**, 33–39 (2012)
5. Ershov, D., Lukyanenko, I.: Dynamic model of technological equipment drive with S DOFs. In: 2019 Proceedings of 14th International Conference on Electromechanics and Robotics “Zavalishin’s Readings”, Smart Innovation, Systems and Technologies, vol. 154, pp. 171–182 (2020)
6. Ershov, D., Zlotnikov, E.: Own fluctuations of technological systems. J. Phys. Conf. Ser. **1399**(22055), 1–6 (2019)
7. Panovko, G.: Resonant adjustment of vibrating machines with unbalance vibroexciter. Problems and solutions. In: 2020 Smart Innovation, Systems and Technologies Proceedings, vol. 154, pp. 51–62 (2020)
8. Fedorenko, I.: Dynamic properties of two-mass vibration technological machine. Bull. Altai State Agric. Univ. **137**, 179–183 (2016)
9. Maksarov, V., Olt, Y.: Dynamic stabilization of machining process based on local metastability in controlled robotic systems of CNC machines. J. Min. Inst. **226**, 446–451 (2017)
10. Alekseeva, L.: Determination of transient behavior characteristics in drive control of machines. Min. Inf. Anal. Bull. **4**, 18–25 (2016)
11. Vasilkov, D.: Dynamic system stability when machining with cutter. IOP Conf. Ser. Earth Environ. Sci. **194**(022045), 1–6 (2018)
12. Shvetsov, I.: Vibration stability criterion in assessing the dynamic quality and limiting capabilities of machine tools. IOP Conf. Ser. Mater. Sci. Eng. **656**(012049), 1–6 (2019)

System Representations of Dynamics of Mechanical Oscillatory Structures Based on Frequency Function and Damping Function

Sergey Eliseev, Andrey Eliseev, and Nikolai Kuznetsov

Abstract The issues of assessing the dynamic properties of structural formations were considered in a number of works; interesting results were obtained, which predetermined the interest in assessing the features of dynamic interactions, when structural formations are included in the structure of real systems interacting with the base or motion limiters. The oscillation of a system with several degrees of freedom is considered from the point of view of identifying structural formations and taking into account the features of their interactions. It is proposed to consider a mechanical oscillatory system with the allocation of an object in the form of a structural formation. The frequency function, the damping function, and the parametrizing function are considered as mathematical tools for taking into account the dynamic properties of structural formations. Made the transition from a task object, the dynamic properties of which are characterized by their own numbers in the form of a concentrated mass, to the task of assessment of interaction of elements of vibrating systems with several degrees of freedom with the object in the form of the structures dynamic and structural properties resulting in an estimated frequency response function, damping function, and parametrizing function.

1 Introduction

At present, the general theory of systems finds its applications not only in objects that have developed control systems based on information processing using computational tools; the properties of systems are also manifested in objects that are mechanical oscillatory systems. Such mechanical oscillatory systems represent a wide class of computational schemes that are used to solve various problems of

S. Eliseev · A. Eliseev (✉)
Irkutsk State Transport University, 15, Chernishevsky str., Irkutsk 664074, Russia
e-mail: eavsh@ya.ru

N. Kuznetsov
Irkutsk National Research Technical University, 83, Lermontov str., Irkutsk 664074, Russia
e-mail: knik@istu.edu

A. Ronzhin and V. Shishlakov (eds.), *Electromechanics and Robotics*, Smart Innovation, Systems and Technologies 232, https://doi.org/10.1007/978-981-16-2814-6_29

dynamics of machines, mechanisms, robotic devices, and mechatronics. The development of mathematical apparatus or methods of system analysis is an urgent task, since mechanical oscillatory systems have a wide range of dynamic properties, which are largely predetermined by the structure and specifics of the connections between the constituent elements in the form of structural formations.

Currently, the use of mechanical oscillatory systems in engineering practice as calculation schemes for evaluating the dynamic properties of technical objects operating under intense vibration loads is widespread [1–4]. Among the most well-known approaches to assessing the dynamic properties of mechanical oscillatory systems are methods based on energy relations [5, 6]. In turn, approaches to the use of extreme properties of the ratio of potential energy to kinetic energy have been developed in the concept of the frequency energy function of the argument of the coefficient of connectivity of the forms of motion of elements of mechanical oscillatory systems [7]. In this case, the associated elements are interpreted as structural formations, and the frequency function, which reflects in its form its own and partial frequencies, is a dynamic characteristic of the object under consideration in the form of a structural formation.

The use of connectivity coefficients of motion forms determines the possibility of tincture and correction of vibration fields of transport and technological machines. The physical interpretation of the coefficient of connectivity of motion forms is the lever connection between the elements of mechanical oscillatory systems [6].

However, the methods of evaluation of dynamic characteristics of structures and mechanical oscillatory systems taking into account forces of viscous friction based on the frequency response functions and damping require detailed consideration, because, in particular, that for systems with viscous friction, the character of the movement varies considerably and depends on the ratio of system parameters, including characteristics of forces of viscous friction [8, 9].

The proposed work is devoted to the development of methods for evaluating the features of structural formations included in mechanical oscillatory systems with viscous friction, based on the frequency function and the damping function.

2 Some General Provisions. Statement of the Problem

An elastic-dissipative mechanical oscillatory system with two degrees of freedom with built-in motion transformation devices L_1, L_0, L_2 is considered. The schematic diagram of the mechanical system is shown in Fig. 1.

The generalized coordinates y_1, y_2 denote the position of mass-inertial elements m_1, m_2 relative to the position of static equilibrium. The kinetic energy, potential energy, and scattering function have the form:

$$T = \frac{1}{2}m_1\dot{y}_1^2 + \frac{1}{2}m_2\dot{y}_2^2 + \frac{1}{2}L_1\dot{y}_1^2 + \frac{1}{2}L_0(\dot{y}_2 - \dot{y}_1)^2 + \frac{1}{2}L_2\dot{y}_2^2, \quad (1)$$

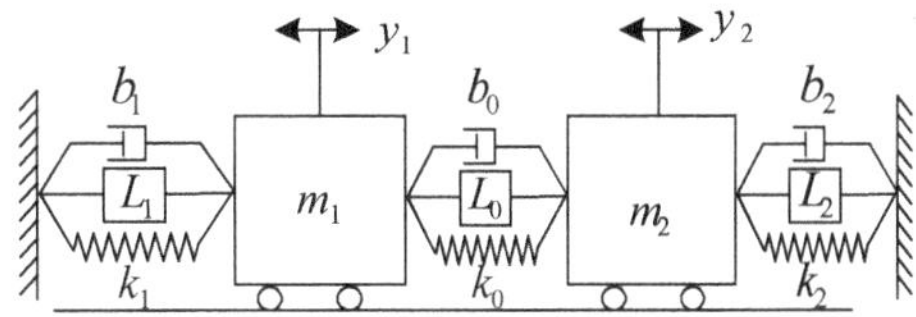

Fig. 1 Mechanical oscillatory system including devices for converting motion L_n, elastic k_n, and dissipative elements b_n, $n = 0, 1, 2$

$$\Pi = \frac{1}{2}k_1 y_1^2 + \frac{1}{2}k_2 y_2^2 + \frac{1}{2}k_0(y_2 - y_1)^2, \tag{2}$$

$$F = \frac{1}{2}b_1 \dot{y}_1^2 + \frac{1}{2}b_0(\dot{y}_2 - \dot{y}_1)^2 + \frac{1}{2}b_2 \dot{y}_2^2. \tag{3}$$

The system of Lagrange equations of the second kind after substituting (1)–(3) is reduced to differential equations:

$$\begin{cases} (m_1 + L_1 + L_0)\ddot{y}_1 - L_0\ddot{y}_2 + (b_0 + b_1)\dot{y}_1 - b_0\dot{y}_2 + (k_0 + k_1)y_1 - k_0 y_2 = 0; \\ (m_2 + L_2 + L_0)\ddot{y}_2 - L_0\ddot{y}_1 + (b_0 + b_2)\dot{y}_2 - b_0\dot{y}_1 + (k_0 + k_2)y_2 - k_0 y_1 = 0. \end{cases} \tag{4}$$

It is assumed that the system performs free movements specified by the initial conditions and has certain dynamic characteristics such as natural and partial frequencies. In general, system (4) has a solution that depends on the initial data.

Along with the solution of the system, the motion $y_1(t)$, $y_2(t)$ of mass-inertia elements is considered in the form of:

$$\vec{y} = \vec{Y}\,\mathrm{e}^{pt},$$

where $\vec{y} = \begin{bmatrix} y_1(t) \\ y_2(t) \end{bmatrix}$ is a vector-function, $\vec{Y} = \begin{bmatrix} Y_1 \\ Y_2 \end{bmatrix}$ is a numerical vector, $p = \sigma + \mathrm{j}\omega$ is a complex constant, and is a time variable. In this case, we can assume that the forms of motion of individual coordinates of the system are connected by a lever relation $Y_2 = \alpha Y_1$, where α is the coefficient of connectivity of the forms.

On the basis of the parameters of the particular case of the system, using a formal procedure, the so-called frequency function of the argument of the connectivity coefficient can be constructed, which reflects such parameters as natural oscillation frequencies and partial frequencies in the features of its form [7].

The task is to develop a method for constructing frequency functions and damping functions that reflect the dynamic features of a mechanical oscillatory system in extreme values, for evaluating structural formations considered as special links with lever properties.

3 Frequency Function for a Mechanical Oscillating System Without Friction

A special case of a mechanical oscillatory system with two degrees of freedom is considered (Fig. 1). The system is formed by two mass-inertia elements m_1, m_2, connected by springs with stiffness k_1, k_0, k_2. Dampers and devices for converting motion are not taken into account. The schematic diagram of the system is shown in Fig. 2.

The frequency function for the system in Fig. 2 takes the form of the following fractional-rational expression, depending on the coefficient of the form [7]:

$$\omega^2(\alpha) = \frac{(k_0 + k_2)\alpha^2 - 2\alpha k_0 + k_0 + k_1}{m_1 + m_2\alpha^2}.$$

Extreme characteristics and values of the frequency function display the dynamic features of a particular mechanical oscillatory system. Along with the extreme values that coincide with the natural frequencies of the system, the limit values and the values of the frequency function at zero that coincide with the partial frequencies are of interest.

Table 1 shows the frequency functions of the considered mechanical oscillatory system (Fig. 2) and its particular variants. Each row in Table 1 includes a set of system parameters, a circuit diagram, a frequency function, and the corresponding partial frequencies.

The presented variants of the forms of frequency functions (Table 1) display the features of dynamic connections between the elements of the considered mechanical oscillatory systems. On the other hand, the frequency function can be interpreted as a dynamic characteristic of some structural formation that is part of a mechanical oscillatory system.

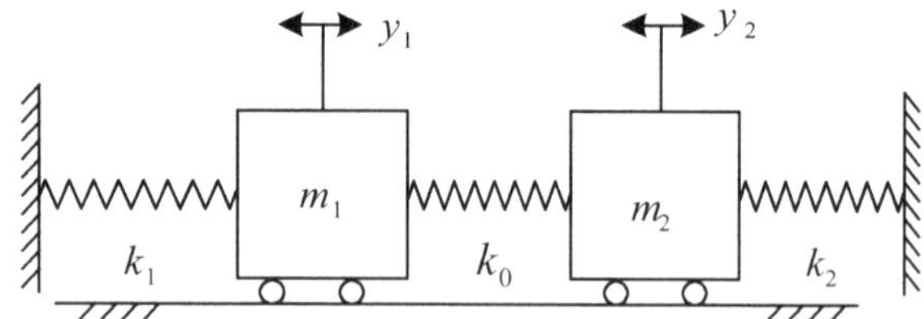

Fig. 2 Mechanical oscillatory system with two degrees of freedom (a special case of the system in Fig. 1) without taking into account friction forces and devices for converting movements

Table 1 Frequency function options

Num.	System Parameters	Schematic diagram	Frequency function	Partial frequencies
1	k_0, k_1, k_2, m_1, m_2		$\omega^2(\alpha) = \frac{(k_0 + k_2)\alpha^2 - 2\alpha k_0 + k_0 + k_1}{m_1 + m_2\alpha^2}$	$n_1^2 = \frac{k_1 + k_0}{m_1}, n_2^2 = \frac{k_2 + k_0}{m_2}$
2	$k_1 = 0, k_2 = 0$		$\omega^2(\alpha) = \frac{k_0(\alpha - 1)^2}{m_1 + m_2\alpha^2}$	$n_1^2 = \frac{k_0}{m_1}, n_2^2 = \frac{k_0}{m_2}$
3	$k_1 = k_2 = k_0 = k$, $m_1 = m_2 = m$		$\omega^2(\alpha) = \frac{2k}{m}\frac{(\alpha^2 - \alpha + 1)}{(1 + \alpha^2)}$	$n_1^2 = \frac{2k}{m}, n_2^2 = \frac{2k}{m}$
4	$m_1 = m_2 = m$, $k_2 = 0, k_1 = k_0 = k$		$\omega^2(\alpha) = \frac{k}{m}\frac{\alpha^2 - 2\alpha + 2}{1 + \alpha^2}$	$n_1^2 = \frac{2k}{m}, n_2^2 = \frac{k}{m}$.
5	$k_0 = 0$	k_1 m_1 m_2 k_2	$\omega^2(\alpha) = \frac{k_2\alpha^2 + k_1}{m_1 + m_2\alpha^2}$	$n_1^2 = \frac{k_1}{m_1}, n_2^2 = \frac{k_2}{m_2}$

4 Damping Function for a Mechanical System with Aperiodic Motion

We consider a mechanical system, which is a special version of the system in Fig. 1, formed by mass-inertia elements m_1, m_2, and dampers b_0, b_1, b_2. The schematic diagram of the system is shown in Fig. 3.

For the system in Fig. 3, the damping function reflecting the features of Eigen forms of motion in extreme values can be constructed on the basis of a formal procedure and presented as a fractional-rational expression depending on the coefficient of the form of connectivity of forms in the form [8]:

$$\sigma(\alpha) = -\frac{(b_0 + b_2)\alpha^2 - 2\alpha b_0 + b_0 + b_1}{m_1 + m_2\alpha^2}.$$

The specific form of the damping function is determined by the characteristics of the elements of the considered mechanical system, which performs free movements under the influence of initial conditions. The features of the damping function include extreme values and values taken when the shape coefficient tends to zero and infinity.

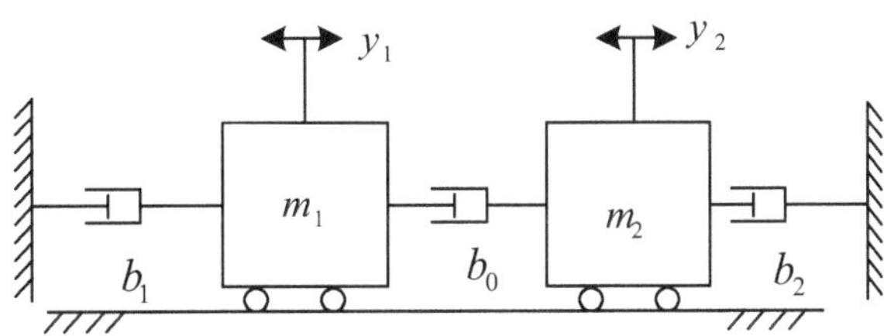

Fig. 3 Mechanical system taking into account the forces of viscous friction

Table 2 Options for damping functions

Num.	System Parameters	Schematic diagram	Frequency function	Partial frequencies
1	b_0, b_1, b_2, m_1, m_2	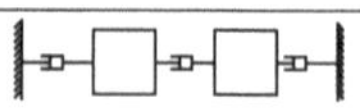	$\sigma(\alpha) = -\frac{(b_0+b_2)\alpha^2 - 2\alpha b_0 + b_0 + b_1}{m_1 + m_2\alpha^2}$	$\sigma_3 = -\frac{b_1+b_0}{m_1}, \sigma_4 = -\frac{b_2+b_0}{m_2}$
2	$b_1 = 0, b_2 = 0$	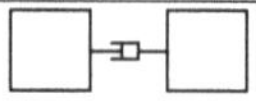	$\sigma(\alpha) = -\frac{b_0(\alpha-1)^2}{m_1 + m_2\alpha^2}$	$\sigma_3 = -\frac{b_0}{m_1}, \sigma_4 = -\frac{b_0}{m_2}$
3	$b_1 = b_2 = b_0 = b$, $m_1 = m_2 = m$	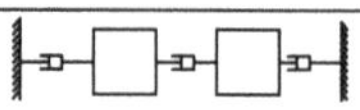	$\sigma(\alpha) = -\frac{2b}{m}\frac{(\alpha^2-\alpha+1)}{(1+\alpha^2)}$	$\sigma_3 = -\frac{2b}{m}$, $\sigma_4 = -\frac{2b}{m}$
4	$b_2 = 0$, $m_1 = m_2 = m$	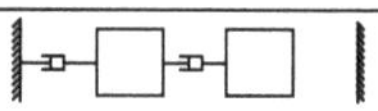	$\sigma(\alpha) = -\frac{b}{m}\frac{\alpha^2-2\alpha+2}{1+\alpha^2}$	$\sigma_3 = -\frac{2b}{m}, \sigma_4 = -\frac{b}{m}$.
5	$b_0 = 0$		$\sigma(\alpha) = -\frac{b_2\alpha^2 + b_1}{m_1 + m_2\alpha^2}$	$\sigma_3 = -\frac{b_1}{m_1}$, $\sigma_4 = -\frac{b_2}{m_2}$

Table 2 presents the characteristic variants of the damping functions of mechanical systems for different values of mass-inertia and dissipative parameters, reflecting the structural features of the corresponding systems.

Thus, for a mechanical system capable of performing periodic movements, the analytical characteristics of the damping functions, which can include, for example, the parity properties, reflect the dynamic features of the structural formations included in the mechanical oscillatory system.

5 Frequency Function and Damping Function for a Mechanical Oscillatory System Taking into Account Viscous Friction

A special version of the mechanical oscillatory system is considered (Fig. 1) with elastic damping elements $b_1 = b_2 = b_0 = b$, $k_1 = k_2 = k_0 = k$, which have symmetry properties. The schematic diagram is shown in Fig. 4.

The frequency function and the damping function for the considered elastic-dissipative mechanical system can be determined taking into account the conditions that characterize the "smallness" of the viscous friction forces [9]:

$$B_\alpha^2 < 4A_\alpha C_\alpha,$$

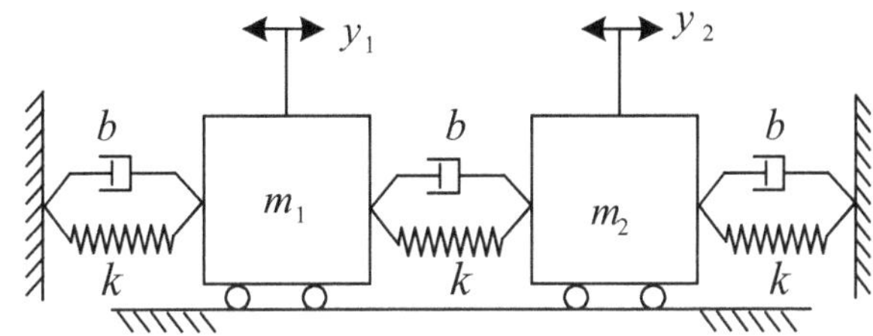

Fig. 4 Mechanical oscillatory system taking into account the forces of viscous friction

where $A_\alpha = m_1 + m_2\alpha^2$, $B_\alpha = 2b(\alpha^2 - \alpha + 1)$, $C_\alpha = 2k(\alpha^2 - \alpha + 1)$ are auxiliary functions of the argument of the connectivity coefficient of motion forms α.

For "small" friction forces, the frequency function and the damping function have the form:

$$\begin{cases} \omega^2(\alpha) = \dfrac{2k(\alpha^2 - \alpha + 1)}{m_1 + m_2\alpha^2} - \left(\dfrac{b(\alpha^2 - \alpha + 1)}{m_1 + m_2\alpha^2}\right)^2; \\ \sigma(\alpha) = \dfrac{b(\alpha^2 - \alpha + 1)}{m_1 + m_2\alpha^2}. \end{cases}$$

For "large" friction forces, the frequency function is zero, and the damping function has two components:

$$\begin{cases} \omega^2 = 0; \\ \sigma_1(\alpha) = -\dfrac{b(\alpha^2 - \alpha + 1)}{m_1 + m_2\alpha^2} - \sqrt{\left(\dfrac{b(\alpha^2 - \alpha + 1)}{m_1 + m_2\alpha^2}\right)^2 - \dfrac{2k(\alpha^2 - \alpha + 1)}{m_1 + m_2\alpha^2}}; \\ \sigma_2(\alpha) = -\dfrac{b(\alpha^2 - \alpha + 1)}{m_1 + m_2\alpha^2} + \sqrt{\left(\dfrac{b(\alpha^2 - \alpha + 1)}{m_1 + m_2\alpha^2}\right)^2 - \dfrac{2k(\alpha^2 - \alpha + 1)}{m_1 + m_2\alpha^2}}. \end{cases}$$

The conditions of "small" forces in the construction of frequency functions and damping functions for a family of mechanical oscillatory systems can be represented using a generalized system parameter and a parametrizing function in the form of an inequality:

$$\gamma_0 < M(\alpha),$$

where $\gamma_0 = \frac{b^2}{4k}$ is the generalized viscoelastic parameter, $M(\alpha) = \frac{1}{2} \cdot \frac{m_1 + m_2\alpha^2}{\alpha^2 - \alpha + 1}$ is the generalized mass-inertia coefficient depending on the shape coefficient α.

The graph of the function M_α defines for each fixed γ_0 set of values in which the "smallness" of the forces of friction, which allows you to build a variant of frequency function and damping.

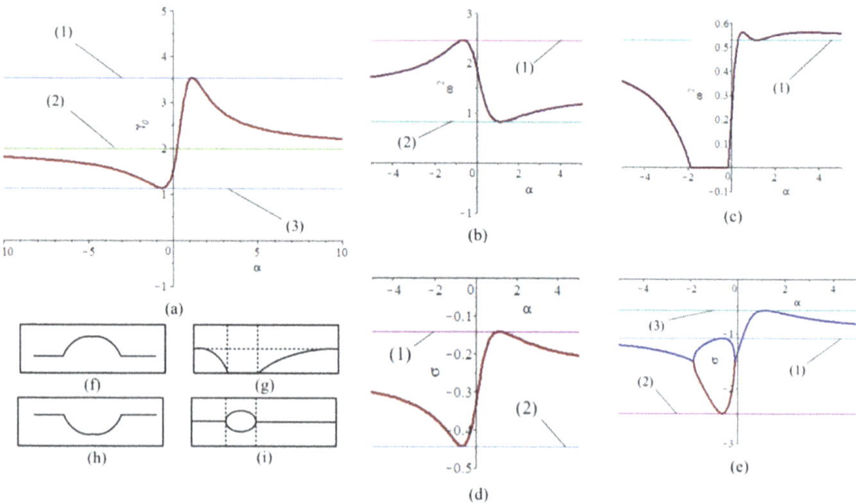

Fig. 5 Features of the formation of the frequency function and the damping function: **a** graph of the parametrizing function of a family of mechanical systems; **b**, **c** frequency functions; **d**, **e** damping functions; **f**, **g** pictograms of the corresponding frequency functions (**b**), (**c**); **h**, **i** pictograms of the corresponding damping functions (**d**), (**e**)

Figure 5 shows the parametrizing function (a), variants of the frequency function and damping function (b)–(i), pictograms (j)–(q), reflecting the essential features of the graphs (b)–(i).

The achieved extreme values of the frequency function and the damping function, constructed taking into account the conditions of "smallness" of the friction forces, are associated with the dynamic characteristics of the mechanical oscillatory. In turn, the variants of the frequency function and damping functions that reflect the essential features of the dynamic characteristics of mechanical oscillatory systems are determined on the basis of the parametrizing function.

6 Parametrizing Functions of Mechanical Oscillatory Systems with a Device for Converting Movements

The set of boundary parameters separating the regions of "small" and "large" friction forces for a mechanical oscillatory system (Fig. 1) is determined by the equation:

$$B_\alpha^2 = 4A_{L,\alpha}C_\alpha,$$

where $A_\alpha = m_1 + m_2\alpha^2$, $L_\alpha = (L_0 + L_2)\alpha^2 - 2\alpha L_0 + L_0 + L_1$, $B_\alpha = (b_0 + b_2)\alpha^2 - 2\alpha b_0 + b_0 + b_1$, $C_\alpha = (k_0 + k_2)\alpha^2 - 2\alpha k_0 + k_0 + k_1$, $A_{L,\alpha} = A_\alpha + L_\alpha$ are auxiliary functions of the argument.

The frequency function and the damping function, the forms of which are determined by the parametrizing function, can determine the dynamic features of structural formations of mechanical oscillatory systems.

If the conditions $B_\alpha^2 < 4A_{L,\alpha}C_\alpha$ for the "smallness" of the viscous friction forces are met, the frequency function $\omega^2(\alpha)$ and the damping function $\sigma(\alpha)$ can be represented as fractional-rational expressions that depend on the shape parameter α:

$$\begin{cases} \omega^2(\alpha) = \dfrac{(k_0 + k_2)\alpha^2 - 2\alpha k_0 + k_0 + k_1}{m_1 + m_2\alpha^2 + (L_0 + L_2)\alpha^2 - 2\alpha L_0 + L_0 + L} \\ \qquad - \left(\dfrac{1}{2}\dfrac{(b_0 + b_2)\alpha^2 - 2\alpha b_0 + b_0 + b_1}{m_1 + m_2\alpha^2 + (L_0 + L_2)\alpha^2 - 2\alpha L_0 + L_0 + L}\right)^2. \\ \sigma(\alpha) = -\dfrac{1}{2} \cdot \dfrac{(b_0 + b_2)\alpha^2 - 2\alpha b_0 + b_0 + b_1}{m_1 + m_2\alpha^2 + (L_0 + L_2)\alpha^2 - 2\alpha L_0 + L_0 + L} \end{cases}$$

For "large" viscous friction forces $B_\alpha^2 > 4A_{L,\alpha}C_\alpha$, the frequency function is identical to zero, and the damping function is two-digit:

$$\begin{cases} \omega^2 = 0; \\ \sigma_1(\alpha) = -\dfrac{1}{2} \cdot \dfrac{(b_0 + b_2)\alpha^2 - 2\alpha b_0 + b_0 + b_1}{m_1 + m_2\alpha^2 + (L_0 + L_2)\alpha^2 - 2\alpha L_0 + L_0 + L_1} \\ \quad - \sqrt{\left(\dfrac{1}{2} \cdot \dfrac{(b_0 + b_2)\alpha^2 - 2\alpha b_0 + b_0 + b_1}{m_1 + m_2\alpha^2 + (L_0 + L_2)\alpha^2 - 2\alpha L_0 + L_0 + L_1}\right)^2 - \dfrac{(k_0 + k_2)\alpha^2 - 2\alpha k_0 + k_0 + k_1}{m_1 + m_2\alpha^2 + (L_0 + L_2)\alpha^2 - 2\alpha L_0 + L_0 + L_1}}; \\ \sigma_2(\alpha) = -\dfrac{1}{2} \cdot \dfrac{(b_0 + b_2)\alpha^2 - 2\alpha b_0 + b_0 + b_1}{m_1 + m_2\alpha^2 + (L_0 + L_2)\alpha^2 - 2\alpha L_0 + L_0 + L_1} \\ \quad + \sqrt{\left(\dfrac{1}{2} \cdot \dfrac{(b_0 + b_2)\alpha^2 - 2\alpha b_0 + b_0 + b_1}{m_1 + m_2\alpha^2 + (L_0 + L_2)\alpha^2 - 2\alpha L_0 + L_0 + L_1}\right)^2 - \dfrac{(k_0 + k_2)\alpha^2 - 2\alpha k_0 + k_0 + k_1}{m_1 + m_2\alpha^2 + (L_0 + L_2)\alpha^2 - 2\alpha L_0 + L_0 + L_1}}. \end{cases}$$

At the same time, for mechanical oscillatory systems considered within the family (Fig. 1), the variety of possible forms of frequency functions and damping functions is determined by the features of parametrizing functions that can be used to evaluate the structure of a mechanical oscillatory system or classify structural formations. Figures 6, 7, 8, and 9 present variants of mechanical oscillatory systems. In particular, Fig. 6 shows a mechanical oscillatory system formed by unrelated mass-inertial elements. Figure 7 shows a diagram of a structural formation formed by two

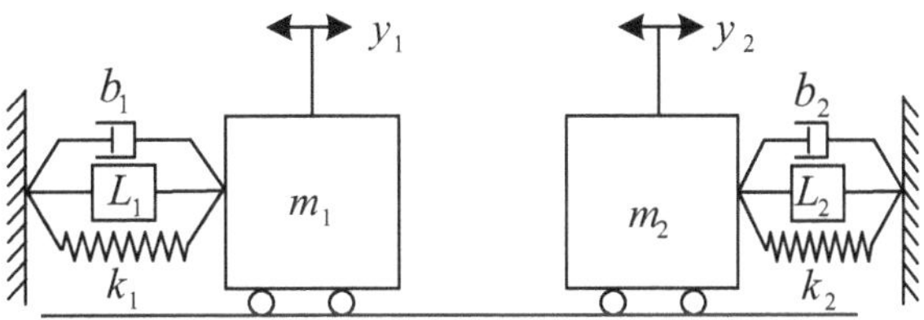

Fig. 6 "Unbound" system

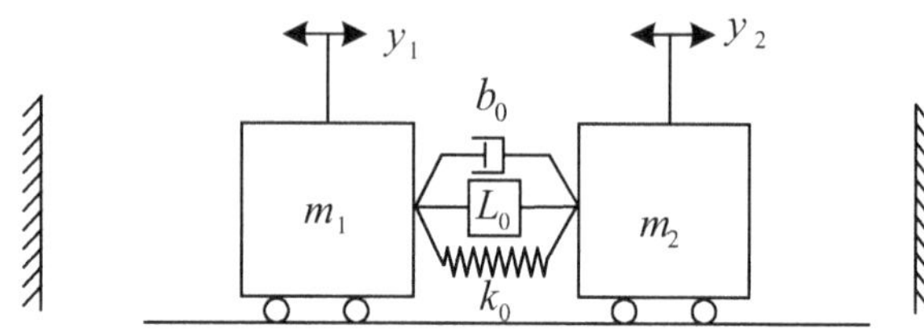

Fig. 7 Dyad

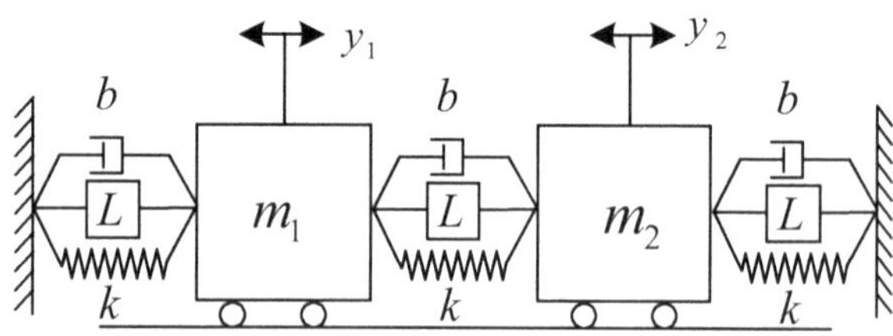

Fig. 8 "Symmetric" system

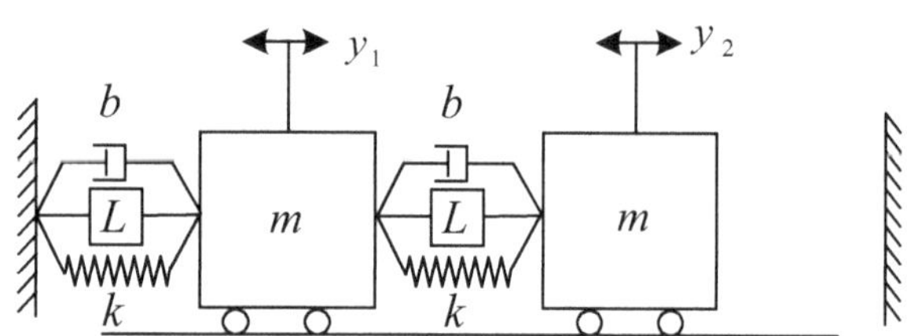

Fig. 9 "Asymmetric" system

mass-inertia elements "connected" to each other by means of an elastic element, a damper and a device for converting motion; mass-inertia elements of the dyad are not connected to any of the support surfaces. In Fig. 8 presents a system, the so-called "symmetric" mechanical oscillatory system, in which elastic elements, dampers and devices for converting motion, connecting mass-inertia elements with each other and with the support surfaces, have the same characteristics.

At the same time, in Figs. 10, 11, 12, and 13, the parametrizing functions of the corresponding mechanical oscillatory systems are presented in Figs. 6, 7, 8, and 9. It can be assumed that these parametrizing functions reflect the structural features of the corresponding mechanical oscillatory systems. Thus, the parametrizing function in Fig. 10 has the property of parity, the parametrizing function in Fig. 11 corresponding to the dyad in Fig. 7 has a discontinuity of the 2nd kind, the parametrizing function in Fig. 13 takes negative values at some points, etc.

Thus, the analytical features of parametrizing functions that define the forms of frequency functions and damping functions of families of mechanical oscillatory systems reflect the features of the considered structural formations.

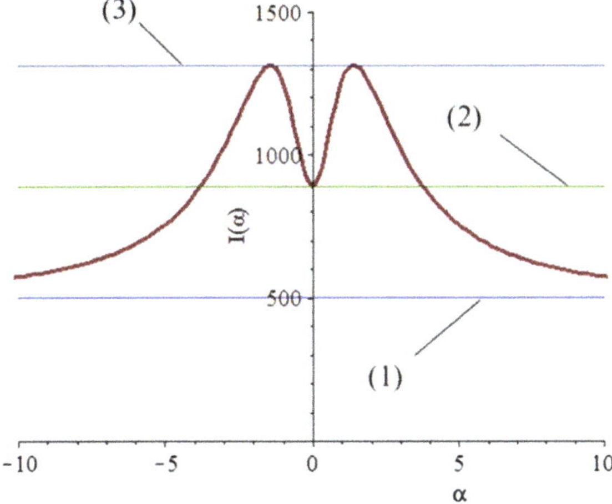

Fig. 10 Parameterizing function for the system in Fig. 6

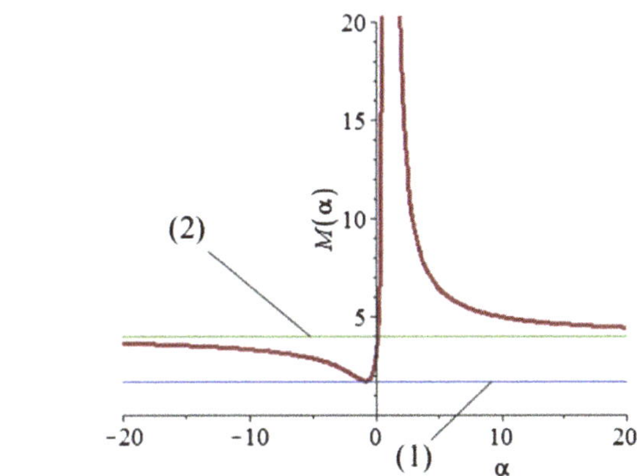

Fig. 11 Parameterizing function for the system in Fig. 7

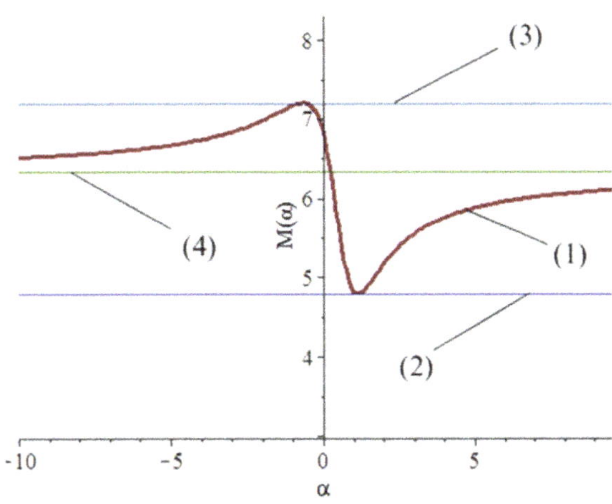

Fig. 12 Parameterizing function for the system in Fig. 8

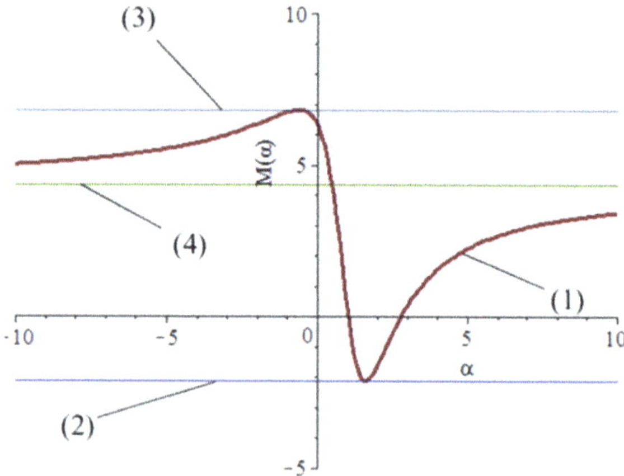

Fig. 13 Parameterizing function for the system in Fig. 9

7 Conclusion

A number of conclusions can be drawn as the results of studies of the features of frequency functions and damping functions for mechanical elastic-dissipative systems formed by two mass-inertia elements containing devices for converting movements.

A method for constructing the frequency function and the damping function of the argument of the connectivity coefficient of motion forms is developed. Based on the developed method shows that shape of graphs of frequency functions and damping reflect at their extreme values of natural frequency and dissipative factors of the motions of mechanical oscillatory systems taking into account forces of viscous friction.

It is shown that the frequency function and damping functions of the argument of the connectivity coefficient of motion forms can be represented in two forms, for "small" and "large" viscous friction forces, respectively. In particular, for "small" viscous friction forces, the frequency function takes positive values, and the damping function has one negative component; while for "large" viscous friction forces, the frequency function takes zero values, and the damping function has two negative components.

It is shown that to determine the variants of the forms of frequency functions and damping functions corresponding to the family of mechanical systems of a given structure, a parameterizing function can be used that compares the parameters of the system of the forms of frequency functions and damping functions that have a number of topological features that represent a criterion for classifying structural formations.

Made the transition from objectives to the object in the form of simple entities, or lumped masses, the dynamic properties of which are characterized by their own numbers, to the task of assessment of interaction of elements of vibrating systems with several degrees of freedom, interpreted as a system with the objects in question in the form of structures representing the links of a new type, dynamic properties which are valued frequency function and damping function, and structural features with parametrizing functions.

References

1. De Silva, C.W.: Vibration: Fundamentals and Practice. CRC Press (2006)
2. Eliseev, S.V., Eliseev, A.V.: Theory of Oscillations. Structural Mathematical Modeling in Problems of Dynamics of Technical Objects. Series: Studies in Systems, Decision and Control, vol. 252. Springer International Publishing, Cham (2020)
3. Eliseev, A.V., Kuznetsov, N.K., Moskovskikh, A.O.: Dynamics of Machines. System Views, Block Diagrams, and Element Relationships. Innovative Mechanical Engineering, Moscow (2019)
4. Eliseev, S.V., Artyunin, A.I.: Applied Theory of Oscillations in Problems of Dynamics of Linear Mechanical Systems. Nauka, Novosibirsk (2016)
5. Strett, J.V.: Theory of Sound. GITTLE, Moscow (1955)
6. Eliseev, S.V.: Applied System Analysis and Structural Mathematical Modeling (Dynamics of Transport and Technological Machines: Connectivity of Movements, Vibration Interactions, Lever Connections). Irkutsk State Transport University, Irkutsk (2018)
7. Khomenko, A.P., Eliseev, S.V.: Development of the energy method for determining the frequencies of free vibrations of mechanical systems. Modern technologies. System analysis. Modeling **1**(49), 8–19 (2016)
8. Eliseev, A.V.: Connectivity of movements in systems with energy dissipation: system approaches. Inf. Math. Technol. Sci. Manag. **3**(19), 4–56 (2020)
9. Eliseev, A.V.: Frequency function and damping function in the evaluation of dynamic processes in mechanical oscillatory systems with symmetry. Adv. Eng. Res. **20**(4), 360–369 (2020)

Mathematical Model of the Electrothermal Process of Heating the Formation by the Ultra-Frequency Electromagnetic Waves

Albert Khalikov, Ilgiz Yangirov, and Ruzil Safiullin

Abstract The article discusses a mathematical model of the electrothermal process of heating the formation with a microwave electromagnetic wave, which is caused by electrical devices capable of operating under high pressure. The issues of using microwave dielectric heating to reduce the viscosity of oil-bearing deposits, which will increase the efficiency of their production, are considered. The article presents an in-depth analysis of various existing and promising methods of ensuring the production of high-viscosity oil and shows the advantages of the thermal method, and on the basis of calculations using mathematical and computer models—the prospects for using microwave radiation. Models of the electromagnetic field are shown that provide uniform heating of the sediment to a sufficient depth. The results of the work create and substantiate the basis for the development of a new technology for thermal increase in the efficiency of oil production, primarily in combined oil fields and northern regions, which are characterized by high oil viscosity.

1 Introduction

Currently, most of the technological processes for the production of products are associated with heating materials. Microwave heating technologies have begun to be actively used not only in domestic conditions, but also in many industries, as effective, multiply accelerating production processes [1–8].

When servicing a gas and oil pipeline system, one of the important tasks remains safe operation and uninterrupted pumping of product raw materials through product pipelines that provide the required rheological properties of viscous media. In works [8–12], it was indicated that the most advanced and promising methods used for this are electrophysical methods of thermal impact on gas and oil pipeline systems. Along

A. Khalikov (✉) · I. Yangirov
The Faculty of Avionics, Energy Engineering and Infocomm Technology, Ufa State Aviation Technical University, K. Marx St., 12, 450000 Ufa, Russia

R. Safiullin
Bashkir State University, St. Zaki Validi, 32, Ufa, Rep. Bashkortostan, Russia

A. Ronzhin and V. Shishlakov (eds.), *Electromechanics and Robotics*, Smart Innovation, Systems and Technologies 232, https://doi.org/10.1007/978-981-16-2814-6_30

with other systems, an electrothermal method for solving this problem was proposed in the works, a mathematical model of a portable spiral inductor was developed and described, which was successfully tested in the field.

The issues of depleting energy resources are relevant along with the problems of development, production, and processing of depleted formations of high-viscosity oil and bitumen (HVOB). The scientific novelty of this work is due to the fact that the issues of depleted energy resources are relevant along with the problems of development, production, and processing of depleted layers of high-viscosity oil and bitumen.

One of the ways to solve this problem is considered in works [6, 7]. They propose safe microwave heating of the bottom of an oil well to enhance the production of hydrocarbons from the formation using a spiral microwave heater (SMH). In the course of the work, a spiral antenna operating at a frequency of 0.8 GHz was designed. The sector of the simultaneous viewing angles of the antenna is not less than $\pm 65°$, and the standing wave ratio does not exceed 3.5 in the entire frequency range [12–16]. The variety of physical properties of the materials that makes up the components of the HVOB mixture requires the development of nonlinear heat transfer problems associated with phase transformations in the reservoir volume [8].

The aim of the work is to evaluate the effectiveness of microwave exposure as a method of intensifying HVOB production in vertical and horizontal methods of drilling wells.

A highly viscous mixture of oil is composed of many chemical components of hydrocarbon chemical elements, waste rock, paraffin, water, and oil. Suppose that the mixture contains only waste rock (clay), water, oil, the process of uniform heating of the mixture proceeds with the formation of steam, heating the clay, and decreasing the oil viscosity.

The aim of the work is to evaluate the effectiveness of microwave exposure as a method of intensifying oil production in vertical and horizontal methods of drilling wells.

2 Model Description

The electromagnetic characteristics of a mixture of HVOB components will be described by the coefficients of the complex dielectric and magnetic permeabilities, for which:

$$\varepsilon_{\mathrm{mix}} = \varepsilon_0(\varepsilon_{\mathrm{R}} - \mathrm{j}\varepsilon_{\mathrm{Im}}) = \varepsilon_0\varepsilon_{\mathrm{a}}, \tag{1}$$

$$\mu_{\mathrm{mix}} = \mu_0(\mu_{\mathrm{R}} - \mathrm{j}\mu_{\mathrm{Im}}) = \mu_0\mu_{\mathrm{a}}. \tag{2}$$

The main factor in converting the energy of an electromagnetic wave into heat is the reactive component of the dielectric constant $\varepsilon_{\mathrm{Im}} = tg\delta^{\mathrm{E}} \cdot \varepsilon_{\mathrm{R}}$ and the dielectric loss factor of the dielectric $tg\delta^{\mathrm{E}} = \varepsilon_{\mathrm{Im}}/\varepsilon_{\mathrm{R}}$, where $\varepsilon_{\mathrm{Im}} = \sum_{i=1}^{3} c_i \varepsilon_{ia} tg\delta_i^E f_i |E_i|^2 / f_{\mathrm{mix}} E_{\mathrm{mix}}^2$ is the complex dielectric mixture constant; other coefficients are presented in works [2, 6, 7].

Let us write down the basic differential equations of Maxwell's electrodynamics to describe electromagnetic waves and Fourier thermal conductivity:

$$\begin{cases} \mathrm{rot}\overrightarrow{H} = \partial \overrightarrow{D}/\partial t + \sigma \overrightarrow{E} + \overrightarrow{j}_{\mathrm{c}}; \\ \mathrm{rot}\overrightarrow{E} = -\partial \overrightarrow{B}/\partial t; \\ c\rho \partial T/\partial t = \lambda \Delta T + q(\overrightarrow{r}, t); \end{cases} \tag{3}$$

where c, ρ, λ—coefficients of heat capacity, density, and thermal conductivity; $q(r;t) = \pi \sum_{i=1}^{3} c_i \varepsilon_{ia} tg\delta_i^E f_i |E_i|^2 / f_{\mathrm{mix}} E_{\mathrm{mix}}^2 = -\mathrm{div}\overrightarrow{\Pi}(r;t)$ is the specific absorbed power of the mixture during microwave heating; $\overrightarrow{\Pi}(r;t)$—Poynting–Umov vector; $\vec{D} = \varepsilon\varepsilon_0 \vec{E}$—electric field strength and induction; $\vec{B} = \mu\mu_0 \vec{H}$—induction and magnetic field strength; σ, t, j_{c}—surface charge density, time, and conduction currents.

From formula (3) follow the characteristics of the electromagnetic field, such as electric and magnetic energy, described by Maxwell:

$$W_{\mathrm{E}} = \int_V \frac{(\overrightarrow{E}, \overrightarrow{D})}{2} \mathrm{d}V, \quad W_{\mathrm{M}} = \int_V \frac{(\overrightarrow{H}, \overrightarrow{B})}{2} \mathrm{d}V, \tag{4}$$

where V is where the volume of space in which the electromagnetic wave propagates, and the integrands characterize the energy densities of the electric and magnetic fields.

After simplifying formulas (3) and (4), taking into account formulas (1) and (2) for electromagnetic waves, we obtain the Helmholtz differential equations (5), which are the basis of microwave emitters [2]:

$$\begin{aligned} \nabla^2 E(\overrightarrow{r}) &= \omega^2 \cdot \mu_{\mathrm{mix}} \cdot \varepsilon_{\mathrm{mix}} \cdot \overrightarrow{E}(\overrightarrow{r}), \\ \nabla^2 H(\overrightarrow{r}) &= \omega^2 \cdot \mu_{\mathrm{mix}} \cdot \varepsilon_{\mathrm{mix}} \cdot \overrightarrow{H}(\overrightarrow{r}). \end{aligned} \tag{5}$$

The energy of electromagnetic waves emitted by the microwave heater is determined by Maxwell's equations (3) from which the formula for the time-average power of emissions or energy transfer by an electromagnetic wave according to the Umov–Poynting theorem in the frequency range follows:

$$P_{\mathrm{EMIS}} = \frac{1}{2} \int_S \left[\overrightarrow{E}, \overrightarrow{H} \right] \mathrm{d}S = \frac{1}{2} \int_S \Pi(r,t) \mathrm{d}S. \tag{6}$$

According to (6), the Umov–Poynting theorem means that the average energy of P penetrated into the envelope surface S, following the modulus and phase, depends on the amplitude, distribution, and corresponding phase of the electric and magnetic fields. The transformation of the surface integral in (6), according to the Gauss theorem, into an integral over the volume, leads to the determination of the loss of active energy due to the finite conductivity of the medium, losses for polarization reversal and magnetization reversal of the environment, i.e., power loss of electromagnetic radiation:

$$\begin{aligned} P_{\mathrm{POL}} &= \frac{\sigma}{2}\int_V \left|\vec{E}\right|^2 \mathrm{d}V + \frac{\omega}{2}\int_V \left(\varepsilon_{\mathrm{Im}}\left|\vec{E}\right|^2 + \mu_{\mathrm{Im}}\left|\vec{H}\right|^2\right)\mathrm{d}V \\ P_{\mathrm{POL}} &= 2\omega\varepsilon_0 \iiint_V \varepsilon_{\mathrm{Im}} E^2 \mathrm{d}V. \end{aligned} \tag{7}$$

Formula (7) expresses the Joule–Lenz law, from which it is possible to obtain a 3D heating source of the electromagnetic field distribution density in non-magnetizable and magnetizable materials [3, 4].

3 Solving Research Problem

The scheme of the microwave effect on the bottom of an oil well and the scheme of the installation (emitter) are presented in [6, 7].

With dielectric heating, the specific power released in the reservoir for a mixture of high-viscosity oil is $q(r;t) = \pi \sum_{i=1}^{3} c_i \varepsilon_{ia} tg\delta_i^E f_i |E_i|^2 / f_{\mathrm{mix}} E_{\mathrm{mix}}^2 = -\mathrm{div}\,\vec{\Pi}\,(r;t)$.

To carry out a thermal calculation, consider in (see Fig. 1) layer HVOB in the form of a rectangular figure, having a width w, length l, height h and under the influence of a microwave heater [8].

The electromotive force (EMF) induced in the circuit is equal to:

$$E_{\mathrm{i}} = -\frac{\mathrm{d}\Phi}{\mathrm{d}t} = -\frac{\mathrm{d}B_{\mathrm{n}}S}{\mathrm{d}t} = -S\frac{\mathrm{d}B}{\mathrm{d}t} = -wl\frac{\mathrm{d}B}{\mathrm{d}t}. \tag{8}$$

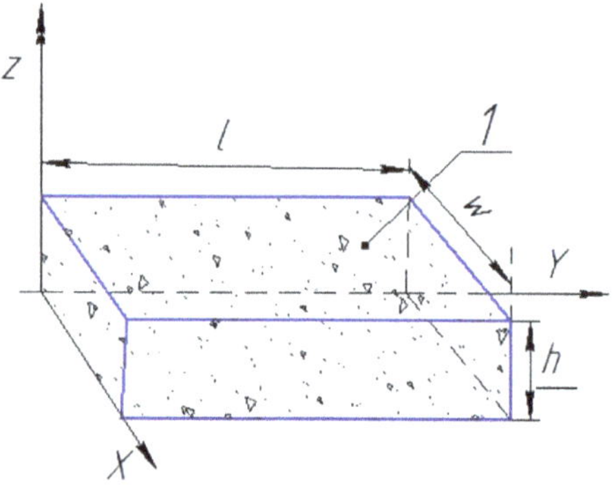

Fig. 1 Representative element of the formation: 1-multicomponent mixture of HVOB

The representative element of the formation has the resistance of the section:

$$R = \rho_{\text{mix}}^{E}\frac{l}{S} = \rho_{\text{mix}}^{E}\frac{l}{wl} = \rho_{\text{mix}}^{E}/w, \tag{9}$$

where ρ_{mix}^{E} is the resistivity of the mixture formation; $V = whl$—the volume of the treated section of the formation with a mixture of high-viscosity oil (Fig. 3).

As a result, a surface current (short-circuited eddy current) is formed, penetrating into the reservoir volume and leading to its heating:

$$Q = I_T^2 R \cdot t = \left(-\frac{w^2 l}{\rho_{\text{mix}}^{E}} \cdot \frac{\mathrm{d}B}{\mathrm{d}t}\right)^2 \cdot \frac{\rho_{\text{mix}}^{E}}{w} \cdot t = \frac{w^3 l^2}{\rho_{\text{mix}}^{E}} \cdot \left(\frac{\mathrm{d}B}{\mathrm{d}t}\right)^2 \cdot t. \tag{10}$$

In this case, the amount of heat is released, determined by the formula:

$$Q_{\text{c}} = c_{\text{mix}} m_{\text{mix}} \Delta T = c_{\text{mix}} \rho_{\text{mix}} V_{\text{mix}} \Delta T \tag{11}$$

where c is the specific heat of the mixture; m is the mass of the treated surface; ΔT—change in processing temperature.

Equating the formula for the amount of heat to the magnitude of the alternating magnetic field, we obtain the formula [3–6]

$$\begin{aligned} c_{\text{mix}} \rho_{\text{mix}} wlh \Delta T &= \frac{w^3 l^2}{\rho_{\text{mix}}^{E}} \cdot \left(\frac{\mathrm{d}B}{\mathrm{d}t}\right)^2 \cdot \Delta T, \\ \Delta T &= \frac{w^3 l^2}{\rho_{\text{mix}}^{E}} \cdot \left(\frac{\mathrm{d}B}{\mathrm{d}t}\right)^2 \cdot \Delta T / c_{\text{mix}} \rho_{\text{mix}} wlh; \end{aligned} \tag{12}$$

where ρ_{mix} is the density of the formation mixture; t—processing time. From here it is possible to determine the instantaneous temperature value, which follows from the Laplace differential equation:

$$\Delta T = \frac{w^2 l}{c_{\text{mix}} \rho_{\text{mix}}^{E} \rho_{\text{mix}} h} \cdot \left(\frac{\mathrm{d}B}{\mathrm{d}t}\right)^2 \cdot \Delta T. \tag{13}$$

An alternating magnetic field is generated by a changing electric:

$$\frac{\mathrm{d}B}{\mathrm{d}t} = \omega B \delta_{\text{m}} \cos\left(\omega t - \frac{\pi x}{\tau_1} v\right) + \frac{1}{2}\omega B v_{\text{m}} \cos \omega t + \frac{1}{2}\omega B v_{\text{m}} \cos\left(\omega t - \frac{2\pi x}{\tau_1} v\right). \tag{14}$$

What allows you to determine the effective value of the square of the induction SMH:

$$\left(\frac{\mathrm{d}B}{\mathrm{d}t}\right)^2 \approx \omega^2 B^2 \delta_\mathrm{m} \cos^2\left(\omega t - \frac{2\pi x}{\tau_1}\nu\right),$$
$$\frac{\sqrt{2}}{2}\left(\frac{\mathrm{d}B}{\mathrm{d}t}\right) = \frac{\sqrt{2}}{2}4\pi^2 f^2 B^2 \delta_\mathrm{m} = 2\sqrt{2}\pi^2 f^2 B^2 \delta_\mathrm{m}. \tag{15}$$

Hence, we obtain an expression for the effective value of the surface temperature of the HVOB formation:

$$\Delta T_\partial = \frac{2\sqrt{2}\pi^2}{c_\mathrm{mix}\rho^E_\mathrm{mix}\rho_\mathrm{mix}} \cdot \frac{w^2 l}{h} \cdot B^2 \delta_m \cdot \Delta T. \tag{16}$$

As a result, you can calculate the efficiency of the microwave heater installation:

$$\eta = \frac{Q}{I \cdot U} \cdot 100\%. \tag{17}$$

4 Results and Discussion

Figure 2 shows the electromagnetic distribution of the heating source in the SolidWorks software, the density of the field strength in the 3D model, as well as the density of the distribution of the temperature field near the spiral coil. Here, the electromagnetic field is created in a cylindrical microwave chamber, which has a longitudinal waveguide arrangement.

As can be seen from Fig. 4, volumetric heating occurs inside the ceramic chamber and then propagates into the formation. Heat-resistant ceramic mirror furnace body (microwave chamber), with a focusing plate, serves to direct microwave radiation in the desired direction. For the HVOB reservoir volume system, according to [6, 7], the position of microwave installations can be large in size both for vertical and inclined well drilling. The very duration of heating the formation should be interval

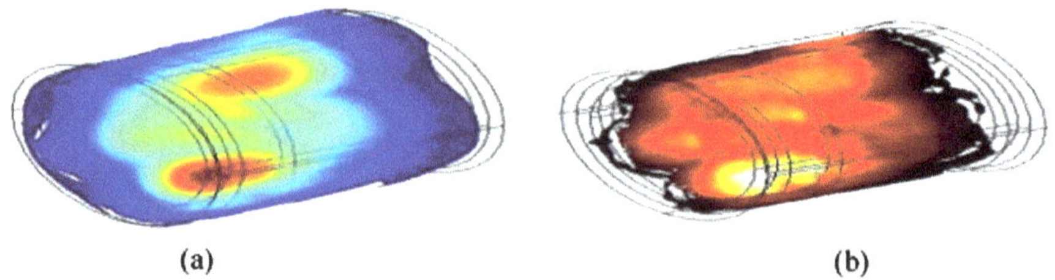

Fig. 2 Density of the temperature field distribution near the spiral coil: **a** at the beginning of work and **b** over time

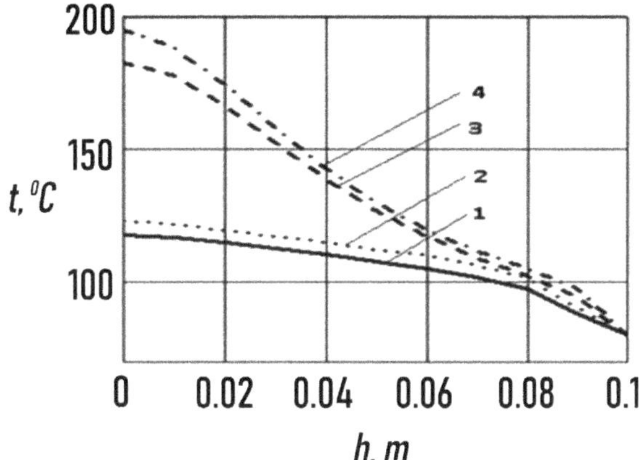

Fig. 3 Curves of temperature distribution in HVOB for data: 1.2–400 MHz; 3.4–900 MHz

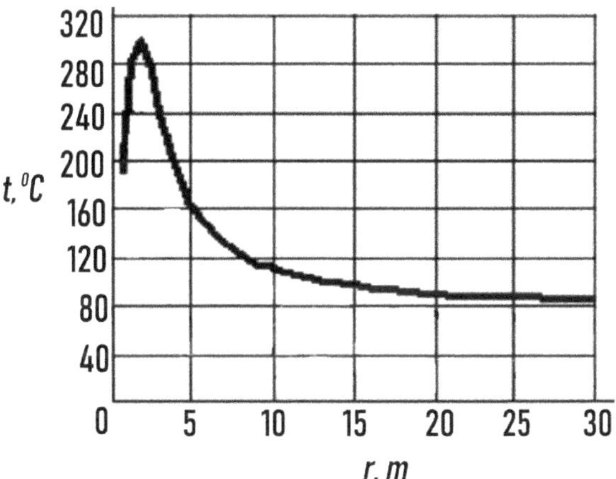

Fig. 4 Distribution of reservoir temperature with distance

in time, with its relaxation to prevent the process of sintering of the mixture and its coking.

The heat flow from system (3) has the form $q(z;\, t) = 2\alpha F \Pi_0 \exp(-2\alpha z)$—the function of the power density of the heat loss source; α—attenuation coefficient; F is the transmission coefficient; Π_0 is the Poynting–Umov vector in vacuum. The initial and boundary conditions for the differential equation (3) have the form $T(z > 0, 0) = T_0$, where T_0 is the reservoir temperature at the initial moment of time.

The solution of the differential equation of heat conduction taking into account the boundary and initial conditions is known [2] and has the form:

$$T(z, t) = T_0 + \frac{F_e \Pi_0}{2\alpha\lambda} \exp(-2\alpha z)\left[\exp(4\alpha^2 a^2 t) - 1\right], \tag{18}$$

where a is the thermal diffusivity; F_e is the energy transmission coefficient.

For a multicomponent mixture of HVOB, taking into account the expression for the power density function of the heat loss source $q(r;t) = \pi \sum_{i=1}^{3} c_i \varepsilon_{ia} tg\delta_i^E f_i |E_i|^2 / f_{\text{mix}} E_{\text{mix}}^2$, we obtain solution (18) in the form:

$$T(r,t) = T_0 + \frac{\Pi_0}{3} \sum_{i=1}^{3} \frac{c_i F_{\text{ei}}}{2\alpha_i \lambda_i} \exp(-2\alpha r)\left[\exp(4\alpha^2 a^2 t) - 1\right]. \tag{19}$$

Figure 3 shows the temperature distribution curves according to formula (19) at complete melting (during the transition from solid to liquid phase) of HVOB for the given values of frequencies and microwave heating power of 1.5 and 3 kW.

As can be seen from the graphs, as the thickness of the formation increases, the microwave heating temperature decreases. The energy of a microwave electromagnetic wave with a frequency of 900 MHz is partially spent on heating the formation in the liquid phase formed during its melting. The reason for the dissipation of electromagnetic energy in the formation is associated with the fact that the depth of its penetration at this frequency is significantly less than at a frequency of 400 MHz. Thus, the energy consumption for heating and melting at 400 MHz will be less than at 900 MHz.

From formula (19), we find the time period before the start of melting at the upper boundary of the formation of high-viscosity oil $z = 0$, depending on the moisture content and porosity

$$t_{\text{mel}}(n,k) = \frac{c\rho(n,k)}{\lambda 4\alpha^2(n,k)} \ln\left[1 - \frac{2\alpha(n,k)\alpha T_0}{F\Pi_0}\right]. \tag{20}$$

The research results allow us to draw the following conclusions—with an increase in the moisture content of an oil reservoir, the time period before the start of melting decreases, and with a decrease in porosity, it increases.

These results allow us to conclude that when determining the requirements for the electrical properties of a dielectric material during microwave melting, special attention should be paid to its moisture content and porosity. To reach the liquid phase as quickly as possible when designing microwave devices, it is necessary to provide a subsystem for preliminary humidification of the microwave heating object.

Figure 4 shows the reservoir temperature distribution curve with distance according to formula (20).

As can be seen from the graphs, there is a gradual heating of the HVOB formation to a certain temperature, it reaches saturation—the critical temperature of 290 °C, and then it melts. In this case, a significant portion of the microwave heating energy is spent on heating the layer of the multicomponent mixture in the liquid phase, which is formed during the melting of the initial solidified layer. This is due to the fact that the depth of penetration of microwave heating differs for different radiation frequencies and significant energy losses occur during its absorption by the formation. Therefore, for optimal operation of this installation, it is necessary to initially test it in the field. As you know, with increasing temperature, the viscosity of oil decreases and the light fractions of oil and formation water evaporate. Hydrocarbon and carbon dioxide gases, which are formed during evaporation, displace oil from the formation. The process of heat transfer and extraction of HVOB can be accelerated by introducing water together with an oxidizing agent.

5 Conclusion

This paper presents based on the analysis of various existing and promising methods enable the extraction of heavy oil and bitumen are shown the advantages of thermal method, on the basis of calculations on mathematical and computer models, the prospects for application of a microwave electromagnetic field, providing even heat reservoir at a sufficient depth. The created mathematical model of microwave dielectric heating of a reservoir and its thermal calculation allow us to study the technological process of heating, expand the possibilities of reducing the viscosity of VHOB, and intensifying their production and transportation. The results of the work create and justify the basis for the development of a new technology for thermal efficiency improvement of oil production, primarily in depleted reservoirs with VHOB. It is also relevant for the northern regions, which are characterized by high oil viscosity, where the costs of its production increase sharply and the volumes of the hydrocarbon product that are not extracted increase.

References

1. Zhang, J., Fan, Z.B., Chen, H., Peng, J.H.: Retracted article: enhancements of microwave absorbing properties of weakly absorbing materials based on auxiliary absorption of SiC for microwave heating. J. Mater. Sci. Mater. Electron. **27**(8), 8869–8869 (2016)
2. Kubota, K., Kurokawa, M., Araki, H., Okazaki, T., Lu, L.T., Hagura, Y.: Studies on the microwave heating-rate equations of foods. In: Developments in Food Engineering, pp. 507–509 (1994)
3. Itaya, Y., Uchiyama, S., Mori, S.: Internal heating effect and enhancement of drying of ceramics by microwave heating with dynamic control. In: Kowalski, S.J. (eds.) Drying of Porous Materials. Springer, Dordrecht (2006)
4. Itaya, Y., Uchiyama, S., Mori, S.: Internal heating effect and enhancement of drying of ceramics by microwave heating with dynamic control. Transp. Porous Med. **66**, 29–42 (2007)
5. Liu, C., Sheen, D.: Analysis and control of the thermal runaway of ceramic slab under microwave heating. Sci. China Ser. E Technol. Sci. **51**(12), 2233–2241 (2008)
6. Yangirov, I.F., Safiullin, R.A.: Engineering calculation of helical microwave heater of the downhole oil wells. Petrol. Eng. **17**(5), 99–105 (2019) https://doi.org/10.17122/ngdelo-2019-5-99-105
7. Peng, Z., Hwang, J.Y., Park, C.L., Kim, B.G., Onyedika, G.: Numerical analysis of heat transfer characteristics in microwave heating of magnetic dielectrics. Metall. Mater. Trans. A **43**(3), 1070–1078 (2012)
8. Bykov, Y.V., Egorov, S.V., Eremeev, A.G., Kholoptsev, V.V., Plotnikov, I.V., Rybakov, K.I., Sorokin, A.A.: Effects of microwave heating in nanostructured ceramic materials. Powder Metall. Met. Ceram. **49**(1–2), 31–41 (2010)

9. Egorov, S.V., Rybakov, K.I., Semenov, V.E., Bykov, Y.V., Kanygina, O.N., Kulumbaev, E.B., Lelevkin, V.M.: Role of convective heat removal and electromagnetic field structure in the microwave heating of materials. J. Mater. Sci. **42**(6), 2097–2104 (2007)
10. Ehlers, R.A., Metaxas, A.C.: Finite element modelling of thin metallic films for microwave heating. In: Advances in Microwave and Radio Frequency Processing, pp. 217–225 (2006)
11. Loik, D.A., Mamontov, A.V., Nazarov, I.V., Nefedov, V.N.: The idea of constructing microwave devices for the uniform heating of sheet materials. Meas. Tech. **52**(3), 312 (2009)
12. Horikoshi, S., Schiffmann, R.F., Fukushima, J., Serpone, N.: Engineering of microwave heating. In: Microwave Chemical and Materials Processing, pp. 145–182 (2018)
13. Horikoshi, S., Schiffmann, R.F., Fukushima, J., Serpone, N.: Materials processing by microwave heating. In: Microwave Chemical and Materials Processing, pp. 321–381 (2018)
14. Horikoshi, S., Schiffmann, R.F., Fukushima, J., Serpone, N.: Physics of microwave heating. In: Microwave Chemical and Materials Processing, pp. 87–143 (2018)
15. Huang, K., Lu, B.: The precise condition of thermal runaway in microwave heating on chemical reaction. Sci. China Ser. E Technol. Sci. **52**(2), 491–496 (2009)
16. Chernousov, Y.D., Shebolaev, I.V., Ivannikov, V.I., Ikryanov, I.M., Bolotov, V.A., Tanashev, Y.Y.: An apparatus for performing chemical reactions under microwave heating of reagents. Instrum. Exp. Tech. **62**(2), 289–294 (2019)

Research of Performance Characteristics of WPT System Associated with Mutual Arrangement of Coils

Konstantin Krestovnikov and Aleksei Erashov

Abstract The paper presents the derivation of systems of equations, relating the efficiency of the wireless power transmission system (WPT) with the transmitted power and the distance or displacement between the receiving and transmitting spiral coils. The initial data for the derivation of equation systems were obtained on a prototype assembly of a bidirectional WPT. Experiments were performed in which the values of efficiency and transmitted power were obtained for distances in the range up to 25 mm and displacements between the receiving and transmitting coils in the range up to 30 mm. When the distance between the coils with a diameter of 62 mm is changed to 25 mm, the maximum transmitted power of the prototype system is reduced by ~2.8 times compared to the assembly with minimum distance between the coils. With an axial displacement between the coils equal to approximately half the diameter of the coils, the maximum transmitted power drops by ~3.5 times. Several expressions have been proposed for converting the characteristics of the prototype into relative units and connecting them with the geometric parameters of the coil. The presented systems of equations can be used to develop algorithms that provide for identification of various kinds of interference during the operation of WPT, affecting the efficiency and transmitted power. When operating a WPT as part of a mobile autonomous robotic system, the derived systems of equations will make it possible to assess the feasibility of transferring energy between robots or a charging station and a robot at a given relative position.

1 Introduction

In the process of performing tasks by mobile autonomous robotic systems (ARS) [1, 2], using rechargeable batteries as a power source, it is necessary to replenish energy reserves with a certain periodicity. Charging a battery is essentially the transfer

K. Krestovnikov (✉) · A. Erashov
St. Petersburg Institute for Informatics and Automation of the Russian Academy of Sciences, St. Petersburg Federal Research Center of the Russian Academy of Sciences (SPC RAS), 39, 14th Line, St. Petersburg 199178, Russia
e-mail: k.krestovnikov@iias.spb.su

A. Ronzhin and V. Shishlakov (eds.), *Electromechanics and Robotics*, Smart Innovation, Systems and Technologies 232, https://doi.org/10.1007/978-981-16-2814-6_31

of energy from an external power source to the robot. As traditional solutions to this problem, various options can be distinguished for wire and contact methods of energy transfer. Both methods require connectors or pin pairs and often assume a sufficiently high positioning accuracy of the robot. If the robot is operated in an aggressive environment or outdoors, then additional measures are required to protect the connectors and contact pairs from contamination and moisture ingress. It is possible to reduce the human involvement in the operation of the robot and to simplify the process of positioning and subsequent charging of the battery with the use of WPT. The most common solutions in this area are based on the principle of transferring energy between inductively coupled elements and have a number of operational limitations. In contrast to wired power systems, the performance efficiency and the transmitted power largely depend on the distance of the power transmission. Also, at large distances or displacements between the receiving and transmitting parts of the system, the intensity of the stray fields increases, which create interference and negatively affect the operation of electronics located next to the WPT.

The permissible distance and axial displacement between the receiving and transmitting parts of the stationary WPT [3], at which the performance efficiency is optimal, are inferred by the authors empirically. The system is a station for wireless charging of batteries placed in it by a manipulator. The transmitting part is located in a permanently installed dock station, and the receiving part is located in the battery case. At zero bias and the distance between the receiving and transmitting coils, the maximum charging current is 725 mA with an average charging efficiency of 46.4%. With an increase in the distance and axial displacement up to 5 mm, the charging current decreases by 2.233 times, and the battery charging time increases by 1.5 times.

A study of the influence of axial displacement on the efficiency of a wireless power transmission system with two transmitting coils and one receiving coil is presented in [4]. The magnetic coupling between the coils is investigated in a simulation environment. The transmitting and receiving coils of the system are identical solenoid coils made of copper wires with a diameter of 2.75 mm. The wires are wound on wooden cylinders. Winding width is 160 mm, coil radius is 350 mm, number of turns is 5.5. In the simulation, the distance between the transmitting and receiving coils was 1 m and the energy transfer efficiency was 61.01%. This configuration is implemented in real-world settings using solenoid coils and tested to verify the simulation results. Experimental results show a drop in efficiency by a factor of 1.94 when the offset angle of the receiving coil is increased to 35°.

The authors of [5] discuss efficiency of WPT, changing the dimensions of the transmitting coil by increasing its width by 7 times with a step of 7 mm. The system consists of three coils with Litz wire. A feature of this system is the division of the transmitting coil into two, located in the same plane. The use of an intermediate coil makes it possible to increase the effective inductance of the transmitting coil at the resonant frequency and to increase the coupling coefficient. The take-up coil consists of two layers with 9 turns on each layer. The inner diameter of the coil is 21.7 mm, and the outer diameter is 27.02 mm. The distance between the receiving and transmitting coils is 30 mm. With the minimum width of the transmitting coil, the efficiency of the system was 14.43%, with the maximum width it increased by 5.21 times.

Changing the parameters of one of the coils of the WPT to achieve the highest efficiency of the system is used by the authors [6]. The WPT presented in this paper consists of a large rectangular loop, a small square loop and an additional square spiral coil. The large loop has dimensions of 2 × 6 m and is the transmitting part of the system. The authors revealed the influence of the following factors on the efficiency of energy transfer and the resonance frequency: the load connected to the receiving part of the system, the presence of foreign objects in the field of the system, such as conductive materials or the human body. Numerical results show that 50% power transfer efficiency can be achieved when the proposed system is used to charge only one load-optimized device. The transmission efficiency reaches its maximum value at 19.22 MHz. With a deviation of the operating frequency, the efficiency can be reduced by 17 times. The results also show that the efficiency of energy transfer is reduced by 11 times when a non-resonant object (human body) is near the receiving element.

The authors of [7] consider WPT for power transmission to electric bicycles. A copper wire with a cross section of 3 mm^2 is used as a conductor of the receiving and transmitting coils. The outer diameter of the coils is 15 cm. The dependences of the system efficiency on the number of coil windings, the distance between the receiving and transmitting parts of the system and the communication efficiency on the operating frequency, are obtained. It was found that the high efficiency of coupling between the coils is maintained at frequencies of 110–130 kHz. Provided that the air gap between the coils does not exceed 3 cm, the WPT has a power transfer rate of 96 W and an efficiency of 79%.

It is proposed to preserve the high efficiency of the WPT when the distance between the parts of the system is changed [8] by introducing additional matching resonant circuits into the structure of the system and ensuring the optimal distance between them. Consistency in the system is achieved by changing the coupling coefficient between the receiving part of the system and two intermediate resonant circuits. The receiving and transmitting coils have one turn with a diameter of 40 cm. An experiment was carried out in which the distance between the intermediate resonance contours was changed from 15 to 100 cm with a step of 5 cm. The efficiency of the system reaches 90% at a distance of 15 cm and rapidly decreases with an increase in the distance between the intermediate resonance circuits. When the distance between the receiving and transmitting parts of the system is 60 cm, the efficiency drops by a factor of 1.94, at 100 cm the respective factor is 3.07.

The introduction of an additional resonant circuit to prevent a decrease in the efficiency of the WPT is described by the authors [9]. They proposed a structure of the U-WPT system, consisting of transmitting, receiving and intermediate resonant circuits. The influence of the number of turns and the location of the intermediate resonant circuit on the efficiency of energy transfer is considered. A simulation model was developed in the SIMULINK software package with the number of turns for the coils from 1 to 48, with a step of 6. The simulation results show that the maximum efficiency is achieved with 30 turns and is 75.82%. The real maximum efficiency under these conditions is 74.83%. The authors of the paper analyzed the influence of the position of the intermediate coil with 30 windings on the efficiency of energy transfer. Also, a simulation model was developed in SIMULINK with varying distances between transmitting, receiving and intermediate coils in the range from 0 to 600 mm, with a step of 100 mm. The simulation showed that the position of the intermediate resonant circuit directly affects the efficiency of energy transfer. With a distance between the coils of 100 mm and 500 mm, the real efficiency of the system is reduced by 18 and 12 times, respectively.

The authors of [10] investigate the dependences of the efficiency of energy transfer with axial displacement between the receiving and transmitting coils. In the work, it is proposed to make the receiving coil smaller than the transmitting one. This solution leads to a decrease in the coupling coefficient (less than 0.5) and a slight decrease in the energy transfer efficiency. It has been empirically established that the coupling coefficient between the coils should not be lower than 0.25. This is required to avoid heating of the coils and significantly reducing of the efficiency of power transfer. The authors propose to wind the receiving coil without gaps, and the transmitting one with gaps between the turns to ensure an even distribution of the field. During the experiment, the receiving coil was ~1.6 times smaller than the transmitting one and consisted of 5 tightly wound coils, while the transmitting coil consisted of 10. The distance between the coils was 10 mm. When forced cooling was used, the transmission efficiency was 77% with a transmitted power of 295 W, with natural cooling, the maximum transmitted power and energy transfer efficiency fell to 74% and 69 W, respectively.

Based on the analysis of scientific works in the areas of development of WPT systems, obtaining their characteristics and researching operating limitations, it is possible to conclude that the main limitations in the use of these systems are related to the relative position of the receiving and transmitting parts of the system. The distance, angular and axial displacement between the receiving and transmitting parts of the system affect their efficiency, the transmitted power, and the level of electromagnetic interference generated by the WPT. With an increase in the air gap, the coupling coefficient of the system parts decreases, and an adjustment of the operating frequency is required. To maintain the required coupling coefficient, receiving coils with a smaller diameter with a denser winding and transmitting coils of a larger diameter with gaps between the windings are used, which ensures a uniform distribution of the magnetic field and reduces the effect of axial displacement. When the air gap changes, the efficiency of the BSPE can be reduced by 18 or more times. With an angular displacement of the receiving coil relative to the transmitting coil from 0 to 35° in one of the considered prototypes, the energy transfer efficiency decreases by 15.21 times.

From the research works considered in this analysis, it is possible to infer some relationships between the efficiency of the system, the transmitted power and the geometric arrangement of the receiving and transmitting coils. The operation of WPTs as part of mobile ARSs requires an analysis of the operational limitations of the WPTs and their systematization for the subsequent development of algorithms for movement, positioning and control of the battery charging system. To systematize the operational limitations of the WPT associated with the relative location of the receiving and transmitting parts of the system, it is necessary to carry out certain experiments that will make it possible to obtain the characteristics of the system for various conditions of power transmission.

2 Obtaining WPT Prototype Characteristics

The study of the dependencies of the characteristics on the distance and displacement between the receiving and transmitting parts of the system was carried out on a real prototype made on the basis of the circuitry solutions presented in [11, 12]. According to the method proposed in [13] for choosing the optimal number of windings of loop coils, resonant circuits with planar spiral coils with 15 windings and an outer diameter of 62 mm were made for the prototype. To obtain the dependences of the efficiency and transmitted power on the distance and displacement between the receiving and transmitting coils, two types of experiments were carried out.

In the first type of experiments, a number of dependences of the transmitted power and system efficiency were obtained for six fixed distances of energy transmission in the range from the immediate proximity of the coils to each other up to 25 mm. The choice of the upper value of the range is due to a significant drop in the electromotive force (EMF) in the receiving coil and the lack of operability of the system at distances exceeding this. During the experiment, the transmitting part was powered from a stabilized source with a voltage of 8.43 V. At the output of the receiving part of the system, the output DC-DC step-up converter sets the voltage at 8.43 V at idle. The load was an electronically controlled ballast with current stabilization. The graph of the dependence of the efficiency of the system on the load power for various distances L is shown in Fig. 1.

As can be seen, the curves in Fig. 1 have a similar shape; at $L = 0$ and $L = 5$ mm, they practically coincide. An increase in the distance between the receiving and transmitting coils leads to a decrease in the maximum load power. When the distance between the coils is changed to 25 mm, the maximum transmitted power decreases ~2.8 times in comparison with the minimum distance between the coils.

The second series of experiments involved obtaining the dependences of the transmitted power and system efficiency under the same conditions as in the first experiment. The distance between the coils was minimal ($L = 0$), while the axial displacement between the coils H varied in the range up to 30 mm. Dependencies were obtained for five axial displacements and are presented in Fig. 2.

When comparing the graphs in Figs. 1 and 2, it can be concluded that the axial displacements between the coils affect the transmitted power and system efficiency to a lesser extent than the distance between the coils. With an axial displacement between the coils equal to approximately half the diameter of the coils, the maximum transmitted power drops by ~3.5 times.

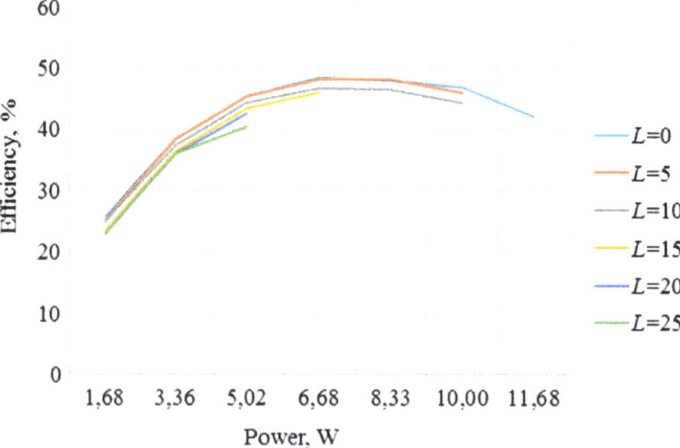

Fig. 1 Dependences of the efficiency of the system on the load power for various distances between the coils

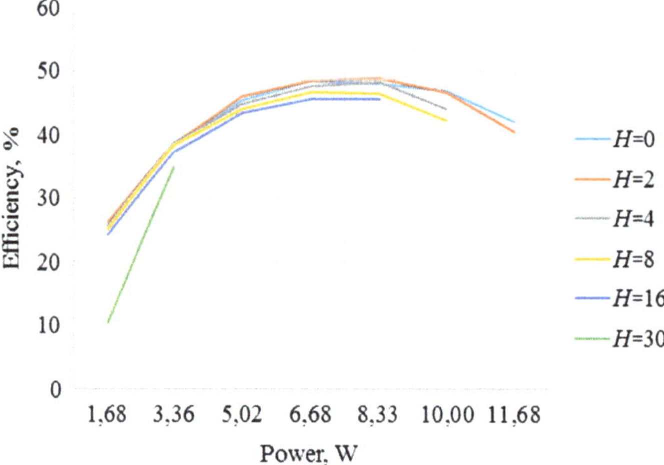

Fig. 2 Dependences of the efficiency of the system on the power of the load for various displacements between the coils

3 Systematization of Characteristics and Obtaining Equations

The data obtained from the experiments in their original form does not enable to calculate the efficiency of the system depending on the distance or displacements between the coils and the transmitted power for WPT with a different size of loop coils. To derive the equations of the dependences of the efficiency of the system and the transmitted power on the distance and displacements between the coils for WPT, it is necessary to express the results obtained in relative values and summarize them in tabular manner.

Experimental data were converted into relative units as follows. The load power P_n corresponding to a certain value of efficiency η_n at a distance L_n or with an offset H_n was converted into a relative value $P_{\text{rel}}(n)$:

$$P_{\text{rel}(n)} = \frac{P_n}{P_{\max}}, \tag{1}$$

where $P_{\max}$ was the maximum load power of WPT.

Relative efficiency $\eta_{\text{rel}}(n)$ corresponding to $P_{\text{rel}}(n)$ was calculated as follows:

$$\eta_{\text{rel}(n)} = \frac{\eta_n}{\eta_{\max}}, \tag{2}$$

where $\eta_{\max}$ was maximum efficiency of WPT.

The influence of the offset parameters and the distance between the coils on the operating parameters of the WPT is related to the dimensions of the looped coils, so it is also necessary to express these parameters in relative units. The distance between the coils is determined using Eq. (3):

$$L_{\text{rel}(n)} = \frac{L_n}{D}, \tag{3}$$

where D is the diameter of the loop coil. The equation will look similar for the axial displacement between the coils:

$$H_{\text{rel}(n)} = \frac{H_n}{D} \tag{4}$$

The results obtained in the above experiments on the transfer of energy with varying distances between the coils, expressed in relative units, are summarized in Table 1.

The data presented in Table 1 were approximated by a third-order power polynomial using a software package for solving computational problems. Constraints on the range of admissible values of variables are added to the obtained surface equation and the following system of equations is obtained:

$$\begin{cases} z = 0.1498 + 0.04462x + 3.14y - 2.147x^2 + 0.3625xy - 3.566y^2 + 3.982x^3 - \\ \quad -0.02121x^2y - 0.5439xy^2 + 1.15y^3 \\ y \le 0.3906x^2 - 1.738x + 1.016 \\ x, y, z \in [0:1] \end{cases}, \tag{5}$$

Table 1 Dependence of efficiency and transmitted power on the distance of power transmission in relative units

$L_{rel} = 0$		$L_{rel} = 0.08$		$L_{rel} = 0.16$		$L_{rel} = 0.24$		$L_{rel} = 0.32$		$L_{rel} = 0.40$	
P_{rel}	η_{rel}	P_{rel}	η_{rel}	P_{rel}	η_{rel}	P_{rel}	η_{rel}	P_{rel}	η_{rel}	P_{rel}	η_{rel}
0.14	0.53	0.14	0.52	0.14	0.51	0.14	0.48	0.14	0.47	0.14	0.47
0.29	0.80	0.29	0.80	0.29	0.77	0.29	0.75	0.29	0.75	0.29	0.75
0.43	0.94	0.43	0.94	0.43	0.92	0.43	0.90	0.43	0.88	0.43	0.70
0.57	1.00	0.57	0.99	0.57	0.96	0.57	0.95				
0.71	0.99	0.71	1.00	0.71	0.96						
0.86	0.97	0.86	0.95	0.86	0.80						
1.00	0.87										

where $x = L_{rel}$, $y = P_{rel}$, $z = \eta_{rel}$. A visual representation of the surface built on the basis of (5) is shown in Fig. 3.

As follows from (1), (2) and (4), the initial data of experiments in the transfer of energy with a displacement between the coils are converted into relative units and presented in Table 2.

Similarly, to the case for the distance between the coils, a polynomial approximation of the surface obtained on the basis of the data for the axial displacement of the coils was carried out. As a result, the obtained second-order polynomial obtained

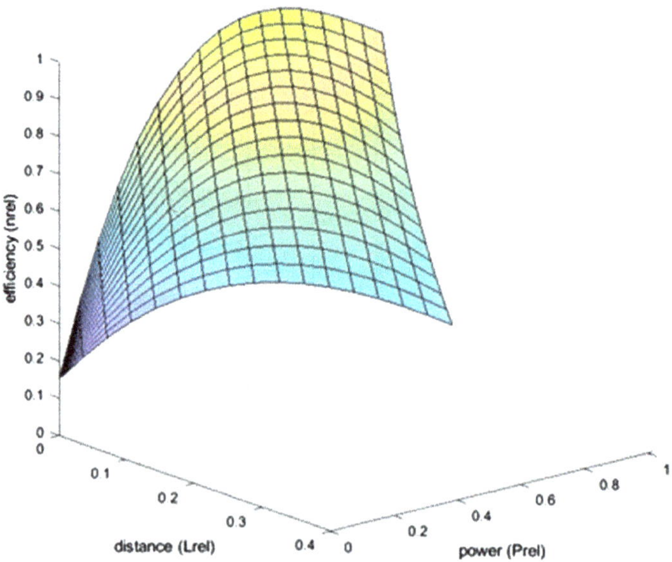

Fig. 3 Dependency of the efficiency of the system from the load power and the distance between the coils

Table 2 Dependency of efficiency and transmitted power on axial displacement between coils in relative units

$H_{rel} = 0$		$H_{rel} = 0.03$		$H_{rel} = 0.06$		$H_{rel} = 0.13$		$H_{rel} = 0.26$		$H_{rel} = 0.48$	
η_{rel}	P_{rel}	η_{rel}	P_{rel}	η_{rel}	P_{rel}	η_{rel}	P_{rel}	η_{rel}	P_{rel}	η_{rel}	P_{rel}
0.14	0.53	0.14	0.54	0.14	0.53	0.14	0.52	0.14	0.50	0.14	0.22
0.29	0.80	0.29	0.80	0.29	0.80	0.29	0.79	0.29	0.77	0.29	0.72
0.43	0.94	0.43	0.95	0.43	0.92	0.43	0.91	0.43	0.89		
0.57	1.00	0.57	1.00	0.57	0.98	0.57	0.96	0.57	0.94		
0.71	0.99	0.71	1.01	0.71	1.00	0.71	0.96	0.71	0.94		
0.86	0.97	0.86	0.96	0.86	0.91	0.86	0.87				
1.00	0.87	1.00	0.84								

additional restrictions on the range of admissible values of variables and the following system of equations was derived:

$$\begin{cases} z = 0.2402 - 0.1936x + 2.322y - 0.4821x^2 + 0.1042xy - 1.774y^2 \\ y \le -1.342x^2 - 0.7919x + 1.03 \\ x, y, z \in [0:1] \end{cases}, \quad (6)$$

where $x = H_{rel}$, $y = P_{rel}$, $z = \eta_{rel}$. A visual representation of the surface built on the basis of (6) is shown in Fig. 4.

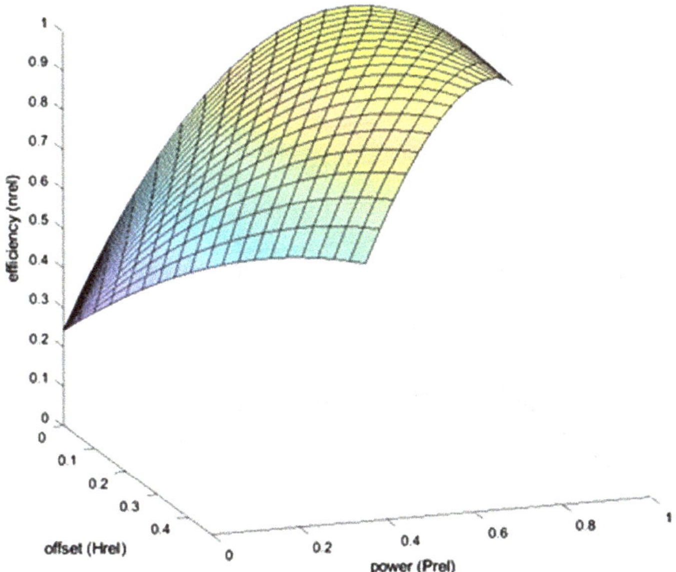

Fig. 4 Dependence of system efficiency on load power and displacement between coils

Comparison of the surfaces obtained on the basis of these experiments, given in relative units, clearly shows a greater dependency of efficiency and transmitted power from the distance of power transmission than on the axial displacement between the coils.

4 Conclusion

The derived systems of equations enable to calculate the efficiency of power transmission depending on the load power and the distance or displacement between the coils for wireless power transmission systems based on inductively coupled resonant circuits with planar spiral coils. The use of algorithms based on the presented systems of equations, with utilization of WPTs as part of mobile ARS, enable to identify various kinds of interference in the transmission of energy. If the calculated value of the system efficiency is lower than the actually measured one, then it is possible to indirectly conclude either about errors in the positioning of the ARS, or about the presence of foreign objects between the receiving and transmitting coils. It is also possible to conduct a preliminary assessment of the feasibility of transferring energy between the swarm robots [14, 15] or the charging station and the robot at a given reciprocal position and make a decision to transfer energy or continue positioning.

The use of energy transfer control algorithms developed on the basis of the proposed systems of equations for powering autonomous sensors can have a significant impact on the development of cyber-physical systems [16, 17].

Further research will be aimed at developing a navigation algorithm for a group of autonomous swarm robots, which will consider the energy resources of agents, distribute the final coordinates and calculate the points for power supply replenishment.

Acknowledgments This work is supported by the RFBR project No. 19-08-01215_A.

References

1. Pavliuk, N., Saveliev, A., Cherskikh, E., Pykhov, D.: Formation of modular structures with mobile autonomous reconfigurable system. In: Proceedings of 14th International Conference on Electromechanics and Robotics "Zavalishin's Readings", pp. 383–395. Springer, Singapore (2020). https://doi.org/10.1007/978-981-13-9267-2_31
2. Pshihopov, V.K., Medvedev, M.: Group control of autonomous robots motion in uncertain environment via unstable modes. SPIIRAS Proc. **5**(60), 39–63 (2018). https://doi.org/10.15622/sp.60.2
3. Zhang, J., et al.: Battery swapping and wireless charging for a home robot system with remote human assistance. IEEE Trans. Consum. Electron. **59**(4), 747–755 (2013). https://doi.org/10.1109/tce.2013.6689685

4. Skaik, T.F., AlWadiya, B.O.: Design of wireless power transfer system with triplet coil configuration based on magnetic resonance. IU-JEEE **17**(1), 3169–3174 (2017)
5. Zhong, W., Xu, D., Hui, R.S.Y.: A method to create more degrees of freedom for designing WPT systems-coil splitting. In: Wireless Power Transfer, pp. 63–73. Springer, Singapore (2020). https://doi.org/10.1109/TAP.2010.2044321
6. Yuan, Q., Chen, Q., Li, L., Sawaya, K.: Numerical analysis on transmission efficiency of evanescent resonant coupling wireless power transfer system. IEEE Trans. Antennas Propag. **58**(5), 1751–1758 (2010). https://doi.org/10.1109/TAP.2010.2044321
7. Pellitteri, F., et al.: Experimental test on a contactless power transfer system. In: 2014 Ninth International Conference on Ecological Vehicles and Renewable Energies (EVER), pp. 1–6. IEEE (2014). https://doi.org/10.1109/EVER.2014.6844092
8. Duong, T.P., Lee, J.W.: Experimental results of high-efficiency resonant coupling wireless power transfer using a variable coupling method. IEEE Microw. Wirel. Compon. Lett. **21**(8), 442–444 (2011). https://doi.org/10.1109/lmwc.2011.2160163
9. Sun, Y., Ye, Z.H.: Power transfer efficiency analysis of U-WPT system. In 2016 Asia-Pacific International Symposium on Electromagnetic Compatibility (APEMC), vol. 1, pp. 858–861. IEEE (2016). https://doi.org/10.1109/apemc.2016.7522890
10. Low, Z.N., Chinga, R.A., Tseng, R., Lin, J.: Design and test of a high-power high-efficiency loosely coupled planar wireless power transfer system. IEEE Trans. Ind. Electron. **56**(5), 1801–1812 (2008). https://doi.org/10.1109/tie.2008.2010110
11. Krestovnikov, K., Cherskikh, E., Pavliuk, N.: Concept of a synchronous rectifier for wireless power transfer system. In: IEEE EUROCON 2019—18th International Conference on Smart Technologies, pp. 1–5. IEEE (2019). https://doi.org/10.1109/eurocon.2019.8861856
12. Krestovnikov, K., Cherskikh, E., Smirnov, P.: Wireless power transmission system based on coreless coils for resource reallocation within robot group. In: International Conference on Interactive Collaborative Robotics, pp. 193–203. Springer, Cham (2019). https://doi.org/10.1007/978-3-030-26118-4_19
13. Krestovnikov, K., Cherskikh, E., Bykov, A.: Approach to choose of optimal number of turns in planar spiral coils for systems of wireless power transmission. Elektron. Elektrotech. **26**(6), 17–24 (2020). https://doi.org/10.5755/j01.eie.26.6.26181
14. Krestovnikov, K., Cherskikh, E., Ronzhin, A.: Mathematical model of a swarm robotic system with wireless bi-directional energy transfer. In Robotics: Industry 4.0 Issues & New Intelligent Control Paradigms, pp. 13–23. Springer, Cham (2020). https://doi.org/10.1007/978-3-030-37841-7_2
15. Endo, T., Maeda, R., Matsuno, F.: Stability analysis of swarm heterogeneous robots with limited field of view. Inform. Autom. **19**(5), 942–966 (2020). https://doi.org/10.15622/ia.2020.19.5.2
16. Vatamaniuk, I.V., Yakovlev, R.N.: Algorithmic model of a distributed corporate notification system in context of a corporate cyber-physical system. Model. Optim. Inf. Technol. **7**(4) (2019). https://doi.org/10.26102/2310-6018/2019.27.4.026
17. Vatamaniuk, I.V., Iakovlev, R.N.: Generalized theoretical models of cyberphysical systems. Proc. Southwest State Univ. **23**(6), 161–175 (2019). https://doi.org/10.21869/2223-1560-2019-23-6-161-175

Modernization of Automatic Excitation Control Systems of Generators in Syrdarya TPP

Tokhir Makhmudov and Obid Nurmatov

Abstract For reliable operation of electrical systems, it is necessary to ensure the stability of parallel operation of power plants and power systems. Violation of their stability leads to the repayment of a large number of electricity consumers, equipment damage and other negative consequences. Excitation systems can improve the quality of voltage control during sudden fluctuations in the load and with a short circuit at the output of synchronous generators. The aim of this work is to verify the automatic excitation controllers installed during the modernization process on the generators of the Syrdarya TPP in the DISILENT power factory software package. The article presents the results of a study in which the effect of the installed system of automatic control of excitation in the process of modernization on the quality of transient indicators is analyzed. Transient analysis showed an improvement in transient quality indicators such as decay time, overregulation and static error.

1 Introduction

Syrdarya thermal power plant (TPP) is the largest power plant in Central Asia. The construction of the Syrdarya TPP began in the fall of 1966 with the Trust «Golodnostep GES stroy». On December 26, 1972, the first power unit with a capacity of 300 MW was put under load. With the subsequent commissioning of the second units in 1973, the third unit in 1974, the fourth and fifth units in 1976, the seventh in 1977, the eighth in 1978 and the ninth in 1979. With the commissioning of the last tenth power unit in 1981, the construction of thermal power plant was completed, and its installed capacity reached 3000 MW [1].

The main stages of development and modernization of the station:

- In 2001–2002, the Syrdarya TPP was reconstructed at power units No. 7 and 8;
- In the period 2014–2015, a full-scale modernization of power units No. 1, 2 of Syrdarya TPP was carried out;

T. Makhmudov (✉) · O. Nurmatov
Tashkent State Technical University, Universitetskaya, Tashkent, Uzbekistan 100095
e-mail: tox-05@yandex.com

A. Ronzhin and V. Shishlakov (eds.), *Electromechanics and Robotics*, Smart Innovation, Systems and Technologies 232, https://doi.org/10.1007/978-981-16-2814-6_32

- In March–December 2020, it is planned to upgrade two more units of the Syrdarya TPP—the fifth and sixth, with a capacity of each up to 325 MW, including through the installation of new automatic excitation control (AEC) systems for synchronous generators.

2 Materials and Methods

The excitation system should contribute to the effective control of voltage and increase the stability of electrical systems [2]. This system should be able to quickly respond to violations, increasing the stability of the transition process [3]. Synchronous generators are the main sources of electricity [4]. AEC systems are widely used in the generator exciter control system [5]. The main way to control the reactive power of a generator is to control the excitation of the generator using AEC. The role of the AEC is to maintain a constant voltage level at the generator terminals under normal operating conditions, as well as transient conditions [6]. In most modern systems, an automatic excitation regulator is a controller that determines the output voltage of the generator (and sometimes the current) and then forms a corrective effect, changing the control of the exciter in the desired direction [7]. As a result of the development of computer technology, it became possible to test virtual devices for automatically controlling the excitation of synchronous generators [8]. Modern programs allow to create models with which can track input and output data, as well as intermediate calculations. Thus, it is possible to control the operation of automatic regulators in real time and check the correctness of their operation under various modes of the electrical system [9].

Regulation of the excitation of generators has a significant impact on transients in the power system; therefore, adequate systems for automatic regulation of excitation are necessary [10]. For the development of such systems at the design stage, as well as for the training of maintenance personnel, it is important to have information about the processes occurring in the system in various operating modes, including emergency ones [11]. A set of models should ensure the reproduction of the main characteristics operated by AECs that affect the quality of electromechanical transients [12].

The level of reliability of the functioning of the energy system is determined, along with other factors, the reliability and operational characteristics of the used electric power equipment, control devices and systems, including the characteristics of microprocessor-based devices for automatic control, protection and automation [13]. At the same time, products offered on the electric power market are not always free from algorithmic and software errors, and adaptation to operating conditions in power systems is not performed properly [14]. All this leads to an increase in the number of technological failures in energy systems and a decrease in the reliability of parallel operation of power plants and power systems [15]. Certification tests of AEC of synchronous generators should contain the following steps [16]:

- assembly of a test model of the power system;
- conducting certification tests;

- analysis of certification test results;
- creation of a digital model of certified strong action AEC systems of synchronous generators.

The test model of the power system should be equipped with [17]:

- AEC systems of synchronous generators and automatic turbine speed controllers;
- devices implementing short circuits of various types;
- devices for simulation the action of relay protection and automation;
- a system for monitoring and recording parameters of the electric regime.

Testing the model of the automatic excitation controller is carried out in the following order [16, 17]:

1. calculation of the parameters of the primary circuit;
2. launch of a customized model of the electric network;
3. calculation and setting of input parameters of the generator voltage and the excitation voltage of the automatic excitation controller;
4. increase (decrease) in the load level at the generator output (test disturbance 1);
5. a single-phase short circuit to ground (0.04 s duration) at the output of the power plant (test disturbance 2);
6. obtaining simulation results and analysis of the operation of automatic excitation regulators.

The relevance of the methods of virtual modeling of the model of the power system of Uzbekistan, as well as the AEC systems of synchronous generators, lies in the possibility of studying the dynamic properties of the system under study under various external disturbances (switching on–off loads, short circuits, etc.).

3 Simulation

Currently, the power system of Uzbekistan has a problem of verification of newly installed power equipment and automation systems, in connection with which a decision was made to conduct a virtual simulation of the power system of Uzbekistan and automatic control systems in the DIgSILENT power factory software package. Figure 1 shows a diagram of the main connections of the Syrdarya TPP with the power system of Uzbekistan, and in addition, the load flow calculation results are shown.

Tables 1, 2, 3 and 4 show the technical parameters of models of generators, transformers and power lines, as well as systems of test AECs.

Since currently there is a modernization of the AEC systems of generators No. 5, 6, Table 4 shows the parameters of the AEC before and after the modernization.

Designations used in Table 1:

P_{nom} rated active power of the generator;
U_{nom} rated voltage of the generator;

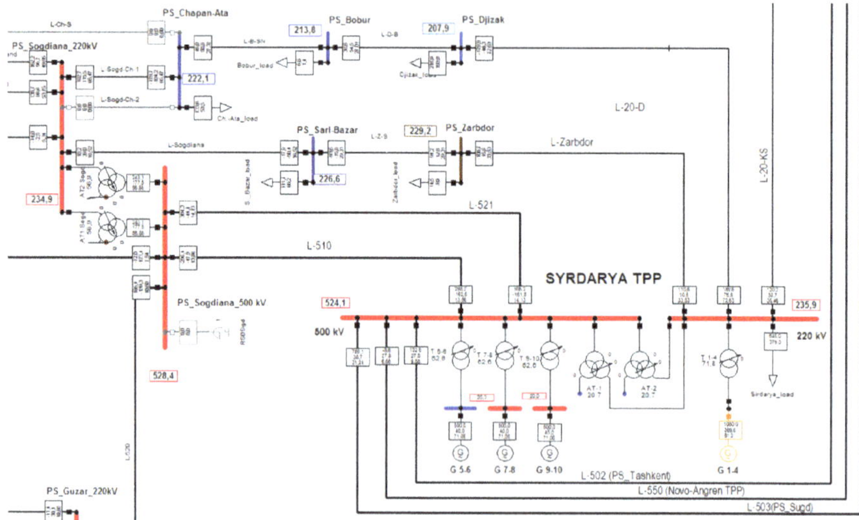

Fig. 1 Scheme of the Syrdarya thermal power plant

Table 1 Parameters of synchronous generators

Generator number	P_{nom}	U_{nom}	cosφ	Reactances[a]			T_{d0}	T_j
				x_d	x'_d	x''_d		
	MW	kV		p.u.	p.u.	p.u.	s	s
G 1-10	300	20	0.85	2.195	0.3	0.195	7	8.5

[a]Relative reactance values are given to rated apparent power and rated voltage

Table 2 Transformer model parameters

Transformer number	S_{nom}	U_H	U_M	U_L	P_0	P_{SC}	U_{SC}, *H-M*/*H-L*/*M-L*	I_0
	MVA	kV	kV	kV	MW	MW	%	%
T 1-4	400	242	–	20	320	880	11	0.4
T 5-10	400	525	–	20	315	790	13	0.45
AT 1-2	167	500	220	38	90	315	11/35/21.5	0.25

x_d longitudinal synchronous reactance;
x'_d transient longitudinal synchronous reactance;
x''_d over transient longitudinal synchronous reactance;
T_{d0} time constant of the field winding with an open stator winding;
T_j mechanical inertial constant of the unit (generator and turbine);
cosφ power factor.

Designations used in Table 2:

Table 3 Parameters of power lines connecting Syrdarya TPP with a power system

Line designation	X_1	X_0	R_1	R_0
	Ω	Ω	Ω	Ω
L-502	47.97	174	3.984	55.33
L-550	39.91	144.768	3.315	46.033
L-503	23.2	97.14	1.853	12.984
L-521	66.47	241.12	5.52	76.67
L-510	67.77	245.83	5.63	78.17
L-20-D	23.95	79.29	4.06	17.3
L-Zarbdor	48.726	161.29	8.27	35.19
L-20-KS	43.189	142.96	7.33	31.19

Table 4 Settings for AEC systems of model synchronous generators

Generator number	Type of excitation system	Type of AEC	Gains					T_{ES}
			K_{0U}	K_{1U}	K_{1If}	K_{0f}	K_{1f}	
G 1-4	Thyristor	AEC-SA	50	4	1	3	1	0.04
G 5-6	Thyristor	AEC-PA (before modernization)	25	–	–	–	–	0.04
		AEC-SA (after modernization)	50	4	1	3	1	0.04
G 7-8	Thyristor	AEC-SA	50	4	1	3	1	0.04
G 9-10	Thyristor	AEC-PA	25	–	–	–	–	0.04

S_{nom} rated apparent power of the transformer;
U_H rated voltage of the high side of the transformer;
U_M rated voltage of the middle side of the transformer;
U_L rated voltage of the low side of the transformer;
I_0 open-circuit current of the transformer;
P_0 transformer no-load losses;
U_{SC} short-circuit voltage of the transformer;
P_{SC} transformer short circuit losses.

Designations used in Table 3:

X_1 reactance of the direct sequence;
X_0 reactance of the zero sequence;
R_1 resistance of the direct sequence;
R_0 resistance of the zero sequence.

Designations used in Table 4:

K_{0U} gain of the proportional channel of the voltage regulator;
K_{1U} gain of the differential channel of the voltage regulator;

$K_{1\text{If}}$ gain of the internal stabilization channel with respect to the derivative of the rotor current;
$K_{0\text{f}}$ gain of the system stabilization channel by voltage frequency;
$K_{1\text{f}}$ gain of the system stabilization channel with respect to the derivative of the voltage frequency;
T_{ES} time constant of the excitation system, s;
AEC-SA strong action automatic excitation controller;
AEC-PA proportional action automatic excitation controller.

During the modernization of blocks No. 5, 6, installation of a system of type ARV-REM is forecasted as AEC-SA. The tuning parameters of the AEC system are given according to the recommendations of the supplier—Concern RUSELPROM.

To represent the AEC system of thyristor excitation of generators at the Syrdarya TPP in a digital model of the power system, the developers of the excitation control system presented the structure of the ARV-REM mathematical model and the ranges of changes in the AEC parameters [18]. Excitation regulation is carried out by the main automatic voltage regulator (AVR), regulation by the first derivative of the rotor current and system stabilization channels.

The transfer function of the AVR is presented in the form:

$$W_1(p) = K_{\text{pU}} + \frac{K_{\max}}{1 + K_{\max} * p * T_{\text{u}}} + K_{1\text{U}} * W_{\text{D}}(p). \tag{1}$$

Here:

K_{pU} the value of the gain at high frequencies;
T_{u} the AVR channel integration time constant;
$W_{\text{D}}(p)$ generator voltage differentiator;
$K_{\max}$ gain limiting factor at low frequencies.

Voltage differentiator:

$$W_{\text{D}}(p) = \mathrm{e}^{-0.01p} \frac{p}{(0.0125p + 1)^2}. \tag{2}$$

Internal stabilizer for the derivative of the rotor current:

$$W_2(p) = K_{1\text{if}} \frac{p}{(0.1p + 1)(0.0128p + 1)} \mathrm{e}^{-0.0033p}. \tag{3}$$

System stabilizer.

Frequency regulator link:

$$W_3(p) = K_{0\text{f}} \cdot \mathrm{e}^{-0.02p} \frac{1}{0.16p + 1} \cdot \frac{2p}{2p + 1} \cdot \mathrm{e}^{-0.0033p}. \tag{4}$$

Derivative frequency link:

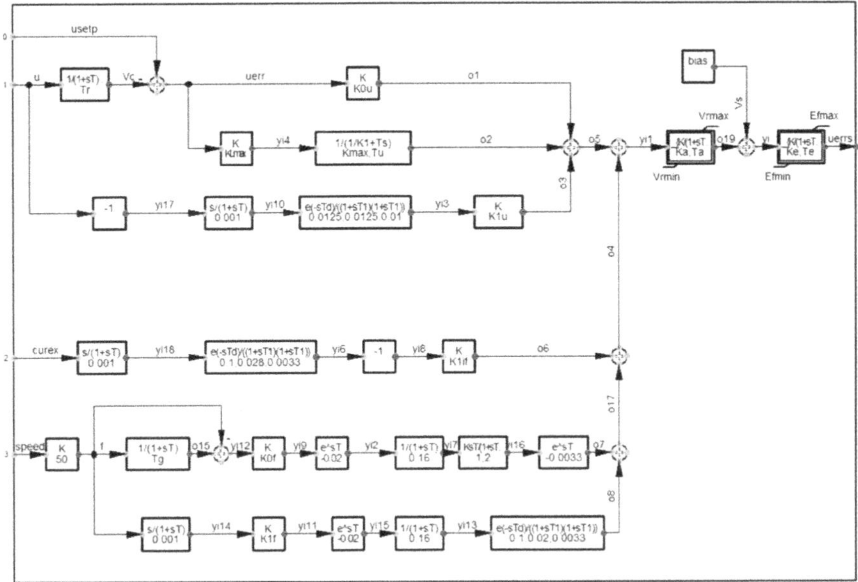

Fig. 2 Structural model of the ARV-REM system

$$W_4(p) = K_{1f} \cdot e^{-0.02p} \frac{1}{0.16p + 1} \cdot \frac{p}{(0.1p + 1)(0.02p + 1)} \cdot e^{-0.0033p}. \quad (5)$$

Based on these materials and the data given in [17], a digital model of ARV-REM was developed for use in the transient calculation program. The block diagram of the digital AEC model is shown in Fig. 2, and its parameters are shown in Table 5.

4 Results

To carry out the work, a digital model was used, created in the program for calculating transients and dynamic stability on the basis of a dynamic model of the unified energy system of Central Asia and South Kazakhstan.

The efficiency of the AEC systems for all settlement modes is evaluated [6, 18]:

- The decay time of electromechanical transients. To determine it, on the graph of the characteristic, two straight lines parallel to the t axis are drawn, which are $\pm 0.05h_{\text{steadystate}}$ from the steady-state value of $h_{\text{steady-state}}$ in either direction (5% tube). t_{tr} is the point in time when the transient response enters the 5% tube and does not exit anymore.
- Overregulation:

Table 5 Values of non-adjustable parameters of the ARV-REM model

Designations	Parameter name	Unit of measurement	Values
T_r	Measurement delay	s	0.028
K_a	Regulator gear ratio	p.u.	1
T_a	Controller time constant	s	0
T_g	Frequency averaging constant	s	0.9
K_e	Exciter transmission coefficient	p.u.	1
T_e	Exciter time constant	s	0.04
The gains of regulation			Customizable gains
K_{0U}	Voltage channel gain	p.u.	
K_{1U}	Derivative of the voltage	p.u.	
K_{1if}	Derivative of the rotor current	p.u.	
K_{0f}	Frequency change channel	p.u.	
K_{1f}	Derivative frequency change channel	p.u.	
Additional channels and restrictions			
K_{Lmax}	Low-frequency gain limit entry key	p.u.	1
K_{max}	Low-frequency boost gain	p.u.	100
T_u	AVR integral channel time constant	p.u.	0.9
V_{rmin}	Excitation regulator limitations	p.u.	−12
V_{rmax}		p.u.	12
E_{fmin}	Field voltage limitation	p.u.	−4
E_{fmax}		p.u.	4

$$\sigma = \frac{\left|h_{\max} - h_{\text{steady - state}}\right|}{h_{\text{steady - state}}} \cdot 100\%. \tag{6}$$

The transients can be aperiodic or oscillatory in nature. The transient response oscillation is determined by the magnitude of overregulation σ. Overregulation σ characterizes the degree to which the system moves away from the oscillatory stability boundary (if the system is on the oscillatory stability boundary, undamped oscillations are observed in the system and $\sigma = 100\%$). The stability margin is considered sufficient if $\sigma = 10\% \div 40\%$.

- Static error (%) is the ratio of the difference between the given $h_{\text{steady-state}}$ and the actual h_{actual} values of the adjustable parameter to the actual value:

$$\delta = \frac{\left|h_{\text{steady - state}} - h_{\text{actual}}\right|}{h_{\text{steady - state}}} \cdot 100\%. \tag{7}$$

Static error characterizes the accuracy of regulation in the steady state and usually $\delta \leq 3 \ldots 5\%$. As the first test disturbance, the regime was intensified by increasing the load at the output of the power plant. Thus, it is possible to obtain the response of AEC systems to this disturbance. The load was increased at the 1st second ($t_0 = 1$ s) after the start of the simulation. Figure 3 shows the results of test disturbance 1.

Designations used in Fig. 3:

U_g generator voltage, kV;
P active power of the generator, MW;
t simulation time, s ($t = 15$ s).

Define the values of the quality indicators of the characteristics of transients shown in Fig. 3a.

Transient decay time:

$$U_g(t) : t_{tr} = t_{tr}^{'} - t_0 = 1.3 - 1 = 0.3\,\text{s}.$$

$$P(t) : t_{tr} = t_{tr}^{'} - t_0 = 3.1 - 1 = 2.1\,\text{s}.$$

Overregulation:

$$U_g(t) : \sigma = \frac{\left|h_{\max} - h_{\text{steady - state}}\right|}{h_{\text{steady - state}}} \cdot 100\% = \frac{|18.85 - 20|}{20} \cdot 100\% = 5.75\%.$$

$$P(t) : \sigma = \frac{\left|h_{\max} - h_{\text{steady - state}}\right|}{h_{\text{steady - state}}} \cdot 100\% = \frac{|702 - 500|}{500} \cdot 100\% = 40.4\%.$$

Static error:

$$U_g(t) : \delta = \frac{\left|h_{\text{steady - state}} - h_{\text{actual}}\right|}{h_{\text{steady - state}}} \cdot 100\% = \frac{|20 - 19.9|}{20} \cdot 100\% = 0.5\%.$$

$$P(t) : \delta = \frac{\left|h_{\text{steady - state}} - h_{\text{actual}}\right|}{h_{\text{steady - state}}} \cdot 100\% = \frac{|500 - 493|}{500} \cdot 100\% = 1.4\%.$$

Quality indicators of transients of the remaining characteristics (Fig. 3) are summarized in Table 6.

From the characteristics of the transition process, it is obvious that the stability of the generators of the power plant was preserved in both versions of the AEC systems.

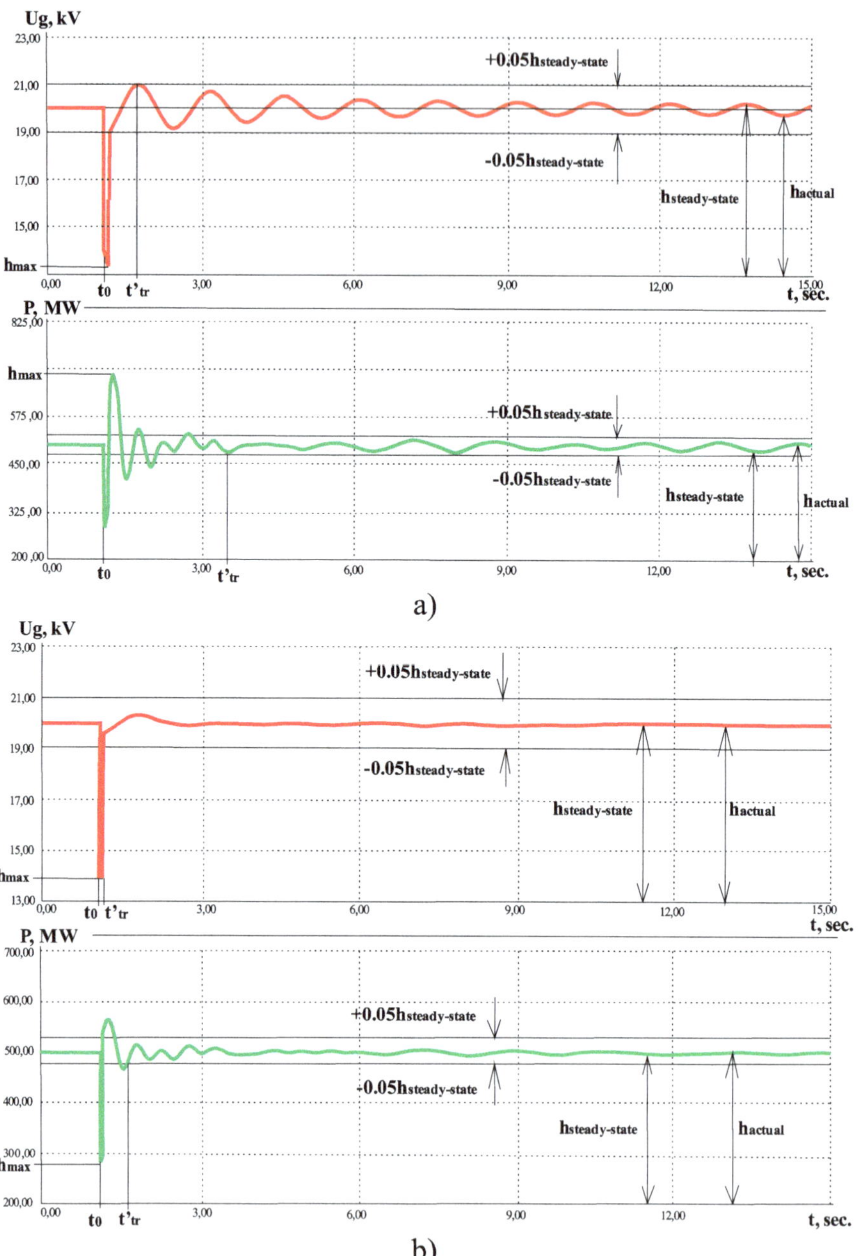

Fig. 3 Characteristics of the short circuit experience **a** AEC—SA; **b** AEC—PA

Table 6 Values of indicators of the quality of transient graphs

Indicators of the quality of the transients	Test disturbance 1		Test disturbance 2	
	Type of AEC system		Type of AEC system	
	AEC-PA	AEC-SA	AEC-PA	AEC-SA
	Characteristics of the transients		Characteristics of the transients	
	$\frac{U_g(t)}{P(t)}$	$\frac{U_g(t)}{P(t)}$	$\frac{U_g(t)}{P(t)}$	$\frac{U_g(t)}{P(t)}$
Decay time t_{tr}, s	$\frac{0.3}{2.1}$	$\frac{0*}{1.1}$	$\frac{0.2}{2.5}$	$\frac{0.1}{0.52}$
Overregulation σ, %	$\frac{5.75}{40.4}$	$\frac{3.9}{30}$	$\frac{33.75}{43}$	$\frac{31}{44}$
Static error δ, %	$\frac{0.5}{1.4}$	$\frac{1.25}{0.6}$	$\frac{0.1}{2}$	$\frac{0}{1}$

[a]The oscillation amplitude turned out to be less than $\pm 0.05 h_{\text{steady-state}}$

However, the introduction of additional control channels into AEC-SA systems provides higher performance and smaller voltage dips during transient conditions.

It is worth noting that the obtained rather large values of one of the indicators of the quality of transients—overregulation, which is associated with suboptimal tuning of the stabilization channels of the AEC systems.

5 Conclusions

The presented simulation results on the mathematical model of the Syrdarya TPP in DIgSILENT power factory showed that the use of AEC-SA can increase the stability margin of the electric power system. The data obtained allow us to count on the successful modernization of the AEC systems of the Syrdarya TPP by installing an AEC system of the ARV-REM type.

A separate study deserves the question of choosing the tuning parameters of the AEC-SA systems, the stabilization channels of which can improve the accuracy of regulation.

References

1. Power machines completes the first stage of modernization of the power units of Syrdarya TPP. https://www.uzdaily.uz/en/post/54456/. Accessed 05 Apr 2020
2. Pavlushko, S.A.: Automatic excitation control of synchronous generators as an effective means to ensure the reliable parallel operation of generation equipment and the united power system as a whole. Power Technol. Eng. **46**, 399–404 (2013)
3. Schaefer, R.C., Kim, K.: Excitation control of the synchronous generator. IEEE Ind. Appl. Mag. **7**(2), 37–43 (2001)
4. Klos, A.: Mathematical Models of Electrical Network Systems: Theory and Applications—An Introduction, p. 412. Springer International Publishing (2017)

5. Özdemir, M.T., Vedat, Ç.: Stability analysis of the automatic voltage regulation system with PI controller. Sakarya Üniv. Fen Bilimleri Enstitüsü Dergisi **21**(4), 698–705 (2017)
6. Anderson, P.M., Fouad, A.A.: Power System Control and Stability, 2nd edn. Wiley-Interscience, Wiley, Hoboken, NJ, USA (2002)
7. Štil, V.J., Mehmedović, M.: Interconnection and damping assignment automatic voltage regulator for synchronous generators. Int. J. Electr. Power Energy Syst. **101**, 204–212 (2018)
8. Ghosh, A., Sanyal, S., Das, A., Sanyal, A.: Automatic electronic excitation control in a modern alternator. In: Modelling and Simulation in Science, Technology and Engineering Mathematics. Advances in Intelligent Systems and Computing, vol. 749, pp. 397–406 (2018)
9. Ghamri, L.Y., Awadh, H., Al Shamsi, N., AlKhateri, S., Khurram, A., Rehman, H.: Robust AVR design for the synchronous generator. J. Eng. **2019**(17), 4111–4115 (2019)
10. Magnus, D., Carbonera, L., Pfitscher, L., Bernardon, D., Tavares, A., Scharlau, C.: Experimental and educational platform for operation tests and parameterization of power system regulators and stabilizers. IEEE Lat. Am. Trans. **17**(01), 54–62 (2019)
11. Allaev, K.R., Makhmudov, T.F.: Analysis of small oscillations of complex electrical systems. E3S Web Conf. **216**, 01097 (2020)
12. Allaev, K., Makhmudov, T.: Research of small oscillations of electrical power systems using the technology of embedding systems. Electr. Eng. **102**(1), 309–319 (2020)
13. Zimmer, H., Niersbach, B., Hanson J.: Optimization of power plant AVR parameters to improve transient voltage stability. In: 11th IEEE International Conference on Compatibility, Power Electronics and Power Engineering (CPE-POWERENG), pp. 71–76 (2017)
14. Komkov, A.L., Popov, E.N., Filimonov, N.Y., Yurganov, A.A., Burmistrov, A.A.: Implementing the system functions of the automatic proportional-derivative excitation control of synchronous generators. Power Technol. Eng. **53**, 356–359 (2019)
15. Chiniforoosh, S., Jatskevich, J., Yazdani, A., Sood, V., Dinavahi, V., Martinez, J.A., Ramirez, A.: Definitions and applications of dynamic average models for analysis of power systems. IEEE Trans. Power Delivery **25**(4), 2655–2669 (2010)
16. Guidelines for Testing Automatic High-Power Excitation Regulators of Synchronous Generators and Algorithms for Their Functioning, Appendix 3 to the Order of the Joint-Stock Company "System Operator of the Unified Energy System", no. 259 (2019)
17. Research and Production Enterprise RUSELPROM-ELECTROMASH. Technical catalog. http://ruselprom-kuzbass.ru/files/elmash.pdf. Accessed 05 Apr 2020
18. IEEE STD 421.5™-2016 (Revision of IEEE STD 421.5-2005): IEEE Recommended Practice for Excitation System Models for Power System Stability Studies. New York, NY, USA

Influence of Automatic Excitation Regulators on Modes of Hydropower Plants

Kakhraman Allaev, Obid Nurmatov, and Tokhir Makhmudov

Abstract The article discusses electrical systems that include pumping stations. Taking into account the fact that the pumping stations are equipped with synchronous motors, it is proposed to use their excitation system in the mode of outputting reactive power to the electrical system. Synchronous motors, due to their design features and high technical and economic indicators, are increasingly used in industry. The expediency of regulating the excitation of a synchronous motor operating on a variable load follows from the following considerations: with large changes in the power on the shaft, to ensure sufficient dynamic stability of the synchronous motor during the period of the load, its excitation must be maintained at a maximum level. To ensure favorable economic and energy performance of the synchronous motor during the period of no load, it is advisable to reduce the excitation current. To ensure the modes of the supply network with high technical and economic indicators, it is required that synchronous motors give maximum reactive energy to the network. The amount of reactive energy delivered, depending on the mode of the supply network and the load of synchronous motors, must change, which is achieved by the automatic excitation controllers (AEC) of synchronous motors according to certain laws. In the work, the results of virtual modeling of the steady state of an electric power system containing a large pumping station are obtained, and the values of voltages in the nodes and the level of losses in the electric network are determined.

K. Allaev · O. Nurmatov (✉) · T. Makhmudov
Tashkent State Technical University, Universitetskaya, 2, Tashkent, Uzbekistan 100095

A. Ronzhin and V. Shishlakov (eds.), *Electromechanics and Robotics*, Smart Innovation, Systems and Technologies 232, https://doi.org/10.1007/978-981-16-2814-6_33

1 Introduction

Synchronous motors, due to their design features and high technical and economic indicators, are increasingly used in industry. Along with such working machines as, for example, pump and fan units, gas and blowers, characterized by an even load curve, synchronous motors are used as a drive for a wide variety of mechanisms with a pronounced variable load, such as: cone, jaw and hammer crushers, ball and rod mills of mining enterprises, rubber mixers in the chemical industry, rocking machines in oil fields, shears and saws for metal, continuous rolling mills, piston compressors, drives for converting units of reversible rolling mills, mine hoisting machines, skip winches of blast furnaces, powerful excavators, and many others [1, 2].

The expediency of regulating the excitation of a synchronous motor operating on a variable load follows from the following considerations: with large changes in the power on the shaft, to ensure sufficient dynamic stability of the synchronous motor during the period of the load, its excitation must be maintained at a maximum level. To ensure favorable economic and energy performance of the synchronous motor during the period of no load, it is advisable to reduce the excitation current. The latter is easily confirmed if we consider the expression of the total losses of a synchronous motor in the following form [3]:

$$\Sigma P = P_{\text{m}} + P_{\text{s}} = I^2 r + i_v^2 r_v + k_{\text{s}} \Phi^2, \tag{1}$$

where are $P_{\text{m}} = I^2 r + i_v r_v$ the losses in the copper of the machine, which are the sum of the heat losses in the stator winding and the excitation winding; $P_{\text{s}} = k_{\text{s}} \Phi^2$ losses in machine steel. With load shedding and constant excitation current, the synchronous motor generally runs at a leading $\cos\varphi$. In this case, the armature reaction flux is magnetizing and the total magnetic flux of the machine increases. Consequently, the losses in the steel of the machine, proportional to Φ^2, increase significantly. Heat losses in the excitation winding under this condition remain unchanged, and the losses $i_v{}^2 r_v$ in the stator winding decrease, although, depending $I^2 r$ on the excitation mode, they can remain at the same level or even increase due to the action of the reactive current. Obviously, a decrease in the excitation current of a synchronous motor during load shedding would reduce the total losses in the machine and, as a consequence, its heating. In case of significant overloads on the motor shaft, in this case, it would be possible to allow large currents in the excitation winding to increase dynamic stability, without fear of overheating. Since the torque of a synchronous motor is determined by the ratio:

$$M = \frac{m U_{\text{H}} E}{\omega_{\text{c}} x} \sin \Theta, \tag{2}$$

it is clear that the regulation of its excitation depending on the load improves the torque utilization of the motor. To ensure the modes of the supply network with high technical and economic indicators, it is required that synchronous motors give maximum reactive energy to the network. The amount of reactive energy supplied, depending on the mode of the supply network and the load of synchronous motors, must change, which is achieved by AEC of synchronous motors according to certain laws. In this regard, AEC of synchronous motors can provide significant assistance in the development of issues of local voltage regulation in networks of industrial enterprises [4, 5].

2 Materials and Methods

AEC devices of synchronous motors interact closely with the entire excitation system. Usually, there are no special requirements for excitation systems in relation to the rate of rise of excitation of motors. The only exceptions are excitation systems for synchronous motors operating on a sharply variable load. In this case, as a rule, the excitation system must provide the possibility of forcing the excitation of the engine; in this case, the voltage and current in the excitation circuit should be maximum, and the time to reach the maximum (ceiling) value of the excitation current should be minimal. The latter is associated with the use of pathogens with a high voltage ceiling and, as a consequence, with a significant increase in the size of pathogens [6, 7]. The following basic requirements are imposed on AEC systems of synchronous motors [8]:

1. Maintaining stable operation when changing modes of the supply network.
2. Stability in ensuring the specified operating mode.
3. Simplicity and reliability of measuring the parameters by which the regulation is carried out and high sensitivity of the measuring elements-sensors.
4. High operational reliability of the regulator and the entire excitation control system.
5. Possibly a smaller delay associated with the inertia that exists both in the controllers themselves and in all elements of the excitation control system [9, 10].

3 Modeling

Let us analyze the effect of a pumping station as a voltage regulator using the example of a diagram of a 9-node electric power system (Fig. 1). The 9-node circuitry follows the IEEE 9-bus system standard of the Institute of Electrical and Electronics Engineers (IEEE), the world's largest professional technical organization [11–13].

The purpose of modeling and calculations here is to analyze the influence of the operating modes of pumping stations on the parameters of the regime of electric power systems.

The sequence of virtual simulation experiments:

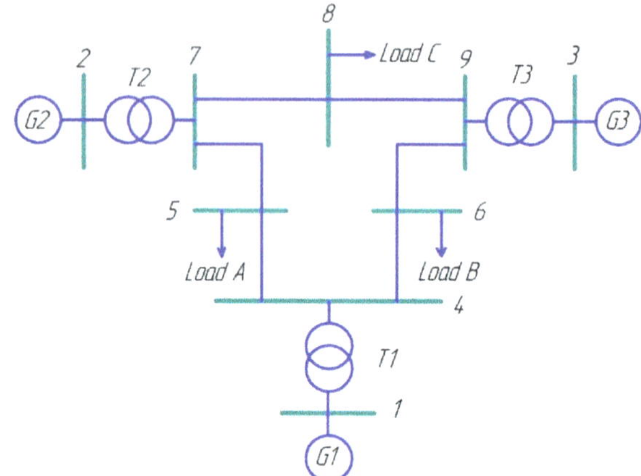

Fig. 1 Schematic diagram of the test 9-node power system

- The system model is implemented in the Power Factory simulator, then the steady state calculation is performed. The analysis of the steady state of a 9-node system is performed using the Newton–Raphson method. In this case, it is assumed that the synchronous motors of the pumping station operate in a static under-excitation mode with $\cos\varphi = 0.8$.
- The synchronous motor of the pumping station is transferred to the overexcitation mode ($\cos\varphi = 0.9$). The steady-state mode is calculated, and the mode parameters are determined.
- The values of active and reactive power losses in the test electrical system are determined.

Tables 1, 2, 3, and 4 show the parameters of the main electrical equipment of the test power system.

Table 1 Synchronous motor parameters

Engine brand	Nominal data					$\frac{M_{max}}{M_{nom}}$	Inertia (kg m^2)
	Active power (kW)	Total power (kVA)	Voltage (kV)	Rotation frequency (rpm)	Efficiency (%)		
SDN-17-89-6U3	4000	4580	6	1000	97,1	1,7	525

Table 2 Parameters of synchronous generators

Generator	Nominal data				Inductance (p.u.)			T_j (s)
	P (MW)	S (MVA)	U (kV)	$\cos\varphi_{nom}$	x_d	x'_d	x''_d	
G1	222	247.5	18	0,9	0,36	0,15	0,1	9,55
G2	163	192	18	0,85	1,72	0,23	0,2	4,165
G3	108	128	13,8	0,85	1,68	0,23	0,2	2,765

Table 3 Initial data of the power system by branches

Branch start	End of branch	R (Ω)	X (Ω)	$Bc \cdot 10^{-6}$ (S)
1	4	0	30,47	0
4	5	5,29	44,965	332,7
4	6	8,993	48,668	298,69
5	7	16,928	85,169	578,45
6	9	20,631	89,93	676,75
3	9	0	31	0
8	9	6,295	53,323	395,08
7	8	4,496	38,088	281,166
2	7	0	33,067	0

Table 4 Initial data of the power system by nodes

Node	U_{nom} (кV)	P_{Load} (MW)	Q_{Load} (MVAr)
Load A	220	125	50
Load B	220	90	30
Load C	220	100	35

In Fig. 2 shows the P-Q diagram of a synchronous motor, thus, the area of admissible operating modes of the motor in the axes of coordinates "PQ" is limited to the right by the rated excitation current, from above by the rated active power and on the left by the minimum consumed reactive power [14].

The results of calculating the steady state for the first simulation experiment (synchronous motor of the pumping station in under-excitation mode) are shown in Fig. 3.

Tables 5, 6, and 7 summarize the main results of the steady state calculation.

The results of calculating the steady state for the second simulation experiment (synchronous motor of the pumping station in overexcitation mode) are shown in Fig. 4.

Tables 8, 9, and 10 summarize the main results of the steady-state calculation.

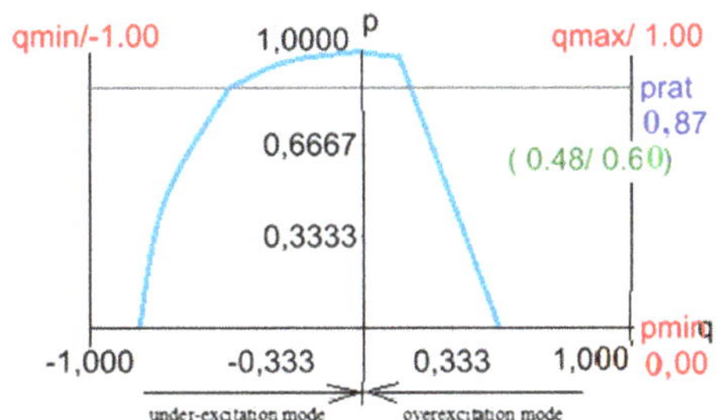

Fig. 2 P-Q diagram of brand synchronous motor SDN-17-89-6U3

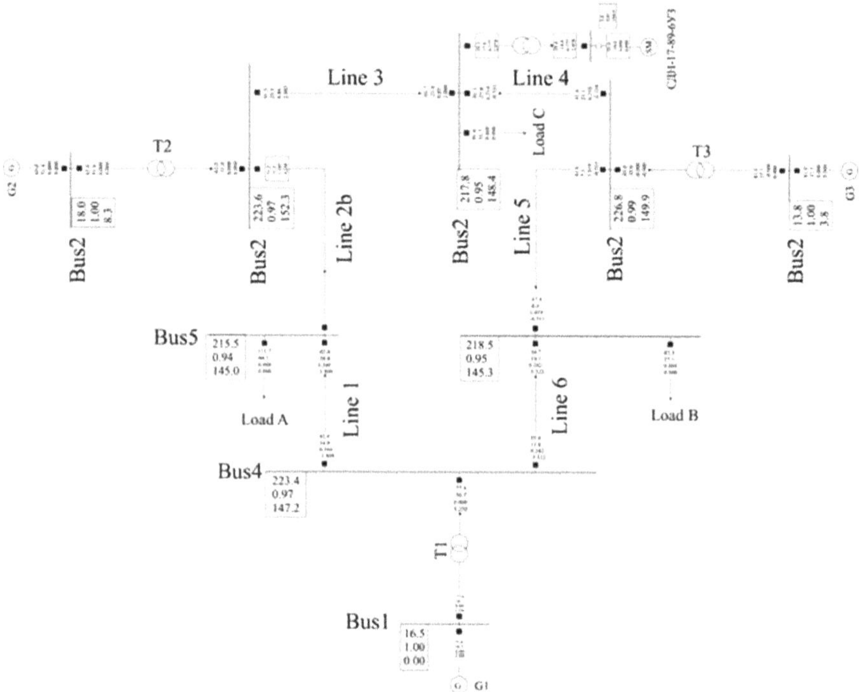

Fig. 3 Results of calculating the steady state for the first simulation experiment

Table 5 Values of **voltages** by nodes, as well as the levels of generation and consumption for the first simulation experiment

Node	U_{nom} (kV)	U_{actual} (kV)	P_{gen} (MW)	Q_{gen} (MVAr)	P_{load} (MW)	Q_{load} (MVAr)
1	18	18	106,9	64		
2	18	18	163	58,2		
3	13,8	13,8	85	41,7		
4	230	222				
5	230	212,5			125	50
6	230	216,1			90	30
7	230	222,9				
8	230	216,9			100	35
9	230	224,7				
SM	6	5,8			35	12,3

Table 6 Levels of losses active and reactive capacity in the electrical network

Branch start	End of branch	ΔP (MW)	ΔQ (MVAr)
1	4	0	8,95
4	5	0,564	0,073
4	6	0,473	−2,239
5	7	1,661	3,613
6	9	0,86	−1,113
3	9	0	5,251
8	9	0,312	−2,237
7	8	0,89	2,7
2	7	0	18,72
8	SM	0,173	2,863

Table 7 Summary indicators of the steady-state calculation for the first simulation experiment

	P (MW)	Q (MVAr)	S (MVA)
Generation	354,93	163,88	390,94
Static load	315	115	335,34
Motor load	35	12,3	37,1
Losses	4,93	36,58	
Charging power Power line		−28,47	

Table 11 shows the comparative results of voltage levels and power losses at different operating modes of the synchronous motor of the pumping station.

4 Conclusions

Based on the analysis of virtual simulation experiments, the following conclusions can be drawn:

1. The transfer of synchronous motors of the pumping station to the overexcitation mode made it possible to increase the voltage level in the nodes of the power system up to 1%, while the total charging power of the lines increased by 0.48 MVAr due to the increase in the voltage in the nodes.
2. The total losses of reactive power in the power system decreased by 4.05 MVAr.

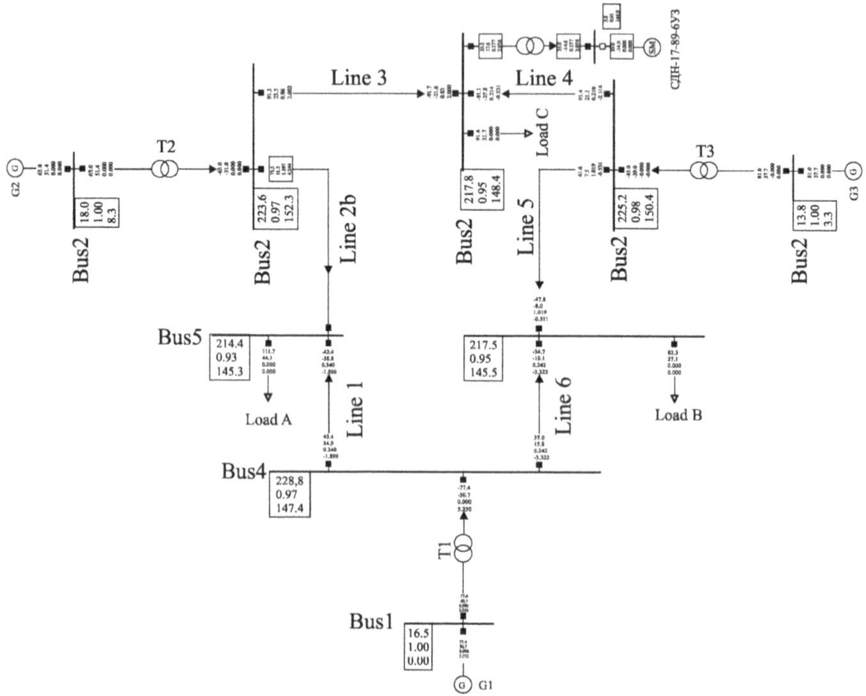

Fig. 4 Results of calculating the steady state for the second simulation experiment

Table 8 Values of **voltages** by nodes, as well as the levels of generation and consumption for the second simulation experiment

Node	U_{nom} (kV)	U_{actual} (kV)	P_{gen} (MW)	Q_{gen} (MVAr)	P_{load} (MW)	Q_{load} (MVAr)
1	18	18	106,7	59,1		
2	18	18	163	43,3		
3	13,8	13,8	85	29,1		
4	230	222,6				
5	230	213,7			125	50
6	230	217,2			90	30
7	230	225				
8	230	221,9			100	35
9	230	226,4				
SM	6	6,2			35	−16

Table 9 Levels of losses of active and reactive capacity in the electrical network

Branch start	End of branch	ΔP (MW)	ΔQ (MVAr)
1	4	0	8,576
4	5	0,537	−0,201
4	6	0,456	−2,367
5	7	1,661	3,542
6	9	0,87	−1,129
3	9	0	4,731
8	9	0,233	−3,052
7	8	0,818	1,938
2	7	0	17,77
8	SM	0,174	2,719

Table 10 Summary indicators of the steady-state calculation for the second simulation experiment

	P (MW)	Q (MVAr)	S (MVA)
Generation	354,75	131,52	378,34
Static load	315	115	335,34
Motor load	35	-16,02	38,49
Losses	4,75	32,53	
Charging power Power line		−28,95	

Table 11 Comparative table of mode parameters at various modes of the synchronous motor

Node	Mode 1*	Mode 2	Branch	Mode 1*	Mode 2
	Voltage (kV)			Losses (MVA)	
1	18	18	1–4	0 + j8,95	0 + j8,576
2	18	18	4–5	0,564 + j0,073	0,537–j0,201
3	13,8	13,8	4–6	0,473–j2,239	0,456–j2,367
4	222	222,6	5–7	1,661 + j3,613	1,661 + j3,542
5	212,5	213,7	6–9	0,86–j1,113	0,87–j1,129
6	216,1	217,2	3–9	0 + j5,251	0 + j4,731
7	222,9	225	8–9	0,312–j2,237	0,233–j3,052
8	216,9	221,9	7–8	0,89 + j2,7	0,818 + j1,938
9	224,7	226,4	2–7	0 + j18,72	0 + j17,77
SM	5,8	6,2	8-SM	0,173 + j2,863	0,174 + j2,719
Total losses (MVA)				4,93 + j36,58	4,75 + j32,53

Note Mode 1—operation of a synchronous motor in under-excitation mode; Mode 2—operation of a synchronous motor in overexcitation mode

References

1. Allaev, K.R., Makhmudov, T.: Analysis of small oscillations of complex electrical systems. E3S Web Conf. **216**, 01097 (2020)
2. Shi, J., Xu, Y., Liao, M., Guo, S., Li, Y., Ren, L., Tang, Y.: Integrated design method for superconducting magnetic energy storage considering the high frequency pulse width modulation pulse voltage on magnet. Appl. Energy **248**, 1–17 (2019)
3. Xu, B., Chen, D., Behrens, P., Ye, W., Guo, P., Luo, X.: Modeling oscillation modal interaction in a hydroelectric generating system. Energy Convers. Manage. **174**, 208–217 (2018)
4. Allaev, K., Makhmudov, T.: Research of small oscillations of electrical power systems using the technology of embedding systems. Electr. Eng. **102**(1), 309–319 (2020)
5. Pimenta, F.M., Assireu, A.T.: Simulating reservoir storage for a wind-hydro hydrid system. Renew. Energy **76**, 757–767 (2015)
6. Allayev, K., Nurmatov, O., Makhmaraimova: Calculastion experimental studies of transition processes in electricity systems with account of hydroenergy installations. J. Crit. Rev. **7**(13) (2020)
7. Pérez-Díaz, J.I., Chazarra, M., García-González, J., Cavazzini, G., Stoppato, A.: Trends and challenges in the operation of pumped-storage hydropower plants. Renew. Sustain. Energy Rev. **44**, 767–784 (2015)
8. Nurmatov, O.: Large pumping stations as regulators of power systems modes. E3S Web Conf. **216**, 01098 (2020)
9. Ghosh, A., Sanyal, S., Das, A., Sanyal, A.: Automatic electronic excitation control in a modern alternator. In: International Conference on Modelling and Simulation, pp. 397–406 (2017)
10. Mirzabaev, A., Isakov, A.J., Mirzabekov, S., Makhkamov, T., Kodirov, D.: Problems of integration of the photovoltaic power stations with the grid systems. IOP Conf. Ser. Earth Environ. Sci. **614**(1), 012016 (2020)
11. Salimi, A.A., Karimi, A., Noorizadeh, Y.: Simultaneous operation of wind and pumped storage hydropower plants in a linearized security-constrained unit commitment model for high wind energy penetration. J. Energy Storage **22**, 318–330 (2019)
12. Yang, W., Yang, J., Zeng, W., Tang, R., Hou, L., Ma, A., Peng, Y.: Experimental investigation of theoretical stability regions for ultra-low frequency oscillations of hydropower generating systems. Energy **186**, 115816 (2019)
13. Chiniforoosh, S., Jatskevich, J., Yazdani, A., Sood, V., Dinavahi, V., Martinez, J.A., Ramirez, A.: Definitions and applications of dynamic average models for analysis of power systems. IEEE Trans. Power Delivery **25**(4), 2655–2669 (2010)
14. Komkov, A.L., Popov, E.N., Filimonov, N.Y., Yurganov, A.A., Burmistrov, A.A.: Implementing the system functions of the automatic proportional-derivative excitation control of synchronous generators. Power Technol. Eng. **53**(3), 356–359 (2019)

Issues of Energy-Efficient Storage of Fuel in Multimodal Transport Units

Eugene Soldatov and Aleksey Bogomolov

Abstract The problem issues of non-drainage storage of cryogenic fuels in modern modular long-term storage systems are considered. The structure of software for remote monitoring the container and small-scale reservoir equipment for storage and transportation of cryogenic fuel is presented. The relationship of software modules for computational modeling of hydrodynamic and thermodynamic processes during storage of cryogenic fuel in stationary and transport conditions, a module for calculating the holding time of non-drainage storage, and individual data transmission modules installed on cryogenic tanks are disclosed. The composition of the software of operators responsible for remote monitoring of the state of multimodal transport units during transportation and long-term storage at the point of consumption is described. An example of energy-efficient technical solution for increasing the holding time of non-drainage storage of liquefied natural gas during storage and transportation in multimodal transport units are considered. In the autonomous power supply system of the multimodal transport unit under consideration, the conversion of solar energy is used to charge accumulator batteries. The results of calculations of electricity consumption for increasing the holding time up to values of 150–200 days, demanded by consumers of cryogenic fuel, are presented.

1 Introduction

During non-drainage storage in stationary reservoirs and multimodal units of cryogenic fuel, in particular, liquefied natural gas (LNG), there is a constant process of evaporation of part of the liquid due to external heat gain [1, 2], because of the impossibility of technical realization of the perfect heat-insulated capacity, even in the case of usage screen-vacuum superinsulation, which is the most effective among all currently used in industry [3, 4]. The power of the heat flow through the insulation,

E. Soldatov · A. Bogomolov (✉)
St. Petersburg Federal Research Center of the Russian Academy of Sciences (SPC RAS), St. Petersburg Institute for Informatics and Automation of the Russian Academy of Sciences, 39, 14th Line, 199178 St. Petersburg, Russia

A. Ronzhin and V. Shishlakov (eds.), *Electromechanics and Robotics*, Smart Innovation, Systems and Technologies 232, https://doi.org/10.1007/978-981-16-2814-6_34

even to tanks of small volume 10–20 m^3, usually exceeds 100 W, and in case of slight losses of the vacuum in the heat-insulating space, it increases significantly.

The development of low-tonnage production of cryogenic LNG fuel was associated with the need to cover peak loads of natural gas consumption, as well as to supply remote areas where it is economically impractical to build gas pipelines. The problem of lack of pipeline gas in remote settlements (up to 50%) in the future can be successfully solved with the help of LNG [5–7].

The study of the problem of modeling heat and mass transfer processes is most interesting when considering cryogenic tanks of small volume: in the case of using cryogenic ISO containers, this is about tanks with a volume of up to 40 m^3. They are mainly used at low-tonnage production facilities and the consumption of industrial gases. The most important thing is the relative average daily losses of the product in small tanks, as a rule, are higher than 1%, even when screen-vacuum superinsulation is used in the construction of cryogenic tanks.

In economically feasible cases, to reduce or completely eliminate losses during storage of cryogenic fuel, various systems for recondensing vapors of cryogenic products are additionally used. In this case, the task is complicated by the need to select an acceptable scheme for the realization of the refrigeration cycle, determine the effective mode of its operation, and ensure monitoring of equipment parameters. Of course, solving these problems requires usage computer modeling methods to support decision making on the development and operation of such technologies.

The study of the efficiency of installations for reverse condensation of cryogenic product vapors is especially important for tanks with liquefied natural gas of small volume, although vapor recondensation systems BOG are also used for large tanks [1, 8–10]. However, large tank systems usually have different refrigeration circuit diagrams and, therefore, use completely different equipment.

In our case, this is mainly about multimodal ISO containers that allow the transportation of a batch of cryofuel by various types of transport (road, river, sea, air) without any refueling at the reference points of transportation [11, 12]. ISO containers are also convenient for loading and unloading operations, since they can be stored in several tiers (see Fig. 1), and also used as a temporary storage base.

This allows ones to offer customers a very convenient solution for organizing a storage base for cryogenic products, while the customer does not need to build a foundation for tank equipment. Road or airfield slabs and reinforced concrete blocks can be used as a supporting surface for temporary mounting of the container on the customer's base.

Fig. 1 Multimodal transport units for LNG storage and transportation [13]

2 Remote Monitoring of the Equipment for Storage and Transportation of Cryofuels

2.1 Structure of Software for Monitoring Remote Equipment for Storage and Transportation of Cryofuel

The main problem issue in the operation of equipment for storage and transportation cryofuel is the study of thermodynamic processes inside a cryogenic tank, for which computer modeling systems are currently used [13–15].

The process of solving the system of equations of hydrodynamic processes and heat and mass transfer processes during storage of cryofuel in a container requires serious computational resources, and the duration of a numerical solution can reach several days or weeks even for simplified models [2, 16–18]. Because of this, it was proposed a separate tasks of determining the main thermophysical parameters of a cryofuel in the vapor space of a tank or container and calculating the holding time of cryofuel [2, 19, 20].

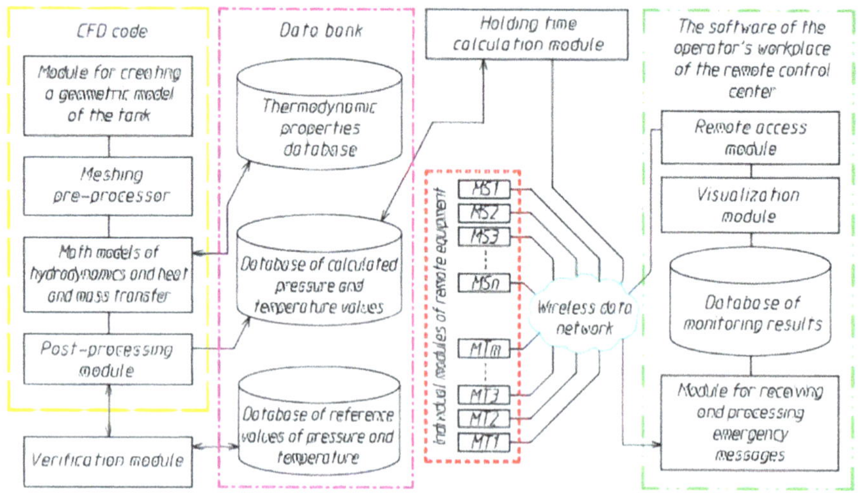

Fig. 2 Structure of software for monitoring remote equipment for storage and transportation of cryoproducts in stationary tanks and ISO containers

Figure 2 shows a block diagram presented the relationship of software systems for solving problems of computer modeling of thermodynamic processes and monitoring of thermophysical parameters of cryogenic equipment of various types. The preparation of computer models was carried out in a simulation software for finite element analysis. When preparing the models, the geometric parameters of the tanks loaded from an external CAD library were used as initial information. The main data for setting the boundary and initial conditions for CFD code were the data on the standard values of heat gain through the superinsulation, determined by the specific manufacturer of the tanks in the passport or another technical documentation [2, 21].

The algorithm of the computational module is based on the lookup table principle for holding time prediction. This solution helps to avoid the need to perform time-consuming calculations of unsteady temperature fields every time on demand. Instead, the interaction of the computational module with the main database of calculated values of pressure and temperature was established. The accumulation of this database was carried out beforehand according to the CFD results of modeling in the form of unsteady temperature fields and pressures in the vapor space of the inner tank [6, 15, 18, 22]. The modeling results can be obtained in both commercial software (ANSYS Fluent, STAR CCM+, COMSOL Multiphysics) and open-source systems (OpenFOAM). Then upon a request from the computing module, an array of pressure and storage time data is returned, from the elements of which the predicted value of the holding time is subsequently determined.

The verification module compares table values of gas pressure and cryofuel temperature values in the vapor phase, after which the models can be corrected.

2.2 About the Software of the Operator's Remote Workplace of the Cryogenic Equipment Control Center

The main feature of application the described software is the ability take into account when making decisions not only the current readings of pressure and level sensors, but also the history of the storage and transportation processes. The application of a wireless data network allow ones to eventually include various tanks and multimodal transport units to the monitoring system, regardless of the distance between tank and data center, as well as regardless of the distance between containers from each other.

The software of the operator's workplace of the remote control center contains a remote access module, a module for visualizing the results of calculations the holding time and the main monitored thermophysical and operational parameters, and a module for receiving and processing emergency messages.

The key difference of individual computer modules for transport cisterns or multimodal transport units MT1, MT2, … from the modules of stationary reservoirs MS1, MS2, … is the technical feasibility to receive information from sensors of mechanical vibrations arising in the process of moving tanks by any means of transport: road, rail, sea, air [1, 23].

The key operational information from the reservoirs and multimodal transport units is sent to the remote control center. First of all, it is data on the pressure and level of liquid cryofuel in each tank, the vacuum level driving the technical state of the thermal insulation. The visualization module displays the current storage and transportation regimes at a given time: regular or emergency, stationary or transport, etc.

It should be noted that when the process of heat and mass transfer approaches equilibrium in the transport mode, it is necessary to take into account the movement of the vapor–liquid interface during the storage process. According to international standards, the maximum liquid filling level of tanks and ISO containers with flammable gases should be such that, after the main volume of liquid has completely warmed up to an equilibrium temperature for maximum operating pressure, the liquid level does not exceed 98% [24]. The software is also used to support decision making when choosing the correct initial level of liquid cryofuel.

3 Technical Solutions and Calculation Results

3.1 Efficiency of a Cryogenerators Based on a Stirling Cycle

Natural gas vapors can also be condensed using gas refrigeration machines [1, 7, 20]. It is advisable to use a modern unit operating according to the reverse Stirling cycle using a helium cycle as a working medium. The main advantages of Stirling cryogenerators are ease of maintenance and reliability.

The refrigerating capacity of a real Stirling refrigeration machine was evaluated using the formula:

$$Q_{\text{real}} = Q_{\text{ideal}} \eta_{\text{rel}}$$

where Q_{ideal} is the refrigerating capacity of an ideal refrigerating machine, η_{rel} is the relative efficiency, which determines the losses due to external heat gains, hydraulic resistances, losses from underrecovery, and other energy losses. At the investigated cryogenic temperature level, the value of the relative efficiency for Stirling cryogenerator does not exceed 0.3. The calculated degree of thermodynamic perfection of the unit based on the Stirling cryogenerator was 0.132 [7].

In stationary systems, the power supply to the refrigeration machine engine can be organized in a standard way from the external electrical grid. For transport installations, an autonomous power supply system should be organized.

3.2 Multimodal Containers with an In-Vehicle Autonomous Power Supply System

The development and wide use of multimodal transport units allows solving problem issues with the transportation of cryogenic products to remote areas of the country by various types of transport without refueling, which in the recent past was an unsolvable or economically impractical task. Based on the results of the analysis of circuit solutions in vapor recondensation systems at various cycles, a number of technical solutions for energy-efficient multimodal units have been developed.

The problem of long-term storage of cryoproducts at the point of consumption is urgent, due to absence of external sources of cold at the point of operation. In addition to delivering cryofuel to remote areas, the tasks of delivering and storing liquid nitrogen for various technological tasks (cooling equipment, inerting pipelines) or liquid argon for filling cylinders on site (providing contractor personnel with inert gas for welding) can also be solved. The use of energy-efficient solutions in modular delivery vehicles has also been made possible by progress in the development of new efficient accumulator batteries [7, 8, 25–27]. Figure 3 presented a transport tank container related for transportation and long-term storage of liquefied natural gas.

The main task of the multimodal unit upgrades is to ensure the LNG transportation conditions over long distances without intermediate refueling by various modes of transport and ensuring long-term storage of the liquid product at the point of consumption without loss to the atmosphere. This is ensured by equipping the tank container with an autonomous vapor recondensation system, which includes a cryogenerator with an electric motor, to which an automatic control system is connected and an inverter connected to the main and backup battery. The pressure in the cryogenic tank is reduced. The switching the electric motor of the gas refrigeration machine occurs upon a signal from the system controller. During operation of the

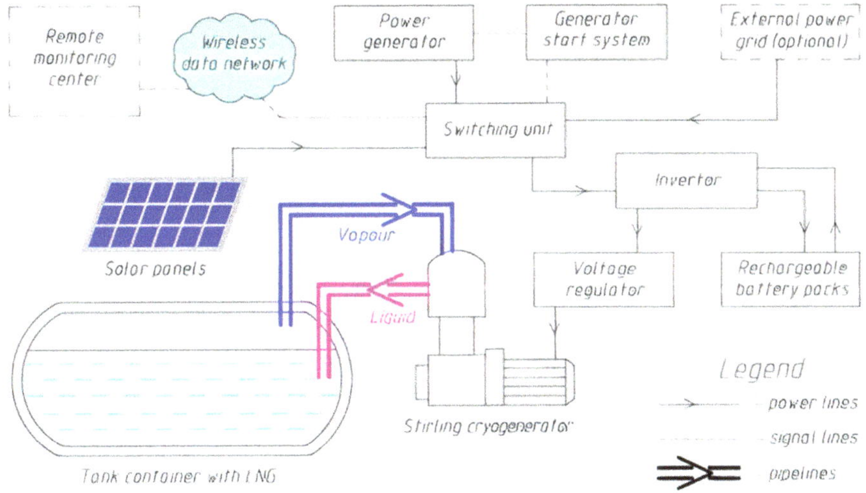

Fig. 3 Multimodal transport unit with in-vehicle autonomous power supply system

system, the pressure and level of the liquid in the tank are constantly monitored. When the upper pressure setpoint is reached, the engine of cryogenerator starts. The pressure gradually decreases to normal pressure value, after which the cryogenerator motor shut off.

The result is the prevention of product losses during the entire period of transportation, and storage of one batch of filled liquefied gas from the moment the production base is released until the end of the process of consumption by the customer. In the absence of unscheduled gas discharges into the atmosphere during normal operation, it is also guaranteed that no explosive mixtures of natural gas with air are formed, which ensures increased production safety. The minimum specific electricity consumption for the recondensation of natural gas vapors, taking into account losses in all units of the refrigeration unit, is 5–10 kW h per 1 m^3 of the gas space of the tank.

If the screen-vacuum thermal insulation is in good working order (vacuum depth 10^{-4}–10^{-3} mm Hg), the discharge of natural gas vapors into the atmosphere is completely excluded during the entire storage period of one batch of cryogenic product, which provides increased safety and eliminates the loss of an expensive product during transportation and storage.

A distinctive feature of the container is the presence of polycrystalline solar panels fixed in the upper part of the container body, which charge the main and backup batteries, and thus ensure complete autonomy of the system. The specific electrical power of the panels is not less than 140 W per 1 m^2 of area, and the solar panels are also protected by a coating of textured impact-resistant glass. The solar-hybrid inverter converts 24 V DC to 220 V AC to power the refrigeration motor and also contains a solar charge controller. The system can be powered from the external electrical grid, if there is access to it in order to recharge the accumulator batteries.

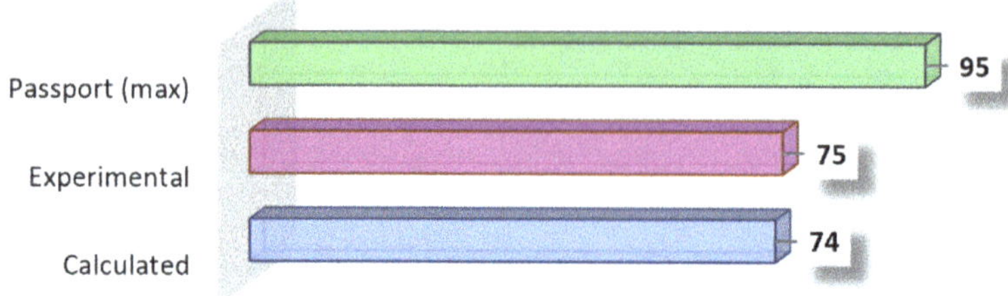

Fig. 4 Theoretical and experimental values of LNG holding time in a multimodal 40 ft tank container

If there is no connection to the external electrical network, the power of the electric motor of the cryogenerator comes from the battery. During the daytime, the batteries are recharged from solar panels. With such a scheme, there is no need for regular recharging from an external power grid: the container is completely autonomous, and the holding time for cryofuel can be increased to values of 150–200 days or more, if necessary.

3.3 Fuel Storage Time Data in Multimodal Transport Units

In Fig. 4, empirical and calculated data on the holding time of LNG in multimodal transport units with a volume of 40 m^3 are presented, showing the example of negligible disparity of the calculation results with experimental data.

The total holding time was taken for multimodal transport units in mixed transport conditions when changing road transport to sea transport, as well as taking into account the long-term stationary storage of the container in the seaport and at the customer's facility.

4 Conclusion

The structure of software for monitoring the processes of transportation and long-term storage of cryofuel, using information from the database of the results of computer modeling of heat and mass transfer processes in a cryogenic tank, is considered. Further areas of research in the field of monitoring the state of equipment for storage and transportation of cryofuel is about an accumulation of a database of reference temperature and pressure values for various types of stationary tanks and multimodal transport units. And, of course, this is about the improvement of software tools for solving the system of differential equations of hydrodynamics and heat and mass transfer in order to reduce time and computational costs.

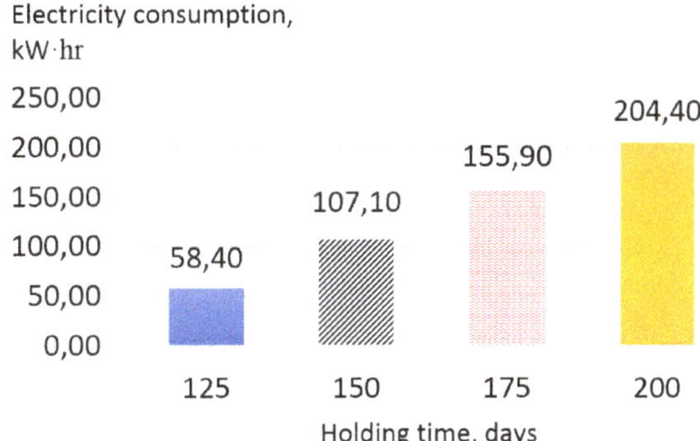

Fig. 5 Estimated values of electricity consumption for increasing LNG holding time to the specified value

In the proposed example for the practical realization of a multimodal transport unit due to the presence of a cryogenerator and an autonomous power supply system, the holding time of a batch of liquefied natural gas can be increased up to 1.5–2 times (see Fig. 5), compared to the theoretical holding time indicated in the passport for a tank for a specific cryogenic product.

Acknowledgements The study was carried out with state support from leading scientific schools Russian Federation, grant No. NSh-2553.2020.8.

References

1. Soldatov, E.S.: Modeling software and monitoring of processes in reservoirs and tanks for long storage of cryogenic products. Bull. Tula State Univ. Tech. Sci. **10**, 385–393 (2019)
2. Yang, H.Q., West, J.: CFD extraction of heat transfer coefficient in cryogenic propellant tanks. In: 51st AIAA/SAE/ASEE Joint Propulsion Conference, pp. 1–27 (2015)
3. Larkin, E., Bogomolov, A., Feofilov, S.: Stability of digital feedback control systems. In: MATEC Web of Conferences, vol. 161, p. 02004 (2018)
4. Wlodek, T.: Prediction of boil-off rate in liquefied natural gas storage processes. In: 17th International Multidisciplinary Scientific GeoConference SGEM 2017, Section Oil and Gas Exploration, pp. 405–413 (2017)
5. Chen, Y.-G., Price, W.G., Temarel, P.: Numerical simulation of liquid sloshing in LNG tanks using a compressible two-fluid flow model. In: Proceedings of the 19-th International Offshore and Polar Engineers, pp. 221–230 (2009)
6. Larkin, E., Bogomolov, A., Gorbachev, D., Privalov, A.: About approach of the transactions flow to poisson one in robot control systems. In: Lecture Notes in Computer Science, 10459 LNAI, pp. 113–122 (2017)
7. Soldatov, E.S., Arkharov, I.A.: Analysis of diagram solutions in the systems of recondensation of liquefied natural gas vapors for transport and stationary tanks of long-term storage. Bull. Tula State Univ. Tech. Sci. **2**, 263–276 (2019)
8. Adom, E., Islam, S.Z., Ji, X.: Modelling of boil-off gas in LNG tanks: a case study. Int. J. Eng. Technol. **2**(4), 292–296 (2010)
9. Domashenko, A.M.: Heat and mass transfer and hydrodynamics in cryogenic fuel systems for ground-based and sea-based objects. Int. Sci. J. Altern. Energy Ecol. **3**, 12–60 (2009). (in Russ.)

10. Larkin, E., Bogomolov, A., Privalov, A., Antonov, M.: About one approach to robot control system simulation. In: Interactive Collaborative Robotics ICR 2018. Lecture Notes in Computer Science, 11097, pp. 159–169 (2018)
11. Bychkov, E.V., Bogomolov, A.V., Kotlovanov, K.Y.: Stochastic mathematical model of internal waves. Bull. South Ural State Univ. Ser. Math. Model. Program. Comput. Softw. **13**(2), 33–42 (2020)
12. Zienkiewicz, O.C. Taylor, R.L., Nithiarasu, P.: The Finite Element Method for Fluid Dynamics, p. 544. Butterworth-Heinemann, Elsevier (2014)
13. Small Scale LNG Solutions. https://www.broadviewenergysolutions.com/lng-container-leasing/. Last accessed 28 Jan 2021
14. Daigle, M.J., Smelyanskiy, V.N., Boschee, J., Foygel, M.: Temperature stratification in a cryogenic fuel tank. J. Thermophys. Heat Transfer **27**(1), 116–126 (2013)
15. Larkin, E., Bogomolov, A., Privalov, A.: Data buffering in information-measuring system. In: 2nd International Ural Conference on Measurements (UralCon), pp. 118–123 (2017)
16. Dobrota, Đ, Lalić, B., Komar, I.: Problem of boil-off in LNG supply chain. Trans. Marit. Sci. **2**(02), 91–100 (2013)
17. Hariti, R., Fekih, M., Saighi, M.: Numerical simulation of heat transfer by natural convection in a storage tank. Int. J. Appl. Innov. Eng. Manag. (IJAIEM) **2**(8), 340–343 (2013)
18. Larkin, E., Akimenko, T., Bogomolov, A., Krestovnikov, K.: Mathematical model for evaluating fault tolerance of on-board equipment of mobile robot. Smart Innov. Syst. Technol. **187**, 383–393 (2021)
19. Larkin, E., Bogomolov, A., Privalov, A.: Discrete model of mobile robot assemble fault-tolerance. In: Lecture Notes in Computer Science, 11659 LNAI, pp. 204–215 (2019)
20. Roh, S., Son, G.: Numerical study of natural convection in liquefied natural gas tank. J. Mech. Sci. Technol. **26**(10), 3133–3140 (2012)
21. Ryou, Y.-D., Lee, J.-H., Jo, Y.-D.: Internal pressure variation analysis and actual holding time test on ISO LNG tank container. J. Korean Inst. Gas **17**(6), 1–7 (2013)
22. Motienko, A.I., Ronzhin, A.L., Basov, O.O., Zelezny, M.: Modeling of injured position during transportation based on Bayesian belief networks. Adv. Intell. Syst. Comput. **451**, 81–88 (2016)
23. Mokhatab, S., Mak, J.Y., Valappil, J.V., Wood, D.A.: Handbook of Liquefied Natural Gas, 1st edn, p. 593. Elsevier, Oxford (2014)
24. European Agreement Concerning the International Carriage of Dangerous Goods by Road (ADR), vol. II. United Nations, New York and Geneva (2018)
25. Polinski, J.: Modeling of multilayer vacuum insulation—complexity versus accuracy. In: Proceedings of the Twentieth International Cryogenic Engineering Conference (ICEC20), pp. 793–796 (2006)
26. Kandoliya, P.D., Mehta, N.C.: Recent research on cryogenic storage tank: a review. Int. J. Res. Appl. Sci. Eng. Technol. (IJRASET) **4**(5), 1681–1686 (2017)
27. Zakaria, M.S., Osman, K., Saadun, M.N.A., Manaf, M.Z.A., Mohd Hanafi, M.H.: Computational simulation of boil-off gas formation inside liquefied natural gas tank using evaporation model in ANSYS fluent. In: Applied Mechanics and Materials, vol. 393, pp. 839–844 (2013)

"Smart Well" Concept in Oil Production

Anton Yashin, Alexander Konev, and Marat Khakimyanov

Abstract Smart well technology is widely implemented in oil production at present. There are about 2 million oil wells around the world, and artificial lift is used for almost 1 million of them. Sucker rod pumps are used in over 750,000 artificial lift wells. A modern well is equipped with a variety of process parameter sensors. These are sensors for pressure, temperature, flow, level, current, voltage, and power. The well controller receives information from sensors. The controller performs preliminary information processing and regulates the operating mode of the electric motor. The article discusses sensors installed on various types of borehole pumping units: electric submersible pumping units, sucker rod pumping units, and progressive cavitation pumping units. The structures of a smart well and a smart field are given. Examples of measurements of electrical parameters at wells are given. To control the operation of the pump, electric drive, current, voltage, active power, reactive capacity, and power factor are measured.

1 Introduction

Downhole oil production is a very difficult and expensive process. The submersible pump is located in the well and is not accessible to personnel. Oil production from wells requires a lot of electricity. The pumping equipment is very expensive. Well workover is very expensive. But oil prices have dropped and remain volatile.

Many factors affect oil production equipment. Such factors include the high temperature, pressure, aggressive chemical environment, particulates, and paraffin formation in the wellbore fluid.

Automation of oil production allows you to avoid equipment breakdowns and optimize well operation. Automation systems are expensive but pay off quickly.

Many scientists were involved in the development of well automation systems, for example Gibbs [1, 2], Aliev et al. [3], Zyuzev et al. [4, 5], Blyuk et al. [6], Vázquez et al. [7], Nederlof and Stephanie [8], Teves et al. [9], and other.

A. Yashin · A. Konev · M. Khakimyanov (✉)
Ufa State Petroleum Technological University, Ufa, Russia

A. Ronzhin and V. Shishlakov (eds.), *Electromechanics and Robotics*, Smart Innovation, Systems and Technologies 232, https://doi.org/10.1007/978-981-16-2814-6_35

Smart well technology frees personnel from traveling to remote fields, reduces energy consumption, and reduces equipment wear.

2 Relevance

Oil producing enterprises have hundreds and thousands of wells. Wells are dispersed over large areas. Distances to wells can reach several hundred kilometers. There are often no roads in such places. It is difficult for operators to reach remote wells. Personnel cannot frequently monitor such wells.

Any accident at a well can cause an environmental disaster. Oil spills and environmental pollution can occur. According to the Ministry of Energy of the Russian Federation, more than 17 thousand accidents with oil spills occurred at the enterprises of the fuel and energy complex in 2019 [10]. According to the Ministry of Emergency Situations and the Ministry of Nature, at least 55 thousand hectares of land in the country are contaminated with oil products [11].

Continuous monitoring of well performance is essential. The well controller detects abnormal operating conditions and shuts down the pumping unit. The well controller changes the pumping rate when the parameters change. The sensors measure well parameters. The information is transmitted to the well controller. The well controller processes the information and adjusts the well operation mode. A smart well can operate on its own without communication with the control room. A smart well enables safe and economical oil production.

3 Oil Well Main Equipment

The main types of pumps for wells are electric submersible pumps (ESP), sucker rod pumps (SRP), and progressive cavity pumps (PCP). Diagrams of these pumps are shown in Figs. 1, 2, and 3.

During well operation with the help of ESP, depth parameters are monitored [12, 13]. For this purpose, a submersible telemetry unit 10 is lowered into the well (Fig. 1), which measures such parameters as the pressure and temperature of the liquid at the pump intake, the temperature of the stator winding of the submersible electric motor, the vibration level, and the voltage at the terminals of the submersible motor. Recently, attempts have been made to measure the liquid flow rate using a specially installed turbine in the lower part of the submersible telemetry unit [14]. The transmission of telemetric information to the surface is carried out through two veins of a three-phase power cable 6 by superimposing a frequency-modulated signal. In the control station 13, the information signal is filtered and decoded. At the wellhead, pressure sensors 1 and temperature at the wellhead 2, annular pressure 3 are also installed, a flow meter 4 and an echo sounder 11 can be installed. Power measurement unit 12 monitors the electrical parameters of the ESP unit [15].

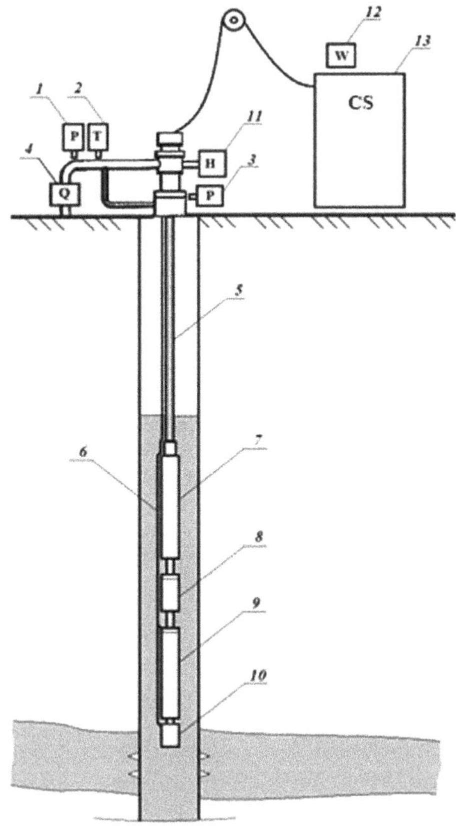

Fig. 1 Installation diagram of process sensors on a well operated by an ESP: 1—wellhead pressure sensor; 2—wellhead temperature sensor; 3—annular pressure sensor; 4—flow meter; 5—tubing string; 6—power cable; 7—ESP; 8—hydro protection unit; 9—submersible electric motor; 10—submersible telemetry unit; 11—echo sounder; 12—wattmeter sensors; 13—control station

It should be noted that submersible telemetry units began to be developed and implemented relatively recently and are not yet capable of providing a long operating time due to severe operating conditions: high temperatures and pressures, vibration, and an aggressive environment.

A diagram of the installation of process sensors in a well operated by a sucker rod pump is shown in Fig. 2.

Wells should contain pressure sensors 1 and temperature at the wellhead 2, pressure in the annulus 3 and, ideally, a flow sensor of the produced fluid 4. However, in practice, the installation of a flow meter at each well is too expensive; therefore, periodic connection of the discharge manifold of each well to a group metering unit is used [16].

A well with a sucker rod pump should additionally be equipped with dynamometer sensors (force sensors on the rod 5 and position 6) and wattmetering 7 [17]. In some oil and gas producing enterprises, pumping units are very worn out; therefore, it is required to equip the units with additional protection sensors: indicators of the condition of the balancer bearing 8, failure of the connecting rod 9, oil level in the gearbox 10, unit roll 11, overheating of the stuffing box 12.

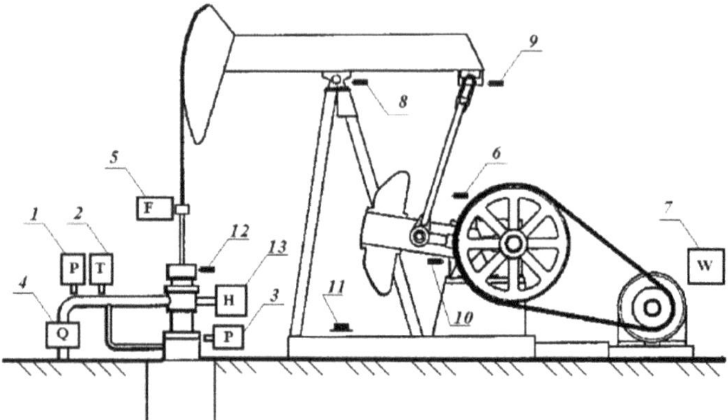

Fig. 2 Scheme of installation of technological sensors on a well operated by a sucker rod pump: 1—wellhead pressure sensor; 2—wellhead temperature sensor; 3—annular pressure sensor; 4—flow meter; 5—force sensor for dynamometry; 6—position sensor for dynamometry; 7—wattmetering sensors; 8—indicator of balancer bearing condition; 9—connecting rod failure indicator; 10—gearbox oil level sensor; 11—unit roll sensor; 12—stuffing box overheating sensor; 13—echo sounder

Fig. 3 Scheme of installation of process sensors on a well operated by PCP: 1—wellhead pressure sensor; 2—wellhead temperature sensor; 3—annular pressure sensor; 4—flow meter; 5—wattmetering sensors; 6—electric motor; 7—tubing string; 8—a column of rods; 9—PCP; 10—force sensor that controls the weight of the column; 11—echo sounder

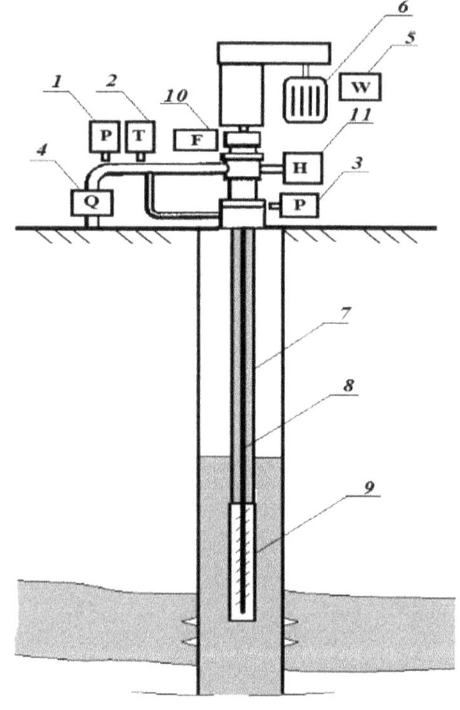

One of the most important parameters controlled during deep pumping oil production is the dynamic fluid level in the well. Recently, automated echo sounders have appeared, which make it possible to make level measurements without human intervention. The echo sounder 13 is mounted on the wellhead and makes measurements at a specified frequency [18].

The use of PCP is one of the new and extremely promising methods of artificial lift of oil [19]. The number of wells with progressive cavity pumps is currently small, but increasing rapidly. Mostly surface-driven units are used, in which a motor with a gearbox is installed at the wellhead, and the rotation is transmitted to the pump through the rod string.

Progressive cavity pumps can be effectively used for the production of high-viscosity oils, as well as when the well fluid has mechanical and abrasive impurities. This provides a higher efficiency in comparison with other types of pumps—up to 60–75% [20].

Wellhead fittings of a well with PCP (Fig. 3) should be equipped with pressure sensors 1 and temperature at the wellhead 2, annular pressure 3, and it can also be equipped with a flow meter 4 and an echo sounder 13. The consumption of electric power by the motor 6 is monitored by means of a wattmeter sensor unit 5. For wells with PCP, a protective sensor 10 can be installed, which measures the weight of the rod string 8.

As can be seen from Figs. 1, 2, and 3, an oil well is a very complex object of management. The well is equipped with a variety of sensors to measure pressure, temperature, flow, force, and other parameters. Information from all sensors must be transmitted to the well controller. Preliminary processing of information is carried out in the well controller. If an emergency occurs, the well controller issues a command to stop the pump motor.

4 Overview of Oilfield Automation Systems

Downhole automation systems began to be developed in the second half of the twentieth century in industrially developed countries such as the USSR and the USA. Dynamometer and wattmeter systems have been developed for sucker rod pumping units. Then there were submersible telemetry systems for electric centrifugal pumps.

In the 1990s, the development of power electronics allowed the creation of borehole pump control stations with a variable frequency drive. The development of telecommunications has made it possible to create oilfield automation systems with remote transmission of information from individual wells and remote control of their electric drive.

Specialists of the Azerbaijan Institute of Petroleum Engineering (AzINMASH) have developed many automation systems: "Gilavar", "NUR", "ChTP", "Aina", and others. Oilfield automation systems have been developed by Lufkin Automation, Weatherford, Baker Hughes, and others.

The oil fields use the Region 2000, Region+, XSPOC, and ARMITS systems. The companies Nefteavtomatika and Izhevsk Radiozavod are also developing automation systems.

But the systems listed above are not smart well and smart field technologies. These systems allow measuring and transmitting a limited set of commands and parameters. These systems do not have artificial intelligence.

5 Oil Well Structure as a Control Object

Figure 4 shows a complex of a downhole oil-producing electric pumping unit. The complex of an electric submersible pumping unit can be considered as a combination of hydraulic and electrical subsystems.

The hydraulic subsystem consists of a well, a centrifugal pump, a tubing string, and wellhead fittings. Oil, water, gas, and solids enter the well from the formation. Then the gas–liquid mixture enters the pump, rises along the tubing string, and through the wellhead fittings enters the collection and preparation system.

The electrical subsystem includes a variable frequency drive, sine filter, transformer, cable line, and submersible motor. The submersible motor is exposed to high temperatures from the bottom of the well [21].

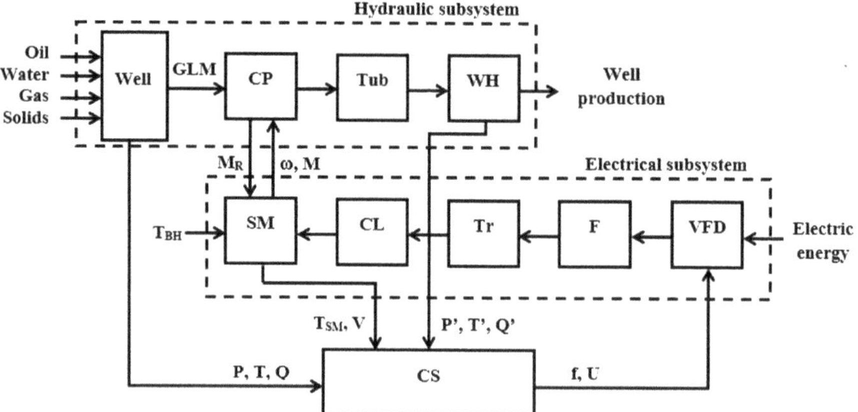

Fig. 4 Scheme of a complex of a downhole oil-producing electric pumping unit: well—oil well; CP—centrifugal pump; GLM—gas–liquid mixture; tub—tubing string; WH—wellhead; SM—submersible motor; CL—cable line; Tr—transformer; F—filter; VFD—variable frequency drive; CS—control system; MR—moment of resistance; M—torque; ω—rotational speed; TBH—bottom hole temperature; P, T, Q—pressure, temperature, and flow rate at the bottom hole; P', T', Q'—pressure, temperature, and flow rate at the wellhead; TSM—submersible motor winding temperature; V—vibration velocity; f—frequency; U—voltage

The connection between the hydraulic and electrical subsystems is expressed in the transfer of torque and speed of rotation from the submersible motor to the centrifugal pump, as well as the moment of resistance from the pump to the motor.

Submersible telemetry transmits to the control system information on pressure, temperature, and flow rate at the pump intake, temperature of windings, and vibration velocity of the motor. Information about pressure, temperature, and flow rate at the wellhead is also received. The control system generates control actions for the variable frequency drive.

The quality of the incoming electricity also influences the operation of electric submersible pump units [22]. Voltage and frequency deviations, higher harmonics, and non-sinusoidality affect the magnitude of losses in the transformer, cable, and motor. These factors reduce the efficiency of the transformer and motor.

6 Measurement of the Insulation Condition of an Oil Well Electrical Equipment

A modern well contains a large amount of electrical equipment. Well electrical equipment includes a transformer, an electric motor, a control station, a cable, a variable frequency drive, a filter, sensors, actuators, and a well controller.

The following methods are used to control the insulation of electrical equipment: high-voltage DC tests, high-voltage AC tests, partial discharges, measurement of insulation resistance, and others [23].

Insulation condition monitoring allows to identify developing defects [24]. Insulation resistance measurement is commonly used today. The partial discharge method is very promising.

7 Smart Well

The smart well is completely controlled by the well controller without human intervention [25]. Automatic control reduces the number of accidents and abnormal conditions. The influence of the human factor is excluded. The well operates as a completely autonomous facility. Energy consumption is reduced.

The structure of the smart well is shown in Fig. 5. The sensors send information to the well controller. The well controller generates a control signal for the variable frequency drive and changes the rotation speed of the electric motor [26].

The objective function of optimizing the well operation can be different: the maximum amount of oil extraction from the well, the minimum specific power consumption, the minimum damage to the equipment, maintaining a given pressure, liquid level, or flow rate:

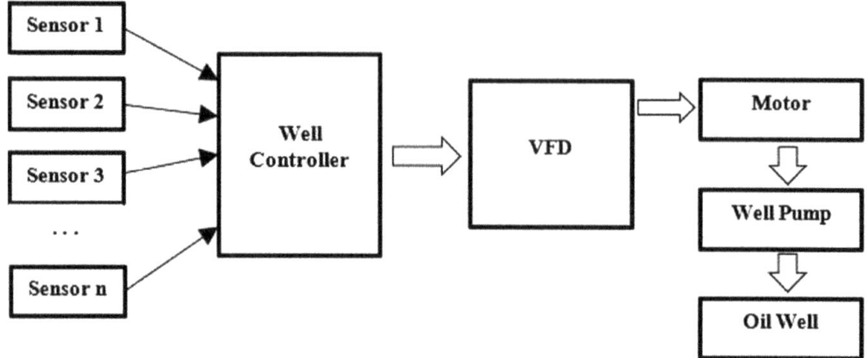

Fig. 5 Structure of the smart well

$$Q(x_1, x_2, \ldots, x_n) \rightarrow \max, \tag{1}$$

$$w(x_1, x_2, \ldots, x_n) = \frac{W}{Q} \rightarrow \min, \tag{2}$$

$$Z(x_1, x_2, \ldots, x_n) \rightarrow \min, \tag{3}$$

$$P(x_1, x_2, \ldots, x_n) \rightarrow P_{\text{ref}}, \tag{4}$$

$$H(x_1, x_2, \ldots, x_n) \rightarrow H_{\text{ref}}, \tag{5}$$

$$Q(x_1, x_2, \ldots, x_n) \rightarrow Q_{\text{ref}}, \tag{6}$$

where

Q is the well production rate;
$x_1, x_2, \ldots, x_n$ well technological parameters;
w specific power consumption;
W consumed electricity;
Z equipment damage;
P pressure;
P_{ref} reference pressure;
H dynamic liquid level;
H_{ref} reference dynamic fluid level;
Q_{ref} reference well production rate.

Smart wells form the "smart field" (Fig. 6) [27].

Information from each well sensor goes to the well controller. The processed information goes to the cluster controller. From the cluster controller, information is

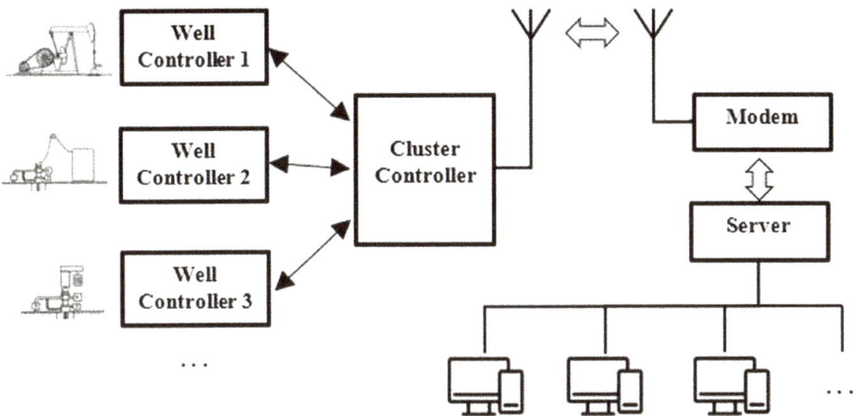

Fig. 6 Scheme of a smart field

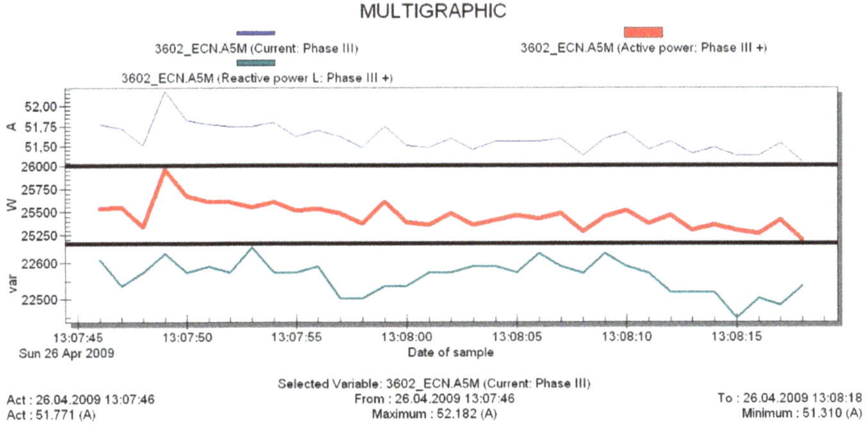

Fig. 7 Current and power of the ESP electric motor

transmitted to the control center. As a result, technologists can get information on their work computer both for each individual well and for the entire field as a whole [28].

Figures 7 and 8 show data on the current and electric power of the ESP and SRP electric motors.

8 Conclusions

Based on the data presented, the following conclusions can be drawn:

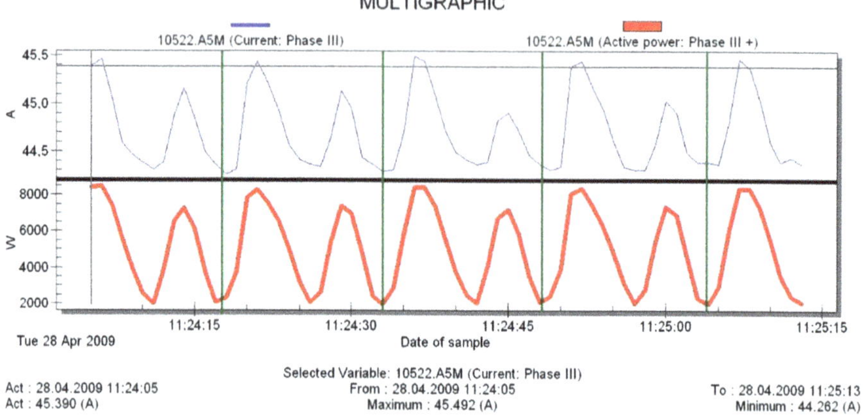

Fig. 8 Current and power of the sucker rod pump electric motor

1. Oil wells are complex objects for automation. It is necessary to control the pressure, temperature, amount of produced fluid, voltage, current, power, power factor, electrical equipment insulation, and other parameters.
2. The smart well controller must itself measure all parameters, process information, and transmit the results to the control room. No human participation is required. The influence of the «human factor» is excluded.
3. The introduction of the smart well technology will reduce the consumption of electricity for oil and gas production, reduce the number of accidents, and increase the production rate of wells. These factors should increase the environmental safety of oil production.
4. The development of automation systems for smart wells and smart fields is an urgent and important task. It is necessary to create new sensors, controllers, information transmission devices and software.

References

1. Gibbs, S.G.: Computing gearbox torque and motor loading for beam pumping units with consideration of inertia effects. J. Petrol. Technol. **27**(09), 1–153 (1975)
2. Gibbs, S.G.: A general method for predicting rod pumping system performance. In: SPE Annual Fall Technical Conference and Exhibition. Society of Petroleum Engineers (1977)
3. Aliev, T., Guluyev, G., Rzayev, A., Pashayev, F., Gadimov, R., Yusifov, I., Sattarov, I.: Noise technology and system for determining of flow rate of oil wells. In: 2012 IV International Conference "Problems of Cybernetics and Informatics" (PCI), pp. 1–3 (2012)
4. Zyuzev, A.M., Bubnov, M.V., Mudrov, M.V.: Sucker-rod pump unit electric drive simulator. In: 2016 2nd International Conference on Industrial Engineering, Applications and Manufacturing (ICIEAM), pp. 1–4 (2016)
5. Zyuzev, A.M., Bubnov, M.V.: Model for sucker-rod pumping unit operating modes analysis based on SimMechanics library. J. Phys. Conf. Ser. **944**(1), 012130 (2018)

6. Blyuk, V., Ershov, M., Komkov, A.: Models and algorithms for quick calculation of electromechanical transition processes of multi-machine electrotechnical systems. In: 2019 1st International Conference on Control Systems, Mathematical Modelling, Automation and Energy Efficiency (SUMMA), pp. 686–689 (2019)
7. Vázquez, M., Suarez, A., Aponte, H., Ocanto, L., Fernandes, J.: Global optimization of oil production systems, a unified operational view. In: SPE Annual Technical Conference and Exhibition. Society of Petroleum Engineers (2001)
8. Nederlof, E., Stephanie, H.: Real-time production optimization for a mature onshore field in Austria. In: SPE EUROPEC/EAGE Annual Conference and Exhibition. Society of Petroleum Engineers (2010)
9. Teves, R., et al.: Production optimization of deep commingled multizone wells with high gas content using ESP and digital solutions. In: SPE Latin American and Caribbean Petroleum Engineering Conference. Society of Petroleum Engineers (2020)
10. Ministry of Energy of the Russian Federation. https://minenergo.gov.ru/en. Last accessed 21 Jan 2021
11. Ministry of Natural Resources and Environment of the Russian. https://www.mnr.gov.ru/en/. Last accessed 22 Jan 2021
12. Khakimyanov, M.I., Shafikov, I.N., Khusainov, F.F.: Electric submersible pumps in oil production and their efficiency analysis. In: Proceedings of the 4th International Conference on Applied Innovations in IT, vol. IV. Bibliothek, Hochschule Anhalt (2018)
13. Glazyrin, A.S., Kladiev, S.N., Afanasiev, K.S., Timoshkin, V.V., Slepnev, I.G., Polishchuk, V.I., Halasz, S.: Design of full order observer with real time monitoring of load torque for submersible induction motors. Bull. Tomsk Polytech. Univ. Geo Assets Eng. **329**(2), 118–126 (2018)
14. Zhang, H., Yu, J., Jiang, Q., Wang, L., Xu, D.: Research on intelligent power supply control based on sensor-less temperature identification of electric submersible motor. In: 2015 9th International Conference on Power Electronics and ECCE Asia (ICPE-ECCE Asia), pp. 2802–2807 (2015)
15. Bolovin, E.V., Glazyrin, A.S.: Method for identifying parameters of submersible induction motors of electrical submersible pump units for oil production. Bull. Tomsk Polytech. Univ Geo Assets Eng. **328**(1), 123–131 (2017)
16. Khakimyanov, M.I., Shafikov, I.N., Khusainov, F.F.: Control of sucker rod pumps energy consumption. In: 2015 International Siberian Conference on Control and Communications (SIBCON), pp. 1–4 (2015)
17. Langbauer, C., Diengsleder-Lambauer, K., Lieschnegg, M.: Downhole dynamometer sensors for sucker rod pumps. IEEE Sens. J. (2020)
18. Tubel, P.S., Mullins II, A.A., Jones, K.R.: Method and apparatus for the remote control and monitoring of production wells, U.S. Patent 6, 176, 312 (2001)
19. Gamboa, J., Olivet, A., Iglesias, J.C., Gonzalez, P.: Understanding the performance of a progressive cavity pump with metallic stator. In: Proceedings of the 20th International Pump Users Symposium, Texas A&M University, Turbomachinery Laboratories (2003)
20. Gamboa, J., Olivet, A., Sorelys, E.: New approach for modeling progressive cavity pumps performance. In: SPE Annual Technical Conference and Exhibition, Society of Petroleum Engineers (2003)
21. Lastra, R.: Electrical submersible pump digital twin, the missing link for successful condition monitoring and failure prediction. In: Abu Dhabi International Petroleum Exhibition & Conference. Society of Petroleum Engineers (2019)
22. Chen, Y., Patil, A., Chen, Y., Bai, C., Wang, Y., Morrison, G.: Numerical study on the first stage head degradation in an electrical submersible pump with population balance model. J. Energy Resour. Technol. **141**(2) (2019)
23. Stone, G.C.: Partial discharge diagnostics and electrical equipment insulation condition assessment. IEEE Trans. Dielectr. Electr. Insul. **12**(5), 891–904 (2005)
24. Montanari, G.C., Hebner, R., Morshuis, P., Seri, P.: An approach to insulation condition monitoring and life assessment in emerging electrical environments. IEEE Trans. Power Deliv. **34**(4), 1357–1364 (2019)

25. Gao, C.H., Rajeswaran, R.T., Nakagawa, E.Y.: A literature review on smart well technology. In: Production and Operations Symposium. Society of Petroleum Engineers (2007)
26. Durlofsky, L.J., Aziz, K.: Optimization of smart well control. In: SPE International Thermal Operations and Heavy Oil Symposium and International Horizontal Well Technology Conference. Society of Petroleum Engineers (2002)
27. Temizel, C., Canbaz, C.H., Palabiyik, Y., Putra, D., Asena, A., Ranjith, R., Jongkittinarukorn, K.: A comprehensive review of smart/intelligent oilfield technologies and applications in the oil and gas industry. In: SPE Middle East Oil and Gas Show and Conference. Society of Petroleum Engineers (2019)
28. Redutskiy, Y.: Conceptualization of smart solutions in oil and gas industry. Procedia Comput. Sci. **109**, 745–753 (2017)

Processing NPP Electromechanical Equipment Diagnostic Signals Using Principal Component Analysis in Hardware–Software Complexes

Elena Abidova, Artem Dembitsky, Alexander Lapkis, Irina Zarochintseva, and Alexander Chernov

Abstract The paper raises the issue of increasing the sensitivity while diagnosing NPP equipment due to a larger time period between overhauls. The current diagnostic systems that use a deterministic component of diagnostic information do not provide sensitivity to every defect. The paper proposes to use an approach based on the principal component analysis for processing diagnostic signals. To confirm the increase in the sensitivity and selectivity of electrically operated valve diagnostics, the results of a full-scale experiment are given, which show reducing the diagnostic error probability thanks to the proposed approach. The processing algorithm is given, and the principle of automatic selection of the main component is described. The proposed approach has been brought to an engineering implementation: the requirements for the hardware–software complex that collects and processes diagnostic information are formulated; the technical characteristics of the pilot prototype of the complex implemented at Novovoronezh NPP are described. It is shown that using the proposed approach, defects (e.g., stem deformation), which cannot be detected by other methods, can be discovered. Similar results have been achieved in diagnostics of other NPP equipment. Improving the diagnostic quality will make it possible to carry out equipment repairs in the required volume under the conditions of the NPP transition to an 18-month cycle.

E. Abidova (✉) · A. Dembitsky · A. Lapkis · I. Zarochintseva · A. Chernov
Volgodonsk Engineering Technical Institute—A Branch of the National Research Nuclear, University "MEPhI", 73/94, Lenin Str., Volgodonsk, Rostov Region 347360, Russia

A. Dembitsky
e-mail: aedembitskii@mephi.ru

I. Zarochintseva
e-mail: IVZarochintseva@mephi.ru

A. Chernov
e-mail: VITIkafIUS@mephi.ru

A. Ronzhin and V. Shishlakov (eds.), *Electromechanics and Robotics*, Smart Innovation, Systems and Technologies 232, https://doi.org/10.1007/978-981-16-2814-6_36

1 NPP Equipment Diagnostic Support

1.1 Objects of Diagnostic Testing

The safety and efficiency of a nuclear power plant is largely ensured by the quality of its equipment diagnostics. Currently, equipment diagnostics at all NPPs of the Russian Federation is performed by specialized subdivisions—technical diagnostics departments. The structure of the departments is different, but there are always groups for diagnostics of electrically operated valve and vibration monitoring.

In Russia, there is no other industrial branch to be covered by systems of diagnostics and control as atomic energy. Modern technical and methodological solutions are being implemented, new innovative directions are being developed. At the same time, NPP features are a source of new challenges for the designers of diagnostic systems and complexes [1–3].

Some of the most critical objects for diagnosis are:

1. electrically operated valve (EOV), i.e., about 2000 units of valves, gag valves, gate valves at each power unit, used to regulate and shut off working fluid;
2. control and protection system drives (CPS), which change the position of the absorbing rods in the reactor;
3. refueling machine;
4. diesel generators, which are the sources of backup power supply for NPP emergency systems.

Failure of any of the listed equipment can cause an accident at a nuclear power plant, the scale of which is comparable to the catastrophes at the Chernobyl and Fukushima nuclear power plants [2].

The largest amount of work is carried out when diagnosing EOV, which includes valves, gag valves, gate valves (about 2000 items per unit) [4]. Also, a large amount of work is carried out when diagnosing the rotating equipment—pumps, compressors, fans—by vibration method. Diagnostics of electromechanical equipment constitutes up to 70% of the work; therefore, this paper considers aspects of valve and pump diagnostics.

Increasing the reliability is a key problem in the methods of diagnosing NPP equipment. A specific feature of NPP equipment diagnostics is the impossibility to carry out diagnostics more often than once a year, when the unit is stopped for refueling (if it is a unit with a WWER-type reactor). The refueling period for nuclear fuel is three to five weeks. During this time, according to the results of diagnostics, it is necessary to identify defects that can lead to a failure until the next shutdown, which will be in a year or later. Currently, some units of Russian nuclear power plants are transferring to an 18-month fuel cycle. With the increase in the period for which the diagnostic results are used to make an operability forecast, the requirements for reliability also increase. The effectiveness of the existing diagnostic methods and systems is justified in a 12-month cycle, but an increase in the overhaul period up to

18 months requires diagnostics quality improvement. In this regard, the improvement of diagnostic methods and tools is a highly demanded task [3].

1.2 Standard Approaches to Diagnosis

The operation of standard diagnostic systems is based on the registration and analysis of vibration or equipment current signals [5–8], less often ultrasound, pressure, temperature, and other signals are used. For example, NPP valves are currently diagnosed by electrical indicators [5]. The approach, of course, allows to cover a large amount of equipment in a short period, but its sensitivity is insufficient. The sensitivity of vibration diagnostics in terms of detecting defects in moving parts is higher [6–8], but the registration of a vibration signal is too long and complicated procedure in the scheduled outage conditions at NPP.

The problem of localizing the place of malfunction (instantiation of the defect) is achieved through spectral analysis of the diagnostic signal. It is shown that in the spectrum of vibration and current signals of electromechanical equipment, harmonics of a defective part are manifested [7]. But in practice, the spectral analysis of the EOV current is a laborious, almost not amenable to automation procedure, and is used to substantiate the presence of a malfunction, and not its specific location. The existing systems determine the malfunction location, if at all, by vibration signals, and not by current. Moreover, standard systems issue a list of diagnoses with varying degrees of probability, which complicates the maintenance of equipment.

Maintenance of NPP equipment under conditions of an 18-month fuel cycle puts forward the following requirements for diagnostic methods and tools:

- sensitivity enhancement;
- instantiation of the defect;
- automation of obtaining results.

It is desirable that these requirements are met when processing electrical signals, not vibroacoustic ones.

2 Improvement of Diagnostic Techniques of NPP Equipment

2.1 Clustering Diagnostic Data in Principal Component Space

The sensitivity of standard methods is limited by the incomplete use of information contained in diagnostic signals. This paper proposes to supplement the classical observability model with an operator of nonlinear transformations, that is, to take into account the influence of non-stationary and nonlinear factors on the formation of diagnostic signals. The development of non-stationary processes in electromechanical equipment is described in the works of many authors who used bifurcation diagrams, Poincaré maps and phase portraits [9, 10].

This paper solves the problem of increasing the sensitivity and clarity in the diagnostic results presentation. After analyzing the methods that ensure the solution of this problem, invariant to the chaotic nature of the information being processed, the choice is made on the principal component analysis (PCA). The application of the method in various fields is widely represented in the works of Russian and foreign authors—Golyandin, Pomerantsev, Al-Bukhrabi, Trendafilova [11–13]. However, the method has not been practically used in the processing of diagnostic signals. The widespread use was previously limited to the problem of principal components (PC) selection.

This method is applied to multidimensional data in order to reduce their dimension. The multidimensionality of diagnostic data is determined by the length of the time series, presented in the form of a trajectory matrix. A typical diagnostic signal (vector of output variables) can be modeled as

$$y(k) = r(k) + h(k),$$

where $r(k)$ is the deterministic component, and $h(k)$ is the variable part and noise.

Singular value decomposition is used to separate the deterministic information contained in the signal from the non-deterministic one. To apply the singular value decomposition, the signal must be converted to the Hankel matrix. The Hankel matrix for a discrete vibration or current signal y_i ($i = 1, 2, \ldots, N$) can be obtained using a sliding window of length m as follows:

$$[A] = \begin{bmatrix} y_1 & y_2 & \cdots & y_n \\ y_2 & y_3 & \cdots & y_{n+1} \\ \vdots & \vdots & \ddots & \vdots \\ y_m & y_{m+1} & \cdots & y_N \end{bmatrix} \quad (1)$$

where $N = m + n - 1$.

The choice of the matrix column dimension n is arbitrary, but within the framework of solving practical problems it is advisable to take $n = N/2$. The Hankel matrix (1) is subjected to singular value decomposition.

The method of principal components makes it possible to obtain a matrix $\overline{A}$ of size $m \times n$ with rank $l < L$, which reflects the contribution of the most significant components:

$$\left[\overline{A}\right] = [U_l][S_l][V_l]^{\mathrm{T}}, \quad (2)$$

where $\left[\overline{A}\right]$—is the reconstructed matrix that usually uses the first numbers from the singular values. The rest of the singular values correspond to the components which contribution to the original signal is weaker.

$$\sigma_i > \varepsilon, \quad i = 1, \ldots, l;$$
$$\sigma_i \le \varepsilon, \quad i = l + 1, \ldots, L,$$

where ε is the significance threshold.

However, when applied to diagnostic signals, "strong components" may not display important diagnostic information. Conversely, a component formed in the original signal under the influence of a defect is likely to have a lower value and a higher number.

Expression (2) can be transformed into the following equation:

$$[A] = \left[\vec{A}\right] + [H] = \left[U_l \; U_0\right]\begin{bmatrix} S_l & 0 \\ 0 & S_0 \end{bmatrix}\begin{bmatrix} V_l^{\mathrm{T}} \\ V_0^{\mathrm{T}} \end{bmatrix},$$

where $\left[\vec{A}\right]$ and $[H]$ correspond to $r(k)$ and $h(k)$, S_l contains large numbers $\sigma_i > \varepsilon$, $i \in 1, \ldots, l$, and S_0—contains smaller numbers.

Therefore, the information in the matrix $[A]$ correlates with $y(k)$ and describes two parts: deterministic data $\left[\overline{A}\right]$ and the matrix $[H]$. The matrix $\left[\overline{A}\right]$ contains information about the harmonics of rotating parts (indirect information about the defect), and the matrix $[H]$ is determined by the specific influence of the defect.

One of the ways to determine the ratio of the deterministic and non-deterministic components of the diagnostic signal is to analyze the eigenvalue spectra (EVS) of the matrix $[A]$. If the signal as a whole is deterministic, then it is described by several first components, and the contribution of the rest will be insignificant. On the other hand, if a signal is formed under the influence of non-deterministic factors, then a larger number of components are required to describe it, so the eigenvalues of the higher decomposition components increase.

By choosing directions that are sensitive to a change in state, it is possible to construct a basis where the difference between the characteristics of serviceable and faulty equipment will be significantly greater than in the original space, i.e., the quality of diagnosis will increase. In this paper, the choice of directions of the diagnostic basis is justified by the defect development patterns. The chaotic signal components important for diagnostics are manifested in the higher components of the singular value decomposition [14]. At the same time, the presentation of diagnostic data in the basis of the first components, as recommended in the literature, cannot provide either sensitivity or visualization.

2.2 Approbation of the Method Processing the Diagnostic Signals from Electrical Equipment

Increasing the sensitivity using the method. To demonstrate the advantages of the proposed approach, experiments were carried out in the laboratory environment. The first series of the experiments was aimed at illustrating the increased sensitivity. The experiment was carried out using a stand, which represents a shaft rotating in bearings, driven by an asynchronous electric motor. As part of the experiment, nine states were simulated: load free, defect free; load free, with a defect in the outer ring; load free, with a defect in the inner ring; under load, defect free; under load, with a defect in the outer ring; under load, with a defect in the inner ring; loaded, defect free; loaded, with a defect in the outer ring; loaded, with a defect in the inner ring.

The signals of the motor current in three phases were recorded during the equipment operation. Then the distribution densities of the parameters of serviceable and defective equipment were compared. The initial distributions of the current parameters practically coincided Fig. 1a, which excluded the detection of a defect. The same signals were mapped onto the higher components of the singular value decomposition. By projecting the parameters, that were selected taking into account the defect development pattern, onto the principle components (PCs), the error probability decreased to an acceptable level (Fig. 1b). Thus, the sensitivity clearly increased.

A quantitative assessment of the degree of increase in the sensitivity of diagnosing a particular defect by the proposed method is shown in Table 1. The initial distributions practically excluded correct diagnosis, and the use of PCA ensured error probability decrease to an acceptable level or eliminated errors.

The result is of great importance for the development of diagnostic systems based on electrical signal processing. The widespread introduction of these systems, despite all the advantages (simplicity, efficiency, the possibility of remote data recording, reduction of current loads), is limited by the low sensitivity of the existing processing methods.

Increasing the selectivity using the method. The next series of experiments was carried out to illustrate the selectivity of the proposed approach in relation to different types of defects. The experiment consisted of registering electrical power signals of a known serviceable EOV and equipment with designed defects: stem breaks, different stages of stem deformation, thread defect.

The received current signals were preprocessed by standard methods. The spectrum view (Fig. 2) does not allow us to assume a malfunction and does not say anything about its cause.

The PCA approach was applied to the same data. Namely, the basis is built on the signal from a serviceable EOV; faulty equipment signals are projected onto it. In this case, the PC selection is justified by the maximum distance from the center of the parameter sample in the serviceable state and six faulty ones in one of the dimensions of the multidimensional space.

As can be seen (Fig. 3), the cluster of the serviceable state is removed from the clusters corresponding to faults, and the clusters of faulty states almost do not

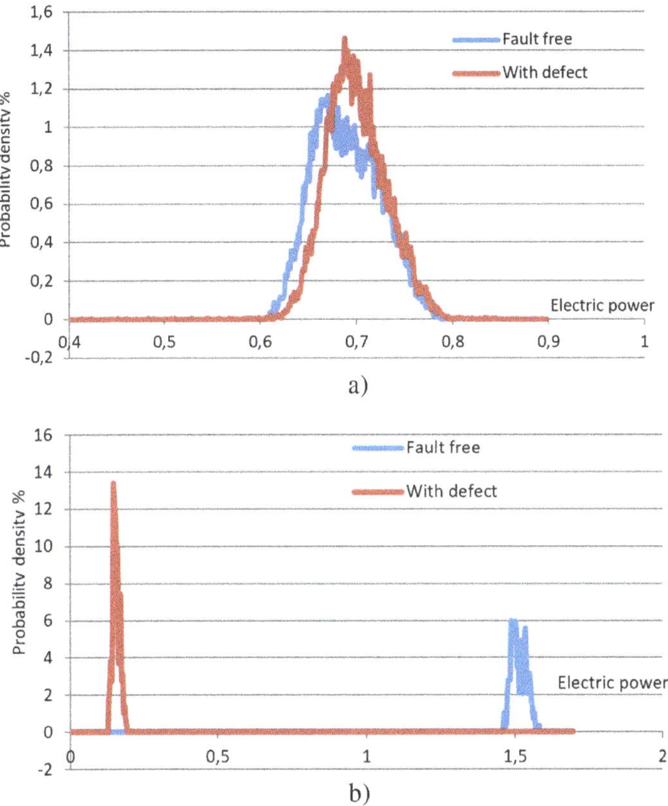

Fig. 1 Functions of the probability distribution density: **a** amplitudes of the current signal envelopes; **b** projections of the amplitudes of the current signal envelopes on the first PC

Table 1 Reduction of the detection errors probability due to the application of PCA to the diagnostic results

Recognizable states		False alarm, %	Skip, %
Load free, defect free	Load free, with a defect in the inner ring	13	12
Load free, defect free	Load free, with a defect in the outer ring	0	0
Loaded, defect free	Loaded, with a defect in the inner ring	0	0
Loaded, defect free	Loaded, with a defect in the outer ring	0	0
Under load, defect free	Under load, with a defect in the inner ring	1	12
Under load, defect free	Under load, with a defect in the outer ring	22	8

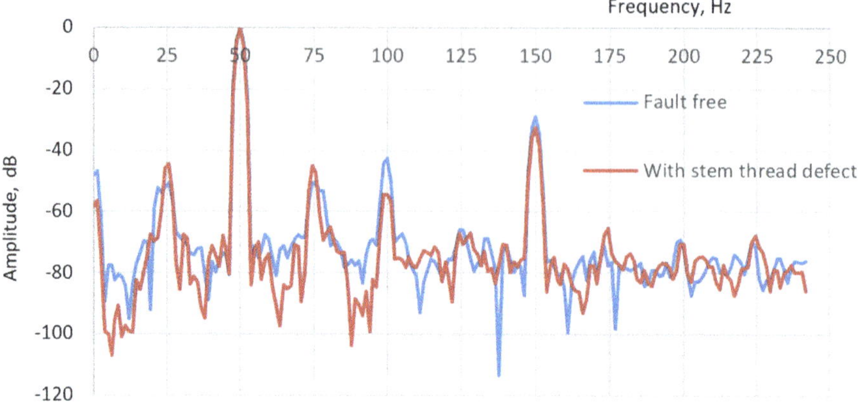

Fig. 2 The spectrum view of the received signals

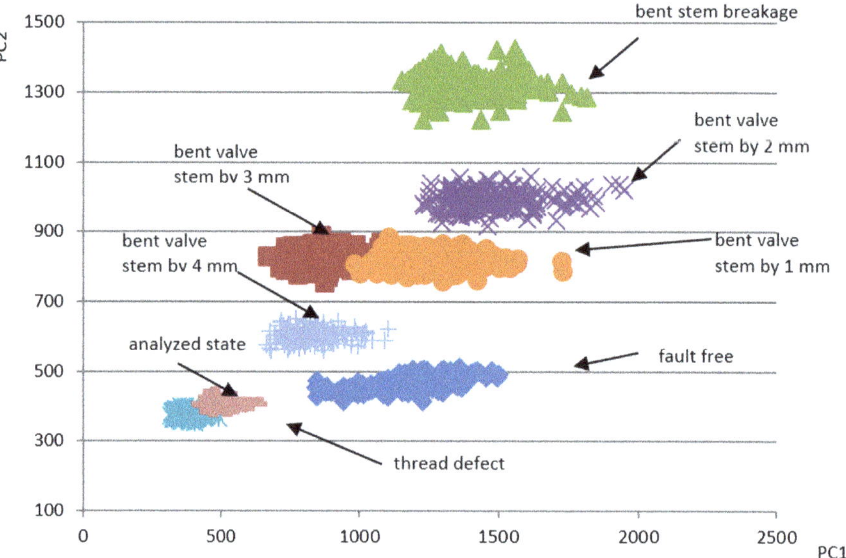

Fig. 3 Projections of electrical power signals onto the PC

coincide. The signal of a conditionally unknown state is projected onto the same basis. In the framework of the experiment, the analyzed state is known a thread defect and it is clearly identified by the coincidence of projections.

The experiment demonstrates that the PCA-based approach in terms of selectivity for defects is a better solution than spectral analysis, which is widely used in existing diagnostic systems. Similar results are achieved when processing other electromechanical equipment of nuclear power plants.

3 Development of Hardware–Software Complex

3.1 Algorithmization of Data Representation in the PC Space

The experimental results formed the basis for the algorithm for comparing the current state with the standard one. The algorithm includes the stages of data preparation, construction of a calibration baseline, projection onto this basis of reference and analyzed data, comparison of projections (Fig. 4). Since the characteristics of faulty equipment are not always available, a simplified algorithm has been additionally developed that compares the current characteristics with the previous ones.

The principle of maximum distance from the center of the parameter sample in the serviceable state and M faulty ones in one of the dimensions of the N-dimensional space is laid in the basis of the automatic PC selection. The center of the parameter projection of the ith state on the jth direction:

$$\bar{x}_{ij} = \frac{\sum_{k=0}^{m_i} x_{ij}^k}{n_i},$$

where x_{ij}^k is the kth parameter coordinate of the ith state to the jth direction, m_i is the sample size for the ith state.

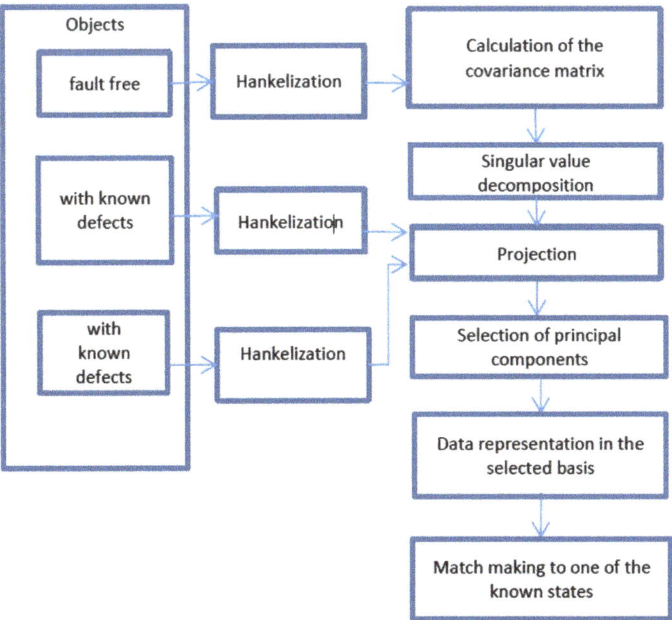

Fig. 4 Block operation scheme "representation in the PC space"

The sum of the distances from the projection center of the serviceable state $\bar{x}_{0j}$ onto the jth direction up to the projection centers of the ith states onto the same direction $\bar{x}_{ij}$ $i \in 1, \ldots, M$:

$$d_j = \sum_{i=1}^{M} \left|\bar{x}_{0j} - \bar{x}_{ij}\right|.$$

Thus, having calculated the sums of the distances $\{d_1, d_2, \ldots, d_L\}$, it is advisable to choose the maximum value to justify the direction, which provides the greatest distance between the parameters of the serviceable and faulty equipment. After sorting the sums of distances in descending order, one can choose directions for calibration baseline construction. The utility of including an additional jth measurement to increase the sensitivity in determining the ith state can be estimated by the following calculation:

$$D_{ij} = \sqrt{(\bar{x}_{01} - \bar{x}_{i1})^2 + \left(\bar{x}_{0j} - \bar{x}_{ij}\right)^2}.$$

The D_{ij} parameter permits to estimate how much an additional measurement inclusion contributes to an increase in sensitivity if the projection onto one direction does not provide the required diagnostic quality indicators. The choice of space directions can also be justified by the selectivity or necessity to demonstrate the degree of a single defect development.

3.2 *Technical Description of the Hardware–Software Complex*

The proposed approach is brought to engineering implementation. A pilot hardware–software complex that collects and processes diagnostic information is developed (Fig. 5).

The modular-type complex consists of: a base module, a module for measuring electric power, a module for measuring voltage, a module for measuring vibroacoustic signals and ultrasound. Measurement modules can be connected to the base module in various combinations by the user's request. The complex is designed as a portable device (dimensions 280 × 187 × 22 mm, weight 1014 g). The complex is intended for diagnostics of electrically operated valves, pumps, and fans at nuclear power plants.

The software part of the complex includes both a block of linear transformations (spectral analysis and statistical data processing) and blocks that analyze the nonlinear part of information: entropy parametrization, representation on a complex plane, representation in the PC space.

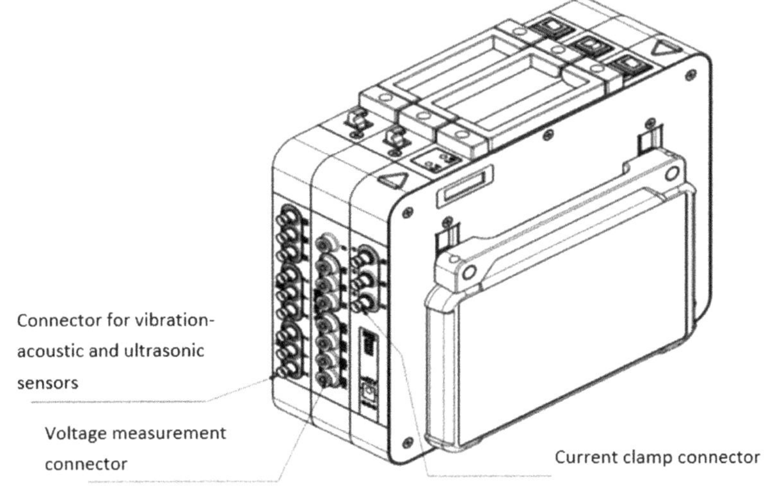

Fig. 5 General view of hardware–software diagnostics of electric drive equipment

4 Conclusion

Technical diagnostics is characterized by a high degree of responsibility in ensuring the safe operation of NPP power units. Improving the efficiency of existing methods of processing diagnostic information is possible through the development of algorithms that provide additional data of power equipment state from signals containing non-stationary and chaotic components.

In order to improve the methods of processing diagnostic information, it is proposed to use the results of multidimensional transformations of the deterministic and stochastic characteristics of the object under test, based on the selection of the principal components of the singular value decomposition that provides a solution to the clustering problem.

Algorithms for constructing reference space with components reflecting the contribution of the deterministic, variable and chaotic components are developed for diagnostics of various NPP objects. Hardware–software system pilot prototypes executing the proposed approach are developed and implemented.

To achieve an increase in the sensitivity of diagnosis and selectivity in relation to the state is planned by introducing the approach based on PCA. The diagnostic quality upgrading makes it possible to carry out equipment repairs as required in the de-sired volume under the conditions of the NPP transition to an 18-month cycle.

References

1. Danilov, A., Povarov, V., Burkovsky, V., Podvalny, S., Gusev, K.: Intellectual decision-making system in the context of potentially dangerous nuclear power facilities. In: MATEC Web of Conferences, vol. 161, p. 02009. EDP Sciences (2018)
2. Bakirov, M., Povarov, V.: Development and implementation of the technology for operational diagnostics of critical equipment damage as a procedure for managing the NPP resource. VSU Bull. Ser. Phys. Math. **1**, 5–17 (2015)
3. Bakirov, M., Povarov, V.: Elaboration and installation of technology of online diagnostics equipment damage as a procedure of NPP lifetime management. In: 3rd International Conference on NPP Life Management (PLIM) for Long Term Operations (LTO), Salt Lake City, UT, USA (2012)
4. Seinov, S., Goshko, A., Adamenkov, A.: Technical Diagnostics of NPP Valves. Mechanical Engineering (Library of the NPP Fitter), Moscow (2012) (in Russ.)
5. Liu, Z., Gao, Q., Niu, H.: The fault diagnosis of electromagnetic valves based on driving current detection. Acta Armamentarii **35**(7), 1083–1090 (2014)
6. Hong, L., Dhupia, J.S.: A time domain approach to diagnose gearbox fault based on measured vibration signals. J. Sound Vib. **33**(7), 2164–2180 (2014)
7. Smith, W.A., Randall, R.B.: Rolling element bearing diagnostics using the case western reserve university data: a benchmark study. Mech. Syst. Signal Process. **64–65**, 100–131 (2015)
8. Kuemmlee, H., Gross, T., Kolerus, J.: Machine vibrations and diagnostics the World of ISO. In: IEEE Industry Applications Society 60th Annual Petroleum and Chemical Industry Conference, Chicago (2013)
9. Li, S., Chen, T., Fahimi, B.: On the occurrence of nonlincar dynamic phcnomcna in the hysteresis-controlled switched reluctance motor drive. In: 44th Annual Conference of the IEEE Industrial Electronics Society (IECON), pp. 5710–5715 (2018)
10. Koutsouvasilis, P., Driot, N., Lu, D., Schweizer, B.: Quantification of sub-synchronous vibrations for turbocharger rotors with full-floating ring bearings. Arch. Appl. Mech. **85**(4), 481–502 (2008)
11. Al-Bugharbee, H., Trendafilova, I.: A new methodology for fault detection in rolling element bearings using singular spectrum analysis. Int. J. Condition Monit. **7**(2), 26–35 (2017)
12. Golafshan, R., Yuce Sanliturk, K.: SVD and Hankel matrix based de-noising approach for ball bearing fault detection and its assessment using artificial faults. Mech. Syst. Signal Process. **70–71**, 36–50 (2016)
13. Abidova, E.A., Chernov, A.V., Lapkis, A.A.: Bearing defects diagnostics using the principal components analysis. IOP Mater. Sci. Conf. Ser. Mater. Sci. Eng. **680**(1), 012005 (2019). Available at: https://iopscience.iop.org/article/10.1088/1757-899X/680/1/012005. Last accessed 26 Feb 2021
14. Abidova, E.A., Lapkis, A.A., Chernov, A.V.: Bearing defects diagnostics using the principal components analysis. IOP Conf. Ser. Mater. Sci. Eng. **680**(1), 012005 (2019, Nov) (IOP Publishing)

Digital Control of Continuous Production with Dry Friction at Actuators

Eugene Larkin, Aleksandr Privalov, Alexey Bogomolov, and Tatiana Akimenko

Abstract Digital control system by complex multi-loop continuous productions is analyzed. Common analytical model of data processing with taking into account time characteristics of von Neumann controllers is obtained. It is shown that polling procedure born time lags, which affect the control parameters of the system as a whole. Common matrix description of a system with dry friction at the actuator and time lags in direct/feedback circuits is worked out. For computer software of arbitrary complexity, the method of time lags calculation, based on the description of control algorithm as a semi-Markov process, is proposed. Theoretical postulates are confirmed by simulation of two-loop digital control system functioning.

1 Introduction

Digitizing of industry is rather modern concept of its development [1, 2], which is linked with utilization of von Neumann controllers for managing by continuous production [3]. Von Neumann computer, as physical device, in comparison with traditional analogue controllers, has new properties which are determined by sequential interpretation of control algorithm operators, unfolding in real physical time [4–6]. Due to that, computers, besides software control principle realization born time lags which, in turn, affects on quality characteristics of control system as a whole. On practice, controller with using its software processes random data, and control algorithm, as a rule, includes decision operators at its branching points. So time of action computation, transmitted to actuator after receiving data from sensors, is a random one. After transmitting data to continuous object under control, real actuators begin move executive unit, which occurs at the presence of dry friction in moving parts (i.e., gate valves) of technology installation [7]. Dry friction at actuators and digital controller time lag qualitatively change the behavior of entire system

E. Larkin (✉) · A. Bogomolov · T. Akimenko
Tula State University, Tula 300012, Russia

A. Privalov
Tula State Lev Tolstoy Pedagogical University, Tula 300026, Russia

A. Ronzhin and V. Shishlakov (eds.), *Electromechanics and Robotics*, Smart Innovation, Systems and Technologies 232, https://doi.org/10.1007/978-981-16-2814-6_37

during control. For taking into account both factors when system design it is necessary to have adequate model, which, from one side, is based on the semi-Markov processes theory [8–11] for evaluation of time lags, and from the other side uses nonlinear control system investigation methodology for description of dynamics of object under control.

Methods of assessing continuous production control system performance at the design stage are not widespread that confirms necessity and relevancy of investigation in the area.

2 Structure of Continuous Production Digital Control System

Structure of the system is shown on Fig. 1. As it follows from the figure, it includes K-loops object under control (OUC) and von Neumann-type digital controller (DC). Digital controller produces vector $\mathbf{F}(s) = [F_1(s), \ldots, F_k(s), \ldots, F_K(s)]$ of desired values of controllable parameters, and on the base of it, and on the base of feedback vector $\mathbf{X}_0(s) = \left[X_{0,1}(s), \ldots, X_{0,k}(s), \ldots, X_{0,K}(s)\right]$ value of control signal vector is calculated $\mathbf{U}(s) = [U_1(s), \ldots, U_k(s), \ldots, U_K(s)]]$. Signals $U_1(s), \ldots, U_k(s), \ldots, U_K(s)$, transmitting through an interface transform to time-shifted signals $U_{sn,1}(s), \ldots, U_{sn,k}(s), \ldots, U_{sn,K}(s)$, which are fed to inputs of

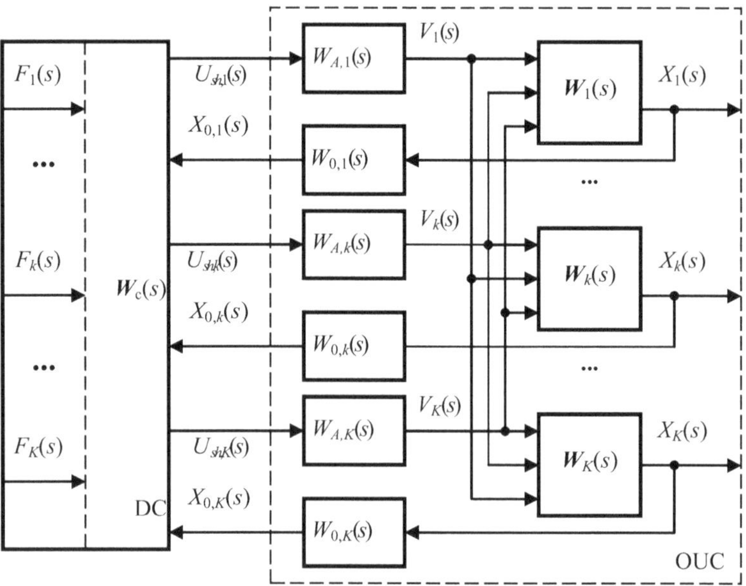

Fig. 1 Flowchart of digital control system

actuators $W_{A,1}(s), \ldots, W_{A,k}(s), \ldots, W_{A,K}(s)$, which, in turn, drive OUC gate valves.

Dynamics of physical processes in the OUC is performed with the matrix:

$$\mathbf{W}(s) = \begin{bmatrix} \mathbf{W}_1(s) \\ \cdots \\ \mathbf{W}_k(s) \\ \cdots \\ \mathbf{W}_K(s) \end{bmatrix} = \begin{bmatrix} W_{11}(s) \ \ldots \ W_{1l}(s) \ \ldots \ W_{1K}(s) \\ \cdots \\ W_{k1}(s) \quad W_{kl}(s) \quad W_{kK}(s) \\ \cdots \\ W_{K1}(s) \ \ldots \ W_{Kl}(s) \ \ldots \ W_{KK}(s) \end{bmatrix}, \tag{1}$$

where $W_{kl}(s)$ describes influence of the k-th actuator on the l-th controllable parameter; s is the Laplace operator [12].

On practice, linear model of technological processes may be obtained from primary analytical description by means of its Taylor expansion.

State of object under control is defined with vector $\mathbf{X}(s) = [X_1(s), \ldots, X_k(s), \ldots, X_K(s)]$, elements of which are measured by sensors $W_{0,1}(s), \ldots, W_{0,k}(s), \ldots, W_{0,K}(s)$. Outputs of sensors form feedback vector $\mathbf{X}_0(s)$, which sequentially, element-by-element, is inputted into the DC. On outputs of digital controller action vector $\mathbf{U}(s) = [U_1(s), \ldots, U_k(s), \ldots, U_K(s)]$ is generated by software. If in DC linear data processing is realized (i.e., digital PID controller [2, 3]), computation of control action may be described by the $K \times K$ matrix:

$$\mathbf{W}_c(s) = \begin{bmatrix} W_{c,11}(s) \ \ldots \ W_{c,1l}(s) \ \ldots \ W_{c,1K}(s) \\ \cdots \\ W_{c,k1}(s) \quad W_{c,kl}(s) \quad W_{c,kK}(s) \\ \cdots \\ W_{c,K1}(s) \ \ldots \ W_{c,Kl}(s) \ \ldots \ W_{c,Kl}(s) \end{bmatrix}, \tag{2}$$

where $W_{c,kl}(s)$ is transfer function for calculation of the l-th part of action $U_k(s)$, putted on the k-th actuator, according difference signal $F_l(s) - X_{0.l}(s)$.

DC processes error signal, performed in the discrete form, which in time domain may be expressed as follows:

$$\mathbf{e}(t) = \sum_{n=0}^{\infty} [\mathbf{f}(t - n\theta) - \mathbf{x}_0(t - n\theta)] = [e_1(t), \ldots, e_k(t), \ldots, e_K(t)], \tag{3}$$

where t is the physical time; θ is the sampling period;

$$\mathbf{f}(t) = [f_1(t), \ldots, f_k(t), \ldots, f_K(t)] = L^{-1}[\mathbf{F}(s)], \tag{4}$$

$$\mathbf{x}_0(t) = \left[x_{0,1}(t), \ldots, x_{0,k}(t), \ldots, x_{0,K}(t)\right] = L^{-1}[\mathbf{X}_0(s)]. \tag{5}$$

$L^{-1}[\ldots]$ is the inverse Laplace transform [12].

Linear processing $e_k(t) \in \mathbf{e}$ to obtain additive component of the action $u_l(t)$, $1 \le l \le K$, before it transmitting to the input of l-th actuator, reduces to calculation of convolution sum:

$$u_{kl}(t - n\theta) = \left[\sum_{n=0}^{J} w_{c,kl}(t - n\theta)\right] * \left[\sum_{n=0}^{\infty} e_k(t - n\theta)\right]$$
$$= \sum_{n=0}^{\infty}\sum_{j=0}^{J} w_{c,kl}(\tilde{t} - j\theta) \cdot e_k[t - (n - j)\theta], \quad (6)$$

where $\sum_{n=0}^{\infty} e_k(t - n\theta)$ is the sampled form of signal $e_k(t)$; $\sum_{n=0}^{J} w_{c,kl}(t - n\theta)$ is the linear difference equation for processing signal $e_k(t)$ to obtain k-th additive component of the signal $\tilde{u}_k(t - n\theta)$; $\tilde{t}$ is the auxiliary argument, which starts from $t = n\theta$ and directed inversely with relation to t; J is the order of linear difference equation, for data processing; * is the convolution sign.

Laplace transform of (6), when $\theta \to 0$ gives the following dependence of data processing in DC:

$$\mathbf{U}(s) = \left[\mathbf{F}(s) - \mathbf{X}_{\text{sh},0}(s)\right] \cdot \mathbf{W}_c(s), \quad (7)$$

where $\mathbf{U}(s)$ is the action vector calculated, but not transmitted to actuators through the interface; $\mathbf{X}_{\text{sh},0}(s)$ is the feedback vector, received through the interface.

Real actuators, which drive gate valves, with taking into account dry friction in moving parts, from input $\mathbf{U}(s)$ till outputs $\mathbf{V}(s) = [V_1(s), \ldots, V_k(s), \ldots, V_K(s)]$ are described with the $K \times K$ diagonal matrix $\mathbf{W}_A(s) = \left[W_{A,kl}(s)\right]$, elements of which are as follows:

$$W_{A,kl}(s) = \begin{cases} 0, & \text{when } k \ne l; \\ W_{A,k}(s), & \text{when } k = l. \end{cases} \quad (8)$$

In turn, $W_{A,k}(s)$ may be obtained from Laplace transform of the second-order differential equations, which describes the drive with dry friction and local valve position feedback, as follows:

$$\begin{cases} s^2 V_k(s) + \xi_k s V_k(s) + \kappa_{k,2} \cdot L\left[\text{sgn}\left(L^{-1}[s V_k(s)]\right)\right] = \kappa_{k,1}\tilde{U}_k(s); \\ \tilde{U}_k(s) = U_{\text{sh},k}(s) - \kappa_{k,3} V_k(s), \end{cases} \quad (9)$$

where $V_k(s)$ and $U_k(s)$ are output and input signals of actuator; ξ_k is the time parameter of drive acceleration characteristic; $U_{\text{sh},k}(s)$ is the signal on the actuator drive; $\kappa_{k,1}$, $\kappa_{k,2}$, $\kappa_{k,3}$ are aspect ratios; $\text{sgn}\left(L^{-1}[s V_k(s)]\right)$ is the dry friction function;

$$\text{sgn}\left(L^{-1}[s V_k(s)]\right) = \begin{cases} 1, & \text{when } L^{-1}[s V_k(s)] > 0; \\ -1, & \text{when } L^{-1}[s V_k(s)] < 0. \end{cases} \quad (10)$$

Values of vector $\mathbf{U}(s)$ in the von Neumann-type controller are calculated sequentially, element-by-element, also both data $\mathbf{U}(s)$ output, and data $\mathbf{X}_0(s)$ input is executed element-by-element too. So, in the system there exist lags, both between input/output elements of the same vector $\mathbf{X}_0(s)/\mathbf{U}(s)$, and between input of k-th element of $\mathbf{X}_0(s)$, $X_{0,k}(s)$, and output the l-th element of $\mathbf{U}(s)$, $U_l(s)$. Due to the fact that time lags between transactions in control contours are essential values for system response, below is considered that time lags are counted from moment of output the $U_1(s)$ element till the input the of $X_{0,k}(s)$. All other data are inputted/outputted in/out of DC with lags, nominated as follows:

- $U_l(s)$ output with lags $\tau_{u,l}$, $1 \le l \le K$.
- $X_{0,k}(s)$ input with lags $\tau_{0,k}$, $1 \le k \le K$.

In accordance with the theorem about shifting in the time domain [12]:

$$L[\varphi(t-\tau)] = \exp(-\tau s)\Phi(s), \quad \tau > 0, \tag{11}$$

where τ is shift value; t is the time; $\varphi(t)$ is a function; $\Phi(s)$ is the Laplace transform of $\varphi(t)$.

From (11) it follows that

$$\mathbf{U}_{\text{sh}}(s) = \mathbf{U}(s) \cdot \mathbf{Q}_u(s), \tag{12}$$

$$\mathbf{X}_{0,\text{sh}}(s) = \mathbf{X}_0(s) \cdot \mathbf{Q}_0(s), \tag{13}$$

where $\mathbf{U}_{\text{sh}}(s)$, $\mathbf{X}_{0,\text{sh}}(s)$, are vectors $\mathbf{U}(s)$, $\mathbf{X}_0(s)$, elements of which are delayed on time after passing through interface; $\mathbf{Q}_0(s) = \left[Q_{0,kl}(s)\right]$, $\mathbf{Q}_u(s) = \left[Q_{u,kl}(s)\right]$ are diagonal lag matrices, in which

$$Q_{u,kl}(s) = \begin{cases} 0, & \text{when } k \ne l; \\ 1, & \text{when } k = l = 1; \\ \exp\left(-\tau_{u,k}s\right), & \text{when } 2 \le k = l \le K; \end{cases} \tag{14}$$

$$Q_{0,kl}(s) = \begin{cases} 0, & \text{when } k \ne l; \\ \exp\left(-\tau_{0,k}s\right), & \text{when } k = l; \end{cases} \tag{15}$$

With taking into account (1) and (2), (12), (13), data processing in the system may be described as:

$$\mathbf{X}_{0,\text{sh}}(s) = \left[\mathbf{F}(s) - \mathbf{X}_{0,\text{sh}}(s)\right] \cdot \mathbf{W}_c(s) \cdot \mathbf{Q}_u(s) \cdot \mathbf{W}_A(s) \cdot \mathbf{W}(s) \cdot \mathbf{W}_0(s) \cdot \mathbf{Q}_0(s), \tag{16}$$

where $\mathbf{W}_0(s) = \left[W_{0,kl}(s)\right]$ is the $K \times K$ diagonal matrix, which elements are as follows:

$$W_{0,kl}(s) = \begin{cases} 0, & \text{when } k \neq l; \\ W_{0,k}(s), & \text{when } k = l. \end{cases} \quad (17)$$

As it follows from (16), time lag in feedback circuit of the system with cross-links is the linear combination of time lags on whole ways, on which feedback signal comes for the DC processing.

3 Semi-Markov Model of DC Operation

For estimation of time intervals, the model of von Neumann computer operation in time domain should be worked out. For simplicity, it may be represented as including transaction operators only. Due to the fact that control process in such model is reduced to vectors $\mathbf{U}(s)$ and $\mathbf{X}_0(s)$ quests, algorithm is the cyclic one, but in it absent a looping effect. A control algorithm may generate transactions in arbitrary sequence, but any quest does not repeat twice consecutively. Also, due to the fact that for control action $\mathbf{U}(s)$ computation of all elements of vector $\mathbf{X}_0(s)$ should be used, the strong connectivity conditions are imposed [13] on the graph, which represents the structure of control algorithm. In common case, such properties have the full oriented graph without loops, shown in Fig. 2a.

With taking into account randomness of time interval between transactions and stochastic transactions sequence for external observer, the adequate approach to algorithm simulation is semi-Markov process [8–11], which states are abstract analogues of transaction operators of algorithm. Semi-Markov process is represented by the semi-Markov matrix:

$$\mathbf{h}(t) = [h_{kl}(t)] = [g_{kl}(t)] \otimes [p_{kl}(t)], \quad (18)$$

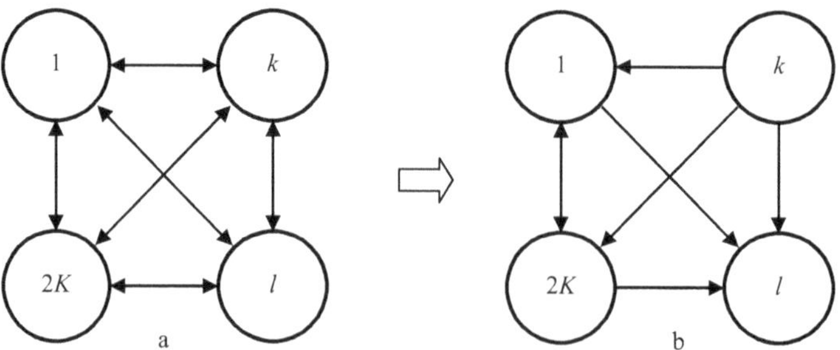

Fig. 2 Common structure of semi-Markov process (**a**) and the model for time interval estimation (**b**)

where $p_{kl}(t)$ is probability of the direct switching from the k-th state to the l-th state; $g_{kl}(t)$ is the pure time density of residence the process (17) in the k-th state before switching into the l-th state; $\otimes$ is the direct multiplication sign; t is the physical time.

Semi-Markov process (18) is the ergodic one and does not include both absorbing and partially absorbing states. Due to semi-Markov process ergodicity on densities $g_{k,l}(t)$ and probabilities $p_{k,l}(t)$, following restrictions are imposed:

$$0 < T_{kl}^{\min} \leq \arg[g_{kl}(t)] \leq T_{kl}^{\max} < \infty, \quad 1 \leq k, l \leq 2K; \tag{19}$$

$$\sum_{k=1}^{2K} p_{kl} = 1; \tag{20}$$

$$\int_{T_{k,l}^{\min}}^{T_{k,l}^{\max}} g_{kl}(t)\mathrm{d}t = 1, \tag{21}$$

where $2K$ is the common quantity of transaction operators; $T_{kl}^{\min}$ and $T_{kl}^{\max}$ are upper and lower bounds of density's $g_{kl}(t)$ domain.

When estimation of time intervals between transactions, it is no matter how semi-Markov process (18) gets l-th state from the k-th one. Determining in the case is that switch is the first, but not second, third, etc. For time interval estimation, initial semi-Markov process should be transformed into the process with the structure, shown in Fig. 2b, in which k-th state is the starting one and l-th state is the absorbing one. For getting such structure:

- k-th column and l-th row of $\mathbf{h}(t)$ should be reset to zeros;
- probabilities $p_{ij}(t)$ in all rows excluding the l-th, and in all columns, excluding k-th, should be recalculated as follows:

$$p'_{ij} = \frac{p_{ij}}{1 - p_{ik}}, \quad 1 \leq i, \ j \leq 2K, \ i \neq k, \ j \neq l. \tag{22}$$

In such a way:

$$\mathbf{h}(t) \to \mathbf{h}'(t) = \left[g_{kl}(t) \cdot p'_{kl}\right]. \tag{23}$$

After recalculation probabilities according (22), partially absorbing states are annihilated, and events of getting the l-th state from the k-th state begin to make up a full group of incompatible events. In such a way, time density of wandering from the first state to the l-th state may be estimated as follows:

$$g_{1,l}^{\Sigma}(t) = \mathbf{I}_k^r \cdot L^{-1}\left[\sum_{j=1}^{\infty} \left\{L\left[\mathbf{h}'(t)\right]\right\}^j\right] \cdot \mathbf{I}_l^c, \tag{24}$$

where $\mathbf{I}_k^r$ is the row vector, first element of which is equal to one, and other elements are equal to zeros; $\mathbf{I}_l^c$ is the column vector, l-th element of which is equal to one, and other elements are equal to zeros.

For time density (24), the expectation and the dispersion may be calculated, as usual [14]:

$$T_{kl}^{\Sigma} = \int_0^{\infty} t \cdot g_{kl}^{\Sigma}(t)\mathrm{d}t; \tag{25}$$

$$D_{kl}^{\Sigma} = \int_0^{\infty} \left(t - T_{kl}^{\Sigma}\right)^2 \cdot g_{kl}^{\Sigma}(t)\mathrm{d}t. \tag{26}$$

Expectations $T_{kl}^{\Sigma}(t) = \tau_{kl}$ give middle estimations of time delays. Also time intervals may be estimated with using "three sigma rule" as follows [14]:

$$\tau_{kl} = T_{kl}^{\Sigma} + 3\sqrt{D_{kl}^{\Sigma}}. \tag{27}$$

Estimations (24), (25), (26) define lags of input/output l-th element with respect to k-th. Changing indices k and l, one can calculate elements of matrices both $\mathbf{Q}_u$ and $\mathbf{Q}_0(s)$.

4 Example of Control System Analysis

As an example, the two-loop system of control the gas mixture pressure and components concentration in the reactor is considered (Fig. 3).

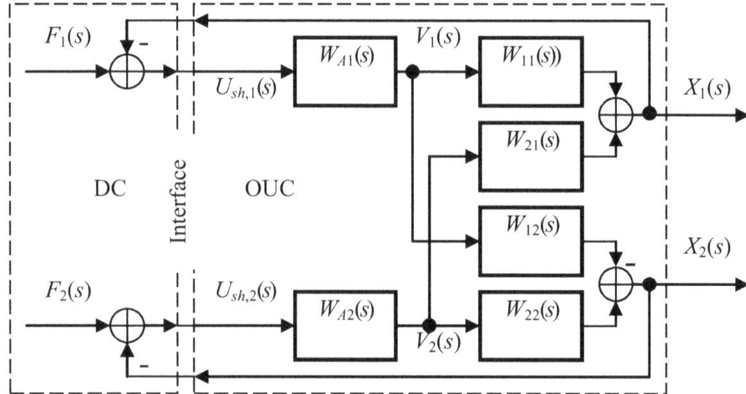

Fig. 3 Two-loop digital control system

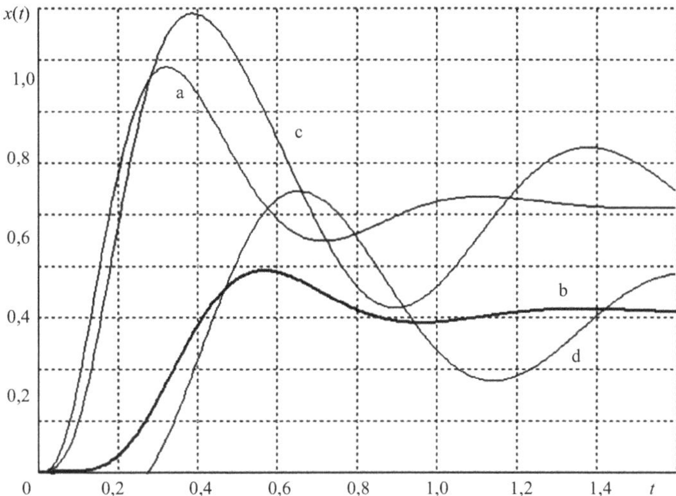

Fig. 4 Performance of the system

Transfer function which defines OUC dynamics is as follows:

$$W_{11}(s) = W_{21}(s) = \frac{3}{0.5s + 1}; \quad W_{12}(s) = W_{22}(s) = \frac{1.5}{0.5s + 1}. \tag{28}$$

In the system proportional feedback is realized. Inputs $F_1(s)$ and $F_2(s)$ are Laplace transform of Heaviside function $L^{-1}[F_1(s)] = 1 \cdot \eta(t)$. Actuators are performed with Eqs. (9), (10), where $\kappa_{1,1} = \kappa_{2,1} = 400$; $\kappa_{1,2} = \kappa_{2,2} = 4$; $\kappa_{1,3} = \kappa_{2,3} = 1$; $\xi_1 = \xi_2 = 20$. Lags at the interface are as follows: $\tau_{0,1} = 0$ c; $\tau_{0,2} = 0.02$ c, $\tau_{u,1} = 0.04$ c, $\tau_{u,1} = 0.06$ c.

Figure 4 shows the performance of the system. Plots a and b show the transient process in the system under analogue control, while plots c and d show the transient process in the system under digital control. Plots a and c show the pressure in reactor installation, while plots b and d show the gas concentration in the reactor.

As one can see from the plots lags in control loops increase both overshooting, and time of control, that one should take into account when design von Neumann computer-based digital control system and working out a software for it.

5 Conclusion

As a result, the mathematical model of von Neumann computer control of continuous production is worked out. It is proved that real-time characteristics of digital controller, namely feedback lags, may be calculated for control algorithm of any complexity. Time characteristics estimated should be taken into account when working out both common configurations of the system as a whole and software of the system. Increasing algorithm complexity for reaching quality characteristics of control process may have opposite effect, due to appearance of excessive lags in control loops.

Further investigations in the domain may be directed to working out methods of practical digital control algorithms synthesis, optimal to complexity–quality ratio.

Acknowledgements The study was carried out with state support from leading scientific schools Russian Federation, grant No. NSh-2553.2020.8.

References

1. Löfving, M., Säfsten, K., Winroth, M.: Manufacturing strategy formulation, leadership style and organisational culture in small and medium-sized enterprises. Int. J. Manufact. Technol. Manage. **30**(5), 306–325 (2016)
2. Landau, I.D., Zito, G.: Digital Control Systems, Design, Identification and Implementation. Springer, Berlin (2006)
3. Aström, J., Wittenmark, B.: Computer Controlled Systems: Theory and Design. Tsinghua University Press, Prentice Hall, Hoboken (2002)
4. Larkin, E.V., Ivutin, A.N.: Estimation of latency in embedded real-time systems. In: 3-rd Mediterranean Conference on Embedded Computing (MECO-2014), pp. 236–239 (2014)
5. Fadali, M.S., Visioli, A. Digital Control Engineering: Analysis and Design. Elsevier Inc., Amsterdam (2013)
6. Auslander, D.M., Ridgely, J.R. Jones, J.C.: Real-time software for implementation of feedback control. In: Levine W.S. (eds.) The Control Handbook. Control System Fundamentals, pp. 16-1–16-32. CRC Press, Taylor and Francis Group, US
7. Karnopp, D.C., Margolis, D.L., Rosenberg, R.C.: System Dynamics: Modeling, Simulation and Control of Mechatronic Systems. Willey, New Jersey (2012)
8. Bielecki, T.R., Jakubowski, J., Niewęgłowski, M.: Conditional Markov chains: properties, construction and structured dependence. Stochast. Process. Appl. **127**(4), 1125–1170 (2017)
9. Ching, W.K., Huang, X., Ng, M.K., Siu, T.K.: Markov Chains: Models, Algorithms and Applications. International Series in Operations Research & Management Science, vol. 189. Springer Science + Business Media NY (2013)
10. Howard, R.A.: Dynamic Probabilistic Systems. Vol. 1: Markov Models. Vol. II: Semi-Markov and Decision Processes. Courier Corporation (2012)
11. Janssen, J., Manca, R.: Applied Semi-Markov Processes. Springer, US (2006)
12. Schiff, J.L.: The Laplace Transform: Theory and Applications. Springer, NY (1991)
13. Larkin, E., Ivutin, A., Esikov, D.: Recursive approach for evaluation of time intervals between transactions in polling procedure. In: 2016 8th International Conference on Computer and Automation Engineering (ICCAE 2016), vol. 56, p. 01004 (2016)
14. Kobayashi, H., Marl, B.L., Turin, W.: Probability, Random Processes and Statistical Analysis. Cambridge University Press, Cambridge (2012)

Automated Product Life-Cycle Control System

Sergej Solyonyj, Aleksandr Rysin, Ilya Voropaev, Oksana Solenaya, and Maria Sozdateleva

Abstract This article describes how to create a product life-cycle tracking system in a continuous production environment. To do this, using the example of the production of electrical equipment, the weaknesses of the production process of the enterprise were identified and the directions of information automation were determined. Methods for tracking the life cycle of products used in different areas were considered, and their advantages and disadvantages were identified. As a result, a system was developed using QR codes applied to products and read by portable devices. The advantages of using the implemented system are increasing production efficiency by tracking the employment of employees and applying disciplinary measures to employees, reducing the percentage of defects in production due to timely receipt of information on the progress of production processes by management, which ensures timely organizational measures, reducing the percentage of utilization by optimizing the logistics system in production—marking of completed production processes and positioning of parts after passing through the processes, reducing the consumption of raw materials.

1 Introduction

The gradual introduction of cyber-physical systems into technological production processes already in the current production conditions, aimed at obtaining benefits due to the scale of production, will increase competition in the market, expand markets, and increase the competitiveness of various sectors of the economy of entire countries [1]. Fundamentally new, breakthrough business models and technologies should become the tools for such a transformation of the economy. This place will be taken by 3D printing technologies, robotization, the Internet of things, digital platforms, virtual spaces for learning, digital ecosystems, advanced analytics of big data, and, of course, development in the field of artificial intelligence [2]. According

S. Solyonyj · A. Rysin (✉) · I. Voropaev · O. Solenaya · M. Sozdateleva
Saint-Petersburg State University of Aerospace Instrumentation, SUAI, 67, Bolshaya Morskaya Str., Saint-Petersburg 190000, Russia

A. Ronzhin and V. Shishlakov (eds.), *Electromechanics and Robotics*, Smart Innovation, Systems and Technologies 232, https://doi.org/10.1007/978-981-16-2814-6_38

to estimates by the McKinsey Global Institute, the Internet of things alone by 2025 will annually bring the world economy a profit of 4–11 trillion US dollars [3].

Combining all these technologies within a single concept allows us to imagine the production of the future: a fully integrated and optimized workflow with significantly increased efficiency at all stages and the relationship between manufacturers and customers at the level of "customization" of manufactured units and maximum compliance of manufactured products with requirements customer, as well as more distant relations between man and machines in heavy production areas: complete replacement of a person with machine labor, replacement of some of the simplest working positions with typical operations by robotic systems capable of independent interaction with each other in real time.

Product life-cycle management (PLM) is a product life-cycle management technology [4]. These are systems that combine the workflow within the company, conducted during production, and technology for keeping a quick record of manufactured products and operations on parts.

In the market of software products for tracking the life cycle of products, there are currently developments of various large companies. Basically, the presented products are complex solutions that can be divided into modules and consist of computer-aided design systems (CAD), enterprise resource planning systems (ERP), product data management systems (PDM), manufacturing execution systems (MES), enterprise content management systems (ECM), and PLM systems [5].

In general, PLM systems are divided into the following components:

- CAD is a computer-aided design (CAD) system designed to carry out design work using computer technology and also allows you to create design and technological documentation for individual products, buildings, and structures [6].
- ERP is a class of systems for the management of production, labor resources, finance and assets, focused on optimizing the resources of an enterprise [7].
- PDM is a module that provides management of complex information about a product [8]. The latter term can be understood as different objects, including technically labor-intensive (ships, missiles, complex computer networks). The PDM system makes it possible to establish interaction between users, control large flows of engineering and technical information, and obtain delimited access to data at any stage of product development/manufacturing. In many respects, therefore, it is considered the main one when choosing a management module [8].
- MES system is a specialized application software designed to solve problems of synchronization, coordination, analysis, and optimization of production within the framework of any production. MES systems belong to the class of shop floor control systems, but they can also be used for integrated production management in the enterprise as a whole. MES is the interface between the ERP layer and the production layer. They are an essential component for creating vertical integration, as shown in Fig. 1. The three layers together can be part of the company management system (for which ERP systems are the most common tools), production management (MES), and the production system. The latter usually contains

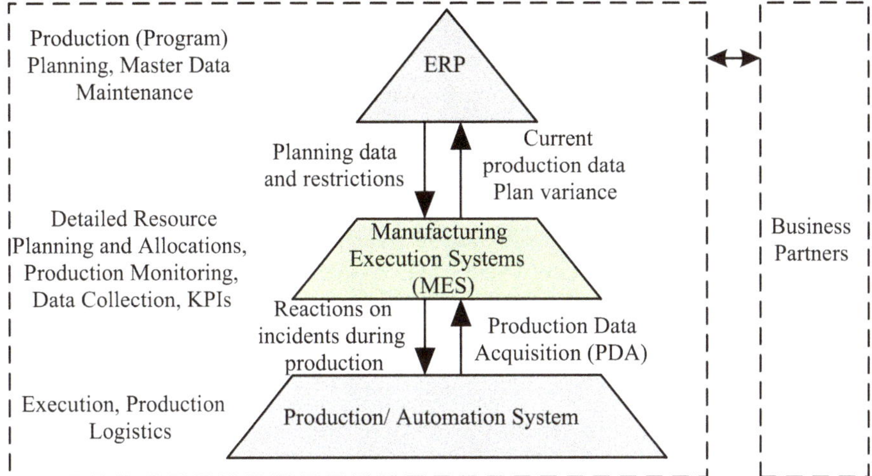

Fig. 1 Place of MES systems in the vertical integration of PLM system components

hybrid software and hardware systems such as: distributed control systems (DCS), programmable logic controllers (PLC), distributed numerical control (DNC), supervisory control and data acquisition (SCADA) systems, and other control systems, designed to automate the process in which products are manufactured [9].

Unlike ERP systems, which usually provide a very wide range of functionality covering all business processes of an enterprise at all stages of production, MES is aimed at providing the ability to quickly respond to events occurring in the production process for the company (situational detailed planning). MES adopts a microscopic, more detailed view of production data (often limited to a single plant or production area), compared to a macroscopic, holistic view of an ERP system, and therefore MES systems are designed to compensate for one of the main disadvantages of system modules of ERP systems in production: the impossibility of ensuring the integration of production data obtained in workshops in real time into the system of general planning and production management. It is this gap that MES systems are trying to fill.

In general terms, the definition of MES systems implies the following characteristics:

- high level of detail (obtaining data directly from production processes);
- relatively short planning horizon (reactive (situational) planning);
- bidirectional communication for interaction with both ERP systems and systems in production [10].

2 Problem Statement

At the enterprise in question, production is organized in four main workshops. Storage of blanks and finished products before shipment is carried out in the internal premises of the enterprise. Storage of blanks, their marking and creation of layouts at the enterprise are not organized. Production of products is divided into five main stages:

1. Production of billets from rolled metal of various thicknesses is carried out in a workshop with partial automation: the workshop is equipped with bending, punching, revolving machines, as well as an automated plasma cutting machine from Amada. After the production of blanks at this site, they are stored on pallets and transported to the next site or stored in the internal premises of the enterprise.
2. Pre-assembly. In this separate workshop, welding and cleaning work is performed with the parts, which is associated with a high probability of mechanical damage to the marking on the parts. Then the workpieces are also transported to the next site or stored in the internal premises of the production of an enterprise for the production of electrical equipment.
3. Mechanized painting of blanks. For this purpose, the blanks are placed on a conveyor with brackets in a suspended state—then the blanks pass in turn a paint shop, a heat-drying shop, and a varnishing shop. These works also do not technologically imply the preservation of marking on the workpieces. Storage of finished products is not provided within this production facility.
4. Laying of wiring cables inside the product. This process is not automated and involves a long painstaking work of the company's employees in laying cables, first on the product model, then installing the created wiring system inside the produced element. The storage of parts after this operation is also organized inside the production premises.
5. Final assembly and packaging. At the stage of final assembly, the manufactured units are supplemented with controls and small parts, then packed on pallets and awaiting dispatch to the customer. Warehousing of parts before shipment is also carried out in the internal premises of the production.

Thus, in the absence of storage facilities for organized storage of parts, as well as the complexity of the logistics scheme, the only possible option to reduce costs for an electrical equipment manufacturing enterprise was to implement a system that would help optimize logistics at the enterprise at the current stage of production development.

Based on the analysis of the company's activities, the current state of production, the need for automation of production in terms of control of the product life cycle was determined.

The product life-cycle control system will allow, even at the stage of cutting metal sheets, to understand which order this or that part will be part of, track the stages of its production, and automatically generate documentation upon completion of the parts processing stages. The main problem in the production process is the movement of parts between workshops and the problem of marking parts: parts are lost after going through production processes due to an incorrectly constructed logistic scheme of production and the lack of an available tool for tracking production processes carried out with the company's products for the production of electrical equipment.

At the production of a separate batch of the company's products, a study was carried out of the flow of business processes, an assessment of the need to implement the system, an economic analysis of the result of the implementation of the system, the development and testing of the MES system.

In an enterprise, the overall infrastructure of a PLM system looks like this:

- CAD system—CREO from PTC;
- PDM system—Windchill from PTC (hereinafter Windchill will be referred to exactly as a PDM system due to insufficient use of Windchill functionality in the conditions of the considered production);
- ECM system—EDMS 1C: UPP (1C: Manufacturing Enterprise Management 7) [11].

The structure of interaction between the specified software products needs optimization, as well as the general structure of the life-cycle tracking system at the enterprise. The main problem identified in the course of the study is the analysis of labor productivity after the release of the product due to the lack of the ability to track the status of the released product at the time of production.

At the current stage of development of the PLM system at this enterprise, the MES system includes the marking of parts and packages of parts in production by applying an order code and various labels on them using permanent markers. This method of labeling and tracking the product life cycle has serious disadvantages: in terms of the durability of marking, the mark is overwritten with a marker in the process of cleaning parts, and in the process of painting, and even in the process of transporting parts, uniformity of marking—marking of parts differs not only between different workshops, but even between different people responsible for marking parts, availability of information about details and design documentation for shop foremen and engineers, the complexity of the formation of documentation—first of all, invoices, the possibility of creating a unified system of marking and tracking of parts. But there are also advantages: low cost, availability, speed of implementation.

To understand the problem, a deep analysis of production processes was carried out. The production of final products involves the production of starting material of various thicknesses, cutting and cutting of parts from the resulting material, initial mechanical assembly of parts, welding of parts, the stage of mechanized powder or enamel painting of parts, final assembly of parts, installation of equipment and wiring, preparation for transportation and transportation itself to the customer. Thus, a life-cycle tracking system is required for implementation at the following stages of production [12]: receiving material, chopping and cutting of blanks, cleaning of blanks, mechanical assembly work, welding works, powder and enamel painting of workpieces, final assembly, transportation, storage of blanks between stages of production.

The system being developed will be tested at a production site with partial automation. The site has machines for automatic plasma cutting, cutting metal, punching technological holes, bending. In this production, all the above processes are presented, which will allow the implementation and testing of the system at the most universal stage of production. The main task of the system in this production area is registration of the processes performed, setting up a shift-daily task for production, tracking the execution of shift-daily tasks by the section employees, tracking the location of workpieces.

For this purpose, it was decided to introduce a marking system into production, as well as a marking tracking system using portable scanners. The introduction of this system will make it possible to build not only a product tracking system at the production stage, but also create a technological platform for the implementation of a piecework wage system, the most effective in the conditions of the specified production.

3 Research Results

In the process of creating the system, the main stages of creating automated systems in accordance with GOST 34.601-90 "Automated systems. Stages of creation." The following strategy was adopted for the development of the system: modeling, the choice of technology, creation of a test version, creation of the core of the program, creation of basic subsystems, implementation in production, further maintenance of the system.

Based on the results of the development and testing of a system for tracking production processes at one production site of the enterprise, it is assumed that the system will be distributed and implemented in all technological processes in production. During the development process, organizational measures were also taken to restructure the established scheme of production processes for the work of the created tracking system: orders to create a working group, appoint responsible persons for the development and implementation of the system, made changes to the flowcharts created in production, job descriptions.

The developed system assumes the possibility of using various technologies for both visual and machine recognition of tags by the system: the proposal for production consists in combining technologies for creating QR tags, server technologies, and RFID-NFC tags.

QR coding has the following disadvantages: limited tag capacity, the need to update the label when tracing the product life cycle, possible loss of marks after painting, and the need for a stable connection to the network.

However, there are also advantages to using such labels: relative ease of tag generation, ease of integration with the existing production infrastructure, low development and implementation costs. RFID tags have certain disadvantages: difficulty in implementation and integration with the existing infrastructure, high price. The advantages of using RFID tags are: autonomy, the ability to rewrite the label, resistance to impact when processing parts, free label format.

Server storage technologies have several disadvantages: the need for a stable connection to the network and high price. Server storage technologies have the following advantages: uniformity (1 tag of any format is used, containing a link to a resource for storage during the entire processing period of the part), the ability to create a system with access control, the possibility of implementing a system of accounting and analysis of working time, creation of a database of statistical data on the processing time of parts and operating time to create a standardization system.

It was revealed that this system can achieve the greatest effect when combining three technologies. At the first stage, a mechanism for automatic creation of QR tags was introduced into production for their further use in the existing infrastructure of the enterprise, which will significantly speed up the marking process, as well as lead to the uniformity of tags in the above-mentioned production and reduce the risk of losing parts during transportation.

As a result of the inspection of production processes and the workflow system, it was decided to orient the tracking system to register the completion of specific technological processes with details. This solution is associated with the need to reflect and document completed processes in the 1C: UPP system. Subsequently, it is planned to connect a separate 1C: UPP application for document flow to the database of the system or create a separate application for electronic document flow within the developed system based on the results of tracking production processes for the required time. It was decided to use the system for registering the produced technological processes at the stage of passing technical control. Thus, the number of operations for technical control of products in production will be increased, which will improve the quality of products, and the document flow of the department of technical control will be simplified.

The use of the system in the workshop for the primary processing of metal products of the enterprise for the production of electrical equipment has transformed the flowchart of production processes as follows:

Working:

1. Places the workpiece in the automated plasma cutting machine;
2. Loads the cutting program to the machine memory;
3. Monitors the process produced by the machine;
4. Receives the workpieces after cutting, attaches the printed QR codes to the workpieces in the places indicated in the technological maps (least involved in further machining and assembly of the workpieces).

Quality control specialist:

1. Checks the compliance of the manufactured part with technological standards;
2. In case of compliance, it scans the QR tag attached by the worker and confirms the status of the manufactured product;
3. Directs the workpieces to the next bending/turret punching/drilling machine in accordance with the product WPS.

Then the process is repeated with each of the machines. Based on the results of scans based on the number of production processes and parts included in the order flowchart, the system automatically assigns the order completed status to the order. Upon completion of the order, the shop manager and quality control department can generate reports on the completion of the order and its technological compliance. It should be noted that in the marking system based on the 1C: UPP functionality considered in the process of market research, there is a certain drawback, corrected in the developed system: the QR code generated by the 1C: UPP system is located on the accompanying documentation of a specific technological process, which implies scanning at the workplace of the shop manager. In the reality of the production under consideration, such a scanning system is associated with the reflection in the reporting documentation of incorrect information due to the non-obligation of employees and the subsequent transfer of incorrect information to top management. The developed system makes it possible to restrict the passage of the next technological process over the part to determine the quality of the previous process, as well as the ability to set the status of the produced process only by direct contact with the part and vising the results of the produced process. In addition, the system assumes the establishment of an employee responsible for scanning before starting work. The system provides for fixing the scanner to a specific production area, which allows reducing the amount of information transmitted to the server part of the system for processing, simplifies the application interface, and also simplifies the interaction between the client and the server (less information is transmitted, which reduces the load on the transmission channel data, makes it possible to quickly transfer information, reduces the number of errors in data transmission) [13].

For the development of the system, a client–server architecture was chosen, where the client is:

- A scanner that sends information about the scanned part, the employee responsible for the registration of the produced technological process, as well as the location of the stored blanks to the server for processing;
- Web client of the system, displaying information on the current status of orders and individual details to authorized users.

As part of the work on choosing a platform, a comparison was made between the two most popular marking technologies in production: development of a system based on the technology of production document circulation as part of the 1C: UPP system (1C programming language), passive QR code scanners DataLogic; development of a system based on the Java language using ready-made frameworks, virtual machine technology, an active ZKC scanner on the Android operating system, an application on the Android platform for scanning tags and transferring data to the system.

The choice was made in favor of system configuration number 2 due to the vast potential of this system configuration for development: active development and support of the listed technologies by platform developers, updating of libraries, the emergence of new frameworks for developing systems based on these platforms, as well as due to the extensive functionality of programs for the development of systems based on these platforms, placed in the public domain. In addition, the initially laid down functional flexibility of the Android platform, as well as a large number of application programs and interfaces embedded in the operating system for reading data using various devices: the ability to quickly connect to the operating system of both an RFID scanning device and QR scanner, the availability of publicly available libraries for the most popular reading devices allows you to develop a system and ensure further development of the system without excessive labor costs of the developers of the mobile application. The popularity of the Android platform and the Java programming language among developers, system analysts and programming specialists will prevent staff shortages in specialists for supporting and developing the developed system in the future.

A database was deployed at the production facilities for testing the developed system. The specified database will allow you to quickly put down a sign of the completion of the technological process of processing a product using a scanner, as well as to quickly receive information about putting this sign down in the database at the manager's workplace (department head/production/section foreman). The database is implemented in Postgre SQL.

In addition, in the Web interface of the program, the functionality for editing technological processes was implemented to ensure compliance with technological maps. Also, a functionality was created for the operative editing of the list of employees who have access to the system.

To ensure the work of QCD specialists in production, a separate application was developed for the ZKC scanner based on the Android 5.1.1 operating system to scan QR and RFID tags and transfer the received data to the server for processing. The main functions of the application have been reduced due to the need to comply with the trade secrets of the electrical equipment manufacturing company in this work.

The scanner has a standard CMOS barcode (QR) code reader and a standard radio frequency (RFID) reader. To ensure scanning by these devices, a special package of functions and methods is installed in the operating system, which allows you to access the scanners by transmitting commands, as well as receive and transmit decoded information from/for the scanner by the application functionality. Information from scanners is received as a string variable with a set of characters. The functionality of the application provides for the activation of the QR code scanner both using soft buttons (using methods called by pressing a button inside the user application) and physically (using a dedicated button on the physical keyboard of the scanner).

The general scheme of data transfer in production after the implementation of the system is represented by the DFD diagram shown in Fig. 2. Data-flow diagrams (DFD) is a methodology for reflecting the movement of data flows. The application of this methodology allows you to graphically reflect the structure of the system, describing the sources and destinations of data within the system, the logical functions of the system, data streams and data stores connected to store the system data.

The system has three data storages, as well as data flow schemes that ensure the integration of information systems with each other. The data packet received during the operation of the scanner application is sent for processing by the server, and then, the converted data is entered into the system database.

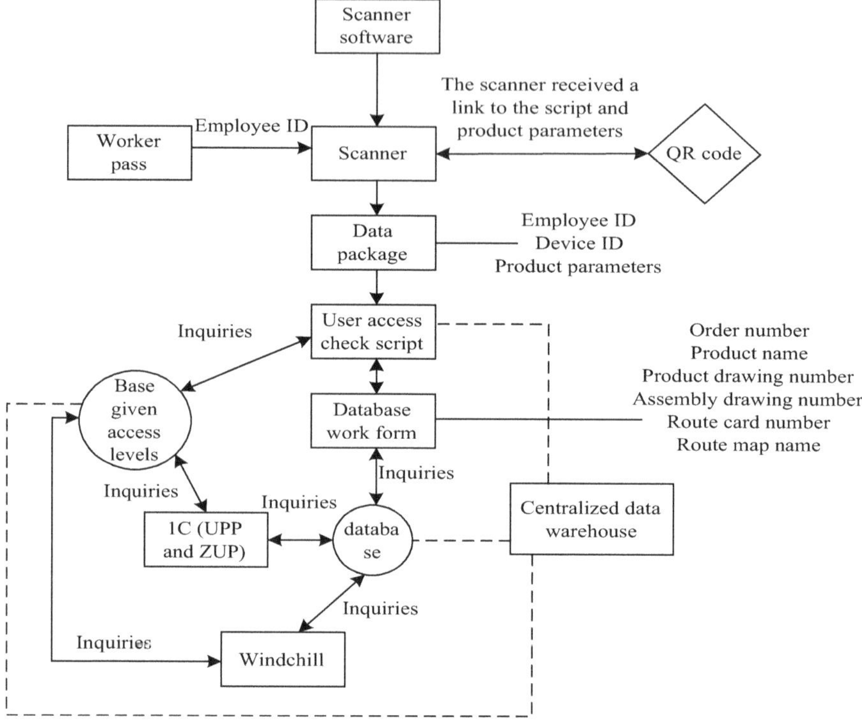

Fig. 2 DFD general scheme of the system after implementation into production

Thus, the MES system was introduced into the general scheme of the product life-cycle tracking system (including the electronic document management scheme), which already exists at the electrical equipment manufacturing enterprise, which allows providing operational data on completed production processes as in the context of whole orders, and in the context of individual parts and production processes [14]. The flowchart of the program is shown in Fig. 3.

4 Conclusion

A partially mechanized section of primary metal processing and primary assembly was chosen as a production process for testing the system. In the main part of the production area, there is a plasma cutting machine. The machine is loaded with a sheet cutting program, formed in the design and programming department of CNC machines—after the blanks leave the working area, the worker sticks marks on the blanks for tracking during the production process. Then the workpieces are sent in accordance with the technological maps further to other production areas in order to carry out additional processing. The workplaces of the specialists of the quality

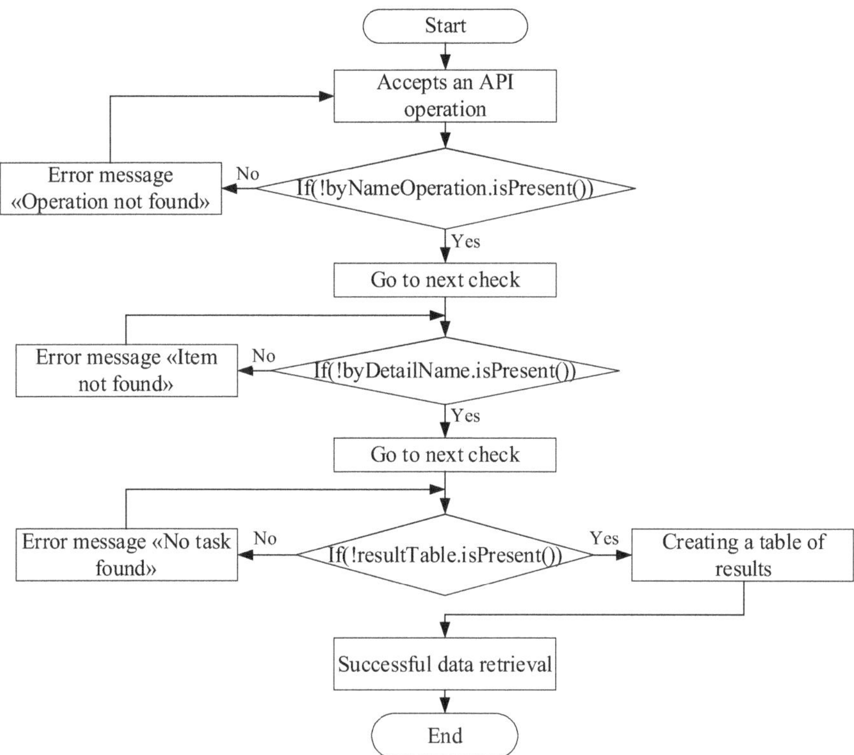

Fig. 3 Program flowchart

control department are located near the workplaces. The staff of the quality control department are constantly in the room and approach the places of production of blanks as the technological processes of production are completed for control. After control and completion of production processes, according to the flowcharts, blanks can be sent either to other production areas, or stored in an intermediate storeroom located on the left in the diagram.

Within the framework of the Lean Manufacturing project, an order was prepared on the use of QR coding parts in production as a marking in the following format: order number (in the format: 3 characters 3 characters); part code in accordance with the flowchart for manufacturing a part (in the format of 3 alphabetic characters, 9 digital characters).

The marking is located in places that are least involved in the mechanical processing of parts due to the instability of any coating for printing a QR code to mechanical stress.

As part of the test operation of the system, a separate order was identified, according to which the parts were produced using the developed system. The infrastructure for using the system was also arranged: a place was determined for storing marked workpieces, a printer was installed for quick printing of QR tags, equipment was installed to provide access to the organization's local network inside the production facility, where the system was tested (access points and repeaters Wi-Fi).

It should be noted that the introduction of the system already at the stage of test operation made it possible to correctly organize the storage system for workpieces in production facilities. As a result of the completion of production processes with blanks, further consumers of blanks from other workshops of the enterprise were able to quickly find the parts needed for further processing, focusing on the system data on the latest technological process produced with blanks [15].

The system has been fully tested. After analyzing the results of testing the system, the advantages and disadvantages were noted.

Advantages: increasing production efficiency by tracking the employment of employees and applying disciplinary measures to employees, reducing the percentage of defects in production due to the timely receipt of information on the progress of production processes by management, which ensures timely organizational measures, reducing the percentage of utilization by optimizing the logistics system in production—marking of completed production processes and the location of parts after passing the processes, reducing the consumption of raw materials. In the long term, the introduction of a life-cycle tracking system based on QR codes or RFID tags creates the basis for global production automation.

Disadvantages: server adjustments are required: creation of a script for automatic transfer of the attribute of the number of parts to the system database, terms of solution implementation, the complexity of integrating the formation of a QR code with a program for the CNC machine tool, it is necessary to integrate the xls data parser from the shop floor into the work of the server.

Acknowledgements This research is supported by the RFBR Project No. 20-08-01056 A.

References

1. Digital Russia: a new reality. https://roscongress.org/materials/tsifrovaya-rossiya-novaya-realnost/. Last accessed 2020/05/01
2. McKinsey: The internet of things in Russia will accelerate the economic effect of the digitalization of the economy. https://iot.ru/promyshlennost/mckinsey-internet-veshchey-v-rossii-uskorit-ekonomicheskiy-effekt-ot-tsifrovizatsii-ekonomiki/. Last accessed 2020/05/01
3. Tarasov, I.V.: Industry 4.0: concept, concepts, development trends. https://cyberleninka.ru/article/n/industriya-4-0-ponyatie-kontseptsii-tendentsii-razvitiya/. Last accessed 2020/04/23
4. Zadorov, I.A.: PDM and ERP systems as a unified information system of enterprise management. In: Materials of the International Scientific and Practical Conference, pp. 145–150. Closed Joint Stock Company "University Book", Kursk (2011)
5. Stages of the product life cycle. http://www.salogistics.ru/students/suai_2011/page3.html. Last accessed 2020/05/01
6. Sinelnikov, A.V., Bachurin, A.V.: Features of integration of PDM system with cad systems. In: Procedures of the XVII International Scientific and Practical Conference, pp. 398–401. Institute for Problems of Ex. Them. V.A. Trapeznikov, Moscow (2017)

7. Classification and characteristics of CIS/Corporate information systems. https://www.sites.google.com/site/korpinfsis/home/klassifikacia-i-harakteristiki-kis/. Last accessed 2020/05/01
8. PDM and PLM Systems. http://asapcg.com/press-center/articles/pdm-i-plm-sistemy/. Last accessed 2020/04/28
9. Schmidt, A., Otto, B., Kussmaul, A.: MES Services in the Automotive Industry. University of St. Gallen—for Business Administration, Economics, Law and Social Sciences (HSG), St. Gallen, Switzerland (2011)
10. Why does a giant need an MES system? Experience of one implementation. https://isup.ru/articles/1/9903/. Last accessed 2020/05/01
11. Borovkov, A.I., Burdakov, S.F., Klyavin, O.I., Melnikova, M.P., Mikhailov, A.A., Nemov, A.S., Palmov, V.A., Silina, E.N.: Computer Engineering: Textbook. Allowance. Publishing House of Polytechnic, Sankt-Petersburg (2012)
12. Vasilyeva, I.A., Kolosova V.V., Sazonov A.A.: Management of the life cycle of products in the conditions of transformation of production. Bull. Moscow State Reg. Univ. Ser. Econ. **3**, 50–58 (2019)
13. Gehrke, I., Schauss, M., Kusters, D., Gries, T.: Experiencing the potential of closed-loop PLM systems enabled by industrial internet of things. Procedia Manuf. **45**, 177–182 (2020)
14. Asl-Najafi, J., Yaghoubi, S.: A novel perspective on closed-loop supply chain coordination: product life-cycle approach. J. Cleaner Prod. **289**, 125697 (2021)
15. He, B., Shao, Y., Wang, S., Gu, Z., Bai, K.: Product environmental footprints assessment for product life cycle. J. Clean. Prod. **233**, 446–460 (2019)

On the Method for Estimation of Pipeline Durability Taking into Account of Technical Condition Diagnostic Results and Safety

Eleonora Zavoychinskaya

Abstract The problem of determining pipeline structure durability based on the results of calculation of its structural element service life at the design stage and the residual service life of elements with defects established by diagnostic explorations and taking into account the requirements of social and environmental safety is formulated and solved. There are proposed relations for the structure failure probability distribution function (the highest hierarchical level) through the failure probability distribution function of its similar structural elements (the lowest hierarchical level). The element durability is proposed to determine according to the theory of failure loading processes and a stochastic model of scale-structural fatigue. Assuming that the failure probability should not exceed its acceptable value (the criterion of structural reliability), the equations for finding of the service and the residual life are written. The ratios for the determination of technogenic and anthropogenesis risks at structure destruction and criteria of durability determining, taking them into account, are given. These criteria are the theoretical generalization of the known relations used in design practice on project and operation stages of pipeline various sections. The proposed approach is original, and the author is not aware of similar works of other researchers.

1 Introduction

The purpose of this study is to develop the experimental and theoretical foundations of the stochastic method for assessing of longevity and diagnostic periods of the technical conditions of various long structures whose elements are at the internal pressure of the pumped product, the action of mass forces, temperature field and natural-climatic and technogenic influences.

The methodological basis for the proposed method of durability estimation is the works of scientists on the fundamental scientific problem of the technogenic safety

E. Zavoychinskaya (✉)
Mechanical and Mathematical Department, Moscow State University named after M.V. Lomonosov, Leninskie Gori, 1, 119991 Moscow, Russia
e-mail: elen@velesgroup.com

A. Ronzhin and V. Shishlakov (eds.), *Electromechanics and Robotics*, Smart Innovation, Systems and Technologies 232, https://doi.org/10.1007/978-981-16-2814-6_39

of structure operation [1–10]. The proposed approach is original, and the author is not aware of similar works of other researchers.

Here, three main technogenic spheres such as people, constructions and the environment are constructed, and accordingly, the concepts of technogenic and anthropogenesis risks are introduced, namely the probability of injury to people (the social risk), the probability of destruction of industrial objects (the industrial risk) and the probability of flora and fauna destruction (the ecological risk) located in a potentially dangerous zone near the structure at destruction. The modern problems of safety estimation are to formulate the criteria of structural reliability, to study of the probabilities of the appearance and spread of injury factors at the structure destruction, to estimate of technogenic and anthropogenesis risks based on the operation data about the section current state on potential and realized risks, to establish the acceptable risks, to develop of numerical modeling for durability and residual life into account the technogenic and anthropogenesis risks, to create of longevity management methods, to calculate of potential economic damage at the construction and operation.

The nature of the change in operating loads, a significant heterogeneity of the mechanical characteristics of materials, a variation in design technological factors, as well as the need to take into account of technological and operational defects, lead to the need to use the probabilistic methods for estimation of the durability and crack resistance of structural elements. The probabilistic parameters of material properties (characteristics of crack opening, Woehler curve, Coffin-Manson equation, Paris relation, yield stress, etc.) are entered in the durability calculation, and random stationary loading processes are considered [2, 4, 5, 11–13]. One of the main modern directions is the creation of algorithms for predicting the residual life of structural elements based on the established patterns of failure processes at the micro-, meso- and macro-levels.

The proposed method for durability estimation for structures such as pipelines is based on a block-hierarchical approach, according to which several hierarchical structural levels are distinguished, and the service life is calculated sequentially from the lowest to the highest hierarchical structural levels [4, 5, 11–16]. The structure (the highest hierarchical level) is conventionally divided into large macrosegments, namely the sections (fragments, etc.) with linear dimensions l_k, $k = 1, \ldots, K$, $\sum_{k=1}^{K} l_k = L_0$ (L_0 is the total structure linear size) on the functional and design principle: linear sections with branches and looping, crossings over natural and artificial obstacles (roads, railways, air crossings over water obstacles, ravines, underwater crossings, etc.), connection nodes of other structures, construction of gas and oil metering stations, gas recovery units, units of starting and receiving of cleaning devices, design of head and intermediate pumping stations, etc. Sections consist of calculated segments such as structural elements (the lowest hierarchical level), for which internal and external loads and effects can be approximately considered as homogeneous. Each section consists of a significant number $n_{k,q}$ ($n_{k,q} \geqslant 5 \times 10^2$) similar q—elements, $q = 1, \ldots, \mathbb{Q}$, with characteristic linear dimensions l_q (base metal, longitudinal and ring welded joints, tee connection, diversion, adapter, bottom).

The durability $t_{f,k,q}$ of the k-section q-element is determined on the theory of failure loading processes [4], the theory of scale-structural fatigue [11–16] and the well-known methods of fracture mechanics. The durability of pipelines at operational loading is described by random processes that take into account the potential stochastic element failure, the random mechanical loading, the random natural environmental influences, etc. Therefore, a stochastic approach and methods of random process theory and statistical analysis are chosen. As a toolkit for numerical experiments and solving of practical problems, finite element methods of the ANSYS software package are used.

The standards of pipeline operation provide for a comprehensive diagnostic of the technical condition during operation. Diagnostics of a section or several sections of a pipeline (middle hierarchical level) are carried out by various methods: photogrammetric, color, multi-zone, infrared and radio frequency surveys including aerospace and helicopter surveys; physical non-destructive diagnostic methods (echo-vibration, infrared, electrometric, X-ray and radioactive radiation); methods of measuring the hardness of the metal surface layers; magnetic and ultrasonic methods (when passing smart projectiles); mechanical potentially destructive diagnostic methods for example 25–50% higher test loading than the working loading method. Databases of detected element defects are created, the k-section q-element may contain j-defects, $j = 1, \ldots, J$, having a mechanical, technological and operational origin, identified at r-diagnostic investigation, $r = 1, \ldots, R$.

Large-scale pipeline destruction is caused by stochastic processes of cracking in the base metal, circular and longitudinal welded joints, corrosion and stress-corrosion cracking, metal corrosion loss, metal delamination, formation of scratches, dents, corrugations in the element walls. Defects can be found by means of internal and external diagnostics with a certain reliability, as well as determined by the results of the section previous operation and by the destruction statistics of similar sections. Each type of defect determines the corresponding flow of element destruction. According to the failure loading process theory [4] and the fatigue scale-structural theory [11], the calculation of the k-section q-element residual life on j-defect $\Delta t_{f,k,q,j}$ is carried out.

2 Structural Reliability Criterion

Here, the function $Q = Q(\tau)$, $0 \le Q \le 1$, $\tau \in [0, t]$ is introduced and is determined as the failure probability distribution function (the structural risk) at a time τ [11–16]. So, there is considered the distribution function of k-section failure probability $Q_k = Q_k(\tau)$, $k = 1, \ldots, K$, and the distribution function of $n_{k,q}$ q-elements failure probability $Q_{k,q} = Q_{k,q}(\tau)$, $q = 1, \ldots, \mathbb{Q}$.

It is assumed that the function $Q = Q(\tau)$ is determined through $Q_k = Q_k(\tau)$ according as follows:

$$\text{optimistic scenario} \quad Q(\tau) = \sum_{k=1}^{K} \left(\frac{Q_k(\tau)}{1 - Q_k(\tau)} \right) \prod_{k=1}^{K} [1 - Q_k(\tau)], \tag{1}$$

$$\text{pessimistic scenario} \quad Q(\tau) = 1 - \prod_{k=1}^{K} [1 - Q_k(\tau)]. \tag{2}$$

Relation (1) determines the sum of independent events, namely k-section failure in the absence of the remaining section failure, and relation (2) is the sum of independent events at least of k-section failure. Analogically we have the following:

$$\text{optimistic scenario} \quad Q_k(\tau) = \sum_{q=1}^{\mathbb{Q}} \left(\frac{Q_{k,q}(\tau)}{1 - Q_{k,q}(\tau)} \right) \prod_{q=1}^{\mathbb{Q}} \left[1 - Q_{k,q}(\tau)\right], \tag{3}$$

$$\text{pessimistic scenario} \quad Q_k(\tau) = 1 - \prod_{q=1}^{\mathbb{Q}} \left[1 - Q_{k,q}(\tau)\right]. \tag{4}$$

It is proposed to describe the failure probability distribution function $Q_{k,q} = Q_{k,q}(\tau)$ by a function of the Poisson distribution type on the first failure of q-element in the form:

$$Q_{k,q}(\tau) = \varphi_{k,q}(\tau)\mathrm{e}^{1-\varphi_{k,q}(\tau)}, \; \varphi_{k,q}(\tau) = \lambda_q l_q n_{k,q} \frac{\tilde{t}}{t_{f,k,q}} \tau, \quad q = 1, \ldots, \mathbb{Q}, \; k = 1, \ldots, K. \tag{5}$$

In the relation (5), a parameter $\tilde{t}$ is the economically and socially acceptable structure life, assigned by the design standards. For example, it lies within $35 \div 45$ years for main pipelines, $60 \div 65$ years for tie pipelines, $15 \div 20$ years for field pipelines. Parameter λ_q is the intensity of the q-element failure flow, namely the number of q-element failure per unit of time (year) and per unit of length (km), known from the statistics of destruction at operation of similar structures in similar natural and climatic conditions. For tee connections, the number of failures per unit time attributed to the total element number is considered as λ_q, and relation (5) does not include the value l_q.

The criterion of structural reliability is formulated as follows:

$$Q(\tau) \le \tilde{Q}, \tag{6}$$

on condition $t_{f,k,q} \ge \tilde{t}$, $k = 1, \ldots, K$, $q = 1, \ldots, \mathbb{Q}$, the function $Q = Q(\tau)$ is determined by (1)–(5), $\tilde{Q}$ is acceptable structural risk according to design standards. The structure life is determined by the equation:

$$Q(t_f) = \tilde{Q}. \tag{7}$$

After carrying out R diagnostics and replacing structural elements with unacceptable defects the failure probability distribution function $Q_R = Q_R(\tau)$, $t_R \le \tau$ (t_R is the total time of all R standard diagnostics) is determined through the failure probability distribution function $Q_{k,R} = Q_{k,R}(\tau)$ for k-section as follows:

$$\text{optimistic scenario} \quad Q_R(\tau) = \sum_{k=1}^{K} \left(\frac{Q_{k,R}(\tau)}{1 - Q_{k,R}(\tau)} \right) \prod_{k=1}^{K} \left[1 - Q_{k,R}(\tau) \right], \tag{8}$$

$$\text{pessimistic scenario} \quad Q_R(\tau) = 1 - \prod_{k=1}^{K} \left[1 - Q_{k,R}(\tau) \right]. \tag{9}$$

Correspondingly for $Q_{k,R} = Q_{k,R}(\tau)$ here can be written the following:

$$\text{optimistic scenario} \quad Q_{k,R}(\tau) = \sum_{r=1}^{R} \left(\frac{Q_{k,r}(\tau)}{1 - Q_{k,r}(\tau)} \right) \prod_{r=1}^{R} \left[1 - Q_{k,r}(\tau) \right], \tag{10}$$

$$\text{pessimistic scenario} \quad Q_{k,R}(\tau) = 1 - \prod_{r=1}^{R} \left[1 - Q_{k,r}(\tau) \right]. \tag{11}$$

where the distribution function of k-section failure probability revealed by r-diagnostics $Q_{k,r} = Q_{k,r}(\tau)$ is expressed through the distribution function of q-element k-section failure probability revealed by r-diagnostics $Q_{k,r,q} = Q_{k,r,q}(\tau)$ in the form:

$$\text{optimistic scenario} \quad Q_{k,r}(\tau) = \sum_{q=1}^{\mathbb{Q}} \left(\frac{Q_{k,r,q}(\tau)}{1 - Q_{k,r,q}(\tau)} \right) \prod_{q=1}^{\mathbb{Q}} \left[1 - Q_{k,r,q}(\tau) \right], \tag{12}$$

$$\text{pessimistic scenario} \quad Q_{\kappa,r}(\tau) = 1 - \prod_{q=1}^{\mathbb{Q}} \left[1 - Q_{k,r,q}(\tau) \right]. \tag{13}$$

The failure probability distribution function $Q_{k,r,q} = Q_{k,r,q}(\tau)$ for q-element of k-section on all defects detected by r-diagnostics, $1 \le r \le R$, is set through the distribution function of $n_{k,q}$ q-element k-section failure on defects of j-type, $j = 1, \ldots, J$, revealed by r-diagnostics, $r = 1, \ldots, R$, $Q_{k,r,q,j} = Q_{k,r,q,j}(\tau)$ as follows:

$$\text{optimistic scenario} \quad Q_{k,r,q}(\tau) = \sum_{j=1}^{J} \left(\frac{Q_{k,r,q,j}(\tau)}{1 - Q_{k,r,q,j}(\tau)} \right) \prod_{j=1}^{J} \left[1 - Q_{k,r,q,j}(\tau) \right], \tag{14}$$

$$\text{pessimistic scenario} \quad Q_{K,r,q}(\tau) = 1 - \prod_{j=1}^{J} \left[1 - Q_{k,r,q,j}(\tau)\right]. \tag{15}$$

And finally, the distribution function $Q_{k,r,q,j} = Q_{k,r,q,j}(\tau)$ should be determined by the following relations:

$$Q_{k,r,q,j}(\tau) = \varphi(\tau)\mathrm{e}^{1-\varphi(\tau)}, \varphi(\tau) = \lambda_{r,q,j} l_q n_{k,q} \frac{\tilde{t}}{\Delta t_{f,k,q,j}} \tau, \quad q = 1, \ldots, \mathbb{Q}, \ k = 1, \ldots, K. \tag{16}$$

In (16), $\lambda_{r,q,j}$ is the coefficients of failure flow, i.e., the number of q-element failure on a defect of j-type revealed by r-diagnostics or known from the failure statistics of failure (in this case $r = 1$) per unit of time (year) per unit of length (km), and $\Delta t_{f,k,q,j}$ is the residual life of q-element of k-section on j-failure.

The criterion of structural reliability after regulatory R diagnostics and replacing structural elements with unacceptable defects is determined as follows:

$$Q_R(\tau) \le \tilde{Q}, \tag{17}$$

where $Q_R = Q_R(\tau)$ is the structural risk in the loading interval $[t_R, t]$, according to (8)–(16), $\tilde{Q}$ is acceptable structural risk according to design standards. Finally, the residual life after regulatory diagnostics and replacing structural elements with unacceptable defects is found in Eq. (7) in the form:

$$Q_R(\Delta t_f) = \tilde{Q}. \tag{18}$$

3 Element Durability Calculation

The loading of the pipeline by the internal pressure drops from the compressor station along the length, random and planned (depending on the consumption volume) pressure fluctuations occur with an amplitude 10–15% of the maximum value. It was found experimentally that these vibrations determine the destruction of structural elements, especially in stress concentration zones. The dependence of pressure on time is proposed to be considered as a finite Fourier series [11–16]. The loading of the q-element k-section by internal pressure can be represented by tangential $\sigma_{\theta\theta,q}(k,\tau) \equiv \sigma_{1,q}(k,\tau)$ and axial $\sigma_{zz,q}(k,\tau) = \sigma_{2,q}(k,\tau)$ stresses on the time interval $\tau \in [0,t]$ as follows:

$$\begin{cases} \sigma_{1,q} = K_{\theta,q}\sigma_\theta f(l_k)\left(\alpha + \sum\limits_{s=1}^{4}\sigma^s \sin\omega_s\tau\right), \\ \sigma_{2,q} = K_{z,q}\left(\dfrac{\nu\sigma_{1,q}(\tau)}{K_{\theta,q}} + \sigma_z(k)\right) \end{cases} \quad k = 1, \ldots, K, \quad q = 1, \ldots, \mathbb{Q},$$

$$\sigma_\theta = \frac{p(R - \delta^*)}{\delta^*},\ f(l_k) = 1 - C_k\frac{l_k}{L_0},\ \sigma_z(k) = E\left[-\alpha T_k \pm R\sqrt{\frac{1}{\rho_{1,k}^2} + \frac{1}{\rho_{2,k}^2}}\right],$$

$$T_k = T_0\left(1 - \beta\frac{l_k}{L_0}\right); \tag{19}$$

p—working (standard) pressure of the pumped product, δ^*—the wall thickness with certain tolerances, R—the radius of the main pipe, T_k, T_0—working and initial temperatures after construction, $0.25 \le C_k \le 0.35$—the coefficient of internal gas pressure average values drop, $0 \le \beta \le 0.55$—the coefficient of wall element temperature drop on the length, α, σ^s—asymmetry parameter and stress amplitude, respectively, ω_s—frequency, $\omega_s = 10^s\left[\frac{\text{cycles}}{\text{year}}\right]$, $s = 1, \ldots, 4$, $\rho_{1,k}$, $\rho_{2,k}$—radii of the section axis bending during its laying, $K_{\theta,q}$ and $K_{z,q}$—coefficients of stress concentration at the tangential and axial directions, respectively.

To determine the distribution function of the q-element failure probability $Q_q = Q_q(\tau)$, $0 \le Q_q \le 1$, $\tau \in [0, t]$, $q = 1, \ldots, \mathbb{Q}$, at an element uniform loading, the following constitutive relation is proposed [11–16]:

$$Q_q(\tau) = \frac{\sigma_\theta f(l_k)\alpha^2}{\Pi_1(\tau, k)} + \frac{|\sigma_z(k) + \nu\sigma_\theta f(l_k)\alpha|\alpha}{\sqrt{\Pi_1(\tau, k)\Pi_2(\tau, k)}} + \frac{(\sigma_z(k) + \nu\sigma_\theta f(l_k)\alpha)^2}{\Pi_2(\tau, k)},$$

$$\Pi_1(\tau) = \alpha\sigma^*(\tau) + \sum_{s=1}^{4}\sigma^s\sigma_{-1}(\tau, \omega_s),\ \Pi_2(\tau) = \nu\sigma_\theta f(l_k)\Pi_1(\tau) + |\sigma_z|\sigma^*(\tau). \tag{20}$$

where $\sigma_{-1} = \sigma_{-1}(\tau, \omega_s)$ is the fatigue curve on the defect levels or complete fracture (taking into account the cracking on fracture mechanics) at symmetric uniaxial loading of the base metal, $\sigma^* = \sigma^*(\tau)$ is the long-term strength curve of the base metal (or $\sigma^* = \sigma_{st}$, σ_{st} is the ultimate strength).

The constitutive relation for the distribution function of the q-element failure probability $Q_q = Q_q(\tau)$ at non-uniform stress state is written as follows:

$$Q_q(\tau) = R_q(\tau, k) + \sqrt{R_{1,q}(\tau, k)R_{2,q}(\tau, k)} + R_{2,q}(\tau, k),$$

$$R_{1,q}(\tau, k) = K_{\theta,q}\sigma_\theta f(l_k)\left(\frac{\alpha}{\sigma_q^*(\tau, K_{\theta,q})} + \sum_{s=1}^{4}\frac{\sigma^s}{\sigma_{-1,q}(\tau, \omega_s, K_{\theta,q})}\right),$$

$$R_{2,q}(\tau, k) = K_{z,q}\left(\frac{\nu R_{1,q}(\tau, k)}{K_{\theta,q}} + \frac{|\sigma_z(k)|}{\sigma_q^*(\tau, K_{z,q})}\right) \tag{21}$$

where $\sigma_{-1,q} = \sigma_{-1,q}(\tau, \omega_s, K_{\theta,q})$ is the experimentally determined failure amplitude of the tangential stress of the q-element with a concentrator on the defect levels or complete fracture (taking into account the cracking on fracture mechanics) at symmetric internal pressure, $\sigma_q^* = \sigma_q^*(\tau, K_{\theta,q})$ and $\sigma_q^* = \sigma_q^*(\tau, K_{z,q})$ are experimentally determined failure tangential and axial stresses of q-element with a concentrator at internal pressure.

The asymmetry parameter α and amplitude σ^s, $s = 1, \ldots, 4$, are random variables. Considering various groups of values $(\alpha, \sigma^0, \ldots, \sigma^4)_g$, $g = 1, \ldots, G$, the k-section q-element service life $t_{f,q,g}$ is found (in this paragraph, $t_{f,q,g}$ is designated for simplicity $t_{f,q} = t_{f,q}(k)$ from the following equations:

$$Q_q(t_{f,q,g}) = 1, \quad q = 1, \ldots, \mathbb{Q}, \; g = 1, \ldots, G \tag{22}$$

where $Q_q = Q_q(\tau)$ is determined by (20), (21), further, the minimum values of the service life $t_{f,q,g}$ are considered as the service life $t_{f,q}$ of the k-section q-element:

$$t_{f,q} = \min\{t_{f,q,g}, g = 1, \ldots, G\} \tag{23}$$

The residual service life k-section q-element on j-defect $\Delta t_{f,k,q,j}$ is found in a similar way. For the distribution function of the probability of k-section q-element on j-defect failure $Q_{q,j} = Q_{q,j}(\tau)$, $0 \le Q_{q,j} \le 1$, $\tau \in [0, t]$, $q = 1, \ldots, \mathbb{Q}$, $j = 1, \ldots, J$, relations (20), (21) are valid, and all experimental curves are plotted for q-element with j-defect.

4 The Safe Operation Criterion

Based on the analysis of the literature and regulatory documents, the following main negative factors of damage from pipeline destruction are identified: toxic effects from the outflow of pumping toxic liquids and gases ($i = 1$); thermal effect from ignition of a gas jet flowing out from a through crack ($i = 2$); thermal effect when a cloud of gas-air mixture ignites ($i = 3$); shock air waves from gas and gas combustion product expansion ($i = 4$); defeat from the scattering of destroyed structure fragments ($i = 5$). These factors arise with probabilities $J_i, i = 1, \ldots, 5$, accordingly, which are determined by the regulatory statistics of the appearance of negative factors at destructions of similar structures.

Here, the known notions of the social risk $I_1 = I_1(\tau), 0 \le I_1 \le 1, \tau \in [0, t]$ (the probability of injury to people and located in a potentially dangerous zone near the failure structure), the industrial risk $I_2 = I_2(\tau)$, $0 \le I_2 \le 1$, $\tau \in [0, t]$ (the probability of industrial objects destruction) and the ecological risk $I_3 = I_3(\tau)$, $I_4 = I_4(\tau)$ $0 \le I_3 \le 1$, $0 \le I_4 \le 1$, $\tau \in [0, t]$ (the probability of flora and fauna destruction accordingly) located in a potentially dangerous zone near the structure at construction and operation are considered. Standards for structures determine the acceptable values of social, industrial and environmental risks $\tilde{I}_m$, $m = 1, \ldots, 4$, during the acceptable life $\tilde{t}$.

According to the developed approach, the risk-adjusted safe criteria at the structure operation are the following:

$$Q(\tau)I_m(\tau) \le \tilde{Q}\tilde{I}_m, \quad m = 1, \ldots, 4, \tag{24}$$

$$\text{optimistic scenario} \quad I_m = \sum_{i=1}^{5}\left(\frac{J_i I_{m,i}}{1 - J_i I_{m,i}}\right)\prod_{i=1}^{5}\left[1 - J_i I_{m,i}\right], \tag{25}$$

$$\text{pessimistic scenario} \quad I_m = 1 - \prod_{i=1}^{5}\left[1 - J_i I_{m,i}\right], \tag{26}$$

$$I_{m,i} = \max\left\{\int_0^R\int_0^{2\pi} \rho_m(r,\theta)I_i(r,\theta,\tau)r\mathrm{d}r\mathrm{d}\theta : 0 \le \tau \le t_i\right\}, \quad m = 2, 3, \tag{27}$$

$$I_{m,i} = max\left\{\int_0^R\int_0^{2\pi} \rho_m(r,\theta,\xi)I_i(r,\theta,\tau)r\mathrm{d}r\mathrm{d}\theta : 0 \le \xi \le t_p; 0 \le \tau \le t_i\right\}, \quad m = 1, 4, \tag{28}$$

$$I_1(r,\theta,\tau) = a_1 * \ln\left\{\left(\frac{D(r,\theta,\tau)}{D_1}\right)^2\frac{\tau}{t_1}\right\}, \quad 0 \le \tau \le t_1,\ a_1, D_1 = \text{const}, \tag{29}$$

$$I_2(r,\tau) = a_2 * \ln\left\{\left(\frac{q(r,\tau)}{q_2}\right)^{4/3}\frac{\tau}{t_2}\right\}, \quad 0 \le \tau \le t_2,\ a_2, q_2 = \text{const}, \tag{30}$$

$$I_3(r,\tau) = a_3 * \ln\left\{\left(\frac{q(r,\tau)}{q_3}\right)^{4/3}\frac{\tau}{t_3}\right\} \quad 0 \le \tau \le t_3,\ a_3, q_3 = \text{const}, \tag{31}$$

$$I_4(r,\tau) = a_4 * \ln\left(\left(\frac{p_0}{p(r,\tau)}\right)^{\alpha} + \left(\frac{I_0}{I(r,\tau)}\right)^{\beta}\right), \quad 0 \le \tau \le t_4,\ a_4, p_0, I_0, \quad \alpha, \beta = \text{const}, \tag{32}$$

$$I_5(r) = a_5 + b_5\left(\frac{mv(r,\tau)^2}{I_0}\right), \quad 0 \le \tau \le t_5,\ a_5, b_5, I_0 = \text{const}. \tag{33}$$

where (r, θ) is the polar coordinate system centered at the point of i—negative factor origin, $\rho_m = \rho_m(r, \theta, t)$ is the distribution function of people, objects, flora and fauna density in the zone $[0, r_0]$ depending on time, r_0 is the radius of i—negative factor action, t_i is time period of i—negative factor action. Function $D = D(r, \theta, \tau)$ in (29) is the concentration (referred to a unit of volume) of toxic substance at a point (r, θ) at a time τ, depending on gas density, average wind speed, intensity and duration of emissions and is determined by hydro-aerodynamics methods; constants (a_1, D_1, t_1) lie in the following range: $0.2 \le \alpha_1 \le 2.5$, $7 < -\alpha_1 \ln(D_1^2 t_1) < 60$. Function $q = q(r, \tau)$ in (30) and (31) is the heat flow (per unit surface) at the point

r at the time τ, t_2 is the total time of jet burning, t_3 is the life of the fireball, constant a_3 is chosen equal $a_3 = 2.5$. In (32), functions $p = p(r, \tau)$ and $I = I(r, \tau)$ are the impulse and the maximum overpressure on the wave front, t_4 is the life of the action of blast wave; constants are chosen as follows: $a_4 = -0.2$, $p_0 = 40$ MPa, $I_0 = 450$ kg m/s, $\alpha = 7.5$, $\beta = 11.5$. In (33), m and $\upsilon = \upsilon(r, \tau)$ are, respectively, the mass and velocity of the fragment, and t_5 is the time of fragment flight, and they are found on the solution of the problem about the shock destruction of pressure vessels, and parameters a_5 and b_5 are chosen in the view $a_5 = 10.5$, $b_5 = -21$.

For gas pipelines of the Russian Federation, the federal acceptable risks of operation during an acceptable life are as follows:

$$\tilde{I}_1 = \left(2 \times 10^{-4} - 2 \times 10^{-5}\right)\left[\frac{\text{number of people}}{\text{km}}\right] * L_0,$$

$$\tilde{I}_2 = \left(10^{-3} - 10^{-4}\right)\left[\frac{\text{number of objects}}{\text{km}^2}\right] * S_0,$$

$$\tilde{I}_3 = \left(10^{-1} - 10^{-2}\right)\left[\frac{\text{number of flora representatives}}{\text{km}^2}\right] * S_0,$$

$$\tilde{I}_4 = \left(10^{-2} - 10^{-3}\right)\left[\frac{\text{number of fauna representatives}}{\text{km}^2}\right] * S_0.$$

L_0 is the pipeline length, S_0 is the area of a potentially dangerous zone.

The time of dangerous operation of structure $t_{f,m}$ is found as a solution of the following equation according to (24):

$$Q\left(t_{f,m}\right) = \frac{\tilde{I}_m}{I_m}\tilde{Q}, \quad m = 1, \ldots, 4.$$

In this case, the following inequality: $t_{f,m} \leqslant t_f$ is fulfilled, where t_f is the structure life, determined by Eq. (7) without taking into account the social safety of its operation. The life $t_{f,m}$, $m = 1, \ldots, 4$, the risk-adjusted longevity.

The risk-adjusted residual life $\Delta t_{f,m}$ is determined as a solution of the equation according to (18):

$$Q\left(\Delta t_{f,m}\right) = \frac{\tilde{I}_m}{I_m}\tilde{Q}, \quad m = 1, \ldots, 4.$$

5 Conclusion

Here, the criterion of structural reliability is represented for a pipeline section consisting of a significant number of similar elements (base metal, longitudinal and ring welded joints, tee connection, diversion, adapter, bottom) as an achievement of failure probability distribution function to the project acceptable structural risk.

Section failure is considered as the sum of independent failure. It is proposed to describe the failure probability distribution function by a function of the Poisson distribution type.

The method [13] for the similar element durability determining at the pressure of the pumped product in the form of a finite Fourier series with a random distribution of amplitudes and asymmetries of cycles and temperature is briefly represented.

Here, the developed safety criteria [14, 15] are expounded as an achievement of failure probability distribution function taking into account the probability of injury to humans, industrial object, flora and fauna destruction at the section failure (as a result of toxic and thermal injuries, from shock effects or scattering of destroyed element fragments) to the project acceptable structural risk taking into account the acceptable values of social, industrial and environmental risks. The equation for the durability determining with considering of technological and anthropogenic risks is proposed.

The proposed safe operation criterion was applied for the oil and gas pipelines longevity predicting. A number of conclusions on the service life and residual life of various designs of gas and oil pipelines with a certain level of accumulated defects were prepared [11–16].

Acknowledgements This work was carried out with financial support of the Moscow Center for Fundamental and Applied Mathematics.

References

1. Makhutov, N.A.: Russia's safety. Law, social, economical, scientific and technical aspects. MGF Knowledge, Moscow, p. 1016 (2018) (in Russ.)
2. Makhutov, N.A., Matvienko, Y.G., Romanov, A.N.: Problems of strength, technogenic safety and structural materials science. URSS, p. 720 (2018) (in Russ.)
3. Makhutov, N.A.: Safety and risks: system research and development. Science, Novosibirsk, p. 724 (2017) (in Russ.)
4. Zavoychinsky, B.I.: Durability of main and technological pipelines (theory, calculation methods, design). Nedra, Moscow (1992) (in Russ.)
5. Klyuev, V.V., Bolotin, V.V., Sosnin, F.R.: Mechanical Engineering: An Encyclopedia in 40 Volumes. Vol. IV-3: Reliability of Machines (2003) (in Russ.)
6. Mazur, I.I., Ivantsov, O.M.: Safety of pipeline systems. ITs ELIMA, Moscow (2004) (in Russ.)
7. Kharionovskiy, V.V.: Reliability and resource of gas pipeline structures. Nedra, p. 464 (2000) (in Russ.)
8. Zainullin, R.S., Alexandrov, A.A., Morozov, E.M.: Criteria for safe failure of cracked pipeline system elements. Federal State Unitary Enterprise «Academic Research and Publishing, Production, Printing and Book Distribution Center» Science, p. 317 (2005) (in Russ.)
9. Makhutov, N.A., Gadenin, M.M.: Technical diagnostics of residual life and safety. P.h. Spectrum (2011) (in Russ.)
10. Lepikhin, A.M., Makhutov, N.A., Moskvichev, V.V., Chernyaev, A.P.: Probabilistic risk analysis of technical systems. Novosibirsk, Nauka (2003) (in Russ.)

11. Zavoychinskaya, E.B.: Fatigue large-scale structural failure and durability of structures under proportional loading processes. Author's abstract. Dr. Dissertation, Moscow, Ltd., «Genesis», p. 46 (2018) (in Russ.)
12. Zavoychinskaya, E.B., Ovchinnikova, N.V.: On the durability estimation of extended structures at complex natural and climatic actions. In: Proceedings of the XXV International Symposium on «Dynamic and Technological Problems of Mechanics of Structures and Continuous Media». A.G. Gorshkov, Moscow, vol. 2, pp. 163–171 (2019)
13. Zavoychinskaya, E.B.: A stochastic theory of scale-structural fatigue and structure durability at operational loading. In: Understanding Complex Systems, pp. 71–89. Springer, Germany (2021)
14. Zavoychinsky, B.I., Zavoychinskaya, E.B., Volchanin, A.V.: Probabilistic assessment of the residual operational safety of long structures. Direct. Mag. **7**, 41–46; Eng. J. **12**, 33–36 (2012)
15. Zavoychinsky, B.I., Giller, G.P., Zavoychinskaya, E.B.: Recommendations for assessing of the safety and durability of gas pipelines during design. IRC Gazprom, Moscow (2002) (in Russ.)
16. Zavoychinsky, B.I., Tutnov, I.A., Zavoychinskaya, E.B.: Recommendations for assessing of the safety and durability of main gas pipelines during design. IRC Gazprom, Moscow (2000) (in Russ.)

Survey on Behavioral Strategies of Cyber-Physical Systems in Case of Loss of Integrity

Ekaterina Cherskikh and Anton Saveliev

Abstract This paper analyzes the strategies for the behavior of cyber-physical systems in case of data integrity loss during transmission by nodes, temporary loss of system connectivity, and failure of individual nodes. Some problems causing the malfunction of cyber-physical systems are considered, as well as the methods, proposed for detecting and eliminating the consequences of such integrity loss. The categorization of these methods is carried out. The considered methods are divided into ontological, reorganizational ones, and methods of assessing and comparing states. When using ontological methods, databases with predetermined failure modes, implicit redundancy, and a system architecture that included several types of agents are commonly used. Most of the considered reorganizational methods are based on system reorganization with the removal of faulty nodes or their clustering with the aid of diagnostic agents. In the case of reorganization, decision-making logic, system diagnostics, and actionable scenarios for failure cases are commonly used. When using methods for comparing system states to detect failures, the following were used: categorization of nodes, controllers and monitors, hash functions, and dynamically updated keys. The results of this survey can be applied in developing of proprietary method for detecting failures and automatic recovery of cyber-physical systems.

1 Introduction

A cyber-physical system (CPS) is a set of logical, physical, transformative components, and components with the operator involvement, implemented at the physical and logical levels of the system [1]. With the increasing use of CPS and their wide variety, security problems have arisen, such as software and physical attacks, the detection and elimination of which requires different methods. Malfunctions of the

---

E. Cherskikh (✉) · A. Saveliev
Laboratory of Autonomous Robotic Systems, St. Petersburg Institute for Informatics and Automation of the Russian Academy of Sciences, St. Petersburg Federal Research Center of the Russian Academy of Sciences (SPC RAS), 39, 14th Line, St. Petersburg 199178, Russia
e-mail: cherskikh.e@iias.spb.su

A. Ronzhin and V. Shishlakov (eds.), *Electromechanics and Robotics*, Smart Innovation, Systems and Technologies 232, https://doi.org/10.1007/978-981-16-2814-6_40

CPS [2] can be caused by such damages as failure of one or more system nodes, loss of connectivity, loss of integrity, or modification of data transmitted by nodes [3]. To solve these problems, various methods and algorithms can be applied [4], capable of fully or partially eliminating violations and allowing the CPS to function as expected. Currently, the most popular way to solve the problems of integrity loss consists in the utilization of the architectures that allow the CPS to retain operability in the case of failure of several nodes [5]. The reliability of separately considered CPS nodes [6] and the methods used for assessing the reliability of nodes and predicting possible errors are also considered. When developing methods for eliminating the consequences of CPS failures, it is important to consider the peculiarities of the system [7], such as its heterogeneity or homogeneity, decentralization or the presence of a central control unit, the routine of data transfer between nodes, and probable types of attacks. Further, we will consider ways to detect attacks and mitigate their impact on the further operation of the CPS and ways to eliminate such threats.

2 Methods for Detecting and Eliminating the Consequences of Failures of CPS Nodes Using Ontology

Elimination of the consequences of CPS failures and methods of preventing the causes of such failures can be implemented in several ways, depending on what type of attack was applied to the system [8]. Physical impact on the system, such as removing the sensor from the system or causing physical harm to the sensor, is proposed to be prevented by installing additional sensors that monitor possible breaches. However, such sensors can also be removed by intruders. The generation of frequent data requests, caused by attackers can overload the system's computing resources. It is proposed to solve this problem, as well as the problems of damage, replacement of sensors, and input of incorrect data into the system, causing a failure in decision-making, by hiding the topology and routing infrastructure of the sensor network and minimizing the number of transfer operations for sending data (reducing the bandwidth). It is possible to prevent an external attack on the CPS without disturbing its operation by assigning the master node of the network cluster, which is equipped with its periodic change and intrusion detection mechanisms, which use statistical time analysis. Mitigation of the consequences of disruption of the CPS operation in the event of data loss can also be ensured by using ontologies, intended to partially eliminate data loss. CPS [9] has a decentralized architecture consisting of three layers: processing layer, network layer, and sensitive executive layer. The agents of the last layer have various purposes: collecting and interpreting data, setting interaction rules, and providing communication between agents based on ontological descriptions, processing data and events, and diagnostics. If data batches are lost during data transfer, the diagnostic agent checks for data inconsistency using an ontology to determine the failure and interpolates the data in the database. Then the ontology agent updates the ontology as the reliability metrics have changed. In this way, network agents can partially eliminate the loss of CPS data.

A system similar to [9], presented in [10], also has an architecture that includes several types of agents performing data processing tasks to infer truth. Agents use an ontological knowledge base with rules that determine their behavior in collaboration and control decision-making processes to detect and prevent failures of hardware components of CPS. In the ontology, several types of failures are predefined, and the corresponding actions to prevent them are outlined. When a failure or service event occurs, agents analyze the ability to communicate with each other by sending and receiving messages to complete their tasks (e.g., detecting a failure, storing data in a database). If such an opportunity exists, then the agents begin to collaborate, otherwise they continue to perform their currently active tasks. If the tasks cannot be completed successfully, the agents try again to establish a new relationship.

Having set up the architecture of the CPS of several types of agents, the authors of [11] use self-healing to ensure the fault tolerance of the system, achieved through structural adaptation and an ontological knowledge base with a description of possible failures. During operation, changes can be made to the system: adding and removing components or changing their interaction patterns. As an example of the system operation, its implementation on a Pioneer mobile robot is given to assess the state of charge of its battery. The assessment of the state is performed due to the ontology, which fixes the relationship between the system components and the possibility of using this information to perform reconfiguration in the event of a component failure (in the described case, an ammeter). When the ammeter fails, the state of charge assessment results in incorrect readings. However, other characteristics are available, measured by sensors or obtained by other components, which are related to the battery current. Since information about the relationship between these properties is stored in the ontology, it can be used to restore a failed component.

Replacing failed CPS components by means of ontology is also possible using implicit redundancy [12]. After a component failure is detected, the presented algorithm determines whether the semantics of the failed component can be replaced by combining the services provided by other system components. The authors provide a practical example of using the proposed approach in a car. Thus, the same system property (for example, rotation angle) can be inferred in different ways from several other system properties. For example, the steering angle can be directly measured by a dedicated sensor on the steering rod, but it can also be obtained from the radius of the car's curve, which in turn can be determined from the lateral acceleration and vehicle speed or from the difference in the angular speed of the inner and outer wheels on a curve. The system ontology represents knowledge about the relationships and interactions of the steering mechanism of a car with its wheels, as well as about various other properties of the car (for example, about speed and acceleration). To find the semantic equivalence of these property concepts when a component fails, the substitution algorithm looks for combinations of the provided property concepts that allow it to compute the required value. Once a combination of such provided property concepts has been found, a mediation service is automatically generated, which can then be used to provide the desired property concept based on the values of other property concepts in the system ontology.

The use of an ontology with predetermined types of failures and actions to prevent such failures is efficient in cases when no other types of failures except predefined ones are possible. The use of structural adaptation and implicit redundancy together with an ontological knowledge base will allow the CPS to retain operability by replacing the failed components and the data they provide. Also, the use of implicit redundancy provides for reducing of the number of system components initially. The proposed methods can both eliminate the consequences of failures by completely replacing the required data and mitigate the consequences of failure by partially replacing the functions of the failed component. The use of several types of agents in the CPS, including diagnostic agents, ensures a prompt response to a failure. Further, we will consider methods and algorithms for structural reorganization of CPS components based on the interaction of several types of agents, as in the systems described in this section above, but without using ontological knowledge bases.

3 Elimination of Failures of CPS Nodes Through Structural Reorganization

The presence of mechanisms allowing the CPS to perform self-organization in real time without interrupting work, restarting and reprogramming individual components is an important aspect of the system operation process. Implementation of such a mechanism is presented by the authors [13] to support the operation of the conveyor system with the ability to add a new conveyor, remove a broken conveyor, or replace two separate conveyors. The general self-organization mechanism consists of several stages and is based on several rules. The first stage of work is the detection stage. During the initial setup of the system, none of the conveyor agents knows its position in the system sequence; a distribution procedure is required to determine the sequence of individual agents. Each of the conveyors has two sensors: at the input and at the output, signaling the appearance of an object. When an object is detected by the agent's output sensor, a message is broadcast to all agents in the system, which is processed by the next unit and ignored by the remaining ones. If the next agent detects that the object has not reached its input sensor, it sends an alert message of a possible reordering and indicates the current position. All system agents, whose sequence number is higher than the received position, become active (ensuring that they are in a valid situation, for example, they do not currently own the object for transmission). Subsequently, when a conveyor agent receives an object on its input sensor, it updates its current sequence number and transmits a message with its previous and new number (the updated number received during the order change message). Other agents who receive this message only update their sequence numbers. This process is repeated every time the agent's order is changed. The use of agents and self-organizing principles allows the conveyor system to be controlled in a distributed and decentralized manner using a simple set of rules. The system does not have a central control unit, such configuration provides for the overall system scalability and helps to eliminate isolated failures.

Management of CPS production through the architecture of holonic control ADACOR (ADAptive holonic COntrol aRchitecture for distributed manufacturing systems) [14, 15] is also carried out using the concept of self-organization. Four types of holons are offered: product holon (PH)—representing the products available in the factory catalog of factories and the knowledge for their production; task holon (TH)—responsible for managing the execution of production orders in the shop in real time; operation holon (OH)—representing system resources such as robots and operators responsible for managing their own agenda, as well as managing the physical connection to a real resource; supervisor holon (SH)—responsible for introducing optimization into the system. The self-organization of the system is achieved through the interaction between local individual holons that spread the emergence and the need for reorganization. The system is in a stationary state until a disturbance is detected. A holon that detects an irregularity in the system and tries to perform local repair through self-diagnosis. If restoration is unsuccessful, the autonomy coefficient of the holon increases, and the need for reorganization spreads to the other holons in the system. At the same time, other holons increase their autonomy in accordance with the coefficient of autonomy of the first holon and extend the emerging reorganization to adjacent SH-holons. Each holon, considering its autonomy coefficient, the ability to learn and the magnitude of the autonomy coefficient of the holon that detected the violation, decides whether it should be reorganized or not. As soon as the coefficient ceases to be active, each individual holon again decreases its autonomy factor and returns to the hierarchical control structure, returning to a steady state and accepting the task schedule from its SH-holon. At this point, the SH-holons collect the updated individual timetables reached during the transient period and continue to synchronize and then optimize the existing timetable. Structural self-organization in this work is a way for restructuring the relationship between holons and their organization.

The reorganization of the CPS [16], consisting of two interdependent networks, is performed algorithmically. Due to interdependence, failures of nodes on one network can cause failure of another network and therefore lead to cascading failures of the entire system. The paper presents a k-reliability model and an algorithm for determining cascade failures. The proposed reliability, defined as the probability that at least k normally functioning nodes make up a functional subnet, explains the probability of the system operation and the possibility of connecting such nodes in the network in the presence of failed ones. The algorithm determines whether random failures of network nodes can lead to reliability failures. The authors assume that the failure cascade is caused by the failure of two random nodes in a network of 12 nodes (6 nodes in network A and 6 nodes in network B). The failed nodes are removed along with the links. The remaining nodes establish separate components of network A and network B. Thus, the system reaches a steady state. Depending on the location of the failed nodes, the system can remove additional nodes to achieve a steady state and the steady state changes.

In case of failure of network nodes, their isolation [17] and clustering [18] can be applied, performed by CPS diagnostic agents. The basis of the proposed isolation methodology is agents for diagnostics of failures of CPS sensors. Each agent is intended for one of the interconnected subsystems of the CPS, while it does not exchange information with other agents. The diagnostic agent is responsible for detecting and fixing failures in a specific group of sensors. Diagnostic agents consist of several modules that monitor groups of sensors to be able to detect multiple types of failures. Isolation of multiple sensor failures in each local diagnostic agent is implemented by combining module decisions and applying decision-making logic based on diagnostic reasoning using a matrix of multiple sensor failure signatures. The mechanism for detecting and isolating CPS failures [18] requires training on an error-free sequence of parameters that determines stationary states, upon deviation from which, a failure is reported. Subsequently, during the life of the CPS, the parameters estimated on the basis of the input data are transferred to the Markov model, and the statistical compatibility of this data is measured in terms of the log-likelihood. If the log-likelihood falls below an automatically determined threshold, the smart layer is activated to match a specific change to a fault affecting one of the sensors in the cluster. Failures and deviations in time are interpreted as deviations from the previously set nominal conditions. This is followed by isolation of multiple faults and detection of changes over time. The system is able to simulate the relationship between the descendants of the CPS sensor data and study the nominal operating conditions of the system, as well as compile a fault dictionary based on the data obtained during the entire CPS service life.

The algorithm of uniform distribution and recovery for wireless sensor networks proposed by the authors of [19] considers network nodes as mobile entity, randomly located in environment. The uniform distribution of sensors in environment is ensured by an algorithm consisting of two parts. The first part, the failure algorithm, prepares the mobile sensor nodes for failures by maintaining half the communication distance between them. If the nodes of the system are not prepared for a failure, then it will take additional time to react—the connection between the nodes and the network coverage may be disrupted. The second part—the algorithm to be used after failure—performs the recovery procedure in the network by recovery nodes. In case of failure of some nodes, the CPS algorithm ensures the constant movement of nodes, making communication between nodes and network coverage continuous without significant time losses. Nodes that cannot maintain constant communication with adjacent nodes are assigned nodes responsible for restoring communication between the existing nodes.

To eliminate the destructive effect of multiple failures on the operation of the CPS, the authors of [20] proposed the use of replication of system nodes. For system self-healing, a recovery-oriented self-healing approach is presented, the MANET approach, capable of assessing the damage caused and repair errors and damage. There are two main types of replication agents: heterogeneous and homogeneous. In heterogeneous replication, all replicas are functionally equivalent, but they could be implemented separately, that is, they are not identical, but are designed to perform the same action. With homogeneous replication, replicas are exact copies of the original agent. There are three types of agents in the structure of the CPS. The agents of the first type perform the necessary intrusion detection. The agents of the second type are replicas (one replica for each agent). The agents of the last type are so-called "replication managers." A replication manager is used to monitor and recover active agents in the event of a failure.

Reconfiguration of CPS [21] upon detection of failed nodes is performed step by step. When a failed node is detected, it is removed from the system, and the nodes, that were associated with it, are moved to other neighboring nodes. Two modes of influence on the system that create an internal threat and contribute to the loss of the integrity of the system are considered: the node capture and the introduction of false data. Byzantine fault tolerance is used to detect the failure, allowing the system to continue to operate even if some of the nodes are down or compromised and broadcast false data. The proposed model is capable of adapting to other failures, such as data breaches, single or part failures.

When reconfiguring the CPS, the removal, isolation, and clustering of failed nodes by diagnostic agents are mainly used to preserve the integrity of the system. Decision-making logic is used, but training on an error-free sequence of stationary states of the system is required. In this case, the knowledge base of failures is replenished during the entire service life of the CPS. Less popular is the method for replicating CPS agents, the disadvantage of which is finding a compromise between the number of replicas, and the possible number of failures. Removing and clustering failed nodes can also be performed using system state comparison techniques, which will be discussed in the next part of the paper.

4 Detection and Elimination of CPS Failures by the Method of Assessment and Comparison of States

Failures and interruptions of communication between the nodes of the CPS can cause serious damage to the network infrastructure. The methodology of security analysis and prevention of CPS failures can be based on the comparison of system states. A combination of state and physical change estimation methods for detecting and eliminating communication faults and failures is presented by the authors [22]. In this approach, each node uses the physical connections between the nodes to estimate certain state parameters of the remote nodes in order to detect faults and also to maintain system stability after a fault occurs. Based on this architecture and approach, a fault-tolerant decentralized voltage control algorithm is presented. CPS nodes are categorized by the number of connected neighbors. Messages transmitted between nodes are classified as command or reporting. Command messages are intended to trigger specific physical changes for specific nodes in the system, and the changes will be monitored in the next cycle. Reporting messages are intended to check the current physical state of the system. When a node failure is detected, it is disconnected from the system and categorized as offline.

The measurement of the physical dynamics of the CPS by checking the data and the state of the system is proposed by the authors of [23, 24]. In [24], the L1Simplex architecture is presented, which, in contrast to traditional simplex architectures designed only for correcting software failures, is able to cope with a class of cyber and physical failures. The L1Simplex architecture for eliminating software and physical failures is presented, which contains a security monitor, L1 controllers (high-performance controller (HPC) and high-assurance controller (HAC)) and decision logic for switching between them. The solution proposed by the authors of [23] can help in detecting the occurrence of false data injection attack and mitigate the impact of an attack on the network, but not eliminate the attack completely. Data validation is carried out using a dynamically updated key and cryptographic hash functions. The key is updated using information about the previous state of the original object that has already been transmitted and confirmed to be true. The purpose of a hash function is to generate a fixed-length identifying value for a given dataset. The generation is done again at the destination node, and the two hash values are compared. A mismatch in the values indicates that the data has been tampered with, which the system informs the network operators or starts a predefined protection procedure. The method is decentralized and does not require information about the graph structure and the tracked object, but requires additional computing resources and bandwidth. An example of the practical use of the method used for three ground vehicles with a communication topology created by a digraph is given (Fig. 1). None of the vehicles are equipped with an absolute positioning system. However, vehicle P_1 (which is the leader of the squad) can measure the relative position (distance and direction) of each of the other two vehicles in relation to itself.

This information is transmitted to the corresponding vehicle through the edges $p_{1,2}$ and $p_{1,3}$. The desired construction is that the P_1 vehicle is ahead of the P_2 and P_3 vehicles with a difference of 1 unit in the x-axis and ± 1 unit in the y-direction. A sine-wave reference speed signal is given to the P_1 leader, and P_2 and P_3 are commanded to maintain their position with respect to P_1. The simulation results for

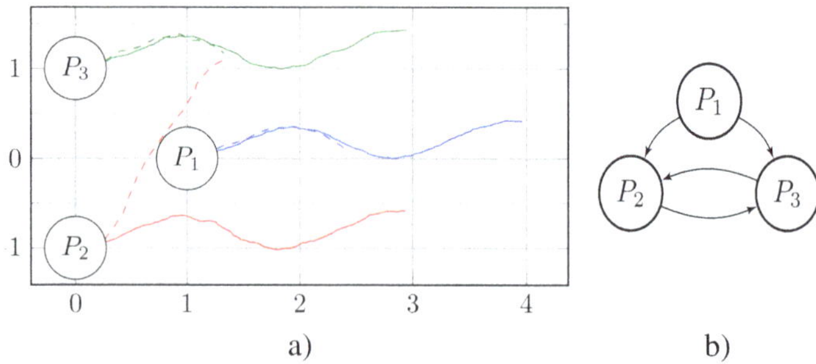

Fig. 1 **a** x–y position plot of robot formation control simulation results under normal operation (solid lines) and under false data injection attack (dashed lines); **b** example communication network among three autonomous agents, [23]

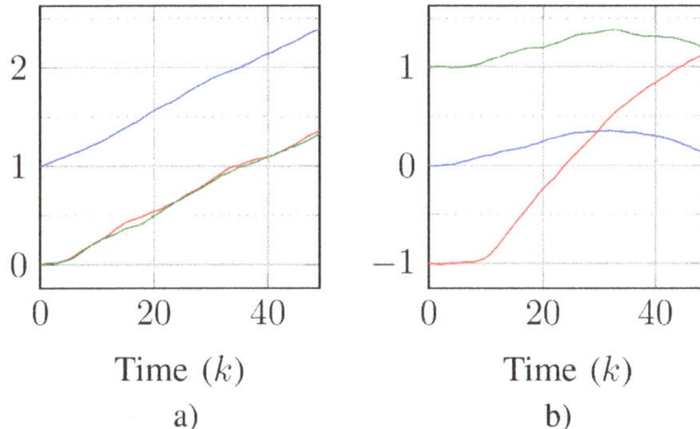

Fig. 2 Time evolution of robot positions under false data injection attack x (**a**) and y (**b**), [23]

both the nominal case and the case when the system is under attack are shown in Fig. 2.

In an attack scenario, the attacker additively violates the measurement of the relative position in the y-direction, transmitted from P_1 to P_2, in an attempt to cause a collision of vehicles P_2 and P_3. From the simulation results presented in the figure, it can be seen that the goal of the intruder is achieved in a collision. The attacker is supposed to replace the corresponding state measurements, while keeping the transmitted hash unchanged. When a P_2 attack is detected through hash mismatch, it queries P_3 for its position relative to P_1. Since the integrity and authenticity of the data received by P_3 is verified using a hash value, it can be trusted and thus transferred to P_2. If it is found that any of the edges $p_{1,3}$ or $p_{3,2}$ is also compromised, then the error appears not to be achievable, since a reliable communication channel from P_1 to P_2 will not be available. After the attack, the characteristics of the control system are corrupted, but the stable operation of the system and the ability to track the path of the vehicle remain.

In [25], the authors present the methodology for the analysis of safety, security, and prevention of failures STPA-SafeSec for CPS used for the electrical network. The network includes a generator, remote and local units for phasor measurement, and a local controller. The controller compares the measurements from the two phasor measuring units and controls the generator. Several scenarios (textual representations of some event of failure during system operation) are presented, indicating possible system failures and solutions for each of them. Scenarios have sub-scenarios, presented as a tree. This structure makes it possible to assess the effectiveness of the applied strategies at the required level of mitigation of the effects of failures.

A decentralized multi-agent method for managing distributed microgrids based on controlling the capacities generated by network nodes is presented in [26]. A certain number of generating capacities (kW) are installed inside each microgrid. CPS nodes are not explicitly associated with a certain type of generation resource, but instead rely on information about the status of local generation assets. Likewise, the amount of installed load in each microgrid is known. The nodes are constantly evaluating the following principal variables: instantaneous total generating capacity, operating level of current generation and current workload. The main goal of a node is to constantly maintain the required level of loads in its microgrid. An emergency instance is defined as either a microgrid failure causing it to leave the system, or a faulty transmission line that results in a loss of that power flow path. The behavior of a microgrid agent in the event of a disaster is first characterized by a disaster response phase followed by a recovery phase. The system enters a normal operating state, except that some microgrid nodes and connections are no longer available to the system. Likewise, the newly reconfigured power system operates according to normal communication protocols.

Assessment and control of the state of the CPS nodes are achieved by categorizing the nodes, checking the transmitted data using a hash function and a dynamically updated key, monitoring the physical changes of the CPS, the controller, and the security monitor. Table 1 shows the types of failures of the considered CPS and ways to eliminate them.

The use of a decentralized structure of the CPS is the most popular way to prevent failures, but in cases where failure was not prevented, methods of eliminating the effects of failures have to be applied. But not all CPSs are ways to completely eliminate the negative impact of a failure; in such cases, methods are used to mitigate the impact of failures on the further operation of the system.

Table 1 Causes and methods of elimination/mitigation of the impact of failures on the operation of the CPS

The causes of the CPS failures	Failure elimination/mitigation methods
Physical impact on the system	Installation of tracking sensors
	Reorganization
	Comparison of system states
Entering invalid data	Minimizing the amount of data transferred
	Comparison of transmitted data
	Using a hash function and a dynamically updated key
Attackers generate frequent requests and external intrusions	Hiding the network topology and routing infrastructure
	Periodic change of the master of the network cluster
Other reasons for failures	Replicating nodes
	Replacing failed nodes with data ontological knowledge base
	Using implicit redundancy
	Removing and clustering failed nodes

5 Conclusion

In most of the cases considered, in order to avoid integrity loss of the CPS, a decentralized structure of the system, several types of agents and various scenarios of the system's behavior are used to eliminate the consequences of predetermined types of failures. When an unknown type of failure of system nodes occurs, it can be added to the ontological database. This knowledge base also provides for replacement of failed components in whole or in part.

Reorganization helps to keep the system in operable condition, but the complete loss of failed nodes and their removal from the CPS is possible. In some cases, clustering and isolation can be utilized. Data losses are mitigated by their ontological replacement or with the use of implicit redundancy performed by normally functioning CPS nodes. The most common methods are those using the structural reorganization of the CPS. Further work will be aimed at developing a custom method for eliminating the consequences and direct violations of the integrity and failures of the CPS units.

Acknowledgements This research is supported by the RFBR Project No. 20-79-10325.

References

1. Vatamaniuk, I.V., Iakovlev, R.N.: Generalized theoretical models of cyberphysical systems. Proc. Southwest State Univ. **23**(6), 161–175 (2019) (In Russ.). https://doi.org/10.21869/2223-1560-2019-23-6-161-175
2. Ronzhin, A.L., Budkov, V.Y., Ronzhin, A.L.: User profile forming based on audiovisual situation analysis in smart meeting room. SPIIRAS Proc. **4**(23), 482–494 (2014). https://doi.org/10.15622/sp.23.28
3. Alguliev, R., Imamerdiev, Y., Suxostat, L.: Ensuring information security of cyber-physical systems. In: 1st Republican Conference On Scientific-practical problems of Software Engineering (2017). https://doi.org/10.25045/NCSoftEng.2017.07
4. Meshcheryakov, R.V., Iskhakov, A.Y., Evsutin, O.O.: Analysis of modern methods to ensure data integrity in cyber-physical system management protocols. Inf. Autom. **19**(5), 1089–1122 (2020). https://doi.org/10.15622/ia.2020.19.5.7
5. Knight, J.C., Sullivan, K.J., Elder, M. C., Wang, C.: Survivability architectures: issues and approaches. In: Proceedings of the DARPA Information Survivability Conference and Exposition. DISCEX'00, vol. 2, pp. 157–171. IEEE (2000)
6. Castaño, F., et al.: Sensor reliability in cyber-physical systems using internet-of-things data: a review and case study. Remote Sens **11**(19) (2019). https://doi.org/10.3390/rs11192252
7. Humayed, A., Lin, J., Li, F., Luo, B.: Cyber-physical systems security—a survey. IEEE Internet Things J. **4**(6), 1802–1831 (2017). https://doi.org/10.1109/JIOT.2017.2703172
8. Karim, M.E., Phoha, V.V.: Cyber-physical systems security. Appl. Cyber-Physical Syst. 75–83 (2014). https://doi.org/10.1007/978-1-4614-7336-7_7
9. Sanislav, T., Zeadally, S., Mois, G.D.: A cloud-integrated, multilayered, agent-based cyber-physical system architecture. Computer **50**(4), 27–37 (2017). https://doi.org/10.1109/mc.2017.113

10. Sanislav, T., Zeadally, S., Mois, G., Fouchal, H.: Multi-agent architecture for reliable Cyber-Physical Systems (CPS). In IEEE Symposium on Computers and Communications (ISCC) (2017). https://doi.org/10.1109/iscc.2017.8024524
11. Ratasich, D., et al.: A self-healing framework for building resilient cyber-physical systems. In: 2017 IEEE 20th International Symposium on Real-Time Distributed Computing (ISORC), pp. 133–140. IEEE (2017). https://doi.org/10.1109/ISORC.2017.7
12. Höftberger, O., Obermaisser, R.: Ontology-based runtime reconfiguration of distributed embedded real-time systems. In: 16th IEEE International Symposium on Object/component/service-oriented Real-time distributed Computing (ISORC), pp. 1–9. IEEE (2016). https://doi.org/10.1109/ISORC.2013.6913205
13. Barbosa, J., Leitão, P., Teixeira, J.: Empowering a cyber-physical system for a modular conveyor system with self-organization. In: Service orientation in holonic and multi-agent manufacturing, pp. 157–170. Springer, Cham (2018). https://doi.org/10.1007/978-3-319-73751-5_12
14. Barbosa, J., Leitão, P., Adam, E., Trentesaux, D.: Dynamic self-organization in holonic multi-agent manufacturing systems: the ADACOR evolution. Comput. Ind. **66**, 99–111 (2015). https://doi.org/10.1016/j.compind.2014.10.011
15. Leitão, P., Restivo, F.: ADACOR: A holonic architecture for agile and adaptive manufacturing control. Comput. Ind. **57**(2), 121–130 (2006). https://doi.org/10.1016/j.compind.2005.05.005
16. Zhang, Z., An, W., Shao, F.: Cascading failures on reliability in cyber-physical system. IEEE Trans. Reliab. **65**(4), 1745–1754 (2016). https://doi.org/10.1109/tr.2016.2606125
17. Reppa, V., Polycarpou, M.M., Panayiotou, C.G.: Decentralized isolation of multiple sensor faults in large-scale interconnected nonlinear systems. IEEE Trans. Autom. Control **60**(6), 1582–1596 (2015). https://doi.org/10.1109/tac.2014.2384371
18. Alippi, C., Ntalampiras, S., Roveri, M.: Model-free fault detection and isolation in large-scale cyber-physical systems. IEEE Trans. Emerg. Top. Comput. Intell. **1**(1), 61–71 (2017). https://doi.org/10.1109/tetci.2016.2641452
19. Jadoon, R.N., et al.: An efficient nodes failure recovery management algorithm for mobile sensor networks. Math. Probl. Eng. (2020). https://doi.org/10.1155/2020/1749467
20. Mechtri, L., Tolba, F.D., Ghanemi, S., Magoni, D.: A twofold self-healing approach for MANET survivability reinforcement. Int. J. Intell. Eng. Informatics **5**(4), 309–326 (2017). https://doi.org/10.1504/IJIEI.2017.087931
21. Mitchell, R., Chen, R.: On survivability of mobile cyber physical systems with intrusion detection. Wirel. Pers. Commun. **68**(4), 1377–1391 (2013). https://doi.org/10.1007/s11277-012-0528-3
22. Abad, F.A.T., Caccamo, M., Robbins, B.: A fault resilient architecture for distributed cyber-physical systems. In: IEEE International Conference on Embedded and Real-Time Computing Systems and Applications, pp. 222–231. IEEE (2012). https://doi.org/10.1007/s11277-012-0528-3
23. Tsiakkas, M., Kolios, P., Polycarpou, M., Panayiotou, C.: Establishing data integrity in networks of cyber-physical systems. In: 2018 European Control Conference (ECC), pp. 350–355. IEEE (2018)
24. Wang, X., Hovakimyan, N., Sha, L.: L1Simplex: fault-tolerant control of cyber-physical systems. In: 2013 ACM/IEEE International Conference on Cyber-Physical Systems (ICCPS), pp. 41–50. IEEE (2013). INSPEC Accession Number: 13769933
25. Friedberg, I., McLaughlin, K., Smith, P., Laverty, D., Sezer, S.: STPA-SafeSec: safety and security analysis for cyber-physical systems. J. Inf Secur Appl **34**, 183–196 (2017). https://doi.org/10.1016/j.jisa.2016.05.008
26. Colson, C.M., Nehrir, M. H., Gunderson, R.W.: Distributed multi-agent microgrids: a decentralized approach to resilient power system self-healing. In: 2011 4th International Symposium on Resilient Control Systems, pp. 83–88. IEEE (2011). https://doi.org/10.1109/ISRCS.2011.6016094

Building, Forming and Processing of Signals of the Electronic Sensor Airspeed Vector's Parameters of Unmanned Aircraft Plane

Vladimir Soldatkin, Vyacheslav Soldatkin, Galina Sokolova, Aleksandr Nikitin, and Elena Efremova

Abstract The analyses of the traditional and developed means of measuring of parameters of airspeed vector, implementing aerometric and vane, vortex or ion-mark methods of parameters of incoming airflow are performed. The article discloses the theoretical foundations of building of the original electronic sensor of direction angle and module of airspeed vector of unmanned aircraft plane with ultrasonic instrumentation channels. The functional scheme of the electronic sensor is presented, the distinctive feature of which is using of two pairs of combined transmitters and receivers of ultrasonic vibrations propagating along and against the direction of incoming flow. Informative signals of ultrasonic instrumentation channels are registered by the measuring circuit including two instrumentation channels. The analytical models are obtained for forming and processing of informative signals, determining the parameters of airspeed vector of unmanned aircraft plane by frequency, pulse time and phase informative signals of ultrasonic instrumentation channels is presented, which are processing with built-in computer. The expediency of using frequency informative signals, excluding methodological errors of measuring airspeed, has been substantiated. The possibility of measuring the direction angle of airspeed vector in range $\pm 180^\circ$ without increasing the number of ultrasonic instrumentation channels is revealed. The competitive advantages and prospects of using the electronic sensor of airspeed vector parameters on small unmanned aircraft plane are considered. The obtained results are the theoretical basis for the development, error analysis and ensuring the accuracy of the electronic sensor of parameters of airspeed vector with ultrasonic instrumentation channels for small-sized unmanned aircraft planes of various classes and purposes.

V. Soldatkin · V. Soldatkin (✉) · G. Sokolova · A. Nikitin · E. Efremova
Kazan National Research Technical University Named After A.N. Tupolev-KAI, 10, Karl Marks Str., Kazan 420111, Russian Federation

A. Ronzhin and V. Shishlakov (eds.), *Electromechanics and Robotics*, Smart Innovation, Systems and Technologies 232, https://doi.org/10.1007/978-981-16-2814-6_41

1 Introduction

The operation of significant class of aircraft planes (AP) for various purposes is carried out in the surface layer of the atmosphere. The main parameters that determine the quality of piloting and safety of aircraft plane's flight in atmosphere are the parameters of airspeed vector. To control manned and unmanned aircraft planes, reliable information is required about the aerodynamic angles of incidence and gliding, module and projections of airspeed vector on the axis of coordinate system is associated with the aircraft plane, which determine the aerodynamics and dynamics of movement of aircraft planes relative to the surrounding air [1–3]. At the same time, the requirements for mass dimensional characteristics, cost and placement of means of measuring of parameters of airspeed vector on unmanned small-sized aircraft plane are increasing. This determines the need to replenish the arsenal of means for measuring aerodynamic angles and airspeed on small-sized unmanned aircraft planes with improved technical characteristics and built using new measurement methods and construction principles.

The creation and expansion of application field of small-sized manned and unmanned AP of various classes and purposes determine the relevance of task of replenishing the arsenal of measuring means of airspeed vector parameters, which have a purely electronic design scheme, low weight and cost, providing measurement of the aerodynamic gliding angle in the entire azimuthal plane, i.e., in the range of $\pm 180°$.

2 Analysis of Measuring Means of Airspeed Vector Parameters of Aircraft Plane

As sources the primary information of the known measuring methods of parameters of vector airspeed of aircraft planes, the module is used (value) of speed vector of incoming airflow and angles, determining the position of speed vector of incoming airflow relative to the axis of the aircraft plane.

Traditional measuring means the parameters of airspeed vector of aircraft, and other aircraft planes implement aerodynamic and vane methods for measuring of the parameters of incoming airflow using autonomous receivers of air pressure and braking temperature, vane sensors of aerodynamic angles distributed over the fuselage of aircraft plane and remote from the calculator, which generates output signals according to the parameters of airspeed vector [4–8]. This complicates the design, increases weight and cost of such measuring means of parameters of airspeed vector, is the reason for the possible clogging and icing of autonomous receivers and sensors, what reduces the reliability of their operation in real operation. All this limits the scope of application of traditional means on small-sized manned and unmanned AP.

The aircraft plane's sensors of aerodynamic angle and airspeed are developed, which implement vortex [9–12] and ion-mark [13–16] measuring methods of parameters of incoming airflow using one (integrated) fixed receiver and built-in computer.

The vortex measuring method of parameters of incoming airflow is based on the effect of forming and periodic disruption of vortices from surface of high-drag bodies installed in flow and forming of so-called Karman's vortex paths behind the bodies. In this case, the frequency of periodic disruption of vortices is proportional to the speed of incoming airflow and size of mid-section of the body. To obtain the dependence of frequency of vortex disruption from direction angle of incoming airflow in the vortex sensor, two wedge-shaped pyramids are used, installed with their bases at an angle of 45° symmetrically to the direction of incoming airflow [8]. The aerodynamic angle and airspeed of aircraft plane in subsonic speed range are determined in the built-in computer using the frequencies of vortex formation parameters behind the wedge-shaped pyramids according to developed analytical models [7]. The absence of moving elements in incoming flow, and use of frequency primary informative signals, simplifies their selection, conversion and processing in the built-in sensor of the aerodynamic angle and airspeed in comparison with traditional means. Using of frequency primary informative signals reduces the errors of their instrumentation channels of sensor. However, the vortex sensor of aerodynamic angle and airspeed [12] provides measuring in range of variation of aerodynamic angle ±25 … 30° what does not allow to use the sensor to measure the gliding angle of the aircraft plane. In the developed ion-mark sensor of aerodynamic angle and airspeed [9], the ion mark with pronounced electrostatic charge is periodically formed into the controlled incoming airflow with using of the spark-gap switch, which moves together with the flow, acquiring its speed and direction. Registration of parameters of incoming airflow is provided by the system of receiving electrodes evenly distributed around the circumference. The receiving electrodes record the time of movement of the ion mark from the spark-gap switch to the circle with the receiving electrodes and angular position of receiving electrode, over which the trajectory of movement of ion mark together with flow passes. With using of multi-channel instrumentation circuit, according to the developed analytical models, the digital output signals are determined and generated in the built-in computer by the measured aerodynamic angle and airspeed of the aircraft plane. The ion-mark sensor does not have elements protruding into the incoming flow that distort the aerodynamics of the aircraft plane. It is the completely electronic device with a built-in computer. Its measuring scheme implements the kinematic method for measuring the parameters of incoming flow without procedural measurement errors. The multi-channel measuring circuit of ion-mark sensor of aerodynamic angle and airspeed [16] provides high-precision measuring in the range of aerodynamic angle variation up to ± 180°. However, stringent requirements for identity and stability of characteristics of countless of instrumentation channels fundamentally complicate the design, increase the weight and cost of sensor, which also limits its use on small-sized unmanned aircraft planes.

The effectiveness of application of the ultrasonic method for measuring of parameters of gas flow and liquid in flow metering and other areas [17–20] determined the direction of development of a wide-range purely electronic sensor of airspeed vector

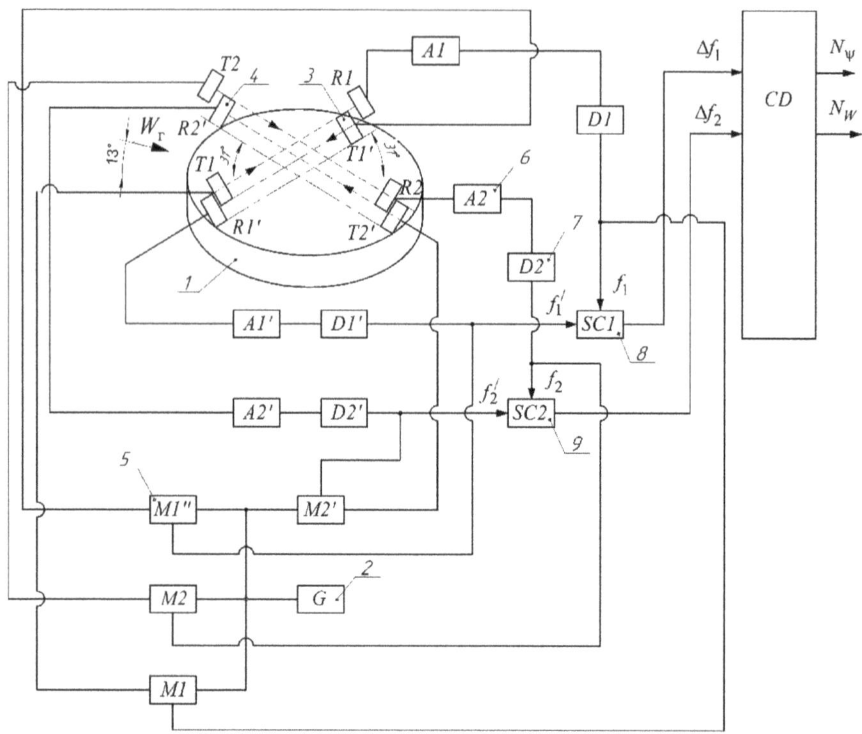

Fig. 1 Functional scheme of the panoramic sensor of aerodynamic angle and true airspeed with ultrasonic instrumentation channels

parameters of small-sized unmanned aircraft plane with one fixed receiver and ultrasonic instrumentation channels, the functional scheme of which is shown in Fig. 1 [21].

3 Building of the Electronic Sensor of Airspeed Vector Parameters with Ultrasonic Instrumentation Channels

The electronic sensor of aerodynamic angle and airspeed implements the ultrasonic method for measuring of parameters of incoming airflow. The sensor implements the kinematic method for measuring of speed and angle of direction of incoming airflow, in which the accuracy of measuring of airspeed vector parameters does not depend on the temperature, pressure and other parameters of surrounding air environment. The working of the instrumentation channels of the electronic sensor is based on recording the difference of propagation time of ultrasonic vibrations in direction of incoming airflow and against the flow, that is, the informative parameters are time–frequency primary informative signals convenient for isolation, transformation and processing.

In the proposed electronic sensor of airspeed vector parameters, for registration of parameters of incoming airflow, pairs of piezoelectric transmitters and receivers of ultrasonic vibrations combined in one design are used. The combined transmitter–receiver pairs are installed on outer streamlined surface 1 of sensor at angle $\theta_0 = 45°$ to axis, relative to which the digitized aerodynamic angle of incidence or gliding of aircraft plane's airspeed vector is measured.

Transmitters 4 $T1$ and $T2$ generate, and receivers 3 $R1$ and $R2$ register ultrasonic vibrations that propagate in the direction of incoming airflow. The transmitters $T1'$ and $T2'$ generate, and receivers $R1'$ and $R2'$ register ultrasonic vibrations propagating against direction of flow.

The inputs of transmitters $T1$, $T2$ and $T1'$, $T2'$ are connected to generator 2, which generates sinusoidal high-frequency oscillations, through control circuits made in form of modulators 5 $M1$, $M2$ and $M1'$, $M2'$, operating in key mode. The outputs of receivers of ultrasonic vibrations $R1$ and $R1'$ through the amplifiers 6 $A1$ and $A1'$, the detectors 7 $D1$ and $D1'$ are connected to inputs of subtraction circuit 8 SC1, at the output of which the informative signal is formed in form of difference $\Delta f_1 = f_1 - f_1{}'$ frequencies f_1 and $f_1{}'$ of ultrasonic vibrations, perceived by receivers $R1$ and $R1'$ in direction of controlled flow. Receivers $R2$ and $R2'$ through amplifiers $A2$ and $A2'$ and detectors $D2$ and $D2'$ are connected to input of subtraction circuit SC2, at output of which the difference $\Delta f_2 = f_2 - f_2{}'$ frequencies f_2 and $f_2{}'$ is formed of ultrasonic vibrations received by receivers $R2$ and $R2'$ against direction of incoming flow. The outputs of the subtraction circuits SC1 and SC2 are connected to the input of calculator 9, which provides processing of informative signals Δf_1 and Δf_2, determining and output of digital signals N_α and N_{V_B} by the aerodynamic angle and airspeed. The obtained information by the parameters of airspeed vector of unmanned aircraft plane is fed to the piloting systems and flight safety systems of the unmanned aircraft plane.

4 Models of Forming and Informative Signals Processing of the Electronic Sensor of Airspeed Vector Parameters of an Unmanned Aircraft Plane

Original analytical models of forming and processing of frequency informative signals of ultrasonic instrumentation channels of the electronic sensor of aerodynamic angle and airspeed have been developed [21]. It is shown that during measuring aerodynamic angle α is found within ±45° (sector I of the figure), time intervals t_1, $t_1{}'$ and t_2, $t_2{}'$ of propagation of ultrasonic vibrations from the transmitters to the receivers in the direction of incoming airflow and against flow are determined by the expressions

$$t_1 = \frac{L}{a + V\cos(\Theta_0 + \alpha)}; \quad t_1' = \frac{L}{a - V\cos(\Theta_0 + \alpha)};$$
$$t_2 = \frac{L}{a + V\cos(\Theta_0 - \alpha)}; \quad t_2' = \frac{L}{a - V\cos(\Theta_0 - \alpha)}, \tag{1}$$

where L is the distance between the transmitters and receivers of ultrasonic vibrations; a is speed of propagation of ultrasonic vibrations in the air; V is speed of incoming airflow; α is the angle of direction of incoming airflow relative to the sensor axis.

Using relations (1), frequencies f_1,f_1' and f_2,f_2' of ultrasonic vibrations coming to the inputs of subtraction circuit SC1 and SC2 are associated with the angle of direction of incoming flow equal to the magnitude and sign of the measured aerodynamic angle α and with the speed of the incoming airflow V, equal to the measured airspeed V, expressions

$$f_1 = \frac{1}{L}[a + V\cos(\Theta_0 + \alpha)]; \quad f_1' = \frac{1}{L}[a - V\cos(\Theta_0 + \alpha)];$$
$$f_2 = \frac{1}{L}[a + V\cos(\Theta_0 - \alpha)]; \quad f_2' = \frac{1}{L}[a - V\cos(\Theta_0 - \alpha)] \tag{2}$$

Obtained relations (2) allow us to determine the frequency differences f_1,f_1' and f_2,f_2' at the output of the subtraction schemes SC1 and SC2 $\Delta f_1 = f_1 - f_1'$ and $\Delta f_2 = f_2 - f_2'$ in the form

$$\Delta f_1 = \frac{2V}{L}\cos(\Theta_0 + \alpha); \quad \Delta f_2 = \frac{2V}{L}\cos(\Theta_0 - \alpha) \tag{3}$$

Analytical models of processing of the input signals of computer and determining of the aerodynamic angle α and airspeed V_B of the unmanned aircraft plane in ultrasonic instrumentation channels of the electronic sensor of airspeed vector parameters are obtained using relations (3) in the form [21]

$$\alpha = \operatorname{arctg}\frac{\Delta f_1 - \Delta f_2}{\Delta f_1 + \Delta f_2} \tag{4}$$

$$V_A = \frac{L}{2}\sqrt{\Delta f_1^2 + \Delta f_2^2}. \tag{5}$$

When the measured aerodynamic angle exceeds the value $\alpha = +45^\circ$ and exits from sector I, the working of the ultrasonic instrumentation channels is similar in sector II. In this case, the value of the aerodynamic angle α, calculated by the ratio (4), should algorithmically increase by $\pi'2$. If the aerodynamic angle exceeds the value $\alpha = +90^\circ$ in sector III, the calculated value of angle should increase algorithmically by π. If the measured aerodynamic angle is negative and exits from the limits of $\alpha = -45^\circ$ and $\alpha = -90^\circ$, the calculated values of the aerodynamic angle should decrease algorithmically by-$\pi'2$ and by-π, respectively.

In this case, the electronic sensor provides a panoramic measuring of the airspeed vector parameters in the entire azimuthal plane in the range of change of aerodynamic angle $\pm 180°$, using analytical models of the form

$$\alpha = K_i \frac{\pi}{2} + \text{arctg}\frac{\Delta f_1 - \Delta f_2}{\Delta f_1 + \Delta f_2}; \quad V_A = \frac{1}{2}\sqrt{\Delta f_1^2 + \Delta f_2^2}. \tag{6}$$

where K_i is the coefficient that takes values for sector I, $K_1 = 0$, for sector II, $K_2 = \pm 1$, and for sector III, $K_3 = \pm 2$.

Panoramic measuring of parameter of airspeed vector of unmanned aircraft plane is provided without increasing the number of ultrasonic instrumentation channels, which simplifies the scheme and construction implementation of the electronic sensor of aerodynamic angle and airspeed of unmanned aircraft plane, including its miniaturization, weight reduction and cost through the use of modern element base integrated.

As can be seen from the ratios (4) and (6), the calculated value of the aerodynamic angle is determined only by the current values of frequency difference Δf_1 and Δf_2, recorded by the corresponding combined pairs of the transmitter–receiver of ultrasonic vibrations in and against the direction of controlled airflow. The expression (5) for calculating the true airspeed includes the distance L between the transmitters and receivers, the technological spread of which is caused of the instrumental error during measuring of speed. However, this error is close to systematic and is easily taken into account when calibrating the sensor.

The implementation of ultrasonic method of measuring of incoming airflow parameters can be achieved by controlling not change frequencies f_1, f_1' and f_2, f_2', recorded by ultrasonic instrumentation channels, and the transit time difference of base distance in and against the incoming flow of short pulses at a given frequency.

In this case, the difference of time intervals, $\Delta t_1 = t_1 - t_1'$ and $\Delta t_2 = t_2 - t_2'$, recorded by the corresponding transmitter–receiver pairs, installed at an angle of $2\Theta_0 = 90°$ symmetrical to relative specified axis of the sensor, is used as informative signals.

The differences Δt_1 and Δt_2 will be determined by the relations

$$\begin{aligned} \Delta t_1 &= t_1 - t_1^{/} = \frac{2LV}{a^2}\cos(\Theta_0 + \alpha); \\ \Delta t_2 &= t_2 - t_2^{/} = \frac{2LV}{a^2}\cos(\Theta_0 - \alpha) \end{aligned} \tag{7}$$

Taking into account the fact that

$$\begin{aligned} \cos(\Theta_0 + \alpha) &= \cos\Theta_0 \cos\alpha + \sin\Theta_0 \sin\alpha \\ \cos(\Theta_0 - \alpha) &= \cos\Theta_0 \cos\alpha - \sin\Theta_0 \sin\alpha \end{aligned}$$

taking $\Theta_0 = 45°$, $\cos 45° = \sin 45° =$, we get

$$\Delta t_1 = \frac{\sqrt{2}LV}{a^2}(\cos\alpha + \sin\alpha);\quad \Delta t_2 = \frac{\sqrt{2}LV}{a^2}(\cos\alpha - \sin\alpha) \tag{8}$$

It is shown that the sum ($\Delta t_1 + \Delta t_2$) and the difference ($\Delta t_1 - \Delta t_2$) will be determined by relations of the form

$$\Delta t_1 + \Delta t_2 = \frac{2\sqrt{2}L}{a^2}V\cos\alpha;\quad \Delta t_1 - \Delta t_2 = \frac{2\sqrt{2}L}{a^2}V\sin\alpha \tag{9}$$

The sum of squares of ${\Delta t_1}^2 + {\Delta t_2}^2$ will be equal to

$$\Delta t_1^2 + \Delta t_2^2 = \frac{2\sqrt{2}L}{a^2}V^2 \tag{10}$$

Using relations (7), the analytical models for calculating of aerodynamic angle α and the true airspeed V_B of the ultrasonic sensor of aerodynamic angle and the true airspeed when using time-pulse informative signals in sector I take the form

$$\alpha = \operatorname{arctg}\frac{\Delta t_1 - \Delta t_2}{\Delta t_1 + \Delta t_2} \tag{11}$$

$$V_A = \frac{a^2}{2\sqrt{2}L}\left(\Delta t_1^2 + \Delta t_2^2\right). \tag{12}$$

The obtained analytical ratios (11), (12) for calculating of aerodynamic angle are same in shape as the ratios (4) and (5) obtained using frequency informative signals. As can be seen from ratio (12), the value of true airspeed depends not only on the measured time intervals Δt_1 and Δt_2, but also on speed a of sound propagation in the air, which depends on ambient temperature, and, consequently, on the altitude, as well as on the distance L between the receivers and transmitters. Changes of sound speed a and distance L cause methodological and instrumental errors of measuring the true airspeed.

When implementing ultrasonic instrumentation channels of the electronic sensor of aerodynamic angle and true airspeed, phase informative signals can be used to register the parameters of controlled airflow.

Due to the difference of the time Δt required for passaging of ultrasonic vibrations of the same distance L along and against the flow, a phase difference occurs on the piezoelectric elements of receivers between the ultrasonic vibrations perceived by them propagating along and against to the direction of the incoming airflow.

For ultrasonic instrumentation channels located at an angle Θ to the direction of the controlled flow, the phase difference $\Delta\varphi$ will be determined by the expression

$$\Delta\varphi = \frac{4\pi f L\cos\Theta}{a^2}V, \tag{13}$$

where f is the frequency of ultrasonic vibrations.

Using relations (13) to the problem of measuring the aerodynamic angle and the true airspeed for the pair of transmitter–receiver located at an angle of $\Theta_0 = 45°$, we obtain

$$\Delta\varphi_1 = \frac{4\pi f LV}{a^2}\cos(\Theta_0 + \alpha); \quad \Delta\varphi_2 = \frac{4\pi f LV}{a^2}\sin(\Theta_0 - \alpha);$$
$$\Delta\varphi_1 - \Delta\varphi_2 = \frac{8\pi f LV}{a^2}\sin\alpha; \quad \Delta\varphi_1 + \Delta\varphi_2 = \frac{8\pi f LV}{a^2}\cos\alpha.$$

Taking into account the values $\cos(\Theta_0 + \alpha)$ and $\cos(\Theta_0 - \alpha)$, using expressions similar to (9) and (10), we obtain analytical models for calculating the aerodynamic angle and airspeed of the ultrasonic sensor of parameters of airspeed vector

$$\alpha = \text{arctg}\frac{\Delta\varphi_1 - \Delta\varphi_2}{\Delta\varphi_1 + \Delta\varphi_2} \tag{14}$$

$$V_A = \frac{a}{2\pi f L}\left(\Delta\varphi_1^2 + \Delta\varphi_2^2\right) \tag{15}$$

Therefore, when using the phase informative signals, as for the time pulse informative signals, calculated by the rations (14) and (15) true airspeed depends from the distance L between the receivers and transmitters, from sound speed a and frequency f of sending ultrasonic vibrations in which change is the cause of instrumental and methodical errors of measurement that also confirms the advantages of the electronic sensor of airspeed vector parameter with frequency informative signals especially when used on small-sized unmanned aircraft planes.

5 Conclusion

Thus, the obtained results show that in comparison with the ion-mark and vortex-integrated sensors, the proposed electronic sensor of parameters of airspeed vector with ultrasound instrumentation channels has the following competitive advantages:

- To register parameters of incoming airflow, only two instrumentation channels are used, built on the basis of two combined pairs of transmitter–receivers of ultrasonic vibrations, which significantly simplifies the design and reduces the weight and cost of the sensor. The use of integrated computer for processing informative signals does not require connecting cables and pneumatic conduits, which is typical for traditional measuring instruments, and also simplifies the construction, reduces weight and cost of the sensor.
- The use of frequency informative signals simplifies their selection, conversion and processing in the sensor's measuring circuit, which simplifies its practical implementation and the formation of digital output signals based on the parameters

of the airspeed vector, as well as the output signals of the electronic sensor of the display system and other consumers.

- The absence of truncation errors of ultrasonic instrumentation channels with frequency informative signals increases the accuracy of measuring of the parameters of air velocity vector. At the same time, the instrumental errors of ultrasonic instrumentation channels associated with the conversion of frequency signals are significantly reduced.
- The absence of moving elements and openings in incoming airflow for the perception of primary information increases the reliability of the electronic sensor of parameters of airspeed vector of unmanned aircraft plane in real operating conditions.

The developed principles of building, models for forming of information signals, their processing and determining of output signals are the theoretical basis for developing, error analysis and ensuring accuracy of instrumentation channels and for the development, manufacture and research in the wind tunnel of experimental samples of variants of the electronic sensor of parameters of airspeed vector with ultrasonic instrumentation channels.

These competitive advantages simplify the design, reduce weight and cost, which expand the scope of application of the electronic sensor on small-sized unmanned aircraft planes of various classes and purposes. The use of the electronic sensor of parameters of airspeed vector with ultrasonic instrumentation channels on unmanned aircraft planes will increase the level of flight safety, improve the quality of piloting and ensure the solution of tactical and technical problems of flight.

References

1. Moiseev, V.S., Gushchina, D.S., Moiseev, G.V.: Fundamentals of the Theory of Creation and Application of Information Unmanned Aircraft Systems. Publishing House of the Ministry of Education and Science of the Republic of Tatarstan, Kazan (2010)
2. Yankevich, Yu.: The use of unmanned aircraft complexes for civil. Aerosp. Courier **6**, 55–57 (2006)
3. Lysenko, N.M. (ed.): Practical Aerodynamics of Maneuverable Aircraft. Voenizdat (1977)
4. Kravtsov, V.G., Alekseev, N.V.: Aerometry of altitude-speed parameters of aircraft. Dev. Syst. Operat. Control. Diagnost. **8**, 47–50 (2000)
5. Kaletka, J.: Evaluation of the helicopter low airspeed system lassie. J. Am. Helicopter Soc. **4**, 35–43 (1983)
6. System zur Bestimung der Fluggeschwindigkeit von Hubschraubern. Burchard Miller. Patenblatt. Patent №0249848 ЕПВ (EP) MK G01P5/00 (1987)
7. Klyuev, G.I., Makarov, N.N., Soldatkin, V.M., Efimov, I.P.: Measuring Means of Aerodynamic Parameters of Aircraft Planes. Publishing house of Ulyanovsk State Technical University, Ulyanovsk (2005)
8. Soldatkin, V.M., Ganeev, F.A., Soldatkin, V.V., Nikitin AV.: Aviation devices, measuring-computing systems and complexes: principles of construction, algorithms for processing information, characteristics and errors. In: Soldatkin, V.M. (ed.) Textbook. Publishing House of Kazan State Technical University, Kazan (2014)
9. Yamasaki, H., Rubin, M.: The vortex flowmeter. Flow Meas. Control Sci. Ind. 975–983 (1974)

10. Kiyasbeyli, A.S., Perelshtein, M.E.: Vortex Measuring Devices. Mashinostroenie (1972)
11. Pankanin, G.L.: The vortex flowmeter: various methods of investigating phenomena. Meas. Sci. Technol. **16**, 1–16 (2005)
12. Soldatkin, V.M., Efremova, E.S.: Features of building and analysis of static accuracy of the vortex air data system of subsonic aircraft. Mechatron. Autom. Control **20**(7), 443–448 (2019)
13. Soldatkin, V.M.: Methods and Means of Measuring of Aerodynamic Angles of Aircraft. Publishing House of Kazan State Technical University, Kazan (2001)
14. Dankert, C.: Messungen der stomungsgesahwindigkeit uberschallschneller gase mittels elektronenstraltehnik. DFLVR-AVA Institut fur Experementelle Stromungstechnik, 154–167 (1982)
15. Barriol, R., Hannoyer, G., Roussean, C.: A new approach for ionic air flow sensors transit time. SAE Technical Paper Series №. 840138, 29–39 (1984)
16. Ganeev, F.A., Soldatkin, V.M.: Ion-mark sensor of aerodynamic angle and airspeed with logometric informative signals and interpolation processing scheme. Izvestiya vuzov. Aviatsionnaya tekhnika **3**, 46–50 (2010)
17. Kremlevsky, P.P.: Flowmeters and Quantity Counters, 3rd edn. (Edited and additional). Mashinostroenie (1975)
18. Novitsky, P.V. (ed.): Electric Measurements of Non-electric Quantities. Energiya, 576 p (1975)
19. Fernandes, D., Gomes, L., Costa, A.: Wind speed and direction measurement based on time of flight ultrasonic anemometer. In: 2017 IEEE 26th International Symposium on Industrial Electronics (ISIE), pp. 1417–1422 (2017)
20. Ghahramani, A., et al.: Measuring air speed with a low-power MEMS ultrasonic anemometer via adaptive phase tracking. Sens. J. IEEE **19**(18), 8136–8145 (2019)
21. Soldatkin, V.M., et al.: Kinematic sensor of aerodynamic angle and true airspeed. Patent RF № 2737518 C1 (2020)

Numerical Analysis of the Near-Resonant Vibrations of a Vibrating Technological Machine with Self-synchronizing Unbalance Vibration Exciters

Alexander Gouskov, Grigory Panovko, and Alexander Shokhin

Abstract The work is devoted to the problem of maintaining near-resonant oscillations of a technological vibration machine with self-synchronizing unbalance vibration exciters under conditions of a variable technological load on its working body. A planar model of a vibration machine with two unbalance exciters driven by asynchronous electric motors is considered. Numerical simulation methods have been used to analyze the dynamics of the system in the near-resonant frequency range at slowly changing mass and position of the mass center of the technological load on the machine's working body. The influence of the rate of change in the mass of the technological load on the oscillations of the system and the self-synchronization of unbalance vibration exciters is established. It is established that the fluctuations in the mass of the technological load lead to fluctuations of mutual phase between the vibration exciters' debalances with the same frequency. In this case, the amplitude of the mutual phase oscillations increases with increasing amplitude of the mass oscillations, practically does not change with a change in their frequency, and significantly increases as one approaches the resonant frequency.

1 Introduction

Technological vibration machines with unbalance vibration exciters (vibrating conveyors, vibrating screens, vibratory crushers, etc.) are widely used in various branches of modern industry. Depending on the purpose and characteristics of the technological process, vibration machines can be used in different ranges of excitation frequencies (rotational speed of the vibration exciter unbalances). These frequency ranges are delimited by the frequencies of natural vibrations of the vibration machine's working body. If the frequency of rotation of the vibration exciters is less than the eigenfrequency of the working body oscillations, then such vibration machines are usually called pre-resonant, in contrast to beyond-resonant machines.

A. Gouskov · G. Panovko (✉) · A. Shokhin
Mechanical Engineering Research Institute of the Russian Academy of Sciences, Moscow, Russia

A. Ronzhin and V. Shishlakov (eds.), *Electromechanics and Robotics*, Smart Innovation, Systems and Technologies 232, https://doi.org/10.1007/978-981-16-2814-6_42

In most cases, vibration machines are used in the beyond-resonant frequency range of excitation [1–4]. This makes it possible to ensure the stability of the required oscillation modes (amplitude and frequency) of the vibration machine's working body, regardless of changes in the parameters of the technological load (composition and inertial characteristics of the processed material). At the same time, the power consumption of the electric drive is mainly aimed at the implementation of the technological process associated with the dissipation of energy in the layer of the processed material and overcoming the frictional forces in the vibration machine's elements. However, to overcome the resonant frequencies when starting the machine, electric motors with excess power are used. This leads to the fact that in the operating mode, the electric motor is significantly underloaded, as a result of which energy consumption increases and the engine's service life decreases. In works [1, 5], using the example of analyzing the power consumption of a technological vibration machine, the calculated values of the required motor power (consumptions associated with vibration impact on the material and friction in bearings) were obtained, which are much less (approximately half) of the installed power. The excess of the installed power (the power required to maintain the required technological mode) in relation to the required one is mainly due to the required power to pass through resonance and raise debalances when starting the machine. The question of choosing the necessary and sufficient installed motor power is important not only for economic reasons. The fact is that in most cases, modern vibration machines are based on the use of two (in some cases even more) self-synchronizing vibration exciters. The operation stability of such self-synchronizing exciters decreases if the installed power of the motors exceeds the required one [1].

The issues of self-synchronization of unbalance vibration exciters in application to technological vibration machines are discussed in various literature; see, for example, [6–9]. The influence of the parameters of the technological load on the dynamics of such machines was investigated, in particular, in [10–12]. Some theoretical and experimental results on the creation of controlled resonant vibration machines are presented in the previous works of the authors of this study [13, 14]. Note that in existing studies, the main attention is paid to the analysis of the stability of self-synchronizing unbalance vibration exciters in the frequency ranges far from resonance, and the issues of the influence of technological load variability on the dynamics of machines in the near-resonant frequency ranges are considered only in a quasi-stationary formulation.

One of the ways to increase the energy efficiency of vibration machines is the use of resonant modes of their working body oscillation, which, however, are usually unstable. In work [5], based on a comparative analysis of energy consumption for the implementation of the beyond-resonant and resonant modes of oscillations of the technological vibration machines' working body with an unbalance vibration exciter, it is shown that the required electrical power of the resonant machine is approximately three times less than that of the beyond-resonant machine. Thus, resonant vibration machines can use less powerful vibration exciters with lighter debalances, which provide a higher stability of their rotation at a significantly lower weight and cost.

However, in a number of cases, the problem of maintenance of stable resonance oscillations mode of vibration machines with self-synchronizing vibration exciters is associated with the dynamic features of the interaction of the machine with a vibration drive and possible fluctuations (variability in time) of the inertial characteristics of the technological load. The latter circumstance is especially important in connection with the need to provide a given type of synchronous debalances rotation.

To adjust and maintain the resonance oscillation mode of vibration machines with the exciters of limited power, it is necessary to use automatic control systems that monitor with a given accuracy the proximity of the synchronous rotation frequency of the debalances to the eigenfrequency of the working body oscillations [14]. Control algorithms for such control systems are usually based on comparing the measured parameters of the machine's motion with the results of the analysis of the steady-state motion modes of their mathematical models. The analysis of such models is carried out by varying the parameters of the system in the given ranges of values under the assumption that for each calculated steady state of motion, these parameters do not change. In this regard, the duration and accuracy of tuning to the required motion mode essentially depend on the correctness and accuracy of measurements of the machine motion's controlled parameters, which, in addition to the technical capabilities of the measuring devices, will be determined by both the decay rate of transient processes and the sensitivity of the measured parameters to fluctuations in the system parameters. For example, in vibratory machines that carry out the technological process due to vibrodisplacement of the processed material along the tray (vibrating screens, vibrating conveyors, etc.), fluctuations of the inertial parameters of the system inevitably arise, due to the uneven supply of material to the working body (tray) and its movement along the tray. A change in the mass of the system leads to a breakdown from the resonant frequency, and a change in the location of the center of mass of the system in such machines can lead to a change in the steady-state mutual phase of synchronous rotation of vibration exciters, and, as a consequence, to a change in the exciting forces and modes of the machine's oscillations [6]. Thus, by measuring the mutual phase of rotation of vibration exciters under conditions of their self-synchronization, it is possible to determine the displacement of the system's center of mass from its initial position and to generate the necessary control signals to correct the machine's oscillation mode. However, the issues of the influence of changes in the inertial parameters of the system on the self-synchronization of vibration exciters near resonance have practically not been considered yet. It is obvious that to create effective algorithms for controlling the resonance oscillations of vibration machines with self-synchronizing unbalance vibration exciters, it is of interest to analyze their dynamics in the near-resonant frequency range depending on the parameters characterizing the change of technological load in time. Thus, the purpose of this work is to identify the effect of changes in the inertial parameters of the technological load on the mutual phase of rotation of inertial self-synchronizing vibration exciters near resonance.

2 Mathematical Model

In this work, a plane model of a vibration machine with two unbalance vibration exciters of limited power is considered (Fig. 1). The model consists of a platform with vibration exciters rigidly fixed on it, which are modeled by a solid with mass m_{pl} and moment of inertia J_{pl} relative to its center of mass at point O. The technological load is modeled by a solid with mass mw and moment of inertia J_{w} relative to its center of mass at point O_{w}, which oscillates together with the platform without separation from it. Features of the model are in taking into account the interaction of the oscillatory system with vibration exciters and the relatively slow compared with the main period of the system oscillations and change in the mass of the process load causing a change in the mass and the center of mass of the oscillatory system.

The exciters have the same mechanical characteristics and rotate in opposite directions. Asynchronous drive motors of vibration exciters have the same torque characteristics, which have a pronounced nonlinear dependence on the rotor speed. Usually, when analyzing the dynamics of mechanical systems, a simplified description of the torque characteristics is used, representing them as a linear dependence obtained as a result of linearization around the operating frequency. In this work, to take into account the features associated with slip of the rotor, the moment characteristics of vibration exciters are described by the Kloss formula [6]:

$$M_j(\dot{\varphi}_{j1}) = 2\sigma_j M_{cj}/(s_{cj}/s_j + s_j/s_{cj}),$$

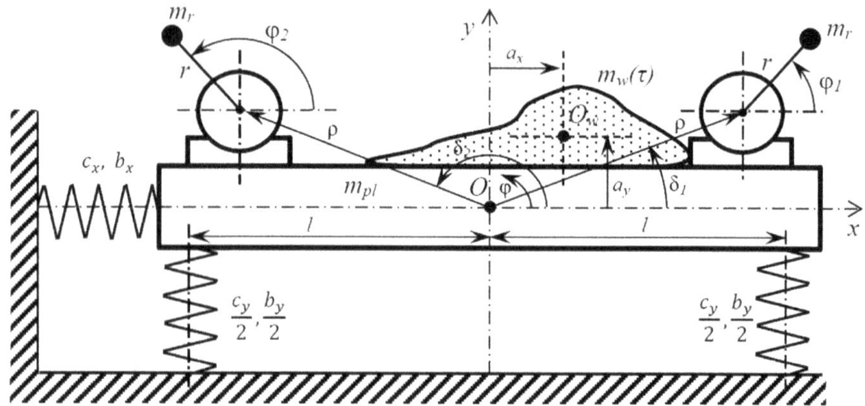

Fig. 1 Design scheme of vibrating machine

where $j = 1, 2$—the vibration exciter number, M_{cj} and s_{cj} are the critical torque and critical slip of the exciter's electric motor, respectively, $s_j = \left(\omega_{0j} - \sigma_j\dot{\varphi}_j\right)/\omega_{0j}$, $\omega_{0j} = 2\pi f_e/P$—motor's synchronous speed, P—number of motor's pole pairs, f_e—power supply frequency in Hz, $\sigma_j = \pm 1$—coefficient taking into account the direction of the torque ($\sigma_j = +1$ corresponds to the action of the motor torque in the counterclockwise direction).

It is assumed that for the implementation of the technological process, it is necessary to excite unidirectional vertical oscillations of the platform, which in a centered system arise during synchronous antiphase rotation of debalances [13, 14]. A change in the mass of the technological load and a shift in its center of mass lead to a violation of the symmetry in the system, which leads to a change in the dynamic properties of the system, a change in the shape of the vibrations, and a possible jump beyond the resonance. In order to identify the features of the machine's dynamics in the resonant frequency range at a slowly changing technological load and then take them into account when developing a control system for the machine's resonant oscillations, this work analyzes the influence of parameters characterizing the change of technological load in time on oscillations of the machine in the resonant frequency range as well as on synchronization of the unbalance vibration exciters.

Motion of the system is described by the system of differential equations:

$$\begin{cases} \dfrac{\mathrm{d}}{\mathrm{d}t}(M(\tau)\dot{x}) + c_x x = \varepsilon\left(-b_x\dot{x} - \dfrac{\mathrm{d}}{\mathrm{d}t}(\mu_1\dot{\varphi}) + \sum\limits_{j=1}^{2} m_\mathrm{r} r\left(\cos\varphi_j\dot{\varphi}_j^2 + \sin\varphi_j\ddot{\varphi}_j\right)\right), \\ \dfrac{\mathrm{d}}{\mathrm{d}t}(M(\tau)\dot{y}) + c_y y = \varepsilon\left(-b_y\dot{y} - \dfrac{\mathrm{d}}{\mathrm{d}t}(\mu_2\dot{\varphi}) + \sum\limits_{j=1}^{2} m_\mathrm{r} r\left(\sin\varphi_j\dot{\varphi}_j^2 + \cos\varphi_j\ddot{\varphi}_j\right)\right), \\ \dfrac{\mathrm{d}}{\mathrm{d}t}(J(\tau)\dot{\varphi}) + c_\varphi\varphi = \varepsilon(-b_\varphi\dot{\varphi} - \dfrac{\mathrm{d}}{\mathrm{d}t}(\mu_1\dot{x} + \mu_2\dot{y}) \\ \quad - \sum\limits_{j=1}^{2} m_\mathrm{r} r\rho\left(\cos(\delta_j + \varphi - \varphi_j)\ddot{\varphi}_j + \sin\left(\delta_j + \varphi - \varphi_j\right)\dot{\varphi}_j^2\right)), \\ J_\mathrm{m}\ddot{\varphi}_1 = \varepsilon(M_1(\dot{\varphi}_1) + m_\mathrm{r} r(\ddot{x}\sin\varphi_1 - \ddot{y}\cos\varphi_1 \\ \quad - \rho\cos\,(\delta_1 + \varphi - \varphi_1)\ddot{\varphi} + \rho\sin(\delta_1 + \varphi - \varphi_1)\dot{\varphi}^2)), \\ J_\mathrm{m}\ddot{\varphi}_2 = \varepsilon(M_2(\dot{\varphi}_2) + m_\mathrm{r} r(\ddot{x}\sin\varphi_2 - \ddot{y}\cos\varphi_2 \\ \quad - \rho\cos(\delta_2 + \varphi - \varphi_2)\ddot{\varphi} + \rho\sin(\delta_2 + \varphi - \varphi_2)\dot{\varphi}^2)), \end{cases}$$

where

$$M(\tau) = m_\mathrm{pl} + m_\mathrm{w}(\tau) + 2m_\mathrm{r},\ J_\mathrm{m} = J_\mathrm{r} + m_\mathrm{r} r^2,$$

$$J(\tau) = J_\mathrm{pl} + J_\mathrm{w} + m_\mathrm{w}(\tau)(a_x^2 + a_y^2) + 2m_\mathrm{r}\rho^2,$$

$$\mu_1 = -m_{\rm w}(\tau)(a_y \cos\varphi + a_x \sin\varphi) - \sum\nolimits_{j=1}^{2} m_{\rm r}\rho \sin(\delta_j + \varphi),$$

$$\mu_2 = -m_{\rm w}(\tau)(a_y \sin\varphi + a_x \cos\varphi) + \sum\nolimits_{j=1}^{2} m_{\rm r}\rho \sin(\delta_j + \varphi),$$

$J_{\rm r}$—inertia moment of exciter's rotor, $\tau = \varepsilon t$, ε—small parameter.

3 Simulation Results

The simulation was carried out numerically with a change in the mass of the technological load according to the harmonic law $m_w(\tau) = m_0(1 + \kappa \sin(\alpha t))$, where m_0—average value of the technological load mass, κ and α—amplitude and frequency of fluctuations in the mass of the technological load, and $\alpha \ll \langle\dot{\varphi}\rangle$, $\langle\dot{\varphi}\rangle$—average value of the debalances' synchronous rotation speed. Note that due to the specifics of the conditions for loading and unloading material onto the working body of machines of the type under consideration, as well as the slow movement of material along the working body (relative to its oscillations), the adopted harmonic law quite satisfactorily describes the slow change in the mass of the material. Numerical simulation was carried out at different frequencies of power supplied to the motors, which changed stepwise with a step of 1 Hz, and exposure at each frequency sufficient to transients decay. In the simulation, the following values of the system parameters have been taken: $m_{\rm pl} = 12.46$ kg, $m_{\rm r} = 0.029$ kg, $J_{\rm pl} = 0.11$ kg m^2, $J_{\rm w} = 0.01$ kg m^2, $J_{\rm r} = 0.8 \times 10^{-3}$ kg m^2, $r = 0.09$ m, $\rho = 0.128$ m, $a_x = a_y = 0.05$ m, $\delta_1 = 22.5°$, $\delta_2 = 157.5°$, $c_x = 580$ kN/m, $c_y = 470$ kN/m, $c_\varphi = 1.8$ kN m/rad, $b_{\rm x} = 300$ N s/m, $b_y = 200$ N s/m, $b_\varphi = 1.5$ N s^2/m, $g = 9.81$ m/s^2, $M_{\rm cr1} = M_{\rm cr2} = 1.19$ N m, $s_{\rm cr1} = s_{\rm cr2} = 0.6$, $P = 2$, $m_0/(m_{\rm pl} + m_0 + 2m_{\rm r}) = 0.1$. The simulations were carried out for various values of parameters κ and α.

Figure 2 shows the change in the absolute values of speed of the electric motors rotation over time when the power frequency changes in the range from 30 to 79 Hz and exposure at each frequency for 5 s, for the case of a constant value of the technological load mass (note that with the chosen display scale, these graphs practically coincide for the left and right electric motors). Near the rotation frequencies $\dot{\varphi}_* \approx 19$ Hz and $\dot{\varphi}_{**} \approx 29$ Hz, corresponding to the resonant frequencies of the system, there are jumps in rotational speed due to the interaction of the oscillatory system with vibration exciters and the limited power of their drives—the Sommerfeld effect.

Figure 3 shows the change in the mutual phase $\Delta\varphi$ between the vibration exciters' debalances corresponding to this case. One can see that in the frequency range up to the first resonant frequency of the system, the value of the mutual phase between the vibration exciters is set near $\Delta\varphi = 0°$, which corresponds to their synchronous in-phase rotation. In this case, planar oscillations of the platform are excited. When approaching the first resonance, there is an increase in the unevenness of the vibration exciters' rotational speed (Fig. 2), as well as an increase in the difference of the phase

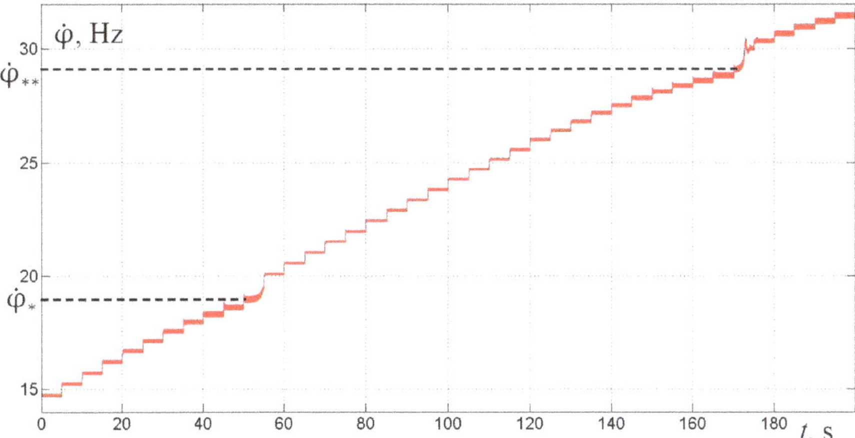

Fig. 2 Motor speed $\dot{\varphi}$ at $m_{\rm w}(\tau) = m_0 = \text{const.}$

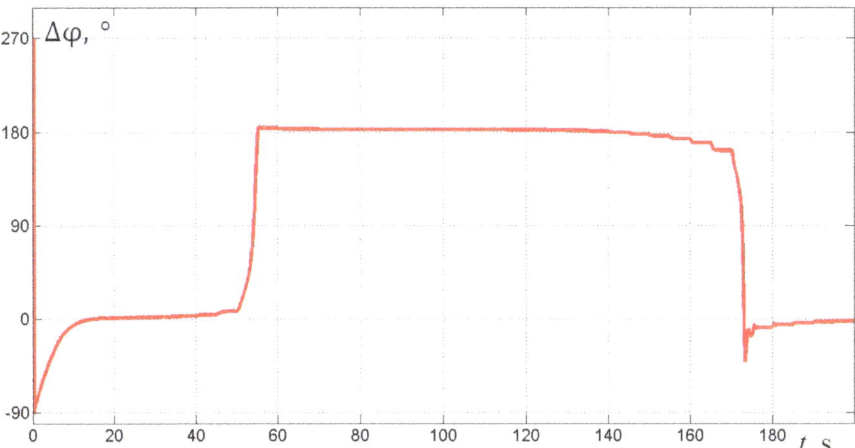

Fig. 3 Phase shift $\Delta\varphi$ at $m_w(\tau) = m_0 = \text{const.}$

shift $\Delta\varphi$ from 0° (Fig. 3), which is a consequence of the interaction of the oscillatory system with vibration exciters of limited power and a significant increase in oscillation amplitudes. After the jump of oscillations into the frequency range after the first resonance, the value of $\Delta\varphi$ changes abruptly and is set near 180°, which corresponds to the required synchronous antiphase rotation of the exciters, and practically unidirectional vertical excitation is realized. In this case, the unevenness of vibration exciters rotation practically disappears due to the establishment of oscillations with small amplitudes. The difference between the oscillation modes of the platform excited in this case from strictly vertical oscillations is due to the displacement of the center of mass of the system from the axis of symmetry of the machine due

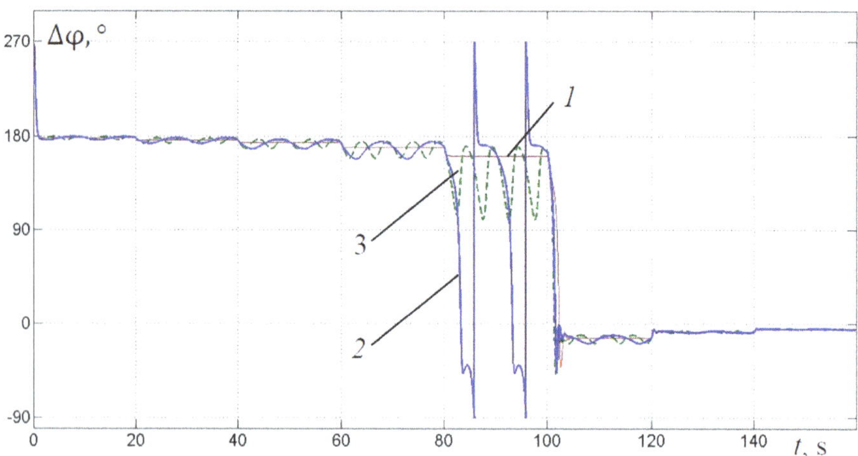

Fig. 4 Phase shift $\Delta\varphi$ near 2nd resonance at: 1—$\kappa = 0$; 2—$\kappa = 0.25$, $\alpha/2\pi = 0.1$ Hz; 3—$\kappa = 0.25$, $\alpha/2\pi = 0.2$ Hz

to the off-center location of the technological load. When approaching the second resonance, the unevenness of the vibration exciters rotation increases, caused by an increase in the vibration amplitudes in all coordinate directions of the platform. The observed shift of the mutual phase $\Delta\varphi$ from 180° is primarily due to an increase in the amplitudes of the angular and horizontal components of the oscillations, and in the case of a strictly centered system, it is not practically appeared. After the jump of oscillations into the beyond-resonant frequency range, the value of $\Delta\varphi$ is set near 0°.

The change in $\Delta\varphi$ near the second resonance at various frequencies of mass change is shown in Fig. 4. In this case, the calculations were carried out with an exposure at each power frequency for 20 s sufficient to observe several periods of change in the mass of the technological load. One can see that a periodic change in mass and, accordingly, in the inertia moment of the system causes oscillations of the mutual phase shift between debalances with a frequency equal to the frequency of the change in mass. For the same amplitude of mass change, the oscillation amplitudes of $\Delta\varphi$ increase as we approach the frequency of jump. In this case, a decrease in the frequency of a change in mass can lead to some earlier jump of oscillations beyond the resonance. So, for example, the change in the mutual phase shift $\Delta\varphi$ of vibration exciters observed at $80\ \text{s} \le t \le 100\ \text{s}$ and $\alpha/2\pi = 0.1$ Hz indicates a desynchronization of their rotational speeds, which inevitably leads to a jump of oscillations into the beyond-resonant mode. At the same power supply frequency in the case of $\alpha/2\pi = 0.2$ Hz, no loss of synchronization of vibration exciters is observed.

The change in $\Delta\varphi$ at various amplitudes of the mass m_w change is presented in Fig. 5. One can see that increase in κ leads to increase in the oscillation amplitudes of $\Delta\varphi$. For the same frequency of mass change, the oscillation amplitudes of $\Delta\varphi$

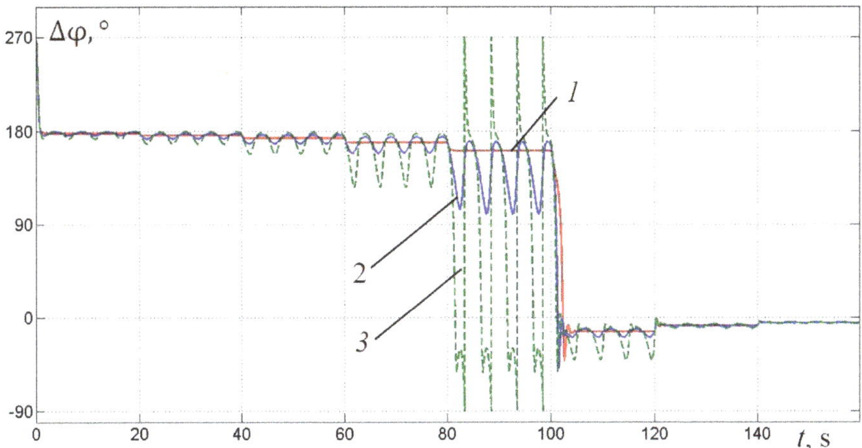

Fig. 5 Phase shift $\Delta\varphi$ near 2nd resonance at: 1—$\kappa = 0$; 2—$\kappa = 0.25$, $\alpha/2\pi = 0.2$ Hz; 3—$\kappa = 0.5$, $\alpha/2\pi = 0.2$ Hz

increase as we approach the frequency of jump. An increase in the amplitude of a change in mass can lead to earlier jump of oscillations beyond the resonance.

4 Conclusion

Thus, as a result of the study, the influence of a low-frequency change in the mass of the technological load on the value of the mutual phase $\Delta\varphi$ between the vibration exciters during their self-synchronization was revealed. It is established that the fluctuations in the mass of the technological load lead to fluctuations of $\Delta\varphi$ with the same frequency. In this case, the amplitude of the $\Delta\varphi$ oscillations increases with increasing amplitude of the mass oscillations, practically does not change with a change in their frequency, and significantly increases as one approaches the resonant frequency. In accordance with this, when developing control systems for resonant vibrations of vibrating machines, the revealed effects must be taken into account in determining the parameters of machines oscillations from measurements of the mutual phase shift between debalances.

Acknowledgements The research was supported by Russian Science Foundation (project No. 21-19-00183).

References

1. Blekhman, I.I., Blekhman, L.I., Vaisberg, L.A., Vasilkov, V.B.: Energy consumption in vibrational transportation and process machines. Obogashchenie Rud **1**, 18–27 (2019)
2. Vaisberg, L.A.: Design and Calculation of Vibrating Screens. Nedra, Moscow (1986)
3. Shah, K.P.: Construction, working and maintenance of electric vibrators and vibrating screens. http://practicalmaintenance.net/wp-content/uploads/Construction-Working-and-Maintenance-of-Vibrators-and-Vibrating-Screens.pdf. Last access 2021/01/31
4. Levendel, E.E. (ed.): Handbook: Vibration in Technics. Vibration Processes and Machines, vol. 4. Mashinostroenie, Moscow (1981)
5. Lyan, I.P., Panovko, G.Y., Shokhin, A.E.: Comparative analysis of energy efficiency in the use of vibration-type process machines in resonant and superresonant operating modes. Obogashchenie Rud **6**, 42–49 (2019)
6. Blekhman, I.I.: Synchronization of Dynamical Systems. Nauka, Moscow (1971)
7. Blekhman, I.I.: Theory of Vibration Processes and Devices. Vibration Mechanics and Vibration Technology. Ore and Metals PH, St. Petersburg (2013)
8. Wagg, D., Neild, S.: Nonlinear Vibration with Control: For Flexible and Adaptive Structures. Springer International Publishing (2015)
9. Balthazar, J.M., et al.: Remarks on passage through resonance of a vibrating system with two degrees of freedom, excited by a non-ideal energy source. J. Sound Vib. **239**(5), 1075–1085 (2001)
10. Fradkov, A., Tomchina, O., Galitskaya, V., Gorlatov, D.: Multiple controlled synchronization for 3-rotor vibration unit with varying payload. IFAC Proc. **46**(12), 5–10 (2013)
11. Yong-Zheng, J., et al.: Influence of load weight on dynamic response of vibrating screen. Shock Vib. **2019**, Art. ID 4232730 (2019)
12. Moncada, M.M., Rodríguez, C.G. Dynamic modeling of a vibrating screen considering the ore inertia and force of the ore over the screen calculated with discrete element method. Shock Vib. **2018**, Art. ID 1714738 (2018)
13. Eremeikin, S.A., Krestnikovskii, K.V., Panovko, G.Y., Shokhin, A.E.: Experimental analysis of the operability of a system to control the oscillations of a mechanical system with self-synchronizing vibration exciters. J. Mach. Manuf. Reliab. **45**(6), 553–558 (2016)
14. Gouskov, A., Panovko, G., Shokhin, A.: To the issue of control resonant oscillations of a vibrating machine with two self-synchronizing inertial exciters. In: Sapountzakis, E.J., Banerjee, M., Biswas, P., Inan, E. (eds.) Proceedings of the 14th International Conference on Vibration Problems. Lecture Notes in Mechanical Engineering, pp. 515–526. Springer, Singapore (2021)

Induction Motor Fault Detection in ESP Systems Based on Vibration Measurements

Arta Mohammad Alikhani, Abolfazl Vahedi, and Pavel Alexandrovich Khlyupin

Abstract Condition monitoring of electrical submersible pumps (ESPs) is vital since they work in a harsh and confined environment. One of the most important parts of ESP systems is the electric motor which is conventionally the induction motor. Therefore, fast and accurate diagnosis of the motor in the ESP system can cut overhaul expenses and prevent further damages and costs. In ESPs, the vibration of the system is usually measured using the downhole sensors. Accordingly, this paper aims to propose a method for bearing fault detection in the induction motor based on vibration measurements. In this method, first, the best features of the vibration signals are selected using a wrapper approach based on the genetic algorithm. Then, the model is trained using the selected features based on *k* nearest neighbor classifier. The proposed method is evaluated using data of an experimental set from Case Western Reserve University (CWRU) Bearing Data Center.

1 Introduction

Many wells do not have enough natural pressure for oil and gas lift. Therefore, to achieve the desired production rate, artificial lift methods are utilized [1]. ESP is one of the popular artificial lift methods. In these systems, a multi-stage pump is connected to the motor through a seal and gas separator to lift the oil from the downhole. In Fig. 1, the detailed configuration of the ESP system is illustrated.

Due to the large costs of the ESP workover, fast detection of ESP failure reduces costs and workover time. Consequently, condition monitoring of ESP is vital in these systems. For monitoring the condition of an ESP system, some downhole sensors are utilized to asset intake pressure, motor temperature, intake temperature, discharge pressure, current leakage, vibration, etc. As is evident in Fig. 1, the gauge is located

A. M. Alikhani · A. Vahedi (✉)
Iran University of Science and Technology, Narmak, 16846 Tehran, Iran
e-mail: avahedi@iust.ac.ir

P. A. Khlyupin
Ufa State Petroleum Technical University (USPTU), 1 Cosmonavtov st, Ufa 450062, Russian Federation

A. Ronzhin and V. Shishlakov (eds.), *Electromechanics and Robotics*, Smart Innovation, Systems and Technologies 232, https://doi.org/10.1007/978-981-16-2814-6_43

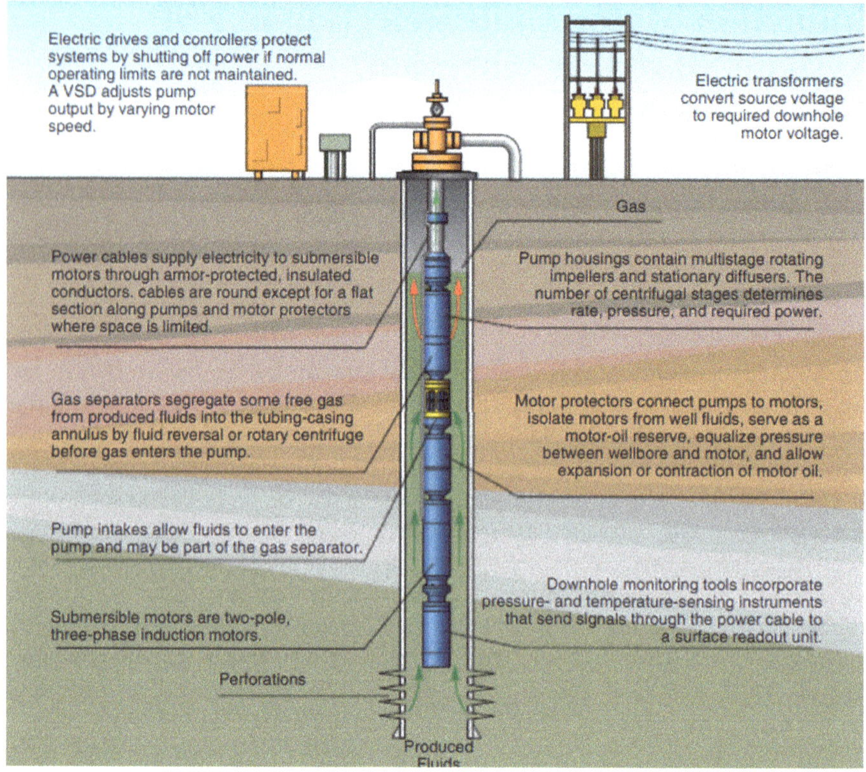

Fig. 1 An illustration of the ESP system [8]

in the bottom of the ESP in the downhole, while the condition monitoring panel is located on the ground surface. Therefore, the measured data by sensors must transmit from downhole to the surface whose distance can be about 5000 ft. There are two types of signal transmission in ESP systems and a separate instrument wire, and also signals can be imposed to the power cable. Despite the presence of some methods for signal transmissions in ESP systems, in the currently used systems, the sample rate is usually low and the sample period is from 1 s to 1 h. Hence, a high-resolution signal cannot be perfectly transmitted to the surface panel without noise. Therefore, the low-frequency features from some signals such as vibration may be useful. Vibration signals are practical for fault detection purposes [2–7].

In addition, in the literature, [9–13] have presented different approaches for condition monitoring in ESP systems. In [9], to investigate the sand wear in ESPs, a sand wear test flow-loop is designed, in which performance degradation, abrasion rate, erosion pattern, and stage vibration of the ESP are recorded. In [10], temperature estimation and vibration monitoring techniques in the literature are studied, and their potential application in ESP motors is proposed. In [11], principal component analysis (PCA) is utilized for the detection of developing ESP faults and prediction of remaining operating time before failure. Motor current signature analysis is used for the condition monitoring of ESP systems in [12]. There are also patents in the field of condition monitoring of ESP systems based on vibration measurements. In [14], based on fast Fourier transform of the vibration, a method is proposed for condition monitoring in ESP systems. The work [15] also presents the main reasons for the failure of wells equipped with installations of electric centrifugal pumps. In 30% of cases, the cause of failures was the effect of vibration on the ESP equipment. Vibration can be caused by technological processes (changes in the properties of the produced hydrocarbon emulsion) and structural changes in the ESP (the appearance of deposits on rotating elements, destruction). Artificial intelligence (AI) and specifically machine learning are an inseparable part of today's world. Therefore, similar to [11, 13], for ESP systems whose reliability is crucial, the application of AI must be assessed to achieve the most appropriate performance of condition monitoring.

Extraction and selection of the measured data play a significant role in the performance of fault detection. Various methods are used to select and extract the most appropriate features from the measured signal for fault detection [5, 16–20]. Overall, the feature selection method can be categorized into wrapper methods, filter methods, and hybrid methods [17]. In wrapper methods, features are selected according to the performance of the model. Since the feature set is usually large, evolutionary algorithms such as genetic algorithm (GA) and particle swarm optimization are utilized for subset selection. However, some direct search methods are proposed for wrapper feature selection [5, 18].

In this paper, a method for fault detection of the electrical motor in the ESP system is based on statistical features of the vibration signal. It is worth noting that the calculation of statistical features is usually more simple than other features and can be also achieved by electronic circuits. However, the best features must be selected for fault detection purposes to achieve the best performance. Therefore, in this study, a wrapper feature selection based on GA is proposed to select the best statistical features. Then, *k* nearest neighbor (kNN) is used for the classification of different states of the system.

2 Methodology

In Figure 2, an overview of the proposed approach is provided. In this method, first, the statistical or time-domain features given in Table 1 are extracted from vibration signals. Then, GA searches for the best features to achieve the least classification error. For this purpose, as is shown in Fig. 2, binary GA is used.

In the binary GA, in each chromosome, the gene with the value of «0» implies that the corresponding feature is not selected while the gene with the value of «1» indicates that the corresponding feature is selected. The cost function of the binary GA is the classification error which is determined by 20-fold cross-validation. The classifier used in this method is kNN with the configuration of Euclidean distance and one nearest neighbor. In each generation of the binary GA, the classification is

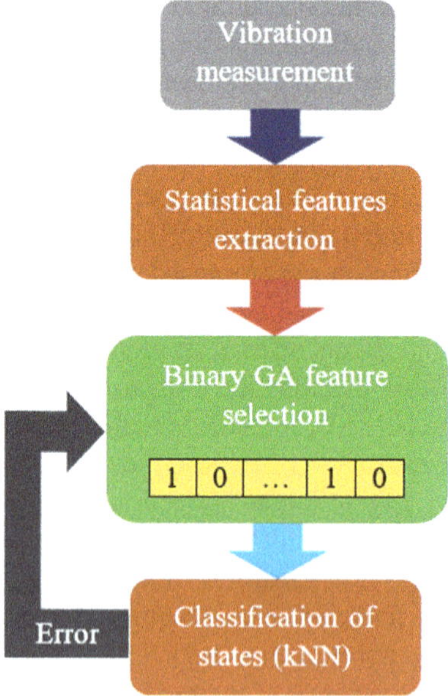

Fig. 2 An overview of the proposed fault detection method

Table 1 Statistical features

Item number	Statistical feature name	Definition	Item number	Statistical feature name	Definition
1	Mean	$\frac{1}{N}\sum_{i=1}^{N} x_i$	9	Kurtosis	$\frac{1}{N}\sum_{i=1}^{N} \frac{(x_i-\bar{x})^4}{\sigma^4}$
2	Mean of absolute	$\frac{1}{N}\sum_{i=1}^{N} \lvert x_i \rvert$	10	Impulse factor	$\frac{\max\lvert x_i\rvert}{\frac{1}{N}\sum_{i=1}^{N}\lvert x_i\rvert}$
3	RMS	$\sqrt{\frac{1}{N}\sum_{i=1}^{N} x_i^2}$	11	Crest factor	$\frac{\max\lvert x_i\rvert}{\sqrt{\frac{1}{N}\sum_{i=1}^{N} x_i^2}}$
4	Median	Median value	12	Energy	$\sum_{i=1}^{N} x_i^2$
5	Mode	Most frequently value	13	3rd moment	$E(x_i - \bar{x})^3$
6	Standard deviation	$\sqrt{\frac{1}{N-1}\sum_{i=1}^{N} (x_i - \bar{x})^2}$	14	4th moment	$E(x_i - \bar{x})^4$
7	Variance	$\frac{1}{N-1}\sum_{i=1}^{N} (x_i - \bar{x})^2$	15	FM4	$\frac{E(x_i-\bar{x})^4}{\left(\frac{1}{N-1}\sum_{i=1}^{N}(x_i-\bar{x})^2\right)^2}$
8	Skewness	$\frac{1}{N}\sum_{i=1}^{N} \frac{(x_i-\bar{x})^3}{\sigma^3}$			

performed on the features determined by each individual. The binary GA comprises initialization (to generate initial population), decoding chromosomes, finding costs for each chromosome, selection, genetic operators (crossover and mutation), and termination [21].

Furthermore, in the kNN method, the class for each instance is determined by its neighbors. The neighbors are selected according to their distance from the intended instance. In the proposed approach, the Euclidean distance is considered which is formulated as follows:

$$d = \sqrt{\frac{\sum_{i=1}^{n}\left(x_i^k - x_i^w\right)}{n}} \tag{1}$$

The equation above calculates the Euclidean distance between two instances with n features as $X^k = \left\{x_1^k, x_2^k, \ldots, x_n^k\right\}$ and $X^w = \left\{x_1^w, x_2^w, \ldots, x_n^w\right\}$.

3 Experimental Evaluation

In order to evaluate the proposed method for time-domain feature selection of the vibration in different states of bearing in induction motor, the data provided by Case Western Reserve University (CWRU) Bearing Data Center is utilized [22]. The vibration signals are measured using the experimental set as in Fig. 3. Although the situation in an ESP system is completely different from this experiment, the outcome can show whether the proposed method is efficient or not.

The data is acquired for different states of the bearing, including healthy, inner race defect, outer race defect, and ball defect. They also include different diameters of fault. Therefore, for this study, 12 states are considered, which include healthy

Fig. 3 Experimental set implemented by CWRU Bearing Data Center [22]

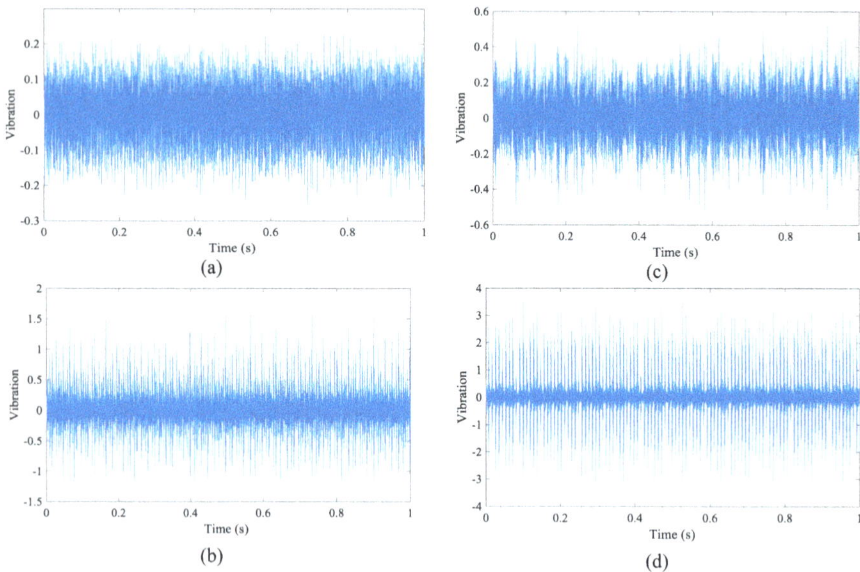

Fig. 4 Vibration signal for states **a** normal, **b** inner race fault, **c** ball fault, and **d** outer race fault with the fault diameter of 0.007″

(#1), inner race bearing fault with fault diameters of 0.007″ (#2), 0.014″ (#3), 0.021″ (#4), 0.028″ (#5), ball bearing fault with fault diameters of 0.007″ (#6), 0.014″ (#7), 0.021″ (#8), 0.028″ (#9), outer race bearing fault with diameters of 0.007″ (#10), 0.014″ (#11), and 0.021″ (#12) with the centered at 6:00 position relative to load zone (load zone centered at 6:00). The data is for the drive end bearing with the rate of 12 kHz. In Fig. 4, the measured vibration signals for normal condition of drive end bearing, inner race fault, ball fault, and outer race fault with the fault diameter of 0.007″ are illustrated.

In the first step, the best statistical features are selected by GA. In Figure 5, the cost functions achieved in 50 generations for different numbers of population are illustrated. According to this figure, GA with 20 individuals results in the best performance, which leads to the best classification accuracy or the least error in the 16th generation. The selected features are mean, mean of absolute, standard deviation, variance, 3rd moment, and 4th moment. The classification accuracy using these features is 98.78%.

In Fig. 6, the confusion matrix for the classification of 12 health condition of drive end bearing in induction motor using the mentioned six statistical features is shown. In this matrix, 100% with true label #1 and predicted label #1 means that all data is accurately classified. Furthermore, 9.8% with true label #7 and predicted label #8 indicates that 9.8% of data of state #7 are misclassified as state #8. As is evident, except for a small proportion of data with states #7, #8, and #10, which misclassified, all other data is correctly classified. More importantly, all data in the

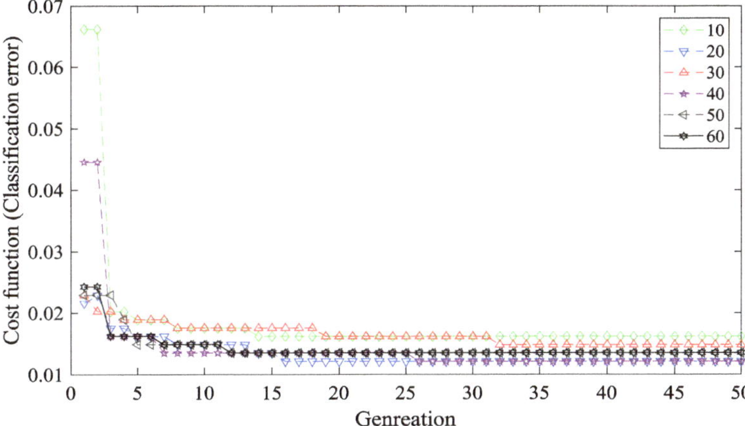

Fig. 5 Mean square error of classification of the selected feature sets in each generation for different population number

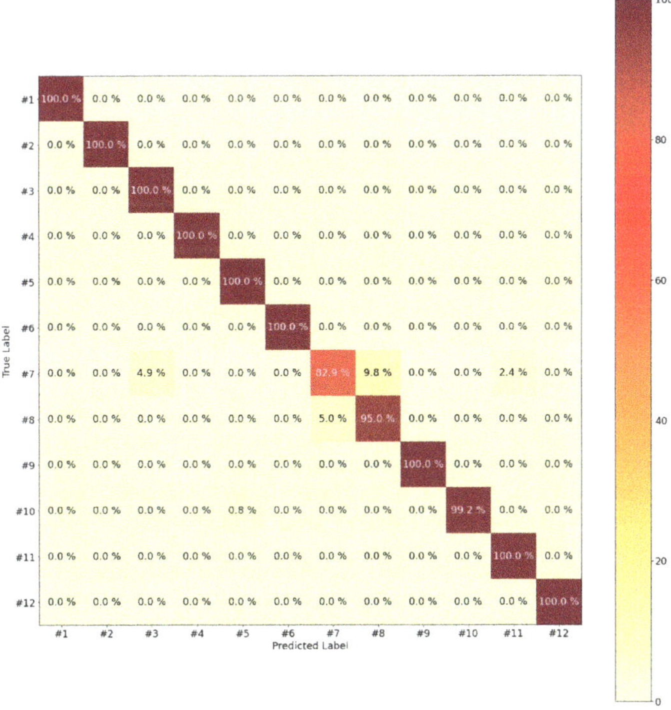

Fig. 6 Confusion matrix for classification of different states of bearing with the accuracy of 98.78%

healthy condition is accurately classified. It means that the accuracy for detection of whether a fault is occurred or not is 100%.

4 Conclusion

In this paper, a method is evaluated for the possible use in the condition monitoring of an induction motor in ESP systems. In this method, statistical features of the vibration signal are chosen for the indication of faults. This is due to the fact that 1—the statistical features are usually easy to calculate and 2—vibration measurements are usually performed in ESP systems. To achieve the best performance of fault detection, a wrapper feature selection method based on GA is utilized to select the best features among 15 statistical or time-domain features. The selected features include mean, mean of absolute, standard deviation, variance, 3rd moment, and 4th moment. These features lead to the classification accuracy of 98.78% for the classification of 12 states of induction motor drive end bearing with one healthy state and 11 faulty states. Therefore, the results show the efficiency of the proposed method for bearing fault detection in terms of accuracy and also simplicity.

References

1. Bafghi, M.B., Vahedi, A.: A comparison of electric motors for electrical submersible pumps used in the oil and gas industry. In: IOP Conference Series: Materials Science and Engineering, vol. 433(1), p. 012091 (2018)
2. Gangsar, P., Tiwari, R.: Multifault diagnosis of induction motor at intermediate operating conditions using wavelet packet transform and support vector machine. J. Dyn. Syst. Meas. Cont., vol. 140(8) (2018)
3. Narendiranath, B.T., Himamshu, H.S., Prabin, K.N., Rama, P. D., Nishant, C.: Journal bearing fault detection based on Daubechies wavelet. Arch. Acoust., vol. 42 (2017)
4. Dubey, R., Agrawal, D.: Bearing fault classification using ANN-based Hilbert footprint analysis. IET Sci. Meas. Technol. **9**(8), 1016–1022 (2015)
5. Rahnama, M., Vahedi, A., Alikhani, A.M., Montazeri, A.: Machine-learning approach for fault detection in brushless synchronous generator using vibration signals. IET Sci. Meas. Technol. **13**(6), 852–861 (2019)
6. Ewert, P., Kowalski, C.T., Orlowska-Kowalska, T.: Low-cost monitoring and diagnosis system for rolling bearing faults of the induction motor based on neural network approach. Electronics **9**(9), 1334 (2020)
7. Junior, R.F.R., de Almeida, F.A., Gomes, G.F.: Fault classification in three-phase motors based on vibration signal analysis and artificial neural networks. Neural Comput. Appl. **32**(18), 15171–15189 (2020)
8. Liang, X., Fleming, E.: Electrical submersible pump systems: Evaluating their power consumption. IEEE Ind. Appl. Mag. **19**(6), 46–55 (2013)
9. Zhu, H., Zhu, J., Zhou, Z., Rutter, R., Forsberg, M., Gunter, S., Zhang, H. Q.: Experimental study of sand erosion in multistage electrical submersible pump ESP: performance degradation, wear and vibration. In: International Petroleum Technology Conference (2019)
10. Liang, X.: Temperature estimation and vibration monitoring for induction motors. In: 2017 IEEE Electrical Power and Energy Conference (EPEC), pp. 1–6 (2017)

11. Abdelaziz, M., Lastra, R., Xiao, J.J.: ESP data analytics: predicting failures for improved production performance. In: Abu Dhabi International Petroleum Exhibition and Conference, Society of Petroleum Engineers (2017)
12. Popaleny, P., Duyar, A., Ozel, C., Erdogan, Y.: Electrical submersible pumps condition monitoring using motor current signature analysis. In: Abu Dhabi International Petroleum Exhibition and Conference, Society of Petroleum Engineers (2018)
13. Yao, C., Li, M.Z., Liu, G.F.: Partial friction fault diagnosis of electrical submersible pump based on support vector machines. In: Advanced Materials Research (2011)
14. Atherton, E.: Plant condition monitoring using vibrational measurements, 2354825 (1999)
15. Dumler, E.B.: Research of a Pneumatic Spring Compensator of Pressure Fluctuations with Quasi-zero Rigidity for a Submersible Electric Centrifugal Pump. In: Cand. Tech. sciences, Ufim. state oil. tech. University (2018)
16. Mohammad-Alikhani, A., Vahedi, A., Mahmouditabar, F., Rahnama, M.: Demagnetization fault diagnosis of FSPM motor based on ReliefF and SVM. In: 2019 International Power System Conference (PSC), pp. 95–99 (2019)
17. Mohammad-Alikhani, A., Rahnama, M., Vahedi, A.: Neighbors class solidarity feature selection for fault diagnosis of brushless generator using thermal imaging. IEEE Trans. Instrum. Meas. **69**(9), 6221–6227 (2020)
18. Mohammad-Alikhani, A., Vahedi, A., Rahnama, M., Bafghi, M.B.: A wrapper-based feature selection approach for accurate fault detection of rotating diode rectifiers in brushless synchronous generators. In: IOP Conference Series: Materials Science and Engineering, vol. 671(1), p. 012045 (2020)
19. Tang, X., Gu, X., Wang, J., He, Q., Zhang, F., Lu, J.: A bearing fault diagnosis method based on feature selection feedback network and improved DS evidence fusion. IEEE Access **8**, 20523–20536 (2020)
20. Patel, S.P., Upadhyay, S.H.: Euclidean distance based feature ranking and subset selection for bearing fault diagnosis. In: Expert Systems with Applications, 154, p. 113400 (2020)
21. Haupt, R.L., Haupt, S.E.: The binary genetic algorithm. In: Practical Genetic Algorithms (2004)
22. Loparo, K.A.: Case Western Reserve University bearing data center. In: Bearings Vibration Data Sets, Case Western Reserve University (2012)

Author Index

A. Ronzhin and V. Shishlakov (eds.), *Electromechanics and Robotics*, Smart Innovation, Systems and Technologies 232, https://doi.org/10.1007/978-981-16-2814-6

Lightning Source UK Ltd.
Milton Keynes UK
UKHW022219050922
408399UK00006B/104